Ancient Irrigation Systems of The Aral Sea Area

The History, Origin, and Development of Irrigated Agriculture

AMERICAN SCHOOL OF PREHISTORIC RESEARCH MONOGRAPH SERIES

The American School of Prehistoric Research (ASPR) Monographs in Archaeology and Paleoanthropology present a series of documents covering a variety of subjects in the archaeology of the Old World (Eurasia, Africa, Australia, and Oceania). This series encompasses a broad range of subjects – from the early prehistory to the Neolithic Revolution in the Old World, and beyond including: hunter-gatherers to complex societies; the rise of agriculture; the emergence of urban societies; human physical morphology, evolution and adaptation, as well as; various technologies such as metallurgy, pottery production, tool making, and shelter construction. Additionally, the subjects of symbolism, religion, and art will be presented within the context of archaeological studies including mortuary practices and rock art. Volumes may be authored by one investigator, a team of investigators, or may be an edited collection of shorter articles by a number of different specialists working on related topics.

American School of Prehistoric Research, Peabody Museum,
Harvard University, Cambridge, MA 02138, USA

ANCIENT IRRIGATION SYSTEMS OF THE ARAL SEA AREA

The History, Origin, and Development of Irrigated Agriculture

Boris V. Andrianov

Edited by
Simone Mantellini

With the collaboration of
C. C. Lamberg-Karlovsky
Maurizio Tosi

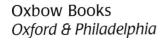
Oxbow Books
Oxford & Philadelphia

The translation of this work has been funded by SEPS

SEGRETARIATO EUROPEO PER LE PUBBLICAZIONI SCIENTIFICHE

Via Val d'Aposa 7 - 40123 Bologna - Italy
seps@seps.it - www.seps.it

A Publication by the American School of Prehistoric Research.

Andrianov, Boris V.
Mantellini, Simone
Lamberg-Karlovsky, C. C.
Tosi, Maurizio

Library of Congress Cataloging-in-Publication Data

Andrianov, Boris Vasilevich, author.
   [Drevnie orositelnye systemy Priaralya. English]
   Ancient irrigation systems of the Aral Sea area : the history, origin, and development of irrigated agriculture / Boris V. Andrianov ; edited by Simone Mantellini, with the collaboration of C.C. Lamberg-Karlovsky, Maurizio Tosi.
        pages ; cm. -- (American School of Prehistoric Research monograph series)
   Includes bibliographical references.
   ISBN 978-1-84217-384-8
   1. Irrigation--Aral Sea Region (Uzbekistan and Kazakhstan)--History. 2. Irrigation farming--Aral Sea Region (Uzbekistan and Kazakhstan--History. 3. Agriculture, Ancient--Aral Sea Region (Uzbekistan and Kazakhstan)   I. Mantellini, Simone, editor. II. Lamberg-Karlovsky, C. C., 1937- III. Tosi, Maurizio. IV. Andrianov, Boris Vasilevich. Drevnie orositelnye systemy Priaralya. Translation of: V. Title. VI. Series: American School of Prehistoric Research monograph series.
   S616.U9A6513 2013
   631.5'8709587--dc23

                                                          2013044185

ISBN 978-1-84217-384-8

Printed in the United Kingdom by Short Run Press, Exeter

Contents

Ancient Irrigation Systems of the Aral Sea Area

B.V. Andrianov

Contributors

Boris V. Andrianov[†]

1919-1993. Professor of Geography, Institute of Ethnology and Anthropology Russian Academy of Sciences. 119991 Moscow Leninsky prospect, 32a, Russia. Ph.D. Historical Sciences, Institute of Ethnography, Russian Academy of Sciences, Moscow.

Simone Mantellini

Ph.D. in Archaeology, Department of History and Cultures, University of Bologna, San Giovanni in Monte 2, 40124 Bologna, Italy

C. C. Lamberg-Karlovsky

Peabody Museum (Director 1977-1990), Stephen Phillips Professor of Archaeology, Department of Anthropology, Harvard University, Cambridge, MA 02138, USA

Maurizio Tosi

Ph.D. in Archaeology, Department of History and Cultures, University of Bologna, San Giovanni in Monte 2, 40124 Bologna, Italy

Irina A. Arjantzeva

Head of Center of Eurasian Archaeology, Institute of Ethnology and Anthropology Russian Academy of Sciences. 119991 Moscow Leninsky Prospect, 32a, Russian Federation

Alison Betts

Associate Professor, Department of Archaeology, Main Quad A14, University of Sydney, Sydney, NSW 2006, Australia

Sergey B. Bolelov

Head of Department of Central Asia Archaeology, State Oriental Museum. 119019, Nikitsky boulevard 12a, Moscow, Russian Federation

Gian Luca Bonora

Visiting Professor, Department of Archaeology and Ethnology "L.N. Gumilev" Eurasian National University, Satpaev Street 2 - 010000 Astana, Kazakhstan

Pavel M. Dolukhanov[†]

1937-2009. Emeritus Professor of East European Prehistory, University of Newcastle upon Tyne.

Henri-Paul Francfort
Directeur de Recherches, Centre National de la Recherche Scientifique (CNRS), UMR7041 "Archéologie et Sciences de l'Antiquité", team "Archaeology of Central Asia", 21 Allée de l'Université, 92023 Nanterre cedex, France

Zamira S. Galieva
Senior Researcher, Department of Central Asia Archaeology, State Oriental Museum, 119019, Nikitsky boulevard 12a, Moscow, Russian Federation

Olivier Lecomte
Directeur de l'UMR 9993, Centre National de la Recherche Scientifique (CNRS), Centre de Recherches Archéologiques Indus-Balochistan, Asie Centrale et Orientale Panthéon Bouddhique, 19, Avenue d'Iéna 75116 Paris, France

Vadim N. Yagodin[†]
Institute of History, Archaeology and Ethnography of the Academy of Sciences of Uzbekistan Republic, Karakalpak Branch (Director 1992-2011,58/31), Tortkul Avenue, 230100 Nukus, Karakalpakstan, Uzbekistan

TRANSLATED BY:
Gian Luca Bonora
Visiting Professor, Department of Archaeology and Ethnology "L.N. Gumilev" Eurasian National University, Satpaev street 2 - 010000 Astana, Kazakhstan

Giò Morse
Ph. D. Candidate in Near Eastern Studies, University of California, Berkeley, USA

Dimitri Kostushkin
MA Degree in Foreign Languages held at the Institute of Foreign languages, Samarkand, Uzbeksitan. Freelance/Independent translator and interpreter. Uzbek-Italian Archaeological Project "Samarkand and Its Territory"

TECHNICAL ASSISTANCE PROVIDED BY:
Andrey Shestakov
MA Degree in Physics held at the Samarkand State University, Uzbekistan, Freelance/Independent Designer and Computer Graphic. Uzbek-Italian Archaeological Project "Samarkand and Its Territory"

COPYEDITING ASSISTANCE PROVIDED TO THE ASPR BY:
H. Dunn Burnett

[†] Deceased

List of Figures

List of Tables

Abbreviations

AEE   1967 *Aerofotograficheskoe etalonirovanie i ekstrapolyatsiya (metodicheskoe posobie)*. Leningrad.

AKK   1960 *Arkheologicheskaya karta Kazakhstana*, K. A. Akishev. Alma-Ata.

DAT   1963 *Dictionnaire Archéologique des Techniques*, Vol. I. Editions de l'Accueil, Paris.
              1964 *Dictionnaire Archéologique des Techniques*, Vol. II. Editions de l'Accueil, Paris.

EOZU   1914 *Ejegodnik otdela zemelnykh uluchsheniy*. Sankt-Peterburg.

ICBIP   **(India. Central Board of Irrigation and Power)**
              1965 *Development of Irrigation in India*. India: Central Board of Irrigation and Power, New Delhi.

IKSSR   1957 *Istorya Kazakhhskoy SSR*. Tom 1. Alma-Ata.

ITN   1963 *Istoriya Tadjikskogo naroda*. Tom I, *C drevneyshikh vremen do V v. i. e.*, B. A. Antonenko, A. M. Belenitskiy, B. G. Gafurov, B. I. Iskandarov, B. A. Litvinskiy, A. M. Mukhatarov, Yu. A. Nikolaev, Z. Sh. Radjabov, S. A. Radjabov. Moskva.
              1964 *Istoriya Tadjikskogo naroda*. Tom II, *Vozniknovenie i razvitie feodalnogo stroya (VI–XVI vv.)*, B. A. Antonenko, A. M. Belenitskiy, B. G. Gafurov, B. I. Iskandarov, B. A. Litvinskiy, A. M. Mukhatarov, Yu. A. Nikolaev, Z. Sh. Radjabov, S. A. Radjabov. Moskva.

IUZ   1955 *Istoriya Uzbekskoy SSR*, Tom I, Kn. 1, S. P. Tolstov, R. N. Nabieva, Ya. G. Gulyamov, V. A. Shishkina. Tashkent.

KhDM   1950 *Khrestomatiya Drevnego mira*. Moskva - Leningrad.

MIKK   1935 *Materialy po istorii karakalpakov*. Moskva - Leningrad.

MITT   1938 *Materialy po istorii Turkmen i Turkmenii*. Tom II, XVI–XIX vv. Iranskie, Bukharskie i Kivinskie Istochniki, V. V. Struve, A. K. Borovkova, A. A. Romaskevicha i P. P. Ivanova. Moskva - Leningrad.
              1939 *Materialy po istorii Turkmen i Turkmenii*. Tom I, *VII–XV vv. Arabskie i Persidskie Istochniki*, S. L. Bolina, A. A. Romaskevicha i A. Yu. Yakubovckogo. Moskva - Leningrad.

MRSA   1926 *Materialy po rayonirovaniyu Sredney Azii*, Kn. 1-2. Tashkent.

OIKK   1964 *Ocherki po istorii Karakalpaskoy ASSR. Tom 1.* Tashkent.

OIT   1940 *Ocherki po istorii Drevnego Vostoka.* Leningrad - Moskva.

OITD   1940 *Ocherki po istorii tekhniki Drevnego Vostoka.* Moskva - Leningrad.

OITN   1954 *Ocherki iz istorii turkmenskogo naroda i Turkmenistana v VIII-XIX vv.* Ashkhabad.

OOI   1966 *Obshchee i osobennoe v istoricheskom razvitii stran Vostoka. Materialy diskussii ob obshchestven-nykh formatsiyakh na Vostoke (Aziatskiy sposob proizvodstva).* Moskva.

PASA   1968 *Problemy arkheologii Sredney Azii. Tezisy dokladov i soobshcheniy k soveshchaniyu po arkheologii Sredney Azii (1-7 aprelya 1968).* Leningrad.

TsGA TSSR   *Tsentralniy gosudarstvennyy Turkmenskoy SSR.* Ashkhabad.

VIR   *Vestnik irrigatsii.* Tashkent.

VRZ   1967 *Vozniknovenie i razvitie zemledeliya.* Moskva.

ZJ   *Zemledelcheskiy jurnal.* Sankt-Peterburg.

Glossary

Gian Luca Bonora

This glossary, brief and without any claim of completeness typical of a dictionary, is only intended to help the reader understand the translation of several specific terms and concepts appearing in Andrianov's book. The Russian words and the archaeological expressions included here are not widely diffused and lack proper translation into English. The glossary is divided into two sections. The first, includes an explanation of typical words and specific expressions, related to agricultural and irrigation practices and to the Central Asian vegetal and animal world, belonging to Russian, Kazakh, Karakalpak, and Uzbek languages. These lexemes are not translated within the text, but only transliterated and italicized. The second section is represented by certain Russian and Central Asian terms that have been translated into English, although they require a short explanation because of their cultural complexity. It must be noted that some words typical of Central Asia physical and cultural landscape, such as kurgan, barchan, wadi, etc., are not included in this glossary because their meanings are internationally known and are included in the typical contemporary dictionary of English language.

Section 1: Explanation of the terms transliterated and not translated

Biyurgun (Биюргун)
Anabasis salsa, a semi-shrub 5-25 cm high of the *Chenopodiaceae* family, one of the species of *anabasis*. It is very widely diffused in Central Asia in saline soils in semi-deserts and deserts from Southern Saratov Oblast to Mongolia, occupying large areas in many places. It is an important fodder crop, especially for camels.

Chigir (Чигирь in Russian; Шигир –*shigir*- in Karakalpak and Kazak)
Primitive hydraulic device formed by a wheel with buckets for lifting water and irrigating small areas.

Farsakh (Фарсах, from Persian *Parasang*)
It is a historical Iranian unit of itinerant distance, usually estimated at 3.4 or 3.5 miles (5.5 or 5.6 kilometers). In antiquity, the term was used throughout much of the Middle East but the Old Iranian language from which it derives can no longer be determined. There is no consensus with respect to its etymology or literal meaning. In addition, to its appearance in various forms in later Iranian languages (e.g., Middle Persian *farsang* or Sogdian *fasukh*), the term also appears in Greek as Παρασάγγης, in Latin as *parasanga*, in Armenian as *hrasakh*, in Georgian as *parsakhi*, in Syriac as *prsha*, in Arabic as *farsakh* and in Turkish as *fersah*.

Gryad (Гряд), see *karyk*.

Irrigator (Ирригатор)
This term is widely used by Andrianov and it can have a double meaning: on one hand, it can be pertinent to either a technician, engineer, or specialist in irrigation, involved in the study or the construction of irrigation and hydraulic works; on the other hand, it is often used to refer to the farmer who provides irrigation to fields.

Itsitek (Итцитек or Итсегек)
Anabasis aphylla, plant of the *Chenopodiaceae* family. A short half-shrub with small branched scale-like opposite leaves, flowers in spicate inflorescences. Fruit are bacciform with yellow or pink wing-like appendages. It grows in saline and clay deserts and semi-deserts of the Near East, Middle and Central Asia, as well as in the Southern European part of Russia and Ukraine, Caucasus, and Southern Siberia.

Kair (Каир)
A type of agriculture without irrigation, practiced in river deltas, on moist, sandy and silt soils (Khazanov 1992, note 3). Rivers in the desert lose their water to evaporation, infiltration in the adjacent soil loosens a large amount of sediments which accumulate and which eventually generate floods, usually in spring or early summer. Stagnating waters from these floods saturate the ground to such an extent that it remains wet until autumn. Fresh river silt is an excellent soil in which to sow that it provides an abundant harvest even with the most primitive cultivation. In the lower reaches of the Amudarya River there were stretches where the population used the areas of natural river floods or close occurrence of groundwater for sowing. This so-called *kair* agriculture is a direct descendant of the earliest types of desert farming. In *kair* agriculture man often had to protect their crops from excessive high water floods by deviating embankments. On *kair* lands it is impossible to cultivate cereals because wheat, barley and other cereals have predominantly surface roots impossible to nourish with the onset of heat beginning from the second half of May. Then, under the heat of the sun, moisture evaporates quickly and it is insufficient to allow the cereals to grow and develop. Thus in *kair* lands it is best to cultivate melons, pumpkins and other cucurbitaceous because of their relatively deep roots (Fedorovich 1948; Gulyamov 1957:59; Lewis 1966:484-485; Andrianov 1995).

Karyk (Қарық, from Kazakh and Karakalpak)
This word can have several meanings: 1) Оросительная канава для бахчевых культур - *Orositelnaya kanava dlya bakhchvykh kultur* = irrigation ditch for melon fields; 2) Арык - *Aryk* = irrigation ditch; 3) Грядка - *Gryadka* = ridged, or raised field; 4) Обилие, Изобилие - *Obilie, Izobilie* = wealth, abundance. In Andrianov's book, the third meaning (raised furrows for cultivation, mainly of melons, watermelons, cucumbers and other cucurbitaceae) fits best with the context.

Ketmen (Кетмень)
Agricultural implement such as a hoe, used in Central Asia for tilling crops, digging ditches, etc.

Keurek (Кеурек or Кеyrek сасыр)
Ferula assafoetida, it is an herbaceous perennial plant, growing to 1-4 m tall, with stout, hollow, somewhat succulent stems, native to the Mediterranean Region and east to Central Asia, mostly growing in arid climates. The leaves are tripinnate or even more finely divided, with a stout basal sheath clasping the stem. The flowers are yellow, produced in large umbels.

Khum (Хум)
Large (up to 1.5 m) earthen jar, tapering downward and with or without a neck, to store water and/or other food supplies. This type of container was widely distributed among the settled farming communities of Central Asia from Neolithic times onwards. The outer surface of the large vessel can be decorated with painted patterns or high relief clay figures. Contemporary Central Asian large containers have handles and a glazed inner surface.

Khutor (Хутор)
Farm, separate farm in association with the land and the estate of the owner.

Mazar (Мазар)
A *mazār* is a tomb or mausoleum. The word deriving from the Arabic verb *zāra* "to visit", whence also comes the noun *ziyārah* "a visit", or "visiting the tomb of a saint for blessings". Though the word is Arabic in origin, it has been borrowed by a number of eastern languages, including Persian and Urdu. The mausoleums of Sufi saints are often places of pilgrimage for Muslims. The city of Mazār-i Sharīf in Northern Afghanistan is so called because it is also famous as a pilgrimage site.

Pakhsa (Пахса)
Pisè, rammed clay, usually with addition of chalk, lime, straw, and gravel, widely used in ancient times for the construction of adobe buildings, structures and dwellings in Central Asia. Today it is still used by local populations mostly in rural environments.

Poisk (Поиск)
Any place in which a moderately brief survey sweep was made. Generally speaking, it is an archaeological site and for this reason a *poisk* could be represented by: a low-quantity surface scatter of material (Точка - *Tochka*, in Russian); by a camp-site or encampment or station (where the collection of artifacts is poor - Стоянка - *Stoyanka*, in Russian); by a settlement, where a dense collection of surface material have been identified (Поселение - *Poselenie* and/or Городище - *Gorodishe*); by a funerary mound or barrow (Курган - *Kurgan*); and by a generic archaeological monument (Памятник - *Pamyatnik*).

Rustak (Рустак, from Middle Persian *rotastak*)
Cornfield, worked or cultivated field.

Sajen (Сажень)
Old Russian measure of length equal to seven feet (2.13 m).

Saxaul / Saksaul (Саксаул in Russian, Сексеуил - *Sekseuil* - in Kazakh)
Haloxylon ammodendron, plant belonging to the *Amaranthaceae*. The *saxaul* is distributed in Middle and Central Asia (Iran, West Afghanistan, Western Turkestan), from the Aralo-Caspian region to the Amudarya River valley, in the lowland areas of Central Asia and China (Mongolia, Xinjiang, Kansu). It is a psammophyte, which grows in sandy deserts, on sand dunes, and in steppe up to 1,600 m a.s.l. In Central Asia, it often forms '*saxaul* forest', while in Middle Asia it usually grows scattered. White *saxaul* is known as *Haloxilon persicum*; black *saxaul* is known as *Haloxilon Aphyllum*.

Sai / Say (Сай, in Russian from Kazakh; see also the synonym Ложбина - *lojbina*)
Dry bed of temporary drainage, seasonal water course.

Sengir (Сенгир, from Kazakh; Сеңгір - *Sengir*, Вершина - *Vershina* and Высокий *Vysokiy*)
Place located on top of a plateau or mountain and visible from a distance.

Solonchak (Солончак)
1) Почва, насыщенная солями, легкорастворимыми в воде - *Pochva, nasyshchennaya polyami, legkorastvorimymi*; 2) Озеро или ключ с соленой водой, солонец - *Ozero ili klyuch s solenoy vodoy, solonets*; a type of soil formed usually by the salinization of soil in steppe, desert, and semi desert regions having

xx   Ancient Irrigation Systems

an exudative water regime in which salts rise to the upper soil layers due to groundwater evaporation from the surface. The profile of *solonchak* soils is differentiated into poorly defined horizons. Below the surface usually there is a swollen and suberose saline horizon; farther down there is a weakly defined or residual humus horizon with streaks and patches of salts. Salinized rock or a water-bearing level occurs more deeply. *Solonchak* contains a substantial amount of highly soluble salts (from 1-3 to 10-15%). A distinction is made between *solonchak* of primary and secondary salinization. The latter forms as a result of improper irrigation. There are semi-desert and sierozem *solonchaks*; the basis for this classification is the residual features of soils, from which the soils were formed. *Solonchak* are found in Central Africa, Asia, Australia, and North America. In Eurasia they occur in the Caspian Lowland, Northern Crimea, Kazakhstan, and Middle Asia. Any agricultural crop of those regions is suitable to be cultivated on *solonchak*. In preparation for cultivation, *solonchak* are desalinized by washing and by lowering the groundwater level (desalinating drainage).

Takyr (Такыр, from Turkic 'smooth', 'even', 'bare')

It is an alkaline soil formation, generally containing only algae and lichens, which are formed by the accumulation of dry elutriated alluvium in natural depressions. Physically, they form smooth, bare, thin, and hard parquet-like or cracked structures which are the result of the rapid drying of silt suspensions and the cementing of surface layers by calcium carbonate crusts. They are distributed over large waterless tracts throughout the deserts of Central Asia, providing convincing evidence of former drainage patterns and the retraction or shift in water courses. Since large *takyr* deposits generally reflect former riverine courses, the occurrence of a *takyr* formed during post-Pleistocene times may indicate a potentially rich area for archaeological research. The *takyr* is almost entirely devoid of vegetation; the flora consists exclusively of algae and lichens. *Takyr* becomes vegetated only when watered by the runoff of spring rains. *Takyr* zones are located outside the Tedjen and Murgab deltas, along the Amudarya and Syrdarya, and around the oases of Northern Bactria. These basins serve as seasonal (springtime) storage places for water where temporary wells are dug. They have a distinctive flora and fauna that attracts grazing animals and predators, and they provide important seasonal plants and animals in the deserts for caravans and herders.

Thalweg (Тальвег - Talveg - in Russian)

English loan word from German (*tal* = 'valley'; *weg* = 'way'), in geography and fluvial geomorphology it means the deepest continuous slope within a valley or watercourse system.

Tugai (Тугай, from Turkic)

Floodplain forest in the deserts of Middle and Central Asia. It is a type of fringing, or gallery, forest. *Tugai* thickets and forests are found in river valleys where the groundwater is close to the surface. Various species of trees are represented, including variable-leaved poplars (*Populus pruinosa* and others), willow (*Salix wilhelmsiana*), tamarisk (*Tamarix* sp.), salt tree (*Halimodendron halodendron*) and buckthorn (*Rhamnus* sp.). Some of these woods have little economic significance (some could be burned as fuel) but are important for retaining water. Typical *tugai* extend along river channels and narrow islands. *Tugai* on rich alluvial soils form dense stands of trees and shrubs entwined by lianas (*Clematis, Calystegia*). The herbaceous cover includes species of reed, dogbane, and, in some places, plum grass (*Erianthus*). In the floodplains of the Amudarya and Syrdarya there is a predominance of variable-leaved poplars; on salinized soils thickets of *Tamarix ramosissima* (2-4 m or, sometimes, 5-6 m in height) and *Tamarix hispida* (up to 1 m in height) predominate. *Tugai* are inhabited by boars, Bukhara deer, Turan Tiger, swamp lynx, rabbit, water rat, mice, many birds, amphibians, and reptiles.

Uldruk (Ульдрук)
Anabasis aphylla, see *Itsitek*.

Uy (Уй, from Karakalpak)
The first and most widespread meaning of this word is Дом - *dom*, Жилище - *Jilishche* = house, dwelling, and any other adobe construction. The second meaning refers to a low-raised mound or hillock between *takyr* areas, like an island between *takyr* lowlands, and on which camelthorn (*Alhagi*) sporadically grows.

Verst (верста)
Old Russian measurement, equivalent to approximately 1.067 kilometers.

Yantak (Жантақ or Янтак, верблюжья колючка - *verblyujya kolyuchka*)
Alhagi, is a genus of Old World plants of the Fabaceae family. They are commonly called camelthorns or manna trees. There are three to five species. *Alhagi* species have proportionally the deepest root system of any plants: a 1 m high shrub may have a main root more than 15 m long; due to their deep root system *Alhagi* species are drought-resistant plants that utilize ground water, thus adapting perfectly to a hyper-arid environment.

Section 2: Supplementary short explanation of some Russian and Central Asian words translated in the new edition

Antichnost – Античность
Classical Antiquity or classical period. It is a broad term for a long period of cultural history centered on the civilizations of ancient Greece, ancient Rome, and others on the Mediterranean Sea. In B. V. Andrianov's work, this term refers mainly to archaeological monuments and historical events dating back between the 4th century BCE and the 3rd or 4th century CE. We therefore refer to them by the adjective 'Antique' (see also Yagodin and Betts 2006: 6).

Arkhaizm – Архаизм
Archaic Period, from the 7th to 5th centuries BCE (see also Yagodin and Betts 2006: 6). In Andrianov's book, Архаизм (*Arkhaizm*) and Архаик (*Arkhaik*) refer to the period of the Khorezm Civilization blossoming in the lower Amudarya, dating to the 6th and 5th centuries BCE, when irrigated agriculture development, connected with the construction of numerous and big diversion canals, was one of the important factors contributing to the formation of the so-called 'Khorezm State'. During this period numerous fortified city-type settlements and multiple farmsteads appeared.

Aryk – Арык
Local term loaned from Turkish and meaning either a major canal for irrigation or a small furrow supplying water to fields.

Aul – Аул (in Russian, *Awıl* in Karakalpak)
Small hamlet, village.

Gyr – Гыр (from Kazakh қыр – *kyr*)
It can have a double meaning: 1) горный хребет, гребень горы - *gornyy khrebet, greben gory* = mountain

range; 2) слегка возвышенная местность с пастбищами и посевными угодьями – *slegka vozbyshennaya mestnost s pastbishchami i posevnymi ugodyami* = a slightly elevated area of pasture land and sowing fields.

Janadarya – Жанадарья or Жаныдарья – and *Inkardarya* – Инкардарья
The Syrdarya Delta consists of six main deltaic branches, here labeled from north to south: Syrdarya; pra-Kuvandarya (in the past called Eskidrjalyk with Kurauly, Eskydaryalik, and Ajikhansaidarya as tributaries); Kuvandarya (with Otakaly and Madenuet as tributaries); Zhanadarya (or Zhanydarya according to the Kyrgyz-Kazakh pronunciation widespread in the 19th and in the first half of the 20th century); Inkardarya (with two main courses, the Upper and the Middle); and lastly Karadarya (or Lower Inkardarya). It must be noted that in all the ethnographical, archaeological, and historical literature produced by the Khorezm Expedition, and thus even in Andrianov's book, the hydronym Karadarya is never mentioned because this ancient riverbed was recognized as the Lower Inkardarya.

Kala – Кала
From Russian, otherwise *qala* in Karakalpak, Kazakh, and Uzbek = a town, usually with a fortified citadel. In the past also referring to an enclosure of yurts fortified by earthen ramparts.

Kel – Көл (in Karakalpak and Kazakh)
Lake, water basin; Кул – *Kul* (in Karakalpak and Kazakh) = ash, cinder. B. V. Andrianov did not distinguish between the two words, because of their pronunciation being similar but dissimilar in meaning.

Lepnaya Keramika – Лепная Керамика
Hand-made rough pottery.

Limannoe oroshenie – Лиманное орошение
A type of estuary irrigation; small-scale gravity soil watering, in spring, by means of local water resources. In this technique surplus water is used from reservoirs, canals and/or melting ice and snow flow from an area higher than farms and fields. These are surrounded by a more or less complex system of dams and embankments, thus appearing similar to an "estuary - semi-enclosed body of water" = Лиман (*liman*, in Russian). In recent times, this technique was mainly widespread in some regions of Central Asia, in the Volga River valley and in the Northern Caucasus where it was used for cereals production.

Meridionalnyy – Меридиональный
Направленный по меридиану, с севера на юг – *napravlennyy po meridian, s severa na yug* = along the meridian, from north to south and vice versa, in a north–south or south–north direction.

Solyanka – Солянка
Salsola, is a genus of the subfamily Salsoloideae in the Amaranthaceae family. A common name of various members of this genus is saltwort, because of its salt tolerance.

Stankovaya Keramika – Станковая Керамика
Wheel-made fine pottery.

Staritsa – Старица

Old stream or riverbed. This term is generally used for a body of water typically found in flat, low-lying areas, and can refer either to an extremely slow-moving stream or river (often with a poorly defined shoreline), or to a marshy lake or wetland.

Там – Там
A single story flat-roofed house, traditionally built of mud bricks but now often with cement.

Zemlekopalka – Землекопалка
Digging stick.

References

Andrianov, B. V.
  1969 *Drevnie orositelnye sistemy priaralya (v svyazi s istoriey vozniknoveniya i razvitiya oroshaemogo zemledeliya)*. Moskova.
  1995 The History of the Economic Development in the Aral Region and its Influence on the Environment. *GeoJournal* 35.1:11-16.

Bilyalova G., Yarygin S., Minardi M., and Bonora G. L.
  2014 *Multilingual Dictionary of Archaeological Terms (Kazakh-Russian-English-Italian)*. L.N. Gumilev ENU. Astana.

Fedorovich, B. A.
  1948 *Lik pustyni*. Moskva.

Gulyamov, Ya. G.
  1949 *Istoriya orosheniya Khorezma s drevneyshikh vremen do nashikh dney*. Tashkent.

Khazanov, A. M.
  1992 Nomads and Oases in Central Asia. In *Transition to Modernity: Essays on Power, Wealth and Belief*, edited by John A. Hall and I. C. Jarvie. Cambridge University Press. Cambridge.

Lewis, R. A.
  1966 Early Irrigation in West Turkestan. *Annals of the Association of American Geographers* 56:484-485.

Yagodin V. N., and A. V. G. Betts
  2006 *Ancient Khorezm*. Tashkent.

Preface: Boris V. Andrianov
and the Archaeology of Irrigation

Simone Mantellini

As Boris Vasilevich Andrianov stated in the very beginning of his book (Andrianov 1969:3, 5), life in Central Asia cannot be possible without artificial irrigation. This is particularly true today as in the past, and it is particularly true in Khorezm, where, without appropriate water management, the harsh climate condition can easily turn a green oasis into a desert.

This is one of the main reasons why many scholars engaged in Central Asian studies have devoted most of their attention to ancient irrigation and hydraulic systems, and their relationship with ancient human settlements (see an overview in Bartold 1965:95-233; Lewis 1966; Andrianov 1995; Lecomte and Francfort 2002). Among these writings, Andrianov's book *Drevnie orositelnye sistemy priaralya* (Ancient Irrigation Systems of the Aral Sea Area) can probably be considered the major output on the subject for several reasons. First, because this work summarizes the results of studies on ancient irrigation achieved during 15 years of research around the Aral Sea with the Khorezm Archaeological-Ethnographical Expedition (hereafter KhAEE). Secondly, this work is fundamental in understanding historical changes in settlement dynamics and environmental transformations which occurred in Khorezm over the last three millennia. Lastly, Andrianov and his archaeological-topographical unit carried out intensive field work aimed at collecting and analyzing data according to an innovative and multidisciplinary approach combining traditional archaeological methods and techniques with those provided by other disciplines such as geography, ethnography, and geology. Before Andrianov, other scholars dealt with the study of ancient irrigation systems in Central Asia, specifically around the Aral Sea (Gulyamov 1957; Voevodskiy 1938). However,

no one addressed this issue through a multidisciplinary approach and a systematic way over such a vast area, and with a long-term perspective as Andrianov did.

In spite of its scientific value, *Drevnie orositelnye sistemy priaralya* was written in Russian and therefore its diffusion was limited to the former USSR countries and among the few Western scholars dealing with this specific research topic. In the last decades, the increase in international archaeological expeditions to Central Asia, as well as the growing interest with issues of desertification and water archaeology, have made the work of Andrianov central to landscape and environmental archaeology projects currently in progress in this region.

The universities of Bologna and Harvard joined to translate Andrianov's book into English to make it widely available. Adding to the initial idea of translating the Russian text was the addition of papers centered on the figure of Andrianov and his contribution to the study of ancient irrigation. The volume is also enriched with a map published by O. Lecomte and H.-P. Francfort (2002) which summarizes the main archaeological discoveries of the KhAEE: major settlements, their chronology and function, graveyards, the main irrigation networks, as well as the aerial and car routes used during field surveys.

The first introductory article is by Pavel V. Dolukhanov and it provides a general overview on Russian, and later Soviet archaeology in Central Asia, beginning with the Russian conquest in the second half of the 19th century. Dolukhanov describes the main archaeological investigation carried out in the former Soviet republics of Central Asia, from the first excavation at Afrasiab (ancient Samarkand) and Merv to the multidisciplinary expeditions in Khorezm, under the directorship of S. P. Tolstov, and in Southern Turkmenia (YuTAKE) headed by M. E. Masson. Furthermore, specific attention is devoted to the important study of ancient irrigation in Central Asia and the fundamental contribution in this field by Andrianov in the Aral Sea and by G. N. Lisitsyna in Southern Turkmenia. The short description on the research history at Samarkand-Afrasiab was provided by Frantz Grenet, Director of the Mission Archéologique Franco-Ouzbeke (MAFOuz).

Sergey B. Bolelov focused his paper on the importance of research on ancient irrigation systems carried out by Andrianov in Khorezm and Lower Syrdarya. Bolelov also recalled a brief history of the KhAEE, its different topics and targets, and remarks on the most advanced multidisciplinary approach given to the expedition by S. P. Tolstov. As soon as the KhAEE started research in the Aral region, a specific archaeological-topographical unit was established, under the direction of Andrianov, in order to map and study the so-called 'lands of ancient irrigation', i.e. the abandoned ancient settlements, cultivated areas and irrigation works. According to Bolelov, the pioneering and innovative research, that combined the use of aerial photos with field surveys and excavations, made Andrianov the founder of the 'archaeology of irrigation' in Central Asia.

The article by Zamira S. Galieva mostly deals with the person of Boris V. Andrianov, who accepted to be her main supervisor when, in the early 1980s, Galieva moved from Tashkent to Moscow to obtain her Ph.D. in Historical Sciences. Even at that time, more than 40 years since its beginning, the fame of the KhAEE was still so high that Galieva considered it an honor to have been trained by Andrianov in uncovering and mapping archaeological evidence through aerial photographs. Following these first experiences, Galieva would improve these methods with other scientific projects and throughout the most advanced applications of informatics. Finally, Galieva remarked how Andrianov was able to connect his human qualities to scientific skills so that he is still remembered today.

The article written by Vadim N. Yagodin and Alison V. G. Betts provides an updated archaeological view of the area formerly investigated by the KhAEE in light of recent discoveries by the Karakalpak-Australian Archaeological Expedition (KAAE). After a general overview of results achieved by the KhAEE and the study of ancient irrigation systems in that area, the authors focus their attention on key sites investigated by the KAAE, in particular on Tash-Kirman-tepe, now interpreted as a ritual center associated with the veneration of fire, and not a fortified manor as Andrianov supposed after his preliminary investigation.

The final article is a theoretical essay written by C. C. Lamberg-Karlovsky on the role of irrigation, and water management more generally. Based on archaeological data, the written sources, and the different schools of thought on this matter, Lamberg-Karlovsky provides a comprehensive analysis on the role of water in ancient civilizations with numerous references to the contemporary situation.

The major problems encountered during the translation process was the presence of many specific terms and technical words belonging to Russian thought and archaeological school. Sometimes these terms do not have an exact corresponding translation in English (see for example the question of *poisk* and *irrigator* in Bonora, *infra*). Thus, these terms have been only transliterated within the text, and then included and explained in detail in a short glossary edited by Gian Luca Bonora at the beginning of the book. Given that this book wants to be also a mean of spreading the archaeology of Central Asia to Western scholars, the glossary includes the explanation of some others terms and concepts typical of the Soviet-Russian archaeological school.

Finally, the whole bibliography at the end of the book includes all the references cited in the text (on this matter, see also the notes on references, *infra*). One part of the bibliography, edited by Irina A. Arjantseva, has been specifically devoted to all the work published by Andrianov throughout his scientific career. It includes the work mentioned in the book and in addition all the geographical-archaeological research, particularly on the subject of irrigation and water management, which he published later. Andrianov's complete scientific writings consist of more than 50 publications, the major part of which are dated to the 1960s culminating in 1969 with the release of *Drevnie orositelnye sistemy priaralya*. Almost ten years later, in 1978, Andrianov also published a further, but less known, essay on this subject entitled, *Zemledelie nashikh predkov* (Agriculture of our ancestors). Of all of Andrianov's writings, four are published in English. The first paper, entitled, *Some Aspects of the Problem of the Interplay of Nature and Society* (Andrianov 1966a) is the translation of an article published earlier in Russian (Andrianov 1966b). In this work, mainly focused on the 18th-20th centuries, Andrianov supports the idea that the major environmental changes occurred recently in the Lower Amudarya due to human activities rather than physical-geographical factors (Andrianov 1966a:3). The second (Andrianov 1976) is a very short comment on the article, *Canal Irrigation and Local Social Organization* (Hunt and Hunt 1976). The discussion focuses on whether or not irrigation was the major cause of development of early states and civilizations. Based on his work in Khorezm, Andrianov declared that the "State power was an important condition, but not the result, of the successful development of irrigation" (Andrianov 1976:756). Although he considered this study interesting, he criticized the approach because the article "... has little specific information on regional irrigation. Furthermore, the question involved in the linkage of the development of irrigation and the rise of state power is not clearly elucidated" (Andrianov 1976:756). The third article (Andrianov 1978b) is a theoretical essay on the concept of the hydraulic society. It has been treated in detail in the introductory paper by Lamberg-Karlovsky (see *infra*). The last paper (Andrianov 1995), is the most interesting from the perspective of archaeology of water management. Referring to the case of the Aral area, the article is actually a general and updated historical overview of irrigation and agriculture in Central Asia based on data collected throughout this region in the last decades. In this work, Andrianov describes the main development steps of irrigated oases and farming practices, from the early agricultural communities of Southern Turkmenia in the 6th-3rd millennia BCE to modern times. Moreover, Andrianov takes into consideration all the historical regions of Central Asia, from the foothill of the Kopet Dag to the West Pamir, and from Bactria to the Lower Syrdarya. In agreement with other eminent scholars, such as Ya. G. Gulyamov (1974) and A. R. Mukhamedjanov (1975, 1994), Andrianov considers the heyday of irrigation development in Central Asia in the first centuries CE, at the time of the Kushan Empire, in connection with the development of urban areas, flourishing of trades, progress in craft, and development of hydraulic engineering (Andrianov 1995:13; see also Andrianov 1969:124; Tolstov 1948a:32, 1948b:113ff.).

The present book is divided in two parts. The first part, chapters 1 and 2, is more general and theoretical, and deals with the ancient irrigation study methods, as well as the origin and development of irrigated agriculture. The second part, chapters 3, 4, and 5, regards specific field work and results achieved by Andrianov and the archaeological-topographical unit, in the so-called 'lands of ancient irrigation around the Aral Sea'.

The book begins by commemorating D. D. Bukinich, the engineer and irrigation specialist whose work, according to Andrianov, marked the beginning of the study of Central Asia ancient irrigation systems.

Then, the author briefly describes the book's aims and structure. He underlines the importance of artificial irrigation in areas with arid climates, and the difficulty in studying ancient irrigation systems and hydraulic devices because of their poor state of preservation. In this regard, the lower reaches of Amudarya and Syrdarya represent an excellent case study due to the amount of field data collected and the work of the many scholars involved with the KhAEE.

In the *Introduction*, the author focuses on the main aspects of ancient irrigation systems and their main socio-economic implications. In particular,

Andrianov underlines the double aspects of irrigation systems: on one hand, they play a major role in the development of arid regions, on the other hand they are difficult to study and date.

The author presents previous studies on ancient irrigation in Khorezm, started in the early 1930s with M. V. Voevodskiy. Ya. G. Gulyamov continued this research during the prewar and postwar years and published the result in the well-known book *Istoriya orosheniya khorezma s drevneyshykh vremen do nashik dney* (History of Irrigation of Khorezm from the Antique Period to Our Day; Gulyamov 1957), that Andrianov considered as "the most important step in studying the history of irrigation in Khorezm. However, some of his conclusions "must be amended and expanded on the basis of new material" (Andrianov 1969:8). The study of Khorezm ancient water management was widely considered also by Tolstov in his monumental work *Drevniy Khorezm* (Ancient Khorezm, Tolstov 1948a), where the author advanced some important conclusions frequently noted and accepted by Andrianov himself. In his book, Tolstov attempted a general historical reconstruction of the population of Khorezm from the earliest period up to the 19th century. After the progress in irrigation techniques in Antiquity and during the Kangju-Kushan periods, Tolstov "convincingly proved that the main reason for the formation of the 'lands of ancient irrigation' was not natural catastrophic changes, but, above all, socio-historical factors, political and economic crises, and that "only a centralized Oriental despotism could create the great canals of Khorezm" (Andrianov 1969:10)

The *Chronology of Research* describes the field work carried out season by season and major interesting points and areas investigated by the archaeological-topographical unit. This team, headed by Andrianov, specifically addressed the systematic survey and mapping of the ancient irrigation system of the Aral Sea area in order to reconstruct in detail settlement patterns developed in this region, from the Bronze Age to their abandonment by the Karakalpaks and Turkmens in the 18th–19th centuries. Andrianov highlights

how this work was widely based on the use of aerial reconnaissance and then field surveys. The results of his unit were then combined with those provided by the other teams forming the KhAEE: archaeology (leader team S. P. Tolstov), geography and ethnography (B. V. Andrianov), geomorphology (A. S. Kes), botany (L. E. Rodin), engineering-geodesy (N. I. Igonin), soil scientist (N. I. Bazilevich). Finally, the author summarizes the data collected during 14 seasons of field work carried out by his team between 1952 and 1964: 1640 *poisk* were investigated on the right bank of Amudarya, 981 *poisk* on the left bank of Amudarya, and 1000 *poisk* on the Lower Syrdarya.

In the first chapter the author focuses on the *Ancient Irrigation Study Methods*. The author describes the approach employed by the archaeological-topographical unit in researching the historical dynamics of irrigation systems and their relation to settlement pattern. The major problems concern the poor state of preservation of canals, hydraulic devices, and field layouts, which many times were buried under sand dunes, or highly damaged by agricultural works and anthropogenic transformations. Andrianov stated that "the cultural landscape is a complex natural-historical formation, in which the effects of influences from different historical periods are gradually accumulated" (Andrianov 1969:16). Therefore he argues that such a study "requires an interdisciplinary approach combining natural geography and human sciences" (Andrianov 1969:16). In this regard the modern desert areas around the Aral Sea investigated by the KhAEE represents a unique case study where the irrigation systems form a sort of skeleton of ancient and modern oases. Andrianov deals also with the difficult task of dating ancient irrigation, mostly because the major canals might have supplied water for a very long time, and digging a few settlements along a canal could not provide a reliable chronology for the canal itself.

Part of the chapter is dedicated to the method of detecting and mapping irrigation networks through aerial methods. Andrianov provides a summary of aerial archaeology history and its application in studying ancient irrigation systems and hydraulic devices. The author summarizes the main publications and pioneer scholars in aerial archaeology, such as G. A. Beazeley, O. G. S. Crawford, A. Poidebard, and R. Chevallier in the West, and the early experience in Russian archaeology and the KhAEE. Andrianov shows a wide knowledge of aerial archaeology history, even of publications outside the former USSR. In particular, he agrees with J. Bradford that aerial archaeology cannot be separated from field archaeological work, and it must be combined with historical research, written sources, geography, and the geology of an investigated area.

The second chapter deals with the *Origin and Development of Irrigated Agriculture*, in different areas of the world. Even in this case, Andrianov shows a very good knowledge of the main research achieved on this matter throughout the world, from the main ethno-archaeological studies of American Indians in the New World by J. H. Steward, C. D. Forde, and E. W. Haury, to the archaeological survey in Mesopotamia by R. McC. Adams, to excavations in the Near East (Jericho, Jarmo, Ali Kosh, Çatal Hüyük, Hacilar, etc.), which allowed to date the appearance of irrigated agriculture in the Old World as early as the 8th-6th millennia BCE.

In *Irrigation and Ancient Civilizations*, Andrianov recalls the theory of 'hydraulic societies' advanced by K. A. Wittfogel, where the development and the maintenance of large-scale irrigation systems were possible only through a centralized and strong state, with a bureaucratic structure and the wide use of forced labor. On this matter Andrianov agrees that "... the slave character of collective irrigation works is not in doubt" (Andrianov 1969:67). After an overview on the development of irrigated civilization in Egypt ('homeland of irrigated agriculture'), Mesopotamia, and China, attention is focused on Central Asia. The author noticed the poor consideration given by western scholars to this region, except for the work of R. A. Lewis (1966), who provided a remarkable outline of West Turkestan early irrigation. Among the research on the history of water management in Central Asia, the work by G. N. Lisitsyna in

Southern Turkmenia is particularly outstanding. Through a comprehensive approach similar to the KhAEE, she was able to identify, in the foothill of the Kopet Dag, early agricultural communities of Central Asia (4th-3rd millennia BCE), which were based on water exploitation from mountain brooks (*sai*) and *kair* irrigation (for *sai* and *kair* see Bonora, *infra*).

In concluding the first part of the book, Andrianov argues that irrigation skills are highly dependent on local and geographical conditions, and closely connected with the technical and socio-economic development of ancient societies.

The third chapter concerns *The Southern Delta of the Akchadarya*, the first area investigated by the archaeological-topographical KhAEE unit. Irrigation works are described according to their chronology and location. Although the data available for the Bronze Age are poor, it is highly possible that the inhabitants of Khorezm practiced irrigated agriculture at that time. Archaeological data proved, during this period, the introduction of some important devices in the development of irrigation technologies. First are the 'head works' at the Tazabagyab settlement, which were used to control the level of flood water in the former riverbed adapted for irrigation. Second, at the settlement of Bazar 8, was the distributor, i.e. an intermediate canal allowing a ramification of the network of canals to irrigate a wider area. The Archaic period (6th-5th centuries BCE) is the building time of massive irrigation systems both on the right and on the left banks of the Amudarya. In that period, canals heads were moved into the major river channel rather than in one of its lateral branches showing the great ability of Khorezmians in building 'artificial rivers' and small 'artificial deltas'. The size and section of canals increased and irrigation networks display a 'sub-rectangular' layout. The Kangju, and especially Kushan periods (4th century BCE - 4th century CE) represent the construction heyday of large fortifications and towns and the development of irrigation techniques. This is connected with the increase in field size and cultivated crops. Considering as an example the 90 km long

Kyrk-Kyz canal, the author suggests that at least 15,000 workers were required for two months for its construction, and 6,000-7,000 people for its seasonal cleaning and maintenance. The Medieval Age is characterized by a socio-economic crisis leading to the decline of many urban centers and the abandonment of settlements and irrigation works. The Khorezmshah period (12th-early 13th centuries) is a time of radical reconstruction of old irrigation systems and further development of Medieval irrigation techniques. The case of the Gavkhore Oasis illustrates the extremely high level of agricultural production reached at that time. Finally, the Mongol invasion in the early 13th century marks the end of Khorezm right bank economic development.

Research in the *Sarykamysh Delta* is presented in the fourth chapter. Here the Bronze Age finds are even poorer than in the Amudarya, thus Andrianov argues that irrigated agriculture and hydraulic facilities appeared in this region somewhat later. However, the ethnographic comparison suggests an integrated economy for that period, where thickets served as pastures for cattle, and the inhabitants fished in the channels and cultivated millet and gourds in the *kair*. In the Archaic period (6th-5th centuries BCE), the construction of important irrigation systems on the Chermen-yab and Daudan was connected with the strong state formation developed in Khorezm. Like on the right bank of the Amudarya, the progress in irrigation technology is considerable, and the water supply pattern was as follows: riverbed-head works-drainage–main canal-feeder -field. The massive construction of large irrigation systems in the Sarykamysh Delta is mainly dated to the Kangju and especially to the Kushan times (4th century BCE-4th century CE), when small canals were combined into a unique greater system. Considering Medieval irrigation works, Andrianov highlights the appearance of the *chigir* (see Bonora, *infra*), i.e. the water-lifting device introduced because of the lowering water in irrigation canals. Research along Medieval Chermen-yab demonstrates how the wide spreading of *chigir* on one hand reduced the surface covered by irrigation facilities but, on the other hand, increased the irrigated land. The traditional scheme of the irrigation network also changed according to this scheme: river-headworks-main canal-distributors of 1st and 2nd order-feeder-*chigir*-field. Also for the Chermen-yab the author provides an evaluation of labor investment required to accomplish its digging: 12,000-14,000 laborers for 50 days and 5,000-6,000 workers for the annual cleaning. Finally, modern era irrigation works are taken into consideration. The 16th-18th centuries were a period of decay for the Sarykamysh Delta and the whole Khorezm, while in the 18th-19th centuries cleaning and reconstruction of Medieval works were implemented in Northern Sarykamysh.

The fifth chapter describes the results achieved during work in *The Lower Syrdarya*, in particular on the left bank of the river. Andrianov provides an overview of the natural conditions of this area. The Lower Syrdarya has less water than the Lower Amudarya and it was a huge deltaic area, with numerous swamps and lakes, before the development of irrigated agriculture. The Bronze Age sites have not been sufficiently studied and irrigation systems of that period are poorly identified. The hydraulic works of Antiquity (4th-2nd centuries BCE), are better preserved, especially in the environs of Babish-Mulla and Chirik-Rabat along the Middle Inkardarya. These systems were based on flood regulation, which is rather primitive if compared to the more complex systems of dikes and head-works developed in Khorezm at the same time. The irrigation scheme was also simple: riverbed–former river-bed (reservoir)–feeder–field. However, quite interesting is the adaption of former riverbeds in reservoir-basins used to maintain the water level required to irrigate fields. This system was particularly widespread in the Djety-asar Oasis between the 1st century BCE and the 9th century CE. For the Medieval period (9th-16th centuries CE), Andrianov mentions the example of irrigation works developed in the so-called 'swamp settlements' and along the Janydarya. In that period the Lower Syrdarya was characterized by a primitive semi-settled economy, combining pastoralism, irrigated agriculture on former riverbeds, and fishing. The situation was different in the Middle Syrdarya, where irrigation was mainly based on gravity systems derived from the main river through head-works and it resembled the contemporary Khorezm systems. Several irrigation works were also found on banks of the Janydarya and Kuvandarya dry riverbeds in connection with abandoned Karakalpak and Kazak farming settlements (17th-19th centuries), who sometimes deepened and rebuilt the canals and the Medieval water works.

In the *Conclusion*, Andrianov retraces the stages of development of irrigation systems and water management after the research of the archaeological-topographical unit in Khorezm and in the Lower Syrdarya. From a methodological perspective, the author recalls the need for a comprehensive study of the landscape, where the single features of irrigation systems must be considered in close connection with the geographical environment. With this assumption, the preserved traces of ancient oases are an excellent source for the study of economy, material culture, and lifestyle of ancient people. However, they can be studied only through an approach combining historical and natural sciences and using different methods, such as field archaeological surveys and deciphering of aerial photographs.

In Khorezm, the development of irrigation techniques started in the Bronze Age, with the first attempts of wetland reclamations, flood controls and primitive forms of *kair* and estuary agriculture. During the Amirabad (9th-8th centuries BCE) and the Archaic (6th-5th centuries BCE) periods regulated riverbeds and former riverbeds began to be turned into small artificial main canals. These water supply improvement methods led to an increase in canal sizes and irrigated land. The Kangju, and especially the Kushan periods (4th century BCE-4th century CE), at the time of the Khorezm State, were characterized by the process of combining local systems into a single, massive system. The next important advance was the Medieval *chigir* (9th-11th centuries), i.e. the

water-lifting device, which allowed an increase of 30-40% of irrigated areas.

The development of water management in the Lower Syrdarya was different, and somewhat slower, than Khorezm. Irrigation appeared only in the mid-1st millennium BCE, and without the complex and extensive systems typical of Khorezm. Andrianov gives a socio-economical explanation of this gap. This vast and wet area required strong efforts in flood control and dike building by a long-time settled population, or perhaps a strong and centralized state. In the 1st millennium BCE this territory was occupied by tribes with an integrated primitive economy of agriculture, herding and fishing. To support this hypothesis, Andrianov notes how the indigenous Karakalpaks of this area lived under patriarchal-kinship ties until the 20th century. Moreover, he also reports labor calculations required for the construction and annual cleaning of irrigation works. Based on some ethnographical studies in the Khiva Oasis, Andrianov evaluates the high cost of labor investment required for such work and introduces his hypothesis on the emergence and development of the slave-owning mode of production. This was typical at the time of the Khorezm State, while in the Medieval period the spreading of the *chigir* reduced significantly the labor cost for constructing and cleaning canals.

In an attempt to consider the origin and development of Khoezm irrigated agriculture in a wider perspective of other Old World arid zones, Andrianov used the most recent ethno-archaeological (V. G. Childe, C. D. Forde) and paleobotanic (H. Helbaek, K. V. Flannery, N. I. Vavilov) studies. He asserts that the spread of irrigated agriculture was not a simple mechanical transfer of skills in farming and irrigation methods from one area to another, but rather a complex historical-cultural process, varied in different ecological conditions of natural vegetation and water resources.

Thenceforth, Andrianov tries to explain the causes of death of the ancient civilizations, and thus the formation of the 'lands of ancient irrigation'. Supporting the theory of geographers L. S. Berg and A. I. Voeykov, and their criticism against the determinism of E. Huntington, Andrianov suggests that the decline of the Khorezmian and Central Asian oases was primarily due to socio-economic factors, such as wars and feudal fragmentation, which contributed to the movement of people, abandonment of cultivated lands and irrigation systems. A similar process can be seen also in the Diyala Basin thanks to the research of R. McC. Adams.

Finally, the author considers his study under a modern perspective. The extension of the ancient irrigation in the Aral Sea covers approximately 5 million ha, that is three times the area covered by the irrigation network at the time of Andrianov. The information available on ancient irrigation and hydraulic works might be extensive and used to plan and sustain modern irrigation projects promoted by the former USSR.

References

Andrianov, B. V.

1966a Some Aspects of the Problem of the Interplay of Nature and Society (as illustrated by the history of the Lower Reaches of the Amu-Dar'ya in the 18th and 19th Centuries). *Soviet Geography* 7:3-14.

1966b Nekotorye aspekty problemy vzaimodeystviya prirody i obshchestva (na primere istorii osvoeniya nizovev Amu-Dari v XVIII–XIX vv.). *Izvestiya Vsesoyuznogo Geograficheskogo Obshchestva*, Tom 98, Vyp. 2:143-156.

1969 *Drevnie orositelnye sistemy priaralya (v svyazi s istoriey vozniknoveniya i razvitiya oroshaemogo zemledeliya)*. Moskva.

1976 On Irrigation and Social Organization. *Current Anthropology* 17:756.

1978a *Zemledelie nashikh predkov*. Moskva.

1978b The Concept of "Hydraulic Society". *Social Sciences Critical Studies Comment* IX(1):193-209.

1995 The History of the Economic Development in the Aral Region and Its Influence on the Environment. *GeoJournal* 35.1:11-16.

Bartold, V. V.

1965 *Sochineniya*, Tom III, *Raboty po istoricheskoy geografii*. Moskva.

Forbes, R. J.

1955 *Studies in Ancient Technology*, Vol. II. E. J. Brill, Leiden.

Gulyamov, Ya. G.

1957 *Istoriya orosheniya khorezma s drevneyshikh vremen do nashikh dney*. Tashkent.

1974 Kushanskoe tsarstvo i Drevnyaya irrigatsiya sredney azii. *Trudy mejdunarodnoy I culture Tsentralnoy azii v kushanskuyu epokhu (Dushanbe, 27 sentyabrya-6 oktyabrya 1968 g.)*, Tom I. Dushanbe.

Lecomte, O., and H.-P. Francfort

2002 Irrigation et société en Asie centrale des origines à l'époque achéménide. *Annales* 57:625-663.

Lewis, R. A.

1966 Early Irrigation in West Turkestan. *Annals of the Association of American Geographers* 56:467-491.

Mukhamedjanov, A. R.

1975 K istorii irrigatsii v kushanskuyu epokhu. *Trudy mejdunarodnoy I culture Tsentralnoy azii v kushanskuyu epokhu (Dushanbe, 27 sentyabrya-6 oktyabrya 1968 g.)*, Tom II. Dushanbe.

1994 Economy and social system in Central Asia in the Kushan Age. In *History of Civilizations of Central Asia. Vol. II, The Development of sedentary and nomadic civilizations: 700 B.C. to A.D. 250*, edited by Jànos Harmatta, pp. 265-290. UNESCO Publishing, Paris.

Tolstov, S. P.

1948a *Drevniy Khorezm. Opyt istoriko-arkheologicheskogo issledovaniya*. Moskva. 1948b *Po sledam drevnekhorezmiyskoy tsivilizatsii*. Moskva-Leningrad.

Voevodskiy, M.

1938 A Summary Report of a Khwarizm Expedition. *Bulletin of the American Institute for Iranian Art and Archaeology* 5(3):235-244.

Notes on Translation, References, and Transliteration

Simone Mantellini

The English edition of *Drevnie orositelnye sistemy priaralya* attempts to be as faithful as possible to the original. The criteria followed in editing are those of *American Antiquity*, while for specific names and styles the *Merriam-Webster's Collegiate Dictionary, 11th ed.*, and the *Chicago Manual of Style, 14th ed.*, were widely used. The introductory papers, notes and captions follow these guidelines as well as the style adopted by Andrianov in the original book edition. All the words in languages other than English have been italicized when not included in the *Webster's Collegiate Dictionary, 11th ed.* (for example *aryk, chigir, poisk* , etc.) and then explained in the Glossary by G. L. Bonora. Titles of Russian publications cited by the author within the text have been italicized, and then translated into English between brackets immediately after the original. For example: *Drevniy Khorezm* (Ancient Khorezm).

However, some changes, corrections and additions to the Russian edition were necessary in order to adapt the translation and the transliteration of many specific terms, names, and concepts belonging to Russian-Soviet archaeology. These changes mostly refer to some inaccurate and unclear citations, especially regarding foreign bibliographies. They are included in the Editor's notes at the conclusion of this volume with a short description of changes while footnotes placed in the original edition remain unchanged in the Endnotes. Question marks are in Andrianov's original Text and Endnotes.

Figures and plates were left as they were in the original edition. Despite efforts to recover original pictures and figures, they were no longer available thus they were scanned at the highest possible resolution in order to maintain good quality images. The tables were completely redone.

The in-text reference citations have been changed only when necessary to provide uniformity with the reference list at the end of the book. Thus, in the multivolume work by V. V. Bartold and N. I. Vavilov, the year of publication was added after the author, and eventually the number of *Tom* in case of two, or more *Tom* in the same year. The number of the *Tom* (volume) and *Vypusk* (issue/number either of a journal or of a monograph in a series) remains numbered in Arabic or in Roman characters according to the original edition.

Reference citations of more authors also agree with the order given by Andrianov, who usually followed a chronological rather than alphabetical sequence.

A final mention concerns the bibliography. References quoted in the text are collected at the end of the volume according to the original book, where they are divided in three sections: 'Proceedings of the founders of Marxism-Leninism', 'References in Russian', and 'Original references in other languages'. To facilitate the reader in searching single references, especially those in Russian, several of the works reported by Andrianov as Abbreviations either of journals or edited volumes had to be spelled out to properly attribute them to the right authors. Despite many efforts, several remained difficult, when not impossible, to be found and thus were left as mentioned in the original book. Abbreviations reported at the beginning of final references concern bibliographical information transliterated from Russian publications (for example, *vyp.* = issue, *izd.* = edition number, etc.).

It must be noted that articles in journals or in edited volumes were reported by Andrianov without specifying singles pages. Checking all the entries would have required a long time so it was done only for those doubtful citations or for those citations requiring additional scrutiny. As mentioned above, all of Andrianov's writings are now listed in I. Arjantseva's additional papers (see *infra*).

The number of abbreviations for journals and series were also reduced considerably after transliteration in order to avoid matching journal abbreviations published either in Russian or in English. For example, *SA* might refer either to *Sovetskaya Arkeologiya* or *Scientific American*. The in-text abbreviations referring to an issue of journal, or series, were replaced, when possible, by the editor of that issue. Thus, *MKhE*, Vyp. 3 becomes 'Tolstov and Itina 1960'. However, some Russian abbreviations remain unchanged and they are reported in the *Abbreviations List*.

The transliteration system used both in the text and in the bibliography is Passport 2003 (see the table at right), which simplifies the names and terms, and makes them more similar to Western standards. For example, the name of orientalist В. В. Бартольд is Bartold, and not Barthold, Barthol'd, or Bartol'd; the culture of Кельтеминар is Kelteminar, and not Kel'teminar, etc. This system considers 'Ж' as 'J', thus Таджикистан is Tadjikistan, and not Tadzhikistan or Tadžikistan; archaeologist Т. А. Жданко is Jdanko, and not Zhdanko or Ždanko; the site of Джейтун is Jeitun, and not Dzheitun or Džeytun, etc.

Place names, archaeological sites, rivers, etc, were always transliterated according to the original edition. Thus, Амударья is Amudarya, and not Amu Darya, Amu-Darya, or Amu-darya; Каракум is Karakum, and not Kara Kum, Kara-Kum, or Kara-kum, etc.

Cyrillic	Passport 2003
А, а	A, a
Б, б	B, b
В, в	V, v
Г, г	G, g
Д, д	D, d
Е, е	E, e
Ё, ё	E, e
Ж, ж	J, j
З, з	Z, z
И, и	I, i
Й, й	Y, y
К, к	K, k
Л, л	L, l
М, м	M, m
Н, н	N, n
О, о	O, o
П, п	P, p
Р, р	R, r
С, с	S, s
Т, т	T, t
У, у	U u
Ф, ф	F, f
Х, х	Kh, kh
Ц, ц	Ts, ts
Ч, ч	Ch, ch
Ш, ш	Sh, sh
Щ, щ	Shch, shch
Ъ, ъ	-
Ы, ы	Y
Ь, ь	-
Э, э	E
Ю, ю	Yu
Я, я	Ya

Acknowledgments

Simone Mantellini

The effort behind the new edition of *Ancient Irrigation Systems of the Aral Sea Area* deserves some important acknowledgments. The first are for C. C. Lamberg-Karlovsky and Maurizio Tosi, who promoted this enterprise and strongly supported the idea to make this book the first issue of a series dedicated to the 'Archaeology of Early Water Management'. Given the excessive waste of water and the problems related to water supply which daily affect a large part of the world's population, the hope is that scientific studies like this might be helpful to raise awareness of an asset as important as it is fleeting. 'Learning from the ancient' is not just a cliché, but a good principle to be followed and implemented in the common interest.

The whole enterprise could never have been done without the support of the Segretariato Europeo per le Pubblicazioni Scientifiche - SEPS in Bologna, which funded the translation of the text. A special thanks to Wren Fournier, who coordinated the editing of the work with patience and consistency, and always provided ready answers and valuable advice. Thanks also to Dimitri Kostushkin, Andrei A. Shestakov, Eugenio Bortolini, Clif Morse for their assistance in the translation into English and the editing of images.

Thanks to Olga N. Inevatkina, Lyuba B. Kircho, and Irina A. Arjantseva who revised the translation into English of the Russian introductory papers. They also provided fundamental aid in solving some issues related to Russian archaeology and some bibliographical references quoted by the author. Mukhamadjan Kh. Isamiddinov has been a precious resource on studies and general aspects of irrigation in Central Asia. Likewise, a special mention for Frantz Grenet, Andrea Gariboldi, Paolo Ognibene, and Andrea Piras for their assistance in the field of Iranian and Avestan questions, terms and names. Thanks also to Gianni Marchesi concerning the cuneiform texts.

A final, but very special, mention is for Serena di Cugno and Morena Agostini who successfully researched all English bibliographical references mentioned in the original edition. Serena also shared arrangement of the final references. Last but not least, this work could not have been realized without the invaluable work and collaboration of Giò Morse and Gian Luca Bonora. Giò was in charge of the huge review of the English form of the volume, while Gian Luca was fundamental in the translation of the book from Russian and other entries concerning Central Asian archaeology and literature. To both goes my deepest gratitude.

I: Central Asian Archaeology: The Russian and Soviet Times

Pavel M. Dolukhanov

A huge, isolated, landmass in the midst of Eurasia, formerly known as Soviet Central Asia, became a pawn in the global power struggle between the Russian and British empires. Contrary to expectations, backward Russia got an upper hand, gradually encroaching into the Transoxian sands and establishing its control over Tashkent (1865), Samarkand (1868) and Turkmenistan (1880s) with the Afghan quagmire and the Persian despotism forming natural borders. By the end of the 19th century this entire area became a Russian colony, with only Khiva and Bukhara retaining a status of formal protectorates. This conquest occurred relatively peacefully, with little or no open resistance. The Russian colonial rule was rather mild. Preoccupied with maintaining 'peace and order, Russian colonial authorities tried to avoid disturbing the traditional way of life and local social networks.

While sharing many common features with the European colonization elsewhere in the world, the Russian conquest of Central Asia had several important distinctions. Unlike other colonial powers, Russia had a stretched and easily penetrable common frontier with its new acquisitions. This facilitated a large scale influx of immigrants which included both peasant folk and skilled workforce from the intelligentsia of Russia's cultural heartlands. The centuries of cultural isolation were rapidly overcome. In 1888 the Trans-Siberian railroad reached Samarkand and by 1905 the Russian railway network reached the Caspian Sea. New European suburbs with Russian educational institutions cropped up some distance from the earlier walled cities. These new urban centers harbored Russian language educational institutions in which the curricula from the outset envisaged training the local elite. The Russian educated diaspora included students of local histories, literature and arts.

The middle and late 19th century was marked by outstanding archaeological discoveries at Central Asia's doorstep, notably in Iran and Mesopotamia. Not surprisingly, soon after the Russian conquest, the first non-professional archaeological digs were reported from that area. Archaeological excavations started already in 1867, when P. I. Lerkh studied the site of Djankent in the Lower Syrdarya (Kohl 1984). Afrasiab (the Old Samarkand) was unsystematically excavated by Russian army officers in the 1860s. Its studies were resumed under more competent direction of N. I. Veselovskiy in 1875. The ancient city of Merv was later explored by V. A. Jukovskiy in 1890.

Concerning the work carried out in Afrasiab, ancient Samarkand, Frantz Grenet kindly commented as follows about the first Russians exploration after the conquest in 1868:

"N. I. Veselovskiy, at that time the leading specialist in Scythian archaeology, excavated Afrasiab more or less like a kurgan, digging deep pits at irregular intervals, with very little results. Between 1904 and 1932, large-scale excavation were resumed by V. L. Vyatkin, Director of the Samarkand Museum. He used a non-destructive approach, mainly searching and following the massive earthen walls. Using also his excellent knowledge of Arabic and Persian records he was able to establish the main features of the pre-Mongol city such as the concentric city walls, the water channels and the pools, the citadel, and the Friday Mosque. But, having no notion of stratigraphy nor of Achaemenid and Greek pottery, he did not recognize the most ancient levels and dated the first phase of the site to the first centuries CE, erroneously rejecting its identity with

Maracanda (the ancient name of Samarkand) mentioned by the historians of Alexander. He published very little. The true chronological and topographical sequence of Afrasiab was established in 1945-48 by A. I. Terenojkin and then by G. V. Shishkina in the 1960-70s".

At the same time General A. V. Komarov, hoping to recover the remains of Alexander the Great, opened up trenches on the northern mound of Anau, the site that played a key role in uncovering Central Asian prehistory (Kohl 1987). Likewise in other European countries, 19th century archaeology was largely dictated by the agenda of antiquarians. Significantly, the first museum was established in Samarkand in the 1870s.

Very soon the antiquities of Central Asia attracted Western archaeologists. In 1904 an international team directed by the American R. Pumpelly, a geologist with vast experience in explorations of Asia, including Siberia, China and Mongolia, started digging the mounds of Anau, 12 km east of Ashkhabad (Pumpelly 1908). Although later criticized for small-size exposures and poor recording techniques, these excavations established the first recognizable cultural sequence, extending from the Eneolithic (northern mound), through the Bronze and Iron Ages (southern mound) and into historic times (city of Anau).

The new stage in the studies of Central Asian prehistory focused on the establishment of Soviet rule in the early 1920s. Soviet policy on 'national republics' vacillated between encouragement of national cultures loyal to proletarian inter-nationalism and ruthless uprooting of the drive for independence branded as bourgeois nationalism. This took the form of establishing Academies of Sciences with numerous research institutes and laboratories in each Soviet Socialist republic (Kazakhstan, Kyrgyzstan, Tajikistan, Turkmenistan and Uzbekistan). As a rule, these academies included either institutes of archaeology or departments of archaeology within the institutes of history. In many cases, and particularly in the republics of Central Asia, archaeological

institutions were initially manned by Russian archaeologists, mostly from Moscow or Leningrad. But eventually they were replaced by national cadres, who were trained either in the local or Russian universities. In those years archaeological investigations in Central Asia were conducted by prominent Russian and local academics, including A. M. Belenitskiy, A. N. Bernshtam, Ya. G. Gulyamov, G. F. Debets, M. M. Dyakonov, B. A. Kuftin, M. E. Masson, A. P. Okladnikov, A. I. Terenojkin, A. Yu. Yakubovskiy among others. These scholars left a legacy in the form of local archaeological schools, which to this day form the backbone of Central Asian archaeology.

Centrally-funded scholarly institutions began to appear as early as the 1920s. Aimed at promoting national cultures in the spirit of proletarian internationalism these institutions sponsored archaeological field projects. In those years the Turkmenkult (Institute of Turkmenian Culture, which organized many archaeological expeditions in Southern Turkmenistan and along the Amudarya between the 1929 and World War II), financed excavations of prehistoric sites in the Merv Oasis and along the Tedjen and Murgab rivers. Significantly B. B. Piotrovskiy, who later became famous for his discovery of Urartian sites in the Caucasus, started there his archaeological career. At the same time, D. D. Bukinich, an irrigation engineer, began digging Namazga-depe, the site he had discovered in 1916.

1937 the year associated in the collective memory of Russians as a synonym of the Great Terror, was a milestone in Central Asian archaeology. During that year the Khorezmian Archaeological Ethnographic Expedition was set up under the aegis of the Soviet Academy of Sciences. Under the efficient and competent directorship of S. P. Tolstov, this institution pioneered numerous novel field techniques, advanced even by today's standards. They in-cluded settlement pattern studies, aerial photo reconnaissance and the use of mechanized digging equipment. Using these techniques the expedition conducted large-scale surveys and minutely recorded excavations of numerous sites in hitherto virtually unexplored areas of Central Asian mesopotamia.

At that time, the first Old Stone Age sites were discovered in Central Asia. One of the most outstanding discoveries was the Neanderthal burial at Teshik-Tash rock shelter in Southern Uzbekistan, made by A. P. Okladnikov in 1938 (Okladnikov 1949). This burial remains the focus of scholarly interest to this day (Krause et al. 2007).

In 1946 the Southern Turkmenistan Complex Archaeological Expedition, better known by the acronym YuTAKE (Yujno-Turkmenistanskoy Arkheologicheskoy Komplesnoy Ekspeditsii) was organized under the leadership of M. E. Masson. The YuTAKE became especially active in 1952-1962 with large-scale excavations of settlements on the Northern Kopetdag piedmont and the Merv Oasis. These excavations included Djeytun - the earliest agricultural settlement northwest of Ashkhabad, and Namazga-Depe - the largest tell-site in that area, and several others. In 1965 the YuTAKE, jointly with the Leningrad (St. Petersburg) Institute of Archaeology under the directorship of V. M. Masson initiated detailed explorations of Altyn-Depe. At about the same time large-scale excavations were undertaken in the Geoksyur Oasis in the Tedjen Delta by I. N. Khlopin and V. I. Sarianidi, respectively. These studies, in which Tashkent University also participated, continued until 1993. In the early 1990s a large multidisciplinary project with the participation of the Institute of Archaeology of the University College of London - UCL, was conducted at Djeytun sites (Harris, Gosden and Charkes 1996). These studies shed a new light on the early development of agricultural systems and the emergence of an urban-type culture in that area. They also resulted in the revision of the older Pumpelly system and the establishment of a new cultural sequence for Eneolithic, and Bronze Ages (Namazga I-VI; Kuftin 1956; Masson 1966, 1971, 1981, 1988).

Multidisciplinary investigations in the Central Asian mesopotamia conducted in the 1930s and 1940s by the Khorezm Expedition under S. P. Tolstov brought to light a panoply of Neolithic and Bronze Age sites, on the basis of which several archaeological cultures, including the Kelteminar, were identified. Later on, they became the object of specific multidisciplinary projects (Vinogradov 1968, 1981; Vinogradov and Mamedov 1975).

Following the early lead of Russian pre-revolutionary Orientalists, Soviet archaeologists attached paramount importance to studies of early irrigation systems. This had both practical and ideological underpinnings. Situated in the area of an extremely arid climate (with less than 300 mm of annual rainfall), agriculture in Central Asia always was and still is dependent on irrigation and hence remained largely restricted to river valleys and oases. In Soviet times, Central Asia became the principal producer of cotton in the USSR which necessitated the expansion of arable land and additional irrigation. Hence the interest in related research and developments. Several scientific institutions were established in national Academies of Sciences aimed at the studies of water resource management. At least theoretically, this kind of research was linked with historical and ethnographic studies, aimed at elucidating the historical experience of Central Asian nations in that area. Yet as often happens, the historical experience was usually ignored by political decision makers. Large-scale irrigation development in the Lower Amudarya Delta was conducted, notwithstanding the historical expert opinion, and eventually wound up in the Aral Sea ecological catastrophe.

The ideological raison d'être of irrigation related studies resided in the concept of 'Asian Mode of Production'. This concept was briefly outlined by K. H. Marx and F. Engels in their paper *British Rule in India* (1853) and much later was (revised) by K. A. Wittfogel in the form of a 'hydraulic civilization' theory (Wittfogel 1957). According to the original Marxist concept, the 'Asian Mode of Production' was based on irrigation and featured a self-sufficient economy; limited degree of labor division; limited private ownership of production means; underdeveloped trade; patriarchal-type slavery; monarchical despotism as the predominant political system. The applicability of this concept to the Central Asian historical realities was a matter of prolonged discussion amongst Soviet historians in the 1950s and 1960s, in the course of which conflicting opinions were advanced but no consensus was reached.

In the 1920s D. D. Bukinich noted that early agricultural settlements on the Northern Kopetdag piedmont strip were located either on the upper portions of alluvial fans of small streams or along the flood plains of the larger rivers. As D. D. Bukinich further noted, these locations were highly favorable for primitive forms of irrigation consisting in small earthen walls along borders of plots, facilitating the collection and storing of flood water (Bukinich 1924:113). Much larger-scale investigations of ancient irrigation systems followed in 1934, when a special project was carried out in order to study the pre-Mongol irrigation network in the cities of Khorezm (Voevodskiy 1938).

Since its creation in 1937 the Khorezmian Expedition had initiated several major projects for the study of early irrigation systems in the lower Amudarya catchment basin. The early stages of these investigations were summarized by Ya. G. Gulyamov (1957) and S. P. Tolstov (1958).

A new stage in the investigations of Central Asian early irrigation was undertaken by Boris Vasilevich Andrianov (1919-1993). Born in Moscow, into an artist family, B. V. Andrianov studied at the Department of Geography at Moscow University, where he graduated in 1944. After 1945, his career became inextricably linked with the Institute of Ethnography of the Russian Academy of Sciences and especially with the Khorezm Expedition. Becoming a professional geographer, B. V. Andrianov headed an archaeological survey branch, which included the processing of aerial photo imagery. In the late 1940s he conducted ethno-geographical studies in the Aral Sea area, which resulted in his Ph.D. dissertation, *Etnicheskaya territoriya karakalpakov v severnom Khorezme XXVIII-XIX vv* (The Karakalpak

ethnic territory in Northern Khorezmia in the 18th-19th centuries; see Andrianov 1951). Since 1952, Andrianov directed his principal multidisciplinary projects targeted at ancient irrigation systems in the Lower Amudarya and Aral Sea area. The years of laborious, minute investigations, resulted in 1969 with his seminal book *Drevnie orositelnye sistemy priaralya* (Ancient Irrigation Systems of the Aral Sea Area) of which an English translation has now became available.

In the 1970s a similar technique which included the processing of aerial imagery was applied by G. N. Lisitsyna in the Geoksyur Oasis on the lower reaches of the Tedjen Basin (Lisitsyna 1978). She concluded that, in the Late Eneolithic, the irrigation system consisted of three parallel canals, drawing water from one of the principal delta branches. A network of minor ditches (*aryks*) branched from each canal.

One of the notable advantages of archaeological studies of the Soviet period was their multidisciplinary character, with the active participation of paleoenvironmentalists, such as I. P. Gerasimov, A. S. Kes, E. D. Mamedov, G. N. Lisitsyna, and many others. These writers' studies have revealed prolonged periods of increased humidity, roughly dated to 8-4 ka BP. During that period perennial fresh-water lakes developed in the present day waterless Karakum desert (Vinogradov and Mamedov 1975). The Amudarya (Oxus) River emptied into the Sarykamysh depression south of the Aral Sea. Its huge delta with numerous prehistoric sites now lies east of this depression. From there the Uzboy River, with its long, sinuous valley, carried the river water further west, into the Caspian Sea. According to more recent estimates, at that time the Aral Sea level stood approximately 100 m asl whereas its present position is 66 m asl (Trofimov 1986).

The spread of agriculture in Europe and Western Asia may be statistically approximated as a gradual expansion from the Levantine center, either enhanced or slowed by environmental factors which created bottlenecks. The earliest signs of agriculture in the Kopetdag piedmont (Sang-i Chakhmak) suggest an age of 7000-6400 cal BC, with 14C dates for early agricultural Djeytun site being 6200-5800 cal BC (Harris, Godsen, and Charkes 1996). Significantly, a network of culturally related pottery-bearing foraging sites arose along the waterways further north (Vinogradov 1981). The stratum with early pottery was dated to the Jebel Cave (Turkmenistan) to 5300-4800 cal BC. Radiocarbon dates of early pottery sites in the Lower Volga and North Caspian Lowland suggest an older age of 8000-6500 cal BC (Vybornov 2008).

A new perspective opens up with the identification of a cool and dry 8.2 ka BP event (6400-6000 BCE), which is observed in a large number of high-resolution climate proxies in the Northern Hemisphere, including Western Asia (Weninger 2006). This event might have triggered, on one hand, a necessity for an artificial regulation of water supply in the form of primitive irrigation schemes and on the other an outflow of surplus population from early farming areas. The expanding population merged with hunter-gatherers further north and transmitted to them traditions of pottery-making. This process encompassed the entire semi-desert and steppe areas during the subsequent Altithermal period.

The last remark concerns the general context in which Soviet era archaeological research in Central Asia was carried out. As one might judge from Andrianov's book, Soviet research at that time involved considerable capital investments on the part of state funding bodies. Andrianov and his associates used aircrafts on a regular basis with expensive aerial photo-equipment and required no less expensive professional aerial-imagery processing. And this was not an isolated case in what Leo S. Kleyn later referred to as the 'Phenomenon of Soviet Archaeology'. Yet this went together with appalling working conditions. I clearly remember our expedition vehicles, ill-fitted for the desert; they were eventually painted dark green. These former army personnel carriers from Central Russia were devoid of any kind of weather protection and with sunrise turned into torture chambers. Often with scarce water and food supplies, baked during the day, frozen during the night, and usually poorly paid, these people achieved outstanding results which, to this day, remain the pride of world archaeology. It is no surprise that many of them died prematurely young. Please remember that when reading this book.

References

Andrianov, B. V.
1951 *Etnicheskaya territoriya karakalpakov v Severnom Khorezme (XVIII-XIX vv.).* Avtoreferat dissertatsii na soiskanie uchenoy stepeni kand. istoricheskikh nauk. Moskva.

1969 *Drevnie orositelnye sistemy Priaralya (v svyazi s istoriey vozniknoveniya i razvitiya oroshaemogo zemledeliya).* Moskva.

Bukinich, D. D.
1924 Istoriya pervobytnogo oroshaemogo zemledeliya v Zakaspiyskoy oblasti v svyazi s voprosom o proiskhojdenii zemledeliya i skotovodstva. *Khlopkovoe delo*, No. 3-4: 92-135.

Gulyamov, Ya. G.
1957 *Istoriya orosheniya khorezma s drevneyshikh vremen do nashikh dney*. Tashkent.

Harris, D. R., C. Gosden and M. P. Charkes
1996 Jeitun: Recent Excavations at an Early Neolithic Site in Southern Turkmenistan. *Proceedings of Prehistoric Society* 62:423-442.

Kohl, P. L.
1984 *Central Asia: Palaeolithic Beginnings to the Iron Age*. Synthèse 14. Edition Recherches sur les Civilisations, Paris.

Krause, J., L. Orlando, D. Serre, B. Viola, K. Prüfer, M. P. Richards, J.-J. Hublin, C. Hänni, A. P. Derevianko, and S. Pääbo
2007   Neanderthals in Central Asia and Siberia. *Nature* 449:902-904.

Kuftin, B. A.
1956   Polevoy otchet o rabote XIV otryada YuTAKE po izucheniyu kultury pervobytno-obshchinnykh osedlozemledelcheskikh poseleniy epokhi medi i bronzy v 1952 g. *Trudy Yujno-Turkmenistanskoy Ekspeditsii AN SSSR. Tom VII, Pamyatniki kultury kamennogo i bronzovogo veka yujnogo turkmenistana.* Ashkhabad.

Lisitsyna, G. N.
1978   *Stanovlenie i razvitie oroshaemogo zemledeliya v yudjnoy Turkmenii.* Moskva.

Marx, K. H. and F. Engels
1957   *Sochineniya,* Tom 9. Izd. 2. Mosvka.

Masson, V. M.
1966   *Srednyaya Aziya v epokhu kamnya i bronzy.* Moskva-Leningrad.
1971   Poselenie Djeytun (Problema stanovlemiya proizvodyashey ekonomiki). *Materialy i issledovaniya po arkheologii SSSR,* No. 180. Leningrad.
1981   *Altyn-Depe.* Nauka, Leningrad.
1988   *Altyn-Depe.* University Museum, University of Pennsylvania, Philadelphia.

Pumpelly, R.
1908   *Explorations in Turkestan. Expedition of 1904. Prehistoric Civilizations of Anau. Origins, growth, and Influence of Environment.* 2 vols. Carnegie Institution of Washington, Washington D.C.

Okladnikov, A. P.
1949   *Teshik-Tash, paleoleticheskiy chelovek.* Moskva.

Tolstov, S. P.
1948   *Drevniy Khorezm. Opyt istoriko-arkheologicheskogo issledovaniya.* Moskva.

Trofimov, G. N.
1986   Paleogidrografiya Uzboya. *Bullyuten Komissii po izucheniyu chetvertichnogo perioda* 55:107-111.

Vinogradov, A. V.
1968   Neoliticheskie pamyatniki Khorezma. *Materialy Khorezmskoy Ekspeditsii,* Vyp. 8. Moskva.
1981   Drevnie okhotniki i rybolovy Sredneaziatskogo Medjdurechya. *Trudy Khorezmskoy arkheologo-etnografiches- koy ekspeditsii,* Tom XIII. Moskva.

Vinogradov, A. V. and E. D. Mamedov
1975   *Pervobytnyi Lyavlyakan.* Moskva.

Voevodsky, M. V.
1938   A Summary Report of a Khwarizm Expedition. *Bulletin of American Institute for Iranian Art and Architecture* 5(3):235-244.

Vybornov, A. A.
2008   *Neolit Volgo-Kamya [The Volga-Kama Neolithic].* Samara Pedagogical University Press, Samara.

Weninger, B., E. Alram-Stern, E. Bauer, L. Clare, U. Danzeglocke, O. Jöris, C. Kubatzki, G. Rollefson, H. Todorova and T. van Andel
2006   Climate forcing due to the 8200 cal yr BP event observed at Early Neolithic sites in the eastern Mediterranean. *Quaternary Research* 66(3):401-420.

Wittfogel, K. A.
1957   *Oriental Despotism: a Comparative Study of Total Power.* Yale University Press, New Haven.

II: Boris Vasilevich Andrianov and the Study of Irrigation in Ancient Khorezm

Sergey B. Bolelov

Once led Boris Vasilevich Andrianov
In the desert the topographical team ...
– (from a song of the Expedition)

The Khorezm Archaeological-Ethnographic Expedition has a special place in Soviet and Russian humanities. It was set up by the outstanding historian, orientalist, ethnographer and archaeologist Sergey Pavlovich Tolstov in 1937. In its first years, the primary goal was to discover and map the archaeological sites in the Southern Aral region which had not been investigated before, either archaeologically or anthropologically. Its exploration tracks included the entire region of desert lands once covered by the irrigation systems of ancient Khorezm, on the lower reaches of the Amudarya. The first archaeological map of this historical and cultural area was created, and the results of this work were published in 1948 by S. P. Tolstov in his *Drevniy Khorezm* (Ancient Khorezm).

In postwar years the work of the Khorezm Expedition began within a new methodological framework, and the research involved not only the Amudarya Delta, but also the Lower Syrdarya (east of the Caspian Sea), the Ustyurt plateau and the central regions of the Karakum and Kyzylkum deserts. The expedition was organized for archaeological and ethnographic work, and it quickly became one of the largest and best-equipped scientific expeditions of the former Soviet Union.

S. P. Tolstov was always convinced of the value of a multidisciplinary approach, and was therefore an adherent of complex research methods. That is why, in addition to archaeologists and ethnographers, specialists such as geologists, geomorphologists, pedologists, geographers, and physical anthropologists worked with the expedition. Airborne research on the large scale used here was a first attempt in world archaeology. The entire Southern and Southeastern Aral Sea region was covered by aerial survey, which produced an archaeological-geomorphological map of the lower reaches of the Amudarya and Syrdarya rivers.

All this work was done by S. P. Tolstov's team of associates that was formed during the years of research. A special place among his pupils belongs to Boris Vasilevich Andrianov. B. V. Andrianov took part in the expedition for the first time in 1946 as a geographer and cartographer, and from then on, all his life and scientific activity were closely connected with the Khorezm Expedition and the Southern Aral Sea region.

Throughout the work of the scholars, all past human activities were studied by the Khorezm Expedition in relation to the water flow of the Amudarya and Syrdarya rivers and the artificial irrigation network derived from them. As Egypt has always been considered a 'gift of the Nile', it is possible to call Khorezm a 'gift of the Amudarya'. For that reason, the history of ancient irrigation drew S. P. Tolstov's attention right from the start of the project. As early as the postwar years it became one of the priority research topics which for a long time defined the direction of research. Thus, in 1952, an archaeological-topographical team headed by B. V. Andrianov was created within the KhAEE, with the aim to study the ancient irrigation systems around the Southern Aral Sea.

In spite of the great importance of his work and results, B. V. Andrianov was not the first to study ancient Central Asian irrigation. In fact, several Russian scientists had already dealt with that subject, as witnessed by *K istorii orosheniya turkestana* (On the history of irrigation in Turkestan), a special section within V. V. Bartold's main work (Bartold 1965:95–233); the works by D. D. Bukinich, primarily those on the irrigation in Afghanistan reported as a chapter in the book *Zemledelcheskiy Afganistan* (Agricultural Afghanistan) written with N. I. Vavilov (see Vavilov and Bukinich 1929); B. A. Latynin's research on the history of irrigation in the Fergana Valley (Latynin 1956, 1957); and many other works that covered to some extent the irrigation of various historical and cultural regions in antiquity.

The beginnings of research on the ancient irrigation systems in Khorezm date back to 1934, when M. V. Voevodskiy surveyed the ancient irrigated land in the dry delta channels of the Daudan and Daryalyk on the left bank of the Amudarya (Voevodskiy 1938). This work was continued by the Uzbek scholar Ya. G. Gulyamov who joined the expedition in Khorezm and studied both the ancient and the modern irrigation systems of Southern Khorezm and the left bank of the Amudarya. The results of Ya. G. Gulyamov's research were published in 1957 in the book *Istoriya orosheniya khorezma s drevneyshikh vremen do nashik dney* (History of the irrigation in Khorezm from its ancient period to present day). All the above mentioned works formed the basis on which the studies of the ancient irrigation systems of the Southern Aral Sea region were developed.

The beginning of work of the archaeological-topographical team of the Khorezm Expedition under B. V. Andrianov mark a new age in the study of the history of ancient irrigation not only in Khorezm, but in the whole of Central Asia, and since then the history of irrigation has become an independent direction of research in historical and archaeological projects.

The study of ancient irrigated lands of Khorezm represented a unique phenomenon in the 1940s to 1960s. The desert fostered the excellent preservation of archaeological monuments which had not yet lost their original form. Extending over a huge area, the remains of ancient irrigation, the ruins of fortified sites and open settlements of various periods represent evidence of early settled farmers in that region. The work of the archaeological-topographical team focused on the ancient irrigated lands of the Aral Sea region with an area of 4.5 million ha, from the Sarykamysh Lake and the Ustyurt plateau to the Middle Syrdarya. Investigations covered the ancient irrigated lands of Khorezm on the right bank of the Amudarya (about 2,000 *poisk* surveyed), the Sarykamysh Delta on the left bank of the Amudarya (about 1,000 *poisk* surveyed), and the enormous area of the Lower Syrdarya. During this work, thousands of settlement sites and the remains of channels and other water works were surveyed and mapped by B. V. Andrianov and his team. Overall it represented a new approach and new methods aimed at reconstructing ancient agricultural settlements and the cultural landscape: "the complex investigation of cultivated landscapes requires an interdisciplinary approach employing natural-scientific, geographic, and humanities disciplines" (Andrianov 1969:16). From the very beginning, the archaeological-topographical team worked in close cooperation with paleogeographers, geomorphologists, and pedologists. In addition, and perhaps for the first time in Central Asian archaeology, the reconstruction of ancient cultural landscapes was based on a wide use of aerial photography which was also applied for identifying large and small irrigation systems in neighboring regions (the Babish-Mulla Oasis in the Syrdarya Delta and the system of ancient irrigation around the ancient settlement of Kalalygyr-kala I). During this work the method of decoding aerial photos of desert landscapes was developed to detect features of various kinds of irrigation systems and underground constructions. Further trace correlations detected on air photos with evidence found during linear survey work made it possible to reconstruct to a large extent the ancient irrigation systems and the landscape transformations which occurred in the Southern Aral Sea region over several millennia, from the Bronze Age up to the Late Middle Age (19th century).

It should be emphasized that the irrigation systems were not considered a separate phenomenon, but a part of the region during a specific historical period. B. V. Andrianov was able to identify some features of ancient irrigation which were typical of certain neighboring regions and differed according to geographical and environmental conditions. All these data were the basis for the reconstruction of paleoeconomic systems of various neighboring communities in the Aral Sea region.

Being a convinced adherent of the interdisciplinary approach to the study of the past, B. V. Andrianov never limited himself to the analysis of ancient irrigation works. During the many years of the expedition, hundreds of known archaeological sites were surveyed, and artifacts characterizing the Khorezmian material culture from the Neolithic to the Late Middle Ages were collected. Some new archaeological monuments were found,

such as the Bronze Age burial at Kokcha 3; the Late Archaic settlement of Dingildje; the oasis of Nurum-depe in the Sarykamysh Delta dated to the Antique period, and many others which have become fundamental for the chronological standardization of certain historical periods of Khorezm.

In conclusion, it is difficult to overestimate B. V. Andrianov's contribution to the study of Ancient and Medieval history of the Aral Sea region and, without doubt, he can be considered the founder of the branch of Central Asian archaeology devoted to the study of ancient irrigation. Without his conclusions and the basic theoretical positions he formulated, the study of ancient history as well as the history of material culture, not only of Khorezm, but of all of Central Asia, would have been impossible.

References

Andrianov, B. V.

    1969   *Drevnie orositelnye sistemy Priaralya (v svyazi s istoriey vozniknoveniya i razvitiya oroshaemogo zemledeliya)*. Moskva.

Bartold, V. V.

    1965   *Sochineniya. Tom III, Raboty po istoricheskoy geografii*. Moskva.

Gulyamov, Ya. G.

    1957   *Istoriya orosheniya khorezma s drevneyshikh vremen do nashikh dney*. Tashkent.12

Latynin, B. A.

    1956   Voprosy istorii irrigatsii drevney Fergany. *Kratkie Soobshcheniya o Dokladakh i Polevykh Issledovaniyakh Instituta Istorii Materialnoy Kultury AN SSSR*, Vyp. 64:15-26. Moskva.

    1957   Voprosy istorii irrigatsii drevney Fergany. *Kratkie Soobshcheniya Instituta Etnografii AN SSSR*, Vyp. XXVI:12-15.

    1959   Nekotorye voprosy metodiki izucheniya istorii irrigatsii Sredney Azii. *Sovetskaya Arkheologiya*, No. 3:19- 28.

    1962   *Voprosy istorii irrigatsii i oroshaemogo zemledeliya drevney Fergany*. Obobshchayushchiy doklad po rabotam, predstavlennym kak dissertatsiya na soiskanie uchenoy stepeni doktora istoricheskikh nauk. Leningrad.

Tolstov, S. P.

    1948   *Drevniy Khorezm. Opyt istoriko-arkheologicheskogo issledovaniya*. Moskva.

Vavilov, N. I. and D. D. Bukinich

    1929   *Zemledelcheskiy Afganistan*. Leningrad.

Voevodsky, M. V.

    1938   A Summary Report of a Khwarizm Expedition. *Bulletin of American Institute for Iranian Art and Architecture* 5(3):235-244.

III: Map of the Main Archaeological Sites of Khorezm published in 2002 after S. P. Tolstov

Henri-Paul Francfort and Olivier Lecomte

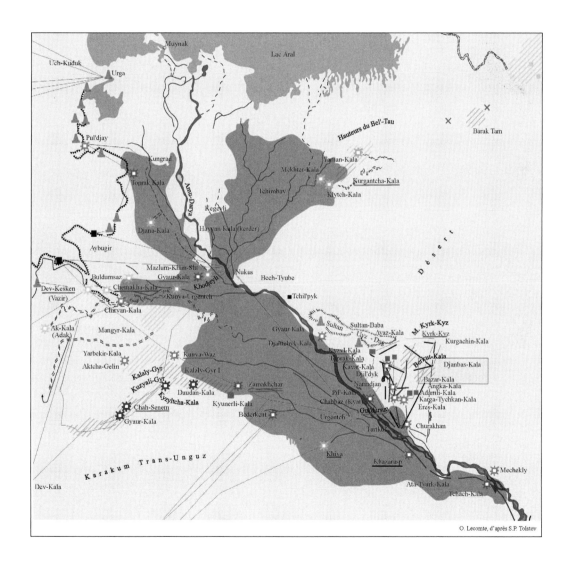

O. Lecomte, d'après S.P. Tolstov

IV: The Karakalpak-Australian Expedition in Khorezm 50 Years after B. V. Andrianov

Alison Betts and Vadim N. Yagodin

B. V. Andrianov and the Tash-Kyrman Oasis

The history of Central Asia is written in the lines of its irrigation canals. In prehistoric times these followed the channels of the ancient river systems, meandering through the landscape, bringing new life to long abandoned desert. Later, they became more formalized, cut according to planned routes to feed the fields that supported towns and cities. In many parts of the region, these systems have long disappeared under more extensive modern ones and can no longer be reconstructed. This is still possible in places where the rivers have changed their course and left the canals dry in the desert, or where political events have caused wide scale abandonment of formerly productive arable land. One of the latter, and also one of the most important areas for study of ancient Central Asian irrigation systems, is Khorezm, the delta of the Amudarya River.

Khorezm is a land of contrasts. Today, contained largely within modern Karakalpakstan, it is a green oasis of irrigated fields planted with sorghum, millet, melons and, above all, cotton. Surrounding the oasis on all sides lie the arid sands of the Kyzylkum, Karakum, and Aralo-Caspian deserts, with only the thin line of the river linking the delta to the outside world. But this life supporting greenery only dates back to the first half of the 20th century when scientific explorations brought the potential of the region to the attention of Soviet planners in Moscow. As the first explorers travelled with great hardship across the sandy wastelands south of the Aral Sea, they were amazed to find the walls of ancient fortresses rising up everywhere above the desert. It was clear that this land had once been rich and prosperous, and therefore could be made so again.

The very first exploration of Khorezm was begun by A. Yu. Yakubovskiy (1928-1929) and M. V. Voevodskiy (1934). These early visits were shortly followed by the establishment of the Khorezm Archaeological-Ethnographic Expedition of the USSR Academy of Sciences, organized in collaboration with several other central and local scientific research institutions. The KhAEE, led by the outstanding archeologist and ethnographer S. P Tolstov, began work in 1937, and was one of the most significant of a number of large archaeological expeditions to various districts of Central Asia launched in the mid-1930s (see especially Tolstov 1948a-b, 1962). When S. P. Tolstov began his research in the lower reaches of the Amudarya, the land was untouched by modern civilization. He noted that it seemed as if a malicious spirit had turned the land to dust, leaving only empty walls, dry canals and barren fields. In the shimmering hot desert air the ruins of ancient cities appeared like mirages on the horizon. Lines of ancient canals stretched for many kilometers, surrounded by innumerable ruins of ancient settlements, and on the cracked dry surface of the *takyr* mudflats the outlines of ancient fields, gardens and vineyards could be clearly seen.

Deeply moved by this magnificent dead land, S. P. Tolstov was reminded of the romantic fairy tale of S. Perro about the sleeping princess in the country where time had stood still, awaiting the return of the handsome prince to kiss the princess and bring the kingdom back to life. S. P. Tolstov himself took on the role of the prince, together with his colleagues and students. One of these was Boris Vasilevich Andrianov, who for many years traversed the harsh, arid landscape, steadily mapping ancient cities, settlements, primary irrigation canals and small distribution canals, fossil field systems and gardens, thus providing a basis for the systematic study of the ancient civilizations that once thrived there.

S. P. Tolstov and his team were the first to create accurate plans of the ancient irrigation systems in the lower reaches of the Amudarya. Over almost half a decade of study the members of the KhAEE identified and plotted the main canals constructed on the right bank of the river in the Antique and Middle Ages. On the basis of this work they came to the important conclusion that the modern system of principal canals on the right bank coincides almost completely with the Antique and Medieval systems (Tolstov 1948a:43–47; Andrianov 1969:99). This research into the ancient irrigation systems of the right bank of the river made it possible to study the wider problem of the ancient and Medieval irrigation systems of Southern Aral as a whole. Its implementation required the ability to work at an interdisciplinary level. B. V. Andrianov combined the skills of topography, geography, ethnography, and archaeology. He became head of a specially created unit within the KhAEE of Academy of Sciences of the USSR. His archaeological-topographical group had the task of mapping and documenting all the regions in the lower reaches of the Amudarya not yet affected by modern civilization, places on the right bank of the river where the desert still preserved intact all the ancient infrastructure.

Research into the problems of the history of irrigation from a geographical perspective permitted S. P. Tolstov, and after him B. V. Andrianov, to establish that at the end of the Bronze Age a branch of the Amudarya flowed towards the Sarykamysh depression, causing the ancient delta branches of the Akchadarya to dry out (Itina 1963:128; Andrianov 1966:111). The inhabitants of the land at the time deepened and cleaned these old river channels, effectively creating artificial canals to lead water to their small settlements and simple fields. In the beginning the canals were modest in scale, but by the end of the Bronze Age, at the time of the so-called Amirabad culture, some already stretched for several kilometers in length. With its incorporation into the world of the Achaemenids around the 7th century BCE (Briant 2002), Khorezm acquired new and advanced

technologies: the potter's wheel, monumental architecture, and also techniques for designing and engineering large irrigation systems. This enabled the full development of Khorezm as a complex agrarian civilization. Based on the material presented by B. V. Andrianov in his book, it is possible to say without exaggeration that the main system of Khorezmian canalization on the right bank of the river was created in Achaemenid times and has existed with certain changes and technical improvements up to the present.

The research of B. V. Andrianov was conducted at a level that was advanced for the time, using both state sponsored air photographs and aerial photography commissioned by the expedition, in conjunction with ground survey. This work, carried out when a considerable part of the territory of ancient Khorezm was desert where all the ancient systems were still preserved, permitted B. V. Andrianov to achieve a very extensive level of reconstruction for almost the whole system of primary canals in ancient and Medieval Khorezm. Andrianov's scientific documentation of the Ancient and Medieval irrigation network and associated agro-irrigation infrastructure is extremely important and timely. In the 1960s and 1970s the lands once under ancient irrigation systems in the south of the Republic of Karakalpakstan, where the main centers of early civilization appeared and where tens of kilometers of ancient canals were recorded, were placed under modern irrigation agriculture and densely occupied. In the course of this development many of the most valuable archaeological features have been lost, among them all traces of the ancient canal and agro-irrigation systems. Now only B. V. Andrianov's work preserves this extraordinary record for scientific study.

After the disintegration of the Soviet Union and the establishment of independent states in Central Asia, the role of scientific study of the archaeology of ancient Khorezm passed to the Institute of History, Archeology and Ethnography (IHAE) of the Karakalpak branch of the Academy of Sciences of Uzbekistan. In 1995, the establishment of the Karakalpak-Australian Archaeological Expedition (KAAE) saw the start of a new chapter in the story of Khorezmian exploration as a collaboration between the University of Sydney and the IHAE. The aim of the Expedition was to follow on from the earlier work of S. P. Tolstov and his team through investigation of the last remaining unstudied oasis in the southern delta region, that of Tash-Kyrman. Full interpretation of the archaeological record at the key sites in the oasis would be impossible without the valuable record left by B. V. Andrianov. His work in Tash-Kyrman Oasis is especially detailed, and much of the evidence on which his data was based has now disappeared. The KAAE has been working at two key sites: Tash-Kyrman-tepe and Akchakhan-kala (Kazakly-Yatkan; see Helms and Yagodin 1997; Helms et al. 2001, 2002; Betts et al. 2009). Tash-Kyrman-tepe was believed by B. V. Andrianov to be a fortified manor, but has now been shown to be a ritual center associated with the veneration of fire. The massive spread of the fortress of Kazakly-Yatkan makes it a likely contender for a regional capital founded around the 2nd century BCE and lasting up into the 1st or early 2nd century CE.

No site in ancient Central Asia could function without its agricultural hinterland, and control of food production was central to the success of the oasis states of Khorezm. Through the vision of S. P. Tolstov and the extensive work of B. V. Andrianov it is possible to reconstruct in large part an archaeological map of the ancient irrigation systems, with their farmsteads and fields, and so to understand in considerable depth the economic and environmental influences that shaped the history of the Tash-Kyrman Oasis and the surrounding lands of ancient Khorezm. There is no doubt that archaeologists working today, with modern methodologies, access to satellite imagery and all the tools of the 21st century, owe a very great debt to the pioneers who trekked across the desert in the early 20th century and recorded so much invaluable data, now long lost to us today.

Acknowledgements

The work of the KAAE is supported by the Australian Research Council (A10009046, DP0556232, DP0877805) as well as National Geographic Society and the generosity of many volunteers.

References

Andrianov, B. V.
1969   *Drevnie orositelnye sistemy Priaralya (v svyazi s istoriey vozniknoveniya i razvitiya oroshaemogo zemledeliya).* Moskva.

Betts, A. V. G., V. N. Yagodin, S. W. Helms, G. Khozhaniyazov, S. Amirov, and M. Negus-Cleary
2009   Karakalpak-Australian excavations in Ancient Chorasmia, 2001-2005: Interim Report on the Fortifications of Kazakly-yatkan and Regional Survey. *Iran* 47:33-55.

Briant, P.
2002   *From Cyrus to Alexander. A History of the Persian Empire.* Eisenbrauns, Winona Lake, Indiana.

Helms, S. W., and V. N. Yagodin
1997   Excavations at Kazakl'i-yatkan in the Tashk'irman Oasis of Ancient Chorasmia: A Preliminary Report. *Iran* 35:43-65.

Helms, S. W., V. N. Yagodin, A. V. G. Betts, G. Khozhaniyazov, and F. Kidd
2001   Five Seasons of Excavations in the Tash-k'irman Oasis of Ancient Chorasmia, 1996-2000. An Interim Report. *Iran* 39:119-144.

Helms, S. W., V. N. Yagodin, A. V. G. Betts, G. Khozhaniyazov, and M. Negus

2002   The Karakalpak-Australian excavations in ancient Chorasmia: the northern frontier of the "civilised" ancient world. *Ancient Near Eastern Studies* 39:3-44.

Itina, M. A.

1963   Poselenie Yakke-Parsan 2 (raskopki 1958-1959 gg.). Materialy Khorezmskoy Ekspeditsii. Vyp. 6, *Polevye Issledovaniya Khoresmoskoy Ekspeditsii v 1958-1961.* Moskva.

Tolstov, S. P.

1948a *Drevniy Khorezm. Opyt istoriko-arkheologicheskogo issledovaniya.* Moskva.

1948b *Po sledam drevnekhorezmiyskoi tsivilizatsii.* Moskva - Leningrad.

1962   *Po drevnim deltam Oksa i Yaksarta.* Moskva.

V: Memories of Boris Vasilevich Andrianov

Zamira S. Galieva

One of the most important periods of my life and my career is connected with Boris V. Andrianov. In the 1970s and 1980s, the scholars dealing with historical landscapes and in the development of monitoring and mapping the archaeological monuments of Central Asia were compelled to use materials that had high visibility and spatial information. However, topographical maps were quite inaccessible for Soviet archeologists because they were classified, and when it was possible to access aerial photos, there was the problem of their decoding.

At that time, in our vast country, the methods of aerial archeology were widely used only within the Khorezm Archeological-Ethnographic Expedition of the Institute of Ethnography of the USSR Academy of Sciences. In 1982, the Institute of Archaeology of the Uzbek SSR Academy of Sciences sent me to Moscow in order to gain a doctorate in History. It was to be with the Khorezm Expedition with Boris Vasilevich Andrianov as personal supervisor. For the younger generation of archaeologists, to which I belong, it was considered an honor to study and to work side by side with B. V. Andrianov and these legendary Khorezmian scholars.

The great prestige of the Khorezmian Expedition lasted for more than seventy years. The important turning period of the expedition was in 1950s when, after returning from World War II, the Khorezmians, unanimous in their enthusiasm and led by the founder and head S. P. Tolstov, resumed their research (Tolstov 1948:3-4). A large-scale multidisciplinary research project developed in the wide region comprising the lower reaches of the Amudarya and Syrdarya and the deserts of Karakum and Kyzylkum. One of the main aims of the research was the reconstruction of the historical landscapes of Khorezm, including the reconnaissance and mapping of archaeological sites and ancient water

systems. The European experience in the use of aerial archaeology methods certainly influenced S. P. Tolstov to expand and to update his project in Khorezm and to extend it along the Lower and Middle Syrdarya Basin.

For many years the multidisciplinary project in Khorezm was one of the most technically advanced expeditions in the former USSR and it introduced aerial methods in complex archaeological, paleogeographical, and cartographical research. Several disciplines were combined to develop the expedition in Khorezm: planning the ground and air routes; carrying out aerial surveys of distinct and wide areas; archeological prospection of discovered sites, ancient roads and water systems. Laboratory decryption of aerial photos, topographical plans of archaeological sites, and irrigation systems were performed before the beginning of each field season (Tolstov 1962).

The successes in studying the environment of settlements and the reconstruction of historical landscapes and water supply were owed, in many respects, to the work of the archeological-topographical team headed by B. V. Andrianov since 1952. For thirteen years (1952-1964) B. V. Andrianov and his associates provided great field-work: considerable areas behind the Aral Sea were surveyed, 1,650 *poisk* were surveyed including a thorough archaeological description of objects and finds (Andrianov, Itina and Kes 1975). All hidden traces of the archaeological landscape revealed by aerial photographs and map analysis were recorded by means of topographical and surveying instruments. Researchers managed and processed more than ten thousand air photos covering 5 million ha of the region, providing important data on the history and changes of the irrigation systems and settlement patterns of ancient Khorezm (Andrianov 1958:311-328; Tolstov 1948:37-62; Tolstov and Andrianov 1957:6-7). B. V. Andrianov generalized the data set of complex archeological-geographical research, providing the analysis and tracing the general character of irrigation systems of Khorezm in each ancient period. He created a method of relative dating of watercourses for different times and established an outline of the irrigation works in the Aral Sea region from the Late Bronze Age up to the 19th century CE (Andrianov 1969:30-41).

From the 1980s, B. V. Andrianov seldom went in the field and spent more time on his scholarly and scientific work. I am personally indebted to him for teaching me geography, geomorphology, cartography, and methods of decoding aerial photos. Thanks to his advice I worked on decoding air photos stored in the archives of the Khorezm Expedition, a unique material created in 1960 by N. I. Igonin on the ancient irrigated lands of Khorezm on the right bank of the Amudarya and the Eastern Aral region (Igonin 1965, 1968).

B. V. Andrianov was an excellent scholar as well as a demanding, responsible, and careful teacher. All my work regarding literature, reports on decoding, results, writing of synopses, and my studies at the Geographical Faculty of Moscow State University were planned and supervised by B. V. Andrianov. Despite his poor health, he decided to train me in the interpretation of aerial photos in the field. Thus, in the summer of 1983,

we took part in the expedition of the Institute of Archaeology of the Uzbek Academy of Sciences which intended to study the ancient settlement of Kavardan, in the district of Tashkent, and the Medieval farmstead of Shakhdjuvar, in the Tian-Shan mountains. I will never forget the extreme mountain travel conditions at an altitude of 3,000 m asl, when B. V. Andrianov commented on all the unexpected adventures of that journey. Although it was my only joint field season with Boris Vasilevich, I remember what I learned from him, as well as his encyclopedic knowledge, his indefatigable diligence and self-control.

Later on, in the 1990s, I tried to improve the methods developed by the Khorezm Expedition while conducting archeological-geographical research on the historical landscape transformations in the Lower Syrdarya Basin of the Eastern Aral region (Galieva 1999, 2002). Seeing the rapid development of computer technology and the appearance of remarkable topographical software, I am often reminded of B. V. Andrianov. He had dreamt of such a time. In each of his lectures, reports and articles he promoted the importance of information obtained by aerial photography and the methods used for that. In that sense B. V. Andrianov was looking at the future, combining a sense for flight and the unknown.

Along the ancient channel near the ancient settlement site of Kavardan (1983). From left to right: Zamira S. Galieva and her son, Kabul A. Alimov (Head of the Kavardan Archaeological Expedition of the Institute of Archeology of the Uzbek Academy of Sciences) and Boris V. Andrianov (Doctor in Historical Sciences of the Institute of Ethnography of the Russian Academy of Sciences).

Shakhdjuvar farmstead in Tian-Shan Mountains (1983). Standing, from left to right: Kabul A. Alimov (Institute of Archeology of the Uzbek Academy of Sciences), Boris V. Andrianov, Mikhail R. Tikhonin (Head of the Shakhdjuvar Archaeological Expedition), Yuri F. Buryakov (Doctor in Historical Sciences, Deputy Director for Research of the Institute of Archeology of the Uzbek Academy of Sciences); sitting, the local people working with the expedition.

Boris V. Andrianov and Kabul. A. Alimov on the ancient settlement of Kavardan (1983).

References

Andrianov, B. V.

1958b Arkheologo-topograficheskie issledovaniya drevney irrigatsionnoy seti kanala Chermen-yab. *Trudy Khorezmskoy arkheologo-etnograficheskoy ekspeditsii. Tom II, Arkheologicheskie i Etnograficheskie raboty Khorezmskoy Ekspeditsii* 1949-1953:311-328. Moskva.

1969 *Drevnie orositelnye sistemy priaralya (v svyazi s istoriey vozniknoveniya i razvitiya oroshaemogo zemledeliya).* Moskva.

Andrianov, B. V., M. A. Itina, and A. S. Kes

1975 Zemli drevnego orosheniya v nizovyakh Syrdaryi i zadachi ikh osvoeniya. *Voprosy Geografii* 99:147-156.

Galieva, Z. S.

1999 *Istoriya kulturnikh landschaftov Vostochnogo Priaralya IX vv. do n.e. XV vv. n.e. (po distantsionnym metodam issledovaniy).* Avtoreferat na soiskanie uchenoy stepeni kandidata istoricheskikh Nauk. Moskva.

2002 Evolyutsiya kulturnikh landschaftov Vostochnogo Priaralya VII v. do n.e. - VII vv. n.e. na primere basseyna Eskidaryalika. Metody rekonstruktsii. *Materialnaya Kultura Vostoka*, Vyp. 3:71-85.

Igonin, N. I.

1962 Ispolzovanie aerometodov v arkheologicheskikh issledovaniyakh. *Sovetskaya Arkheologiya*, No. 1. 1965

1965 Primenenie aerofotosemki pri izuchenii arkheologicheskikh pamyatnikov. In *Arkheologiya i estestvennye nauki*, redaktor B. A. Kolchin, pp. 256-261. Nauka. Moskva.

1968 Issledovaniya arkheologicheskikh pamyatnikov po materialam krupnomasshtabnoy aerofotosemki. *Istoriya, arkheologiya i etnografiya Sredney Azii*:257-266. Moskva.

Tolstov, S. P.

1948 *Po sledam drevnekhorezmiyskoi tsivilizatsii.* Moskva - Leningrad.

1962 *Po drevnim deltam Oksa i Yaksarta.* Moskva.

Tolstov, S. P., and B. V. Andrianov

1957 Novye materialy po istorii razvitiya irrigatsii Khorezma. *Kratkie soobshcheniya Instituta etnografii AN SSSR*, Vyp. XXVI:5-11.

Tolstov, S. P., B. V. Andrianov, and N. I. Igonin

1962 Ispolzovanie aerometodov v arkheologicheskikh issledovaniyakh. *Sovetskaya Arkheologiya* No. 1:3-15.

VI: Irrigation Among the Shaykhs and Kings

C. C. Lamberg-Karlovsky

When you are about to take hold of your field [for cultivation], keep a sharp eye on the opening of the dikes, ditches, and mounds [so that] when you flood the field the water will not rise too high in it. When you have emptied it of water, watch the fields water-soaked ground that it stay virile for you.

-Sumerian Farmers Almanac, ca. 1600 BCE; S. N. Kramer 1963:340

For the whole land of Babylon like Egypt, is cut across by canals. The greatest of these is navigable, it runs where the sun rises in winter, from the Euphrates to another river, the Tigris...

-Herodotus, I, 193; A. D. Godley 1966

Yes, if the kings did not control irrigation, who did?

-J. S. Lansing 1991:4

This is not a technological issue. The technology is easily available. It is a political and organizational issue. Water is a social good.

-P. Gleick (on irrigation); in de Villiers 2000:17

....it is better to be head of the water than head of the people

-C. Bichsel 2009:49

Water is essential to sustain life. Its distribution ranges from abundance to scarcity. Where scarcity prevails, whether as a resource or due to the demands of population, irrigation allows for increased agricultural productivity. For millennia the management of water has been a foundational element within human society. Today we are informed that water shortages, on a continental level, will lead to a crises in sustainability (Solomon 2010). Water management today, as in the past, has been as much an economic reality as a political necessity. In this essay I touch upon a variety of issues that concern irrigation, its management and control. The complex technology of channel, dike, dam, sluice, reservoir, and leveling are all but ignored.

Many of these issues have been addressed in book or monograph length. No pretense is made that any topic addressed is comprehensive, yet, the hope is that the bibliography may allow one to find further readings on the topics addressed.

Vernon Scarborough (2003:11) notes that "water management requires cooperation among people who might otherwise be in conflict". It is of importance to recognize that an entity, whether it be a village, or a network of cities and villages that share a single water system form an ecological unit that are bound by the advantages and necessities of cooperation. While the amount of water remains a limiting factor with regard to populations inhabiting a specific landscape, the availability of water is also a function of organizational skills and institutions that can mobilize labor, technological innovation, and production. Water, specifically its use in irrigation offered a directional arrow to the growth of civilization. It was not the only factor in the evolution of cultural complexity but it facilitated transport, trade, colonial settlement, military conquests, and most importantly the production of an agricultural surplus, and, its consequence, population increase. Water sustains, devastates, and purifies. It is the quintessential resource for sustaining life and a universal presence in origin myths and religious ritual (Frazer 1981:53-56). Water, whether in its feeding of agricultural fields, conversion to energy, or construction of monumental dams has always been a significant actor in transformational events in human history. The historian of cities, Lewis Mumford (1961:71) observed that "the first efficient means of mass transport, the waterway" was the "dynamic component of the city, without which it could not have continued to increase in size and scope and productivity. That the first growth of cities should have taken place in river valleys is no accident....".

Years ago Karl Wittfogel (1957) advanced the thesis that 'irrigation civilization' was characterized by a centralized authoritarian administration. The construction, management and maintenance of irrigation networks, he argued, led to political regimes of 'oriental despotism'. His was a deterministic view for the role of irrigation in stimulating both the origin of the state and its despotic rule. His experience in a Nazi concentration camp (1933-1934) led to personal anguish and his hatred of totalitarian regimes. He advocated a multilinear approach to cultural evolution grounded in Marxist frameworks. In his classic work *Oriental Despotism* (1957) he juxtaposed a 'western' vs. an 'eastern' world of political divergence.* The former is characterized by industrial capitalism in which power and authority was decentralized, innovative, and entrepreneurial. The latter is characterized by highly centralized autocratic sovereigns. Irrigation agriculture allowed these 'despots' to mobilize a large labor force for the control and management of irrigation networks. Stagnation followed.

Wittfogel did not, in fact, propagate the notion, for which he is frequently held accountable,

namely, that irrigation stimulated the productivity of agriculture and totalitarian rule. In his view the importance of 'hydraulic societies' was not in its contribution to agricultural productivity, but in its contribution to a 'revolution in organization'.

In his words (Wittfogel 1957:18):

If irrigation farming depends on the effective handling of a major supply of water, the distinctive quality of water - its tendency to gather in bulk - becomes institutionally decisive. A large quantity of water can be channeled and kept within bounds only by the use of mass labor, and this mass labor must be coordinated, disciplined and led. Thus a number of farmers eager to conquer arid lowlands and plains are forced to invoke the *organizational devices* which - on the basis of a premachine technology - offer the one chance of success; they must work in cooperation with their fellows and subordinate to a directing authority (emphasis mine).

Irrigation involves a multiplicity of tasks including construction, allocation, maintenance, conflict resolution, and the conduct of ritual, all invariably connected to aspects involving water as a limited resource (Scarborough 2003). The importance of irrigation was in its establishment of a new system for the organization and control of labor. In this vein irrigation is seen as a public works program, much as the pyramids of Egypt as argued by (Mendelssohn 1986), that establishes a new organization and control over an institutionalized labor force. Wittfogel maintained that the origin of the 'revolution in organization' was the concentration of power in the hands of a single individual whose organization of labor allowed for the construction of irrigation works, monumental architecture, and a military organization. To substantiate such a perspective he gathered ethnographic data, rarely archaeological, from Asia, Africa and the America's. Wittfogel's

---

*   We note that Boris V. Andrianov (1978) was deeply hostile to Wittfogel's concept of 'Hydraulic Society'. He charged Wittfogel with the misuse of Marxist terminology, rejected the deterministic singularity of irrigation as 'cause', and rejected Wittfogel's notion (promulgated 50 years before Samuel Huntington's 1996 *Clash of Civilizations*) of an insurmountable gulf between a despotic totalitarian East and a democratic West.

emphasis upon 'organization' and 'management' is not dissimilar to that of Henry Wright's (1969, 1977) emphasis on the role of organization in the development of an administrative bureaucracy, so significant in the emergence of the state. Wittfogel's 'hydraulic societies' has given birth to a substantial shelf of books written to counter, to support, and to adjudicate between contending views. For a review of this literature see Fagan 2011; Salomon 2010; Hassan 2006; Scarborough 2003; Mitchell 1973; Hunt and Hunt 1976; and, various citations in this paper. For an extremely interesting and informed critical review of the debate concerning Wittfogel's hypothesis see the engaging paper by Boris V. Andrianov 1978. The 'causal' issue with regard to irrigation, and its role in the emergence of civilization, is well stated by Robert Murphy (1967:29-30) during the period of its central debate:

> When pursuing historical causality, we often end up by chasing our own tails. Does the political requirement of irrigation beget the state, or is a state a necessary precondition for irrigation? Actually, it probably works both ways. The irrigation hypothesis never required us to believe that communities undertook projects beyond their political means and then caught up institutionally to their accomplishments. Of course large-scale irrigation works were built by large-scale polities, but both had antecedents in small communities and small irrigation projects. It would be surprising indeed if significant temporal priorities were to be found, for the two variables probably emerged together. Perhaps our real problem is a mechanistic model of causality that leads us to seek for *the* cause at a point in time distinctly before the effect (emphasis in the original).

The author's proposal that there is a synergistic action regarding irrigation and its centralized coordination, relating scale of irrigation complexity to greater political integration, is an opinion that would have been shared by the author of this book as well as by numerous authors addressing this specific issue, see the seminal works of Bennett

1974, 1976 (new edition 2005); Hunt 1976, 1988; and, Scarborough 2003, for excellent reviews of the debate.

Neither Wittfogel nor Julian Stewart (see below) were the first to propagate the notion that irrigation, and its administered bureaucracy, was implicated in the emergence of both the state and its authoritarian rule. Lev Mechnikov in *Civilization and the Great Historical Rivers* (1889, in French), as pointed out by Andrianov (1978 in a publication following the writing of this book), argued for three stages of cultural evolution: 1) The great riverine civilizations of Egypt, Mesopotamia, India and the Hwang Ho were all resultant from irrigation and shaped despotic slave owning states. A thesis that almost exactly parallels Wittfogel's; 2) Mediterranean oligarchic and feudal states; and, 3) Oceanic... the absence of governmental control.

Mechnikov was followed by Karl Marx and Friedrich Engels in repeatedly pointing out the formative role of irrigation in the emergence of class oppression and the state.

In a letter to Marx Engels wrote:

Artificial irrigation here is the first condition of agriculture and this is a matter of either for the communes, the provinces or the central government. An Oriental government never had more than three departments: Finance (plunder at home), war (plunder at home and abroad), and public works (provision for reproduction) [June 6, 1853] (quoted in Andrianov 1978).

The belief that irrigation is 'causal' to the growth of cultural complexity, or is related to the origin of the state, is today out of fashion; as is its role in population increase (Smith and Young 1972; Boserup 1981). Today's general consensus holds that cultural complexity and state origins precede the centralized control of irrigation practices. Brian Fagan (2011:xxi) summarizes that position: "As I argue, the main interest of rulers and their officials was not in irrigation and water management as such, but in the food surplus they produced, which supported the state". Note: there would be no food surplus without irrigation. This is like saying that modern nation states are not interested in taxes only in what the money can support. The question of a central authorities role in the control of irrigation is directly related to the *scale of complexity* inherent in both the management and the size of the irrigation network. The scale of an irrigation system is directly proportionate to the scale of management. Even within an imagined community of self-sufficient villagers, particularly within an arid environment as the Near East, where the earliest agriculture and irrigation systems are evident by 6000 BCE. Simple furrow irrigation directed water run-off from streams and rivers in canals that proportioned water to allow its benefit to reach the most distant fields (see below). Some individual or group had to be responsible for coordinating the digging of canals, the apportioning of the water, the maintenance of the canals and adjudication in times of land and/or water disputes. In time that individual became the Pharaoh of Egypt whose divine responsibility was the annual inundation of the Nile. One of the earliest pharaohs, referred to as the 'Scorpion King', is depicted on a mace head directing the construction and management of irrigation networks (Kemp 2006). Tension between individual rights to water versus the corporate rights of all sharing a single canal system necessitated some form of management; an authority to adjudicate between disputes.

We note in passing that the Sumerian epic myth *Enki and the World Order* offers some support for the importance of organization and the central control of irrigation systems. This myth writes Richard Averbeck (2003:758a, 2003b) "...consists of imaginative stories that are, in fact, based on reality and/or history, and therefore, reflect foundational understandings of the world that are important to the culture of the composer and those who read his compositions or hear them recited".

Enki, the Lord of the Earth and god of the city-state of Eridu, god of underground waters, fertility and productivity, assigns specific deities to organize and take charge of certain functions within the Sumerian world order. After Enki creates the Tigris and Euphrates, the crops and tools of the farmer, he places Enbilulu as 'Inspector of the Waterways' and appoints Enkimdu as the responsible agent for overseeing the ditches and the dikes of the irrigation networks. Fundamental to the Sumerian 'World Order' is Enki positioning his brother Enlil as the chief god of Sumer. It is Enlil who invents the agricultural hoe, made plants to grow, and organized plows, yokes and oxen teams. Central to this myth is its emphasis upon the maintenance, restoration, and vigilant control over the efficient and prosperous natural order of the Sumerian world.*

Wittfogel's 'hydraulic society' theory, although largely dismissed today, remains of relevance in relating power to the structure of water management. Traditional models advocate hierarchical structures of centralized control, linking economic needs to political power, in which a limited number of individuals maintain a 'despotic' control. One cannot avoid the fact that water management requires cooperation among individuals or groups that otherwise would be in conflict for access to the limited resource. As Robert Netting (1993:12) points out, the inequality within villages necessitates a coordination of access to water which, in turn, require some form of centralized decision making. An alternative interpretation for a degree of centralized control is evident in what Carol Crumley (1987, 1995) has termed *hetarchy*. Unlike hierarchical models in which a top-down linear chain-of-command dominates hetarchies envisage densely settled communities interacting, cooperative, and interdependent.

---

* We note that in Mesopotamia, the Babylonian supreme deity Marduk was responsible for the management of irrigation networks. Marduk, whose symbol was the *marru*, the spade "used for digging canals, was both the supplier of water and the controller of irrigation (Oshima 2007). In like manner, the Scorpion King, pharaoh of Egypt in Dynasty 0, is depicted holding a digging stick and commanding the cleaning and construction of canals. We also note that Yu the Great, founder of the Xia Dynasty of China (ca. 2070 BCE), was accorded divine characteristics taught his people flood control techniques, constructed dikes, and built irrigation canals to distant farm lands (Lewis 2006).

Hetarchy does not imply that coercion and centralization are absent it merely emphasizes the role of interdependency, a corporate management of labor, water, and the production/consumption of agricultural produce. One can readily imagine that hetarchy characterized the management of water resources already in Neolithic times. In this manner we envisage a continuum in the size of irrigations systems related to the *scale* of their management and control, ranging from hetarchical corporate (tribal) to hierarchical (state) control. It is not the presence and/or absence of centralized control that matters it is the *scale* of the irrigation system, on the one hand, and the scale of centralized control, on the other hand, that matters. Hetarchical control is characterized by horizontal economic and political interdependence and co-operation while hierarchical control was vertical in which a single urban center dominated the economic and political structures of a considerable territory.

In discussing the earliest irrigation practices Brian Fagan (2011:119, 31), is influenced by Robert McC. Adams's (1966) comparative study of Mesoamerican and Mesopotamian civilizations. In Adams's study irrigation is relegated to near insignificance in the emergence of cultural complexity. This view was re-affirmed in his seminal archaeological research in Mesopotamia, in which the administrative management of irrigation systems followed, was not 'causal' or related to, the emergence of urban complexity. In his own words the late development of large systems suggest that "if anything, large-scale complex irrigation practices were a derivative of the prior development of urban and state organization rather than vice versa" (Adams 1981:245). One may fairly ask what is meant by 'large-scale' and when are we dealing with 'urban'? Surely, by the middle of the 4th millennium sites in Northern Mesopotamia reached an urban status of over 100 ha (Tell Brak at 130 ha) and are estimated to have a population of 10-15,000 (Ur 2011). In spite of the size and number of such communities, both surveyed and excavated in Northern Mesopotamia, we are

informed by Jason Ur that "Despite the intensive and potentially overextended agricultural economy, and the monumentality of settlement and landscapes, *we should not assume that the hand of a centralized administration lay behind these developments.*" This conclusion follows the authors writing that "*Excavations at these centers revealed remarkable concentrations of political and economic power*: monumental temple, palace institutions, writing and administrative technologies, craft specialization and mass production, and considerable disparities in status and wealth" (Ur 2011a, emphasis mine; see also Ur 2011b). The contradictory implausibility of both statements is self-evident. In spite of such contradiction there is a conclusion "The EBA urban landscape appears to have been the non-planned result of widespread rules and attitudes about land tenure, household based surplus production and the social role of communal meals". One would think that populations in urban environments, numbering in the tens of thousands and constructing monumental temples and palaces, would require the management and centralization of labor in dealing with an "overextended agricultural economy" that reaches far beyond a "household based surplus production". In an urban landscape labor, agricultural production, and surplus, are essential requirements. Similar contradictions, or inherent tensions, exist in Adams (1981:244-245). Prior to his rigid conclusion that "complex irrigation practices were a derivative of the prior development of urban and state organization, rather than vice versa" he emphasizes that the emergence of "Mesopotamian cities can be viewed as an adaptation to this perennial problem of periodic, unpredictable shortages. They provided concentration points for the storage of surpluses, necessarily soon walled to assure their defensibility". Storage, urban supply, he notes are a dominant concern (Adams 1974:3), thus: "It is noteworthy that the objective of urban supply, rather than irrigation, is stressed in the few early royal inscriptions dealing with watercourse maintenance, and that the principal technological distinction, is between navigable and non-navigable channels". Are the navigable rivers not also the major source for irrigation canals? It is relevant to observe that Medieval authors, in discussing the irrigation networks of Mesopotamia and Central Asia clearly emphasize, and offer the names of, the navigable canals while ignoring mention of the omnipresent irrigation canals. The emphasis in the texts is upon the richness and variety of agricultural products rather than the canals that allowed for their production (see Le Grange 1930 and Bartold 1958 for full discussion). What is of relevance is not 'storage' nor even 'surplus' but the organization of *labor* that produced the *surplus* and the management that distributed it! Adams in recognizing that a "narrowly deterministic view of urban genesis as merely the formation of walled storage depots" is insufficient we read:

> The drawing together of significantly larger settlements than had existed previously not only created an essentially new basis for cultural and *organizational growth* but could hardly have been brought about without the development of powerful new means for unifying what originally were socially and culturally heterogeneous groups (emphasis mine).

Thus, it was the 'organizational growth' and, within that context, it was the management of labor that produced the surplus by which cities could come into being! It is not unreasonable to suppose that already within the 4th, if not the 5th millennium that communities that reached the size of 15+ ha (37 ac) were producing a surplus under supervised labor management. Adams projects a perspective that allows for a multiplicity of interacting determinants, favoring an equal opportunity approach for all that might be deemed 'causal' in culture change (see also Adams 1996). In discussing the competition that would have evolved between upstream and downstream access to water, whether riverine or

irrigation, we come to a conclusion that Adams (1981:245) takes to have occurred "no earlier than the 3rd millennium":

> It may well be that differentiation along this [upstream/downstream] and similar lines had as much to do with the appearance of complex, hierarchical state and urban institutions as the much more frequently and explicitly chronicled rivalries that took an interurban, overtly military form... Hence in response to all of them [the challenging ecological conditions] the husbanding of surpluses under hierarchically organized institutions with even a modest, strictly relative assurance of continuity represented the most broadly advantageous course of action that was available. This further implies, of course, that irrigation should be seen not as one of the most important of a group of such forces that tended to pose social challenges of a particular kind".

Au contraire! The control and management of labor as well as the production of a surplus are very much "social challenges of a particular kind".*

The control and management of a labor force, and the production of an agricultural surplus, are as determinist in today's world as they most assuredly were in the ancient world. Whether it is the leader of a tribal entity or a modern state there always was and remains an essential need to control the equitable access to water. Whether it be the Hoover Dam, the Aswan Dam or the Three Gorges Dam the control and management of labor and its relationship to water, are inextricably intertwined with an agricultural surplus. Within the 20/21st century the triumvirate of water/labor/management remains of continental consequence (see Fagan 2011).

The tribal role in the management and control of irrigations systems in the 19th century is nowhere better documented than in Central Asia (MacGhan 1874; Schuyler 1876;

---

*   For a recent discussion on conflicts and violence due to water access, boundaries, and ethnicity (Kirghiz and Tajik) in Central Asia see Bichsel 2009. Her conclusion regarding upstream vs. downstream access to water is that it is "better to be head of the water than head of the people" (p. 49).

O'Donovan 1882; Mukhamedjanov 1978). Southern Mesopotamia, on the other hand, was sparsely settled and as reported by Lady Anne Blunt (1879:442):

> A principal feature of all these [development] schemes seems to be the restoration of fertility to the Babylonian plain south of Baghdad. This, rich as the plain formerly was, could not now be effected without a prodigious outlay in the form of water-works. To reconstruct entirely the Babylonian system of canals is financially impossible, even for the richest country in the world, at the present day; and without irrigation not a blade can grow.

An abundance of 19th century information, gathered during the expansionist Tsarist period on tribes, settlement patterns, social systems, agricultural production and irrigation, followed by the Soviet conquest of Central Asia, offered the foundation and the inspiration for the undertaking of the decades long Khorezmian Expedition of 1937-1945, directed by Sergey P. Tolstov (see below).

The Merv Oasis, in today's Turkmenistan, constitutes the apex of the Murghab deltaic fan which is some 40 miles in width and constitutes a fan of some 1,600 square miles (424,398 ha). In the 19th century Henry Lansdell (1885:476-477) gave the width of the Murghab River as "80-100 paces" and up to "23 feet" in depth. Concerning irrigation he notes the presence of a large dam that:

> Diverts the water among two sections of the oasis by means of two main canals, the Otamish and the Tokhtamish.... each of the two canals distributes the water through about 50 leading arteries, and these in turn feed hundreds of smaller leats.

Each of the above two canals was occupied by different clans of the dominant Tekke tribe, themselves divided into 17 distinctive clans. Additionally, the Merv Oasis was inhabited by the Akhal, Saryk, Salor, Ersari, "and other" tribes totaling "230,000 souls". Significantly, the control of irrigation systems was in the hands of tribal leaders that delegated their authority to lineage heads, which, in turn, managed the control over the distinctive branches along which they resided. The management of irrigation between the different tribes required alliances. The system of tribal and kin control over extensive irrigation networks as described by Lansdell, in both geographical spread and technical sophistication in the construction of dams, dikes, reservoirs etc., is almost identical to that described by A. R. Mukhamedjanov (1978) for the Medieval period in the Bukhara Oasis. Nineteenth century travel literature offers an abundant description of the settlement regimes, a state of constant conflict, an omnipresent irrigation network, and the presence of formidable fortification systems (the *qala*, or *kala*) belonging to distinctive tribes to protect their settlements and their irrigation networks. E. O'Donovan (1882, II:143, 175) observes that the forts functioned, in part, to protect the irrigation networks. In discussing the imposing Baba Kabasi fort, next to the great Dam of Banfi, he states that "without this dam the present cultivated area would be reduced to a condition as bleak and arid as that of the plains that surround it.... the old Sarouk fortress....constituted the central stronghold of Merv and protected the water works".

We note that the forts were also the residences of the tribal leaders of the dominant Tekke tribe. Eugene Schuyler (1876:67-68) in one of the first geographic descriptions of the region wrote:

> The whole of this region shows traces of ancient cultivation, and it is evident that a large population at one time existed here. In various parts there are mounds, now covered with growths of *saxaul* and other shrubs, which are evidently the ruins of former cities. There is an old legend that the whole valley of the Syrdarya was at one time so thickly settled that a nightingale could fly from branch to branch of the fruit trees, and a cat walk from wall to wall and house-top to house-top from Kashgar to the Sea of Aral. From the traces of former culture one can in part believe this... but nothing which could enable the age of the ruins could be ascertained.

Schuyler (1876:284–307) also offers descriptive detail as to environmental and climatic conditions, the management and technology of irrigation networks, the types and quantities of agriculture produced under irrigation (termed *obi*) as opposed to agricultural lands fertilized by spring and autumn rains (termed *lalmi*), details pertaining to the agricultural cycle (fallowing, manuring, seeding, harvesting, watering, etc.), the quantification of lands cultivated for specific harvests, population numbers within specific named settlements, the allocation of the number of laborers assigned to agricultural production and irrigation needs, the traditional accounts of taxation (including that said to be 'most important', the *kosh-pul* – the tax for building and maintaining the irrigation system (Schuyler 1876:305), and finally, the management of the entire Zeravshan irrigation network, and the five types of land tenure, by *both* the centralized authority, i.e. the Emir of Bukhara's administrative management and the local management by the authority of the tribe/lineage.

....the irrigated lands, on account of their richness and fertility, the constancy of their harvests, and the variety of their produce, are by far the most important to the well-being and civilization of the country. The proper regulation of irrigation is, therefore, a matter of the greatest consequence, especially in the valley of the Zeravshan where every drop of water has value, and where without more water there is hardly room for another inhabitant. The worth of land is estimated chiefly by the amount of water to which it has a right, and most of the lawsuits about lands arise out of disputes concerning water (Schuyler 1876:286).

On the waters of the Zeravshan which supplied water to the Bukhara Oasis we read in Arminius Vámbéry (1865:220):

The water flows through a canal, deep enough but not maintained in a state of cleanliness. It is permitted to enter through the Gate of Dervaze Mezar once in intervals of from every eight to fourteen days, according as the height

of the river may allow. The appearance of the water, tolerably dirty when it first enters, is always a joyful occurrence for the inhabitants... It has, it is true, absorbed thousands of elements of miasma and filthiness.

The bilateral control and management of irrigation systems by *both* state and local control, as reported by Schuyler for the 300-year Bukhara Emirate and by Mukhamedjanov for the Medieval period, is similar in many respects to that reported by Robert Fernea for the management of irrigation system in Iraq in the 1960s (see below).

Noting the extreme variability of winter snow in the mountains and spring rain in the Bukhara region Schuyler (1876:292) notes that "It is not surprising then that famines are of not infrequent occurrence". In the famine of 1810 "there was such a famine that men sold their children, their sisters and their mothers, and killed the old people or left them to starve".

Finally, as observed by Adams, conflicts did emerge in the competition for water between upstream and downstream inhabitants (1981:300):

...as an irrigating canal is made for the benefit of all the lands bordering it, the use of water is subject to certain restrictions. Proprietors living near the beginning of the canal have no right to use more than their proper share of water to the detriment of those farther on.

The fact that inhabitants near the "beginning of the canal" routinely took more water is a frequent theme of 19th century travel literature, as well as in the modern era, as noted by Christine Bichsel (2009:49). On the occasion of disputes concerning water rights and access settlements are typically adjudicated at the local level, and are most commonly settled, but, if they are not, the administrative bureaucracy of the Emirate imposes a resolution.

Within Central Asia, by Andrianov in this book, a rich inventory of Bronze and Iron Age settlements have been surveyed and excavated. Today's knowledge of Central Asia has very considerably expanded (Sarianidi 2010; Salvatori and Tosi 2008; Lamberg-Karlovsky 2013) yet the role of irrigation, assuredly present from the beginnings of urban settlement (Masson 2006), remains little examined and less discussed. Again, the earliest evidence, as in Mesopotamia, remains buried under meters of alluvial silt.

Recently, with regard to the Mesopotamian world, Guillermo Algaze (2008) has correctly noted that the low gradient of the Mesopotamian plain, and the ease in which the gravity transport of water facilitates irrigation in agricultural productivity and water transport, forms at least since the 4th millennium a critical element in the 'Mesopotamian advantage'. In recent decades, the use of CORONA, and other satellite images, has facilitated and transformed our understanding of Mesopotamia's ancient landscape – particularly in the study of later periods where the signature of land use is less eradicated by the passing of time (Adams 2005). The anastomosing effects of the Euphrates and Tigris (Pournelle 2003) as well as the very considerable alluviation that buried earlier settlements under meters of silt buried also the evidence of irrigation networks. Dramatic evidence for the extent of alluviation, and its role in burying archaeological sites, was gathered by David Stronach (1961:97, 124) at Ras al 'Amiya in Central Iraq. The top of this 2.3 m high mounded site was found 2 m beneath the modern day surface. The site is dated to the Hajji Mohammed phase of the Ubaid Period (early 5th millennium). Stronach writes:

> At present prolonged irrigation, coupled with long periods of neglect, has increased the degree of salinization in this flat country to a dangerous degree. But in ancient times the rich alluvial soil must have been remarkable productive wherever sufficient water was available. For this reason ever pond, river or water-hole must have been of immense importance in early 'Ubaid times, and the precise location of sites like Ras al 'Amiya must have often depended on the temporary course of some meandering river.... Many sites must lie beneath the alluvial cover of Central and Southern Mesopotamia.

Clearly, in Mesopotamia the very evidence for early irrigation, let alone the nature of its management, lie buried beneath the alluvium. The absence of evidence should not be taken as the evidence for absence! Additionally, contrasting data derived from archeological survey and the 3rd millennium written record, indicate that Mesopotamia was a heartland of villages rather than cities. The presence of a significantly larger number of villages then detected in the land surveys of Adams (1981) has recently been documented by Piotr Steinkeller's study of the written documents. In those texts more than four times the number of villages are recorded when compared to those identified on archaeological survey (Steinkeller 2007). It is entirely within the realm of our present understanding that all of these villages would have to be served by irrigation canals!

The 3rd millennium texts suggest that there was a favoring of barley cultivation over other kinds of cereal. Barley is the most resistant of cereals to saline soil. An emphasis upon the production of barley, reported in the texts, is taken as proxy evidence for intensive irrigation; a well-known agent of salinization. If irrigation water is not drained from the land the water table will rise bringing salts to the surface by capillary action. This, in turn will render the soils increasing less fertile. Thorkild Jacobsen (1982) advanced the controversial thesis that by the end of the 3rd millennium increased salinity, resulting from over irrigation, brought about a shift from wheat to barley production which by the 18th-17th centuries BCE led to the abandonment of many urban centers of Southern Mesopotamia. A detailed critique negating the evidence in support of the 'salinity theory' has been put forth by Marvin Powell (1985).

The verdict implicating irrigation and salinity as involved in both the shift to barley and the abandonment of settlement in Southern Mesopotamia remains inconclusive. There is simply a lack of evidence for 3rd millennium patterns of land use, their relationship to water supply, the shifts in fallow systems, and their technical skills in dealing with (and/or reversing) salinity. The same verdict must be concluded

for the very role of irrigation in the formation of Mesopotamian cultural complexity in the Uruk Period of mid-4th millennium. Although we believe it played some role the scale and degree of its implication simply cannot be determined from the evidence at hand. For this period of time we have little idea on the nature of land ownership, the scale of the irrigation networks, their technological achievements, the organization of labor (a critical component), the relationships and demography that characterized the relations of farmer and pastoral nomad, and indeed, the political and economic systems that structured the communities. Nevertheless, the synergistic action of irrigation and the essential need for its management and allocation was likely involved in the evolution of centralized authority. There is no denying this fact in the later Assyrian/ Achaemenid/Sasanian Empires (Adams 2005; Ur 2011). The detailed analysis of irrigation systems, its technology, labor force, number and types of agricultural items produced, so thoroughly documented for the 19th century by J. A. Barois in his superb *Irrigation in Egypt* (1889), is essential were we to truly understand the nature of Mesopotamia's ancient irrigation practices.

The Medieval Period

There is a considerable library of manuscripts and books from 900 CE to 1200 CE that consider aspects of dynastic rule, geography, social and political conditions, agricultural productivity, and importantly, from our perspective, irrigation networks. The works of such renowned Persian and Arab authors as al-Biruni, al-Idrisi, Ibn Batuta, al-Maqadisi, Samani, al-Masudi, Ibn-Hawkal, al-Istakhri, al-Tabari and many, many others are given detailed review (and bibliography) by Guy Le Strange (1930) and Vasili V. Bartold (1958). The great cities of Baghdad, Isfahan, Samarkand, Merv, Balkh, Bukhara, to mention but a few, as well as their distant countryside are discussed as are the irrigation systems that served them. The fall-off of literature following 1200 CE is directly related to the fact that all of the above cities, and more, were completely destroyed by the Mongol

invasion. In the city of Merv alone over 10 thousand books and manuscripts were publicly destroyed, a task repeated in many of the major cities. Yakut, a resident of Merv, informs us that Merv was completely reduced to rubble by the Mongols and that the dikes, bridges, irrigation networks were all destroyed, the fields reduced to swamps and nine million corpses filled the countryside. Two centuries later Ibn Batuta passed by and reported that Merv consisted of a pile of ruins.

Bartold (1930:89-97, 104-107) tells us that in Samarkand, before its destruction, was inhabited by 10 thousand families and offers a description of the watercourses that fed the city. One canal, the Juybar-Bakar was said to water one thousand gardens while hundreds of surrounding fortified villages, many encircled by moats, were fed by numerous canals dug from the Amudarya (Oxus) River (Bartold 1930:150). The Medieval authors followed a common narrative, referred to as the 'Book of Roads', in which various itineraries, roads and canals detailed the routes and waterways that offered communication between major cities. The names of the principal cities, the towns and villages in-between, their monuments, administrative detail, irrigation networks, industries, and agricultural products formed the substance of their narratives. Bartold (1930:181-182), observes that from this literature one obtains a view of their 'warlike spirit' and the fact that they were "constantly at war with one another". A fact we have seen reaffirmed in the 19th century travel literature.

In the early 8th century the Arab conquest of Central Asia, commanded by Qutayba bin Muslim, and much reported in contemporary literature, avoids discussing the nature, or the extent, of the irrigation works. We note, however, that following the Arab conquests in 733 a great famine is reported while in the anti-Arab revolt of 816-817 the export of grain was terminated. It is entirely possible that the famine was the result of the Arab destruction of the irrigation systems. A contemporary author of the mid-9th century writes of walls built "round the vineyards and cultivated fields of the inhabitants" to protect them from invasion" (quoted in Bartold 1930:211). Even after Arab subjugation of local rule the quarrels over water rights remained a constant factor. Abdullah bin Tâhir, the ruler of Khorasan, took a special interest in agriculture. In attempting to adjudicate irrigation disputes he discovered that the Muslim lawbooks offered neither instruction on their management nor on the resolution of disputes. He summoned the *faqîhs* (religious scholars) of Khorasan, to collaborate with their counterparts in Iraq, and work out the legal principles regarding the use of irrigation water. Deliberations resulted in the writing of the *Kitâb al Quniy* (Book of Canals) that held influence for several centuries. Notable is a concern for the moral issues of equity and justice, the protection of the interests of the peasant class, and an emphasis upon education – all championed concerns of bin Tâhir. He believed that "The autocrat must above all be a good landlord" implying the judicious treatment of the farmer and the maintenance of the irrigation canals (Bartold 1930:213, 226). The absence of being a good 'landlord' was the imposition of excessive taxes. The absence of 'Muslim lawbooks' does not imply the absence of rules governing irrigation networks by the local populations. It might imply that the nascent Arab state had not formalized any colonial rules for dealing with subjugated populations.

Isfarayini writes in the 11th century that ruinous taxation led to disastrous consequences, "the agricultural districts were to a great degree deserted and the irrigation works in some places had fallen into decay, in others had ceased altogether" (Bartold 1930:287). The noted Medieval historian Muqaddasi states that in the 10th century, in the city of Merv, an appointed overseer of irrigation commanded 10 thousand corvée laborers, and horse guards, responsible for the digging and maintenance of the canals. This suggests a high degree of centralized control over both the labor and maintenance of the water networks. As the Murghab River approached Merv major dikes separated the river into four major canals and a large reservoir that nourished fields for miles in every direction, extending "six leagues from the city" to the town of Jîrang "while one league beyond it lay Zark. Here stood the mill where Yazdajird III, the last of the Sassanian kings, fled for shelter, and was murdered by the miller for the sake of his jewels" (Le Strange 1930:400). Sassanian kings were above all responsible for the vast construction of irrigation networks. Among the largest was the construction of the Nahrawan Canal in Iraq that extended from Dur to Madharaya (see Map II, Lestrange 1930) a distance of over 200 miles (for Sassanian irrigation, see Adams 2006). The Sassanian monarchs were not alone in the construction of large irrigation networks. The contemporary Khorezmian Shahs were their equal (see below).

The Ethnographic Evidence

Fagan (2011:119, 31) follows the consensus view: "But there was no central authority" in the management of early irrigation systems. It is relevant to point out that Adams's highly influential conclusion was influenced by the study of his colleague Robert Fernea's (1970) ethnographic study of irrigation among the El Shabana of Southern Iraq. Fernea emphasizes the local management of irrigation systems and sees the El Shabana as 'typical' of tribal groups in Southern Iraq. Several features within the social organization of the tribal El Shabana are ideally suited to irrigation practices. Some aspects of the El Shabana are seen by Fernea as near universals of tribal organization and well suited to the local organization of irrigation. These are:

1. "As in all traditional societies the irrigation technology employed is very simple. Neither survey equipment nor leveling devices are used. Joint ownership of land prevails which helped assure that there were no "plots without access to water" (Fernea's 1970:3) thus avoiding potential conflicts".

2. The El Shebana established alliances with local tribal and nomadic groups having distinctive tribal affiliations. Marriage ties, alliances, shared economic concerns, i.e., permitting pasturage on fallow fields to nomadic groups also alleviated conflict.

3. The El Shabana kept livestock in sheep, goat and cattle which allowed some (or all) to revert to pastoral nomadism should conditions make farming too tenuous. Animals were not only a source of income "but an ultimate insurance against drought, loss of land, or other crises" (Fernea's 1970:6).

4. Tribal hospitality, mutual obligations, kin relations, and ritual ceremony (birth, marriage, circumcision, death, etc.) between different members of the tribe allowed for risk reduction and strengthened the co-operation required in digging and maintaining the irrigation networks. The structured formality in their social relations allowed for both fragmentation and coalescence in time of crises and stability.

5. Fernea argues persuasively that the most adaptive feature of the tribes social organization (and arguably all tribal societies) is its ability to prevent the concentration of wealth and power in the hands of the few. The social organization of tribal societies is generally regarded as egalitarian (Sahlins and Service 1988). Among the El Shabana the *shaykh*, or chief, is regarded as the first among equals. His principal function (typical of all tribal societies) was to lead in combat and act "as a reservoir of tribal law and an astute judge enforcing culturally defined and traditional norms" (Fernea's 1970:9).

6. The *shaykh* lacked political power. The segmentary lineage system of tribal organization inevitably favored a decentralized rule which led, in turn, to frequent revolutions resulting in *shaykhs* emerging from different lineages.

7. Lastly, the *shaykh* did not attempt to build up his land-holdings or invest in his control over new irrigation works. Such efforts would be contested by the tribe. His concern was enhancing his prestige and status. What wealth he acquired he expended "in the form of hospitality, help in crises and the like (Fernea's 1970:10).

The social organization of tribal cultures are well suited to manage and control irrigation networks. Tribal society are clearly not entirely without organization. The organization, however, is not hierarchical as when a state polity takes central control over the entire irrigation network. That form of communal, decentralized control, but nevertheless control, we have referred to as an hetarchy. It is the characteristic structure of decision making in a tribal society.

Did irrigation agriculture in its increasing complexity require ever more bureaucratic management and create the centralized state or vice versa? The debate misses the essential point, namely, irrigation, and its ever increasing technological requirements and consequent population increase are co-evolutionary processes in the development of cultural complexity. Cultivation and irrigation in arid environments go hand-in-hand. The scale of agricultural production and irrigation management trace an evolutionary process that moves from hetarchy to hierarchy, from tribal societies to state formation. Hydraulic societies existed where water resources were both concentrated and essential for agriculture. This concentration, in turn, allowed for the centralized management of the resource for the distribution and allocation of water to distant villages. Although the 'Wittfogel Hypothesis' has little traction today we note that there are some archaeologists that adhere to the belief that irrigation was, in fact, a central factor in the emergence of cultural complexity. Thus, Richard McNeish (1967) contends that irrigation was central to the emergence of civilization in Peru and Mesoamerica while William Sanders (1968) shares the same opinion for Mesoamerica. Almost a decade before Wittfogel proposed his thesis the distinguished anthropologist and ecologist Julian Steward (1949, 1955) proposed that irrigation was a major 'cause' in the emergence of a centralized political authority. Steward proposed that 'irrigation civilizations' (Egypt, Mesopotamia, China, Mesoamerica, and the Andes) had common developmental sequences resulting from their arid environments that required large-scale irrigation. Later, and largely due to the influential criticisms of the 'irrigation hypothesis" by Adams (1966),

Steward (1968:323) drew back from his position and attributed the emergence of Mesoamerican civilization to the centralized control of production and trade rather than irrigation. It is worth pointing out that in 1966 Adams had virtually no evidence as to the role that irrigation played in the emergence of cultural complexity in the early periods in Mesopotamia and scant more for Mesoamerica. His rejection of irrigation as a factor in the emergence of cultural complexity seems to be more against the likely accuracy of Wittfogel's singular emphasis on irrigation as 'cause' or with its political implications for the emergence of 'despotism' than upon any solid evidence that he marshals to oppose its role in the emergence of urban complexity.

The Earliest Irrigation Networks

Rain fed agriculture came into being some 10 thousand years ago in the arid lands of the Eastern Mediterranean. At the Neolithic village of Choga Mami, in Eastern Iraq near the Gangir River, archaeologists discovered the presence of simple gravity flow furrow irrigation dated to the 6th millennium BCE (Oates and Oates 1976). Even earlier simple irrigation practice is argued for the 7th millennium site of Beidha in Southern Jordan (Byrd 2005). The manipulation of water came long before the emergence of social power that controlled the huge irrigation networks of the 3rd millennium. Channeling water to fields must have come soon after the domestication of plants, if so, Choga Mami is a late representative of the manipulation of water to feed agricultural fields. Access to water is tantamount to managing risk whether a hunter-gatherer or a farmer. Where water is a scarce resource and populations are wholly dependent on irrigation for agricultural yield the risks are high and risk management is essential. Thus, the tenuous nature of irrigation necessitates the management of risk for the very survival of the community.

In the late 1960s and 1970s the earliest fully preserved landscape of irrigated agriculture in the Near East was discovered by the Tepe Yahya Project in the Kerman Province of Southeastern Iran. The late Martha Prickett (1986) undertook the survey and excavation of a number of early 6th–5th millennium sites in the Shah Maran-Daulatabad region where a system of spate irrigation was mapped in association with a dense settlement distribution (see also Wilkinson 2003). The village of Dolatabad, in the Shah Maran-Daulatabad basin, is located 24 km from the site of Tepe Yahya in Southeastern Iran. Today the mean rainfall is 150 mm. The extremely restricted rainfall produces an arid environment of sparse xerophytic and halophytic shrub and grass vegetation. In the absence of irrigation there is clearly insufficient rainfall for agricultural production. The Shah Maran-Daulatabad basin is watered by two perennial surface water flows, the Rud-i Gushk and Rud-i Ab Dasht. The Rud-i Gushk channel is a broad anastomosing network of 20–300 m wide braided channels with banks under 2 m (ca. 6.5 ft) in height. Settlements in the immediate region of Tepe Gaz Tavila, form a cluster of 11.57 ha (ca. 43 ac) and date from the late 6th to mid-5th millennium. The *total* settlement within the survey area numbered 313 mounded sites and 227 'scatter sites'. Individual sites were discontinuously settled, however, over the course of 1,200 years the region itself was continuously inhabited. The area of terraced and irrigated agricultural fields encompassed a region of approximately 400 ha, almost 1,000 ac. Irrigation channels were narrow depressions ranging from 30 ft in length to 3.8 mi (6 km). Low stone dams were placed across the rectangular fields to both deflect the water to downslope fields as well as to capture the silt in flood conditions – during the monsoonal period of late summer. The region was totally abandoned in the late 4th millennium, resulting perhaps, from a shift in the pattern of the monsoon (Brooks 2006). The Shah Maran-Daulatabad basin provides the earliest preserved irrigated landscape with an associated settlement regime extending over an area of almost 1,000 ac. The range of over 500 settlements from a maximum of 12 ha (29.65 ac) to six hundredths of a hectare (two-tenths of an acre) represent a likely population in the low thousands living in villages numbering from no more than 50 to several hundred. The settlement and associated irrigation found here is almost two millennia before centralized authority takes over the management of irrigation in Mesopotamia (Adams 1981). The evidence presented here would be contemporary with the Ubaid Period in Mesopotamia where Adams sees little evidence for the centralized management of irrigation. Adams (2007) in discussing Mesopotamia and the role of irrigation has remained wholly consistent in his belief that "Large dendritic systems of canals bringing whole subregions of the plain into coordinated control.... did not exist". If so, Mesopotamia was behind the contemporary agricultural developments of distant Southeastern Iran and, as discussed below, the even more distant irrigated agricultural regions of 5th millennium Turkmenistan.

It is impossible to conceive of the absence of a degree of central management over the Shah Maran-Daulatabad irrigation system that sprawled over 1,000 ac and in which over 500 variably sized settlements were located. Clearly, there had to be some form of management. The scale of settlement density, population, and the expansive irrigated fields would have required an oversight. Without management for scheduling access, amount, construction and maintenance of canals, chaos and combat would prevail in the fight for individual rights, as opposed to corporate needs, for water. The rare evidence derived at Choga Mami and the unique landscape of the Shah Maran-Daulatabad region attest to a scale of irrigation network that required some degree of centralized control. The terraced field systems in the Maran-Daulatabad region consisted of both constructed stone dams (30-80 cm in height) and silt dams to prevent the run-off of sheet-flooding and river flood. Complimenting these water management systems was the construction of canals as well as the construction of dams diverting water from natural widyan into agricultural fields or canals. Excavations of sites in the region of the fields yielded wheat, barley, millet varieties, cultivated grapes, dates, poppy (*Papava* sp.), and legumes (Prickett 1985, 1986a). Within the estimated 300 ha (741 ac) of

ancient terraced fields there are over 30 km (18 mi) of walls lines that were mapped. Many others were too poorly preserved to record. Prickett (1986b:243) is clear in stating that the complexity of the terraced field systems and spate irrigation that bound the agricultural fields into a single network required "Intensive labor input and organization....that supported a large settlement growth for over half a millennium". As nascent as that management might have been the adjoining irrigated fields amid a considerable settlement density was of such a size and complexity as to be beyond the control of individual households. The Shah Maran-Daulatabad region involved over 500 settlements of towns, village and homesteads scattered over 1,000 ac! (For a remarkably detailed analysis and illustrations depicting the environment, ecology, settlement regime and irrigation systems as explored by the Tepe Yahya Project see Prickett 1979, 1986a, 1986b).

The Shah Maran-Daulatabad region may be the best preserved ancient landscape of settlements and irrigation practices within the Near East but it is not the only one. Dr. G. N. Lisitsyna, a paleoethnobotanist affiliated with the Institute of Archaeology, Moscow, was directly involved in the paleogeographical investigations in Southern Turkmenistan. In the eastern part of the ancient delta of the Tedjen Oasis nine archaeological sites were discovered in a region referred to as the Geoksyur Oasis. A general description of this Central Asian world, consisting of dominant rivers forming oases surrounded by forbidding deserts and specific settlement pattern, is well described by Barnard Taylor (1874:17):

> The dry climate, which makes a desert of the greater portion of the land, in fact, allowing habitation only in the neighborhood of the mountains, has given rise to a singular arrangement of the settlements. In the absence of periodic rains the inhabitants are obliged to rely upon the streams which come from the mountains in spring and summer for the fertilizing of their fields. They therefore construct long canals and

ditches from the gorges of the streams to their fields, and thereby, notwithstanding the rudest agricultural implements they obtain regular and excellent harvests, unless there happens to be an unusually scanty snowfall upon the mountains, and the supply of water is diminished in consequence.

The Geoksyur Oasis consists of 400 km² (105 mi²) of irrigated lands in which nine archaeological sites were discovered, ranging from the mid-4th to the early 3rd millennia. Geoksyur 1, a settlement of 8-10 ha (27-36 ac) was fed by Canal 1 that extended over 3 km (1.86 mi). The extent of the irrigation canals dug in the Geokysur Oases amount to 7,500 m³ (264,8593 ft). The 3 km Canal 1 was 3.47 m (11 ft) wide and 1.2 m (3.9 ft) deep. Using Alexander Vaiman's (1961) figures for canal diggers in Sumer it is estimated that the construction of Canal 1 took 2,500 man-days or 100 men working for 25 days. The irrigated lands around the settlement of Geoksyur 1 were estimated to range from 50 to 80 ha (123-197 ac). If these irrigated lands produced one harvest it is estimated that production would amount to 100 metric tons. The possibility for two harvests was considered and deemed possible. This would have produced a substantial surplus. Wheat and barley were sown with a clear dominance of barley. Again, was the preference for barley due to the salinization brought about by irrigation? Adjacent to the settlement of Geoksyur IV, dated to the end of the 4th millennium, a water storage pit, a reservoir, was discovered with a surface of 1,000 m² (10,7642 ft). The Geoksyur surveys and excavations were aided by the pioneering use of aerial photography followed by ground-truthing. Regarding the irrigation network Lisitsyna (1969: 279) concludes:

The changes which have been determined by archaeologists in the distribution of early settlement sites were related, not to those changes in the natural environment which were caused purely by the climate, but to the *organization* of the hydrographic network and the constant oscillations in the water supply of certain regions.

'Organization' implies some degree of control and when one examines the considerable distance traversed by canals, as well as the dams constructed for sheet and river flood control, and the number of villages served by the irrigation network one cannot help but consider some degree of centralized control. It is likely that the Geoksyur and Shah Maran-Daulatabad irrigation systems were managed in a hetarchical manner. Thus, several institutions representing different constituents, i.e. tribes, specific lineages, religious authorities, etc., formed a communal decision-making authority. With the passing of time and under demographic pressure within the ever increasing size of settlements social forces required a new control over the productive forces of the economy. This, in turn, required a greater centralization of authority for securing the necessary labor force, as well as for the design, construction and management of the irrigation systems that characterized the 2nd and 1st millennia BCE. Control over the means of production, whether labor, land tenure, irrigation, trade or craft production, evolved from an earlier hetarchical communal authority to the hierarchical power in the hands of an increasing few. Perhaps, within the Old World the one region with the best evidence for the transition from irrigation as hetarchy, the Geokysur Oasis, to hierarchy, the Khorezmian Expedition, is in Central Asia. The author of this book, Boris V. Andrianov, was involved in that project serving as the director for the specific study of the irrigation systems.

The Khorezmian Expedition

Sergey Yatsenko (2007), on the occasion of the 2006 opening of *The History of One Expedition* at the State Museum of Oriental Art in Moscow, referred to the Khorezmian Expedition as "The Biggest Expedition Studying the Ancient Iranian World". That statement need not have been restricted to the 'Iranian World'. The Khorezmian Expedition was under the leadership of Academician Sergey Pavlovich Tolstov (1907-1976). Born of a family that supported the monarchy his father was an officer in the Emperor's guard, his grandfather a general in the Ural Cossacks. Tolstov was to change allegiance and became a dedicated member of the Communist Party and an academic leader in the Soviet Academy of Sciences. His enthusiasm for the Party whether dedicated, necessary, compromised, or forced is today a subject of debate in Russia. In 1923, Tolstov became a student in the Department of Anthropology and studied with the distinguished environmentalist Professor Vsevelod Aleksandrovich Anuchin (see below). In 1929 Tolstov took part in an expedition to the lower Amu Darya (Oxus) River studying the traditions of the Turkmen Yomud. In 1935 he completed his doctorate and in 1938 became the head of the Khorezmian Archaeological-Ethnographical Expedition. His was a meteoric rise, in both academic and party circles. In 1939 Tolstov became the Director of two leading institutions: the Department of Ethnography in Moscow State University (which he headed until 1952) and the Institute of the History of Material Culture (today the Institute of Archaeology, Russian Academy of Science). In the 1950s he added to these administrative responsibilities the directorship of the Institute of Oriental Studies and the editorship of the magazine, *Soviet Ethnography*. In the 1930s ethnology was being declared a bourgeois science and in 1933 a new system of cultural institutions was established to study regional cultures (*kraevedenie*). Previous institutes with ethnographic interests were closed, their leaders dismissed, some killed (Miller 1956). Tolstov became a new leader, an ideologue supporting the new tenets for the collectivization of farming, the elimination of private property, and the sedentarization of nomads. Following WW II from 1946-1952 he concentrated on the excavation of Toprak-kala and from 1952-1957 on Koy-Krylgan-kala . Both sites are in the Ellik kala, the 'Fifty Fortresses Oasis' in modern day Uzbekistan. At Toprak-kala, capital of the Khorezmian Kingdom and dated to the 1st-5th centuries CE, Tolstov excavated palaces, temples, and wall paintings depicting

Zoroastrian deities, some 14 m in height. At Koy-Krylgan-kala, 400BCE-400CE, excavations revealed a Mazdian Fire Temple. [For photos and accessible text see http://www.heritageinstitute. com/ zoroastrianism/khvarizem/page 3.htm, www.heritageinstitute.com/zoroastrianism/ khvarizem/page 3.htm, one of several web sites pertaining to Tolstov's expedition].

From 1952-1957 Tolstov directed an expedition that was extraordinary in its interdisciplinary perspective (for two seminal publications see Tolstov 1948a-b). We note that the author of this book, Boris V. Andrianov, was both a colleague and a co-author with Tolstov on numerous publications pertaining to the expedition results. The expedition obtained permission to utilize airplanes, carried portable power-stations, photo-labs, and had special caravan automobiles. Excavations were combined with archaeological survey on both the land and through flight. It was among the very first expeditions that combined aerial photography with ground truthing. In 1946 over 9,000 km (ca. 5,500 miles) were covered by air routes with 60 landings and 5,000 air photos taken. More than 250 new sites were recorded. Central to the expeditions concern were the recording of ancient settlements, roads, irrigation systems and the extent of land cultivated. During 10 field seasons, 1952-1961, scientists studied the modern and ancient topography, climate, environment, ethnography and archaeology of the right bank of the Khorezmian Amudarya. In the course of five seasons the field work covered 2,000 km (1,242 miles), recorded and described more than 400 archaeological monuments from the 4th millennium BCE to the 16th century CE. The topographical-archaeological staff responsible for documenting and mapping the ancient irrigation systems was directed by Boris V. Andrianov. The presence of settlements and associated irrigation systems within the Bronze, Iron, and the Medieval Periods were all duly recorded. A dramatic change in the complexity of irrigation is noted in the 6th-5th centuries BCE when irrigation canals were 100-150 km (62-91 mi) long and 2-3 m (6.5-9.8 ft) deep.

Barley, wheat, millet, grapes (for wine production), and a variety of fruits including numerous types of melon (for which the region today remains famed) were all cultivated by irrigated fields. Numerous forts surrounded the settlements. Andrianov (1995:11) in a more recent publication summarizes that within all periods one notes that:

> Periods of political centralization coincided with the rise and spread of irrigation-based farming, and vice versa, the reduction of irrigated areas occurred in troubled war and crises time....in the lower reaches of the Amu-Darya and Syr Darya.... Complex archaeological study of these areas has made possible the reconstruction of complicated history of the development of irrigation and irrigation-based farming.

Tolstov, not including hundreds of his articles, published two highly significant synthetic works summarizing the results of the expedition in 1948 and in 1962. One may fairly ask why is this remarkable expedition virtually unknown in the western literature. Although there are rare instances of its mention, (i.e., Altman 1947) its truly remarkable results, a landmark in the history of 20th century archaeology, are all but ignored. Its 1930s interdisciplinary approach, let alone its independent development of settlement survey analysis, are of historic significance. The answer is perhaps obvious. During the Soviet period travel, let alone collaboration, in Central Asia, even for scientists (maybe particularly for scientists) was not possible. Sadly, political barriers simply prohibited collaborative undertaking and, without such, nations and science become separatist if not even antagonistic. This barrier was only lifted in the mid-1980s with the first USA–USSR collaboration in joint archaeological symposia, five held in both countries, as well

as collaborative field work (Lamberg-Karlovsky 1994) Archaeological collaboration with the USSR was quickly followed by the French, Italian, British, and Germans. By the early 1990s, with the collapse of the Soviet Union, many nations were undertaking collaborative research in Russia and its former 'autonomous republics'.

We mentioned above that Professor Vsevelod Aleksandrovich Anuchin was a mentor to Tolstov. Anuchin, a geographer, was a major figure and much involved in what became known as the 'Anuchin controversy'. It has an important bearing on the entire approach taken by Tolstov. In the 1930s the geographer N. N. Kolosovski was among the most influential scientists in pre-Soviet Russia advocating a unified geographical approach for an appreciation of the importance of the environment on human affairs. Anuchin was his able disciple. The notion that the environment may be a 'determinist' factor in land-man relationships directly countered official Soviet ideology. On this point Stalin's *Short Course* was abundantly clear. In Chapter 4 of the *Short Course*, "Dialectical and Historical Materialism", Stalin states that the chief causal force in social development is not to be found in the geographic environment but in the "mode of productions of material values". In following Marxist-Leninist ideology Stalin asserted that the physical environment can never decisively determine the growth of society.* Such a perception is consistent with Marxist-Leninist views that natural laws and social laws occupy distinctive realms. In sum, man is capable of mastering nature through social development and is not subject to the influence of his natural environment. Anuchin expressed outspoken views in favor of a unified geography that favored an increased emphasis on environmental determinism. He wrote:

---

*   The *Kratkiy kurs* (Short Course) is the *Istorii Vsesoiuznoy Kommunisticeskoy partii (bolshevikov) Kratkiy kurs* 1938 (History of the All Union Party Bolsheviks). This massive volume, written by Stalin (who compiled it from various authors) served as the ultimate and unalterable textbook of Marxism-Leninism-Stalinism. Both the *Kratkiy kurs* and Stalin were denounced by N. S. Khrushchev at the Twentieth Party Congress on February 25, 1956. The *Kratkiy kurs* was withdrawn from circulation and *Pravda* announced that "the cult of the individual leader" was to be replaced by "collective leadership" of the Party (see Avrich 1960 for further details on the *Short Course*).

The interaction between society and nature always bear a clearly defined character, and given any social structure, it will be different in different countries and regions, since it depends on the geographic environment, which cannot be alike all over the surface of our planet.

Anuchin cast down the gauntlet. His criticisms of inadequate environmental planning and conservation, whether the reduction of cropland in favor of urban development, the diversion of rivers, irrigation without soil conservation, the construction of excessively large reservoirs, or the felling of forests without adequate plans for re-growth all have a modern ring. Anuchin's doctoral dissertation *Theoretical Problems in Geography* did not obtain the necessary two-thirds vote and was rejected by his committee. The 'Anuchin controversy' received some attention in the US (Hooson 1962; Chappell 1965) and, significantly, his doctoral thesis, with identical title, was published in the US (Anuchin 1977) but never saw the light of day in the USSR. He did, however, publish parts of this dissertation in literary journals.

In spite of Soviet opposition to environmental determinism the issue of man-land relationship occupied Marx in his recognition that environmental challenges posed specific socio-political economic problems. He took up the notion of 'Oriental despotism' in attempting to explain why some Asian nations, embedded with totalitarian regimes, did not experience the 'normal' transitions from slave to feudal to capitalist, and finally, to socialist conditions. The near universal rejection of Wittfogel's hypothesis concerning 'hydraulic civilizations' overlooks his valuable contribution in focusing on the importance of man-land relationships in Marxist and non-Marxist thought. That relationship was at the very core motivating the research of the Khorezmian Expedition and the contributions of S. P. Tolstov and B. V. Andrianov. In this book Andrianov, collaborating with Tolstov, and directing the study of irrigation for the Khorezmian Expedition is explicit as to the determinative role of environment and the transforming effect of irrigation on both environment and human society.

Irrigation under Colonial Deconstruction

Lastly, we consider an antithesis, namely, rather than states and irrigation being intertwined in the rise of cultural complexity, we consider the thesis that contends that state control of irrigation is ruinous to both environment and agricultural production. McGuire Gibson (1974:15, 31) has this to say about British and Turkish colonialism and rule:

> By supporting and keeping one family [the *shaykh*'s] in position of power, by changing a chief [*shaykh*] to a landlord; by concentrating wealth while inducing individuals to take up small, fixed plots; by imposing yearly taxes and encouraging rents and debts, the central authority brought about widespread violation of fallow. Eventually the selling out by small holders to large landholders did not lead to a reversal of agricultural decline because debt ridden farmers often did not stay on the land as sharecroppers, but became nomads or fled to the cities.

The British administration's construction of dams, the centralization and extension of canal networks, intensification of agricultural production, and absentee landlords led to a predictable increase in salinity and a dramatic decrease in agricultural productivity. Gibson, again notes (1974:15): "Directly, through engineering that promoted water-logging and salinity, the central government acted to undermine agricultural productivity". The reduction of fallow meant a reduction in pasturage and a consequent decrease in livestock. The gradual disintegration of the tribal system, resulting, in part, from a new system of private land ownership, rents, taxes, interest on loans, and a newly constituted market economy all fostered the bankruptcy of the small land-holder as well as labor shortages. Tribal land, previously held communally, became private property transforming the *shaykhs* into landlords who were now concerned "increasingly in their own family interests".

Colonial powers transformed traditional societies. They appointed tribal *shaykhs* as their agents, empowering them as landlords, and investing in them powers which were not traditionally theirs. *Shaykhs* became absentee landlords, their tribal members tenants obligated to taxes and debt, or sold off their lands to speculators. On such occasions Esther Boserup (1974:12-13) observes that:

> When (large-scale irrigation) regions are left in the uncontrolled possession of a landlord class, which is either of foreign origin or partner in precarious alliance with a foreign conqueror, rural investments are in danger of being neglected, because the landlords inevitably go for quick profits and liquid assets. In extreme cases, the result is starvation and depopulation.

To the above voices lamenting the destructive policies of a state control in the management and construction of irrigation systems we can add that of Robert Fernea (1970:37). He studied the traditional irrigation of both the El Shebana tribe and that of Iraq's central government. His conclusion:

> All evidence seems to indicate that localized tribal organization was sufficient to sustain and at times to extend irrigation-based cultivation in the region over many centuries, in the absence of sustained governmental administration and investment. Indeed, contemporary study suggests the extensive patterns of decentralized irrigation agriculture as practiced by the tribes may actually have been better suited to the physical environment of Southern Iraq then the more intensive patterns of land use which have followed technological improvements in the irrigation systems, improvements developed by modern central government in Iraq.

In fact, the advantages of local management over irrigation practices were well recognized by the British colonial administrator, "I am very strongly of the opinion that it is sounder to leave the distribution of [water] from all except the main canals almost entirely to the Arabs themselves.... rather than entrust it to subordinates who will certainly involve our administration in a great

deal of odium" (quoted in Fernea 1970:138). Gibson (1974:7, 15) is even more strident: "In Mesopotamia the intervention of government has tended to weaken and ultimately destroy the agricultural basis of the country". The deleterious effects of state management were recognized at an early date. In the most extensive and well-documented treatise on irrigation in the 19th century we read in J. Barois's exhaustive study, *Irrigation in Egypt* (1889), of the ambivalence and tensions that accompanied the transition from traditional approaches to irrigation in the service of state control.

Barois (1898:99-101) writes:

> As to the distribution of the water amongst the different branches of the canals, as well as for the secondary canals, it is entirely in the hands of the Government agents yet, on the one hand, the farmer enjoys the greatest liberty in opening the ditches or making inlets in the banks of the canals...no [government] rules exist for regulating the discharge...on the one hand the Government assumes all authority over irrigation, and on the other the individuals are subjected for the use of the water to no special regulations...It is the unity of interests which causes the necessary measure to be taken to bring water and to maintain the level and drain it afterwards.

Barois reviews the ancient and modern history of irrigation in Egypt, observing that from time immemorial the sovereign, whether pharaoh or king, laid claim to the ownership of land while in practice the usufruct, the territorial unit, was one in which land was communally held. "This system lasted, with some slight modifications, until the beginning of this century (with the initiation of Mehmet Ali's (1769-1849) developmental programs involving land reform, the construction and control of large scale irrigation networks, the establishment of corvée labor, the employment of 450,000 laborers for four months largely devoted to irrigation construction, and related new laws) and it may be believed that during this long period of time communal rule of property its instability, its insecurity, and especially its character of usufruct, had contributed considerable to the development of a normal system of irrigation based on the rights and interests of individuals".* The binary opposition of state vs. local control of irrigation systems establishes an opposition that functions best as academic debate but not at all in the functioning of irrigation networks. From an evolutionary perspective one concedes that initially there was local control over both communal lands and irrigation systems. Local control does not place the emphasis on either word, 'local' or 'control'. There was control and that control was local. Such control existed under hetarchical principles involving the traditions of tribal rule, lineage authority, and religious and ritual determinants.

The size of the community was directly related to the scale of the irrigation system. Increasing populations require ever increasing agricultural production and both, in turn, require new forms of social organization. There is a linear relationship in the increasing size of community population, agriculture and power. At a certain point (indeterminate?) hetarchical systems give way to an emergence of hierarchical structures of power. That point of transition may be conceived in anthropological terms as what is regarded as the evolution of a chiefdom to a state administered bureaucracy. The irrigation networks of Maran-Daulatabad and the Geoksyur Oasis were assuredly under local control. The control of irrigation systems in the Bronze Age of Mesopotamia and Egypt may mirror that of the El Shabana and Barois's description of 19th century Egyptian irrigation systems. In detailing the small-scale horizontal structures of management and control, those that are hetarchical, as opposed to the vertical structures of hierarchical control, we have, under the enduring influence of Wittfogel, too long favored the later and all but ignored the former. We have missed the fact that:

1) scale matters, both with regard to the irrigation network and population size; 2) of necessity ALL irrigation systems are managed and controlled; 3) and, that the scale of control relates to the scale of social organization; 4) the control of labor comes before the construction of irrigation systems, monumental architecture; and, 5) there are always factors that combine local and central control of irrigation systems whether they be contoured by tribal or state organizations!

Jennifer Pournelle has recently made an important case for natural wetlands, an environment of marshes, rather than irrigated fields, as being the sustainable habitat in which the cities of Mesopotamia emerged. She writes:

In most people's heads – archaeologists, ecologists, environmental scientists – here's the idea that cities happen because somebody invented and managed irrigation. My argument is 'No' – *irrigation is what happened because you had cities*, but the marshes were moving away from them. That's what marshes do. Deltas build up, river mouths migrate and the marshes go with them The city's stick where it is, so it has to start irrigating to raise crop production and replaces all of the marshland resource that have moved too far away (emphasis mine).

Radiocarbon dates were obtained from the site of Eridu, in Southern Iraq, from shells that were embedded in mud bricks confirming the marshland environment during the settlements earliest occupation: "What we found was that in this area, right around where the mound [settlement] is, there were marshes during the bracketed dates". The site was occupied from 8000–4000 BCE. It is her hypothesis, which has considerable merit in the Southern Mesopotamian landscape that cities first emerged in the marshes where aquatic resources were rich and water readily available for agricultural production. This may well be true for Southern Mesopotamia but it is equally true that irrigation was already evident in the Neolithic, prior to such settlements as Eridu in Southern Mesopotamia. Pournelle is quite correct that "irrigation is what happened because you had cities" and such is true not only for Mesopotamia

---

* Private property rights were not fully conceded until 1871 under Khedive Ismail Pasha.

but for Central Asia, China, Egypt, and Peru, to mention a few places in which early cities were dependent on irrigation. Cities contemporary with those in the southern marshes were also present in the Mesopotamian alluvium and would have required irrigation to support the population. In the Late Uruk period Uruk texts record fields of 952 hectares while one from Jemdet Nasr concerns 2,120 hectares of agricultural fields which certainly would have been irrigated (Renger 1995:273). As Steinkeller has observed (1999:302):

> Moreover, as is generally known, in the Mesopotamian alluvium cereal cultivation is impossible without artificial irrigation. Only minimal artificial irrigation can be done in the individual-family level. Under such conditions, irrigation works are indispensable for securing an agricultural surplus, but even a small-scale irrigation system exceeds the labor capacity of a single family. To create such a system, and even more importantly, to maintain it, a suprafamily organizational arrangements be they voluntary or coercive are required.

The increasing complexity of irrigation technology, settlement density, and population increase are best seen as co-evolutionary processes within alluvial environments involved in driving scales of cultural complexity within millennia long histories.

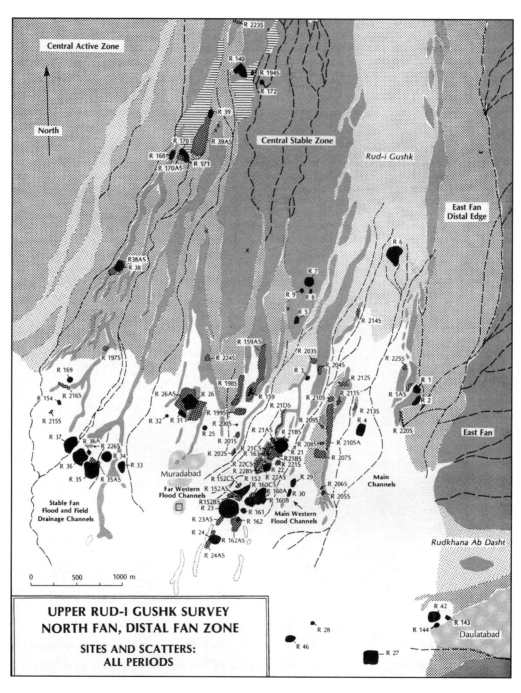

Central Active Zone

North

R 2235

R 140

R 194S

R 172

R 39

Central Stable Zone

Rud-i Gushk

R 170

R 39AS

R 168

R 171

R 170AS

East Fan
Distal Edge

R38AS

R 38

R 6

R 7

R 9   R 8

R 5

R 214S

R 159AS

R 224S

R 203S

R 3

R 204S

R 225S

R 197S

R 212S

R 1

R 169

R 198S

R 159

R 21IS

R 1AS

R 154   R 216S

R 26AS

R 26

R 210S

R 213S

R 2

R 215S

R 199S

R 21DS

R 209S

R 4

R 32   R 31

R 200S

R 21AS

R 21BS

R 208S

R 210SA

R 220S

R 25

R 201S

R 21CS

R 21

R 207S

East Fan

R 37

R 36A

R 226S

R 202S

R 21BS

R21BS

R 34

R 22CS

R 22

R 36

R 33

R 22BS

R 22AS

Main
Channels

R 35

R 35AS

Muradabad

R 152CS

R 152

R 29

R 206S

Far Western
Flood Channels

R 152AS

R 160CS

R 160A

R 30

R 205S

R152BS

R 23

R 163

R 160B

Stable Fan
Flood and Field
Drainage Channels

R 23AS

R 161

R 162

Main Western
Flood Channels

R 24

R 162AS

Rudkhana Ab Dasht

R 24AS

0   500   1000 m

R 42

R 143

R 28

R 144

Daulatabad

R 46

R 27

UPPER RUD-I GUSHK SURVEY
NORTH FAN, DISTAL FAN ZONE

SITES AND SCATTERS:
ALL PERIODS

Settlement sites and principal water channels.

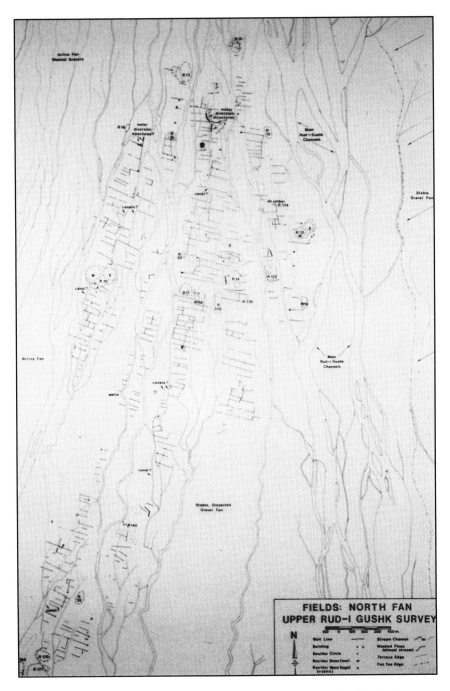

Settlements, water channels, dug canals, and stone walls (perpendicular lines) for alluvial catchment.

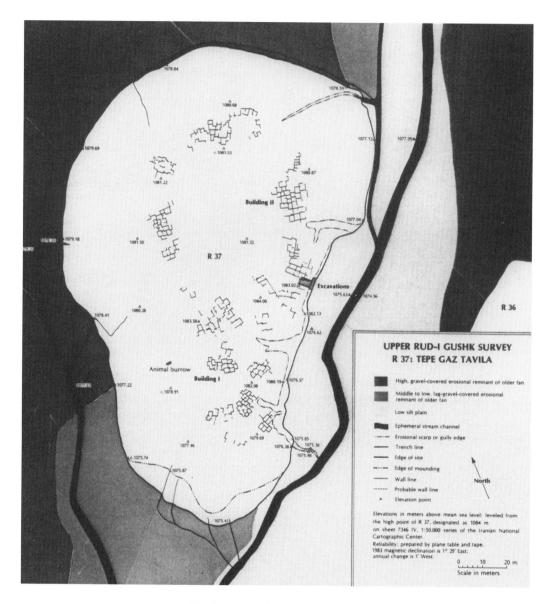

Tepe Gaz Tavila, 5th millenium settlement.

Tepe Gaz Tavila, excavation and surface architecture. The site was bisected by seasonal flooding.

Stone walls for alluvial catchment.

Stone walls for alluvial catchment.

References

Adams, R. McC.

1966   *The Evolution of Urban Society: Early Mesopotamia and Prehispanic Mexico.* University of Chicago Press, Chicago.

1974   Historic Patterns of Mesopotamian Irrigation Agriculture. In *Irrigations Impact on Society*, edited by Theodore Downing and McGuire Gibson. Anthropological Papers of the University of Arizona, No. 25. University of Arizona Press, Tucson.

1981   *The Heartland of Cities: Studies of Ancient Settlement and Land Use. On the Central Floodplain of the Euphrates.* University of Chicago Press, Chicago.

1996   *Paths of Fire: An Anthropologists Inquiry into Western Technology.* Princeton University Press, Princeton, NJ.

2006   Intensified Large-Scale Irrigation as an Aspect of Imperial Policy: Strategies of Statecraft on the Late Sasanian-Mesopotamian Plain. In *Agricultural Strategies*, edited by J. Marcus and C. Stanish. Cotsen Institute of Archaeology, Los Angeles.

2007   *The Limits of State Power on the Mesopotamian Plain.* Cuneiform Digital Library Bulletin.

Algaze, G.

2008   *Ancient Mesopotamia at the Dawn of Civilization.* University of Chicago Press, Chicago.

Andrianov, B. V.

1978   The Concept of "Hydraulic Society". *Social Sciences Critical Studies Comment* IX(1):193-209.

1995   The History of Economic Development in the Aral Region and its Influence on the Environment. *Geojournal* 35(1):11-16.

Anuchin V. A.

1977   *Theoretical Problems in Geography.* Ohio State University, Columbus.

Averbeck, R. E.

2003a   Myth, Ritual and Order. In "Enki and the World Order". *Journal of the American Oriental Society* 123(4):757-771.

2003b   Daily Life and Culture in "Enki and the World Order" and Other Sumerian Literary Compositions. In *Life and Culture in the Ancient Near East*, edited by Richard E, Avereck, Mark W. Chavales and David B. Weisberg, pp. 23-61. CDL Press, Bethesda, MD.

Avrich, P. H.

1960   The Short Course and Soviet Historiography. *Political Science Quarterly* 75(4):539-553.

Bartold, V. V.

1958   *Turkestan Down to the Mongol Invasion, 2nd Edition.* E. J. Gibb Memorial Series. V. Luzac and Co., London.

Bennett, J. W.

1974   Anthropological Contributions to the Cultural Ecology and Management of Water Resources. In *Man and Water: The Social Sciences and Water Management*, edited by L. D. James, pp. 34-81. University of Kentucky Press, Lexington.

1976   *The Ecological Transition: Cultural Anthropology and Human Adaptation.* Pergamon Press, New York.

Bichsel, C.

2009   *Conflict Transformation in Central Asia.* Routledge, London.

Blunt, A. I. N.

1879   *Bedouin Tribes of the Euphrates.* Harper Brothers, New York.

Boserup, E.

1974   *The Conditions of Agricultural Growth.* Aldine, Chicago.

1981   *Population and Technology.* University of Chicago Press, Chicago.

Brooks, N.

2006   Cultural Responses to Aridity in the Middle Holocene and Increased Complexity. *Quaternary International* 151(1):29-49.

Byrd, B.

2005   *Early Village Life in Beidha, Jordan.* Oxford University Press, Oxford.

Chappell, Jr., J. E.

1965   Marxism and Geography. *Problems of Communism* 14:12-22.

Crumley, C.

1987   Dialectical Critique of Hierarchy. In *Power Relations and State Formation*, edited by T. C. Patterson and C. W. Gailey, pp. 155-159. American Anthropological Association, Washington, D.C.

1995   Hetarchy and the Analysis of Complex Societies. In *Archaeological Papers of the American Anthropological Association*, 6 edited by R. N. Ehrenreich, C. L. Crumley and J. E. Levy. Archaeological Papers of the American Anthropological Association, 6.

De Villiers, M.

2000   *Water: The Fate of our Most Precious Resource.* Houghton Mifflin, Boston.

Fagan, B.

2011   *Elixir. A History of Water and Humankind.* Bloomsbury Press, New York.

Fernea, R.

1970   *Shaykh and Effendi Changing Patterns of Authority Among the El Shabana of Southern Iraq.* Harvard University Press, Cambridge.

Gibson, McG.

1974   Violation of Fallow: An Engineered Disaster in Mesopotamian Civilization. In *Irrigation's Impact on Society*, No 25. Theodore E. Downing and MacGuire Gibson. University of Arizona Press, Tucson.

Godley, A. D.

1966   *Herodotus.* Harvard University Press, Cambridge, MA.

Hassan, F.

2006   Archaeology for our Times: Water and Civilization. *The Review of Archaeology* 27:1-25.

Hunt, R. C., and E. Hunt

1976   Canal Irrigation and Local Social Organization. *Current Anthropology* 17(3):389-411.

1988   Size and Structure of Authority in Canal Irrigation System. *Journal of Anthropological Research* 4:335-355.

Huntington, S. P.

1996   *The Clash of Civilizations and the Remaking of the World Order.* Simon and Schuster, New York.

Hooson, D.

1962   Methodological Clashes in Moscow. *Annals of the Association of American Geographers*, 52(4):469-475.

Jacobsen, T.

1982   Salinity and Irrigation Agriculture in Antiquity. *Biblioteca Mespotamica* 14, Malibu.

Kemp, B.

2006   *Ancient Egypt. Anatomy of a Civilization.* Routledge, London.

Kramer, S.

1963   *The Sumerians.* University of Chicago Press, Chicago.

Lamberg-Karlovsky, C. C.

1994   Initiating an Archaeological Dialogue: The USA-USSR Archaeological Exchange. In *Origins of the Bronze Age Oasis Civilization in Central Asia.* Peabody Museum of Archaeology and Ethnology, Cambridge.

2013   The Oxus Civilization (AKA: The Bactria-Margiana Archaeological Complex): A Review. *Reviews in Archaeology*, 30:21-63.

Lansdell, H.

    1885   *Russian Central Asia including Kuldja, Bukhara, Khiva and Merv.* Houghton Mifflin, Boston.

Lansing, J. S.

    1991   *Priests and Programmers: Technologies of Power in the Engineered Landscape of Bali.* Princeton University Press, Princeton.

Le Grange, G.

    1930   *The Lands of the Eastern Caliphate. Mesopotamia, Persia and Central Asia from the Mongol Conquests to the Time of Timur.* Cambridge University Press, Cambridge.

Lewis, M. E.

    2006   *The Flood Myths of Ancient China.* State University of New York Press, Albany.

Lisitsyna, G. N.

    1965   *Oroshaemoe zemledelie epokhi eneolita na yuga Turkmenii.* Moscow.

    1966   Izuchniye geoksyurskoy orositelnoi seti v Yujnoi Turkmenii v 1964. *Kratkiye Soobscheniye Institute Arkheologie* CVII:98-136.

    1996   The Earliest Irrigation in Turkmenistan. *Antiquity* LIII:279-288.

MacGahan, J. A.

    1874   *Campaigning on the Oxus and the Fall of Khiva.* Harper and Brothers, New York.

McNeish, R.

    1967   Mesoamerican Archaeology. *Biennial Review of Anthropology* 326.

Masson, V. M.

    2006   *Kulturogenez Drevnei Tsentralnoi Azii.* Sankt Peterburg.

Mendelsson, K.

    1986   *The Riddle of the Pyramids.* Thames and Hudson, New York.

Miller, M.

    1956   *Archaeology in the USSR.* Praeger, New York.

Mitchell, W. P.

    1973   The Hydraulic Hypothesis; A Reappraisal. *Current Anthropology* 14:5:532-34.

Mukhamedjanov, A. R.

    1978   *Istoriya Orosheniya Bukharskogo Oazisa.* Tashkent.

Netting, R.

    1993   *The Smallholders, Householders: Farm Families and the Ecology of Intensive Sustainable Agriculture.* Stanford University Press, Stanford.

Oates, D. and J. Oates

    1976   Early Irrigation Agriculture in Mesopotamia. In *Problems in Economic and Social Archaeology*, edited by G. De G. Sieveking, et al., pp. 109-152. Duckworth, London.

O'Donovan, E.

    1882   The Merv Oasis. Smith, Elder & Co., London

Oshima, T.

    2007   Marduk, The Canal Digger. *The Journal of the Ancient Near Eastern Society* 30:77-88.

Pournelle, J. R.

    2003   The Littoral Foundations of the Uruk State: Using Satellite Photography Toward a New Understanding of the 5th/4th Millennium BCE Landscapes in the Warka Survey Area, Iraq. In *Chalcolithic and Early Bronze Age Hydrostrategies.* BAR International Series 1123, edited by D. Gheorghiu, pp. 5-23. Archaeopress, Oxford.

2012   Sc.academia.edu/JenniferPournelle/Papers/1086636/MID- HOLOCENE_DATES_FOR_ORGANIC_
RIC_SEDIMENT_PALSTRINE_SHELL_AND_CHARCOAL_FROM_SOUTHERN_IRAQ "Lessons from
Southern Iraq: Urban Marshes and City Survival" HYPERLINK "http://phys.org/news/2012-10- lessons-
iraq-urban-" http://phys.org/news/2012-10-lessons-iraq-urban-marshes-city.html

Powell, M. A.

1985   Salt, Seed and Yield in Sumerian Agriculture. A Critique of the Theory of Progressive Salinization.
Zeitschrift für Assyriologie 75:7-38.

Prickett, M.

1979   Settlement and development of Agriculture in the Rud-i Gushk Drainage, Southeastern Iran. In *Akten
des VII Internationalen Kongresses für Iranische Kunst und Archäologie, Archäeologishe Mteilungen aus Iran*
6:47-56.

1986a Man, Land and Water – Settlement Distribution and the Development of Irrigation Agriculture in the
Upper Rud-I Gushk Drainage, Southeastern Iran. Ph.D. Dissertation, Department of Anthropology,
Harvard University Cambridge, MA.

1986b Settlement During the Early Periods. In *Excavations at Tepe Yahya, Iran 1967-1975*, edited by C. C.
Lamberg-Karlovsky and Thomas W. Beale, pp. 215-247. American School of Prehistoric Research,
Bulletin 38. Peabody Museum, Harvard University, Cambridge, MA.

Renger, J.

1995   Institutional, Communal, and Individual Ownership or Possession of Arable Land in Ancient
Mesopotamia from the End of the Fourth to the End of the First Millennium BC, Symposium on Ancient
Law, Economics, and Society, Part II, *Chicago-Kent Law Review* 71:1, 269-319.

Sahlins, M., and E. Service

1988   *The Evolution of Culture.* The University of Michigan Press, Ann Arbor, MI.

Salvatori, S., and M. Tosi

2008   *The Bronze Age and Early Iron Age in the Margiana Lowlands.* BAR International Series 1806. Archaeopress,
Oxford.

Sanders, W. T.

1968   *Mesoamerica: The Evolution of a Civilization.* Random House, New York.

Sarianidi, V.

2010   *On the Track of Uncovering a Civilization.* Aletheia, St. Petersburg.

Schuyler, E.

1874   *Turkmenistan: Notes on a Journey in Russian Turkmenistan.* Scribner, Armstrong and Co., New York.

Smith, P. E. L., and T. Cuyler Young

1981   The Evolution of Early Agriculture and Culture in Greater Mesopotamia. In *Population Growth:
Anthropological Implications*, edited by Brian Spooner. MIT Press, Cambridge.

Solomon, S.

2010   *Water. The Epic Struggle for Wealth, Power and Civilization.* Harper-Collins, New York City.

Steinkeller, P.

2007   City and Countryside in Third Millenium Southern Babylonia. In *Settlement and Society: Essays Dedicated
to Robert McCormick Adams*, edited by Elizabeth Stone, pp. 185-211. Cotsen Institute, UCLA.

1999   Land Tenure Conditions in Third Millenium Babylonia: The Problem of Regional Variation. In
Urbanization and Land Ownership in the Ancient Near East, edited by Michael Hudson and Baruch A.
Levine. Peabody Museum Bulletin 7, Harvard University, Cambridge, Mass. 289-329.

Stewart, J.

1949   Cultural Causality and Law. *American Anthropologist* 51:1-27.

1955   Introduction: The Irrigation Civilizations, A Symposium on Methods and Result in Cross-cultural Regularities. In *Irrigation Civilizations: A Comparative Approach*. Pan America Union Social Sciences Monograph 1.

Stronach, D.

1961   The Excavations at Ras al 'Amiya. *Iraq* XXIII:95-137.

Taylor, B.

1874   *Central Asia, Travels in Cashmere, Little Tibet and Central Asia*. Scribner, Armstrong, New York.

Tolstov, S. P.

1946   The Early Cultures of Khwarizm. *Antiquity* 78:72-99.

1948a *Drevnii Khorezm. Opyt istoriko-arkheologicheskogo issledovaniya*. Moscow University Press, Moscow. 1948b *Po Slednom Drevnekhoremzmiyskoy Tsivilzatsii*. Moskva-Leningrad.

1962   *Po drevnim delta Oksa i Yaksarta*. Moskva.

Ur, J.

2011a *Physical and Cultural Landscapes of Assyria. The Blackwell Companion to Assyria*, edited by E, Frahm. Blackwell, Oxford (in press).

2011b *Patterns of Settlement in Sumer and Akkad. The Sumerian World*, edited by Harriet Crawford. Routledge, London.

Vámbéry, A.

1865   *Travels in Central Asia*. Harper and Brothers, New York.

Vayman, A. A.

1960   *Dva klinopisnikh documents o provednii orositelnovo kanaa (Two cuneiform documents on the construction of irrigation canals)*. Trude Gosudarstvennovo Ermtaza. Leningrad.

Wilkinson, T. J.

2003   *Archaeological Landscapes of the Near East*. University of Arizona Press, Tucson.

Wright, H.

1969   *Administration of Rural Production in an Early Mesopotamian Town*. Museum of Anthropology Anthropological Papers, Ann Arbor.

1977   Recent Research on the Origin of the State. *American Review of Anthropology* 6, pp. 379-397.

VII: References and Bibliography of Boris V. Andrianov

Irina A. Arjantseva

Publications included in the original text

1951a K voprosu o geograficheskikh izmeneniyakh v delte Amu-Dari. *Voprosy geografii*, Tom 24. Moskva.

1951b *Etnicheskaya territoriya karakalpakov v Severnom Khorezme (XVIII-XIX vv.)*. Avtoreferat dissertatsii na soiskanie uchenoy stepeni kand. istoricheskikh nauk. Moskva.

1952a Ak-Djagyz (k istorii formirovaniya sovremennoy etnicheskoy territorii karakalpakov v nizove Amu-Dari). *Trudy Khorezmskoy arkheologo-etnograficheskoy ekspeditsii. Tom I, Arkheologicheskie i Etnograficheskie Raboty Khorezmskoy Ekspeditsii 1945-1948*, S. P. Tolstov i T. A. Jdanko, pp. 567-585. Moskva.

1952b Materialy k istorii orosheniya v Khorezme. *Glavnyy Turkmenskiy kanal*. Moskva.

1954 K voprosu o prichinakh zapusteniya zemel drevnego orosheniya na Kunya-Dare i Jany-Dare. *Izvestiya Vsesoyuznogo Geograficheskogo Obshchestva*, Tom 86, Vyp. 5:442-447.

1955 Iz istorii zemel drevnego orosheniya Khorezmskogo oazisa. *Pamyati akademika L. S. Berga*, pp. 353-359. Moskva - Leningrad.

1958a Etnicheskaya territoriya karakalpakov v Severnom Khorezme (XVIII-XIX vv.). *Trudy Khorezmskoy arkheologo-etnograficheskoy ekspeditsii. Tom III, Materialy i issledovaniya po etnografii karakalpakov*, T. A. Jdanko, pp. 7-132. Moskva.

1958b Arkheologo-topograficheskie issledovaniya drevney irrigatsionnoy seti kanala Chermen-yab. *Trudy Khorezmskoy arkheologo-etnograficheskoy ekspeditsii. Tom II, Arkheologicheskie i Etnograficheskie raboty Khorezmskoy Ekspeditsii 1949-1953*, S. P. Tolstov i T. A. Jdanko, pp. 311-328. Moskva.

1959a Arkheologo-topograficheskie issledovaniya na zemlyakh drevnego orosheniya Turtkulskogo i Biruniyskogo rayonov Karakalpakskoy ASSR za 1955-1956 gg. *Materialy Khorezmskoy Ekspeditsii. Vyp. 1, Polevye issledovaniya Khorezmskoy ekspeditsii v 1954-1956 gg*, S. P. Tolstov i M. G. Vorobeva, pp. 143-150. Moskva.

1959b Retsenziya na kn. Ya. G. Gulyamova "Istoriya orosheniya Khorezma s drevneyshikh vremen do nashikh dney" (Tashkent, 1957). *Sovetskaya Etnografiya*, 1959 No. 5.

1960a Izuchenie karakalpakskoy irrigatsii v basseyne Jany-Dari v 1956-1957 gg. *Materialy Khorezmskoy Ekspeditsii. Vyp. 4, Polevye issledovaniya Khorezmskoy ekspeditsii 1957 g*, S. P. Tolstov i M. A. Itina, pp. 172–191. Moskva.

1960b Retsenziya na st. *Salinity and Irrigation Agriculture in Antiquity*. Diyala Basin Archaeological Project, report on essential results June 1, 1957, to June 1, 1958. *Sovetskaya Etnografiya*, 1960 No. 2.

1961 Razdel "Zemledelie i irrigatsiya" v istoriko-etnograficheskom atlase Sredney Azii i Kazakhstana (prospekt). *Trudy Instituta Etnografii AN SSSR, nov. ser., Tom X/VIII*. Moskva - Leningrad.

1962 Arkheologicheskaya karta Priaralya. *Po drevnim deltam Oksa i Yaksarta*, S. P. Tolstov, Moskva.

1963 Khozyaystvenno-kulturnye tipy Sredney Azii i Kazakhstana (s kartoy). *Narody Sredney Azii i Kazakhstana, Tom I-II*. Moskva.

1964a Problemy selskokhozyaystvennogo osvoeniya zemel drevnego orosheniya. *Vestnik AN SSSR*, No. 7.

1964b Karakalpaki v nizovyakh Syr-Dari i na Jany-Dare (XVII-XIX vv.). *Ocherki istorii Karakalpakskoy ASSR, Tom I*. Tashkent.

1965 Deshifrirovanie aerofotosnimkov pri izuchenii drevnikh orositelnykh system. *Arkheologiya i estestvennye nauki*. Moskva.

1966a Nekotorye aspekty problemy vzaimodeystviya prirody i obshchestva (na primere istorii osvoeniya nizovev Amu-Dari v XVIII-XIX vv.). Izvestiya Vsesoyuznogo *Geograficheskogo Obshchestva*, Tom 98, Vyp. 2:143-156 c kart.

1966b Some Aspects of the Problem of the Interplay of Nature and Society. *Soviet Geography* 7(10):3-14.

1968a Khozyaystvenno-kulturnye tipy i istoricheskiy protsess. *Sovetskaya Etnografiya*, No. 2.

1968b Problemy proiskhojdeniya irrigatsionnogo zemledeliya i sovremennye arkheologicheskie issledovaniya. *Istoriya, arkheologiya i etnografiya Sredney Azii i Kazakhstana*, A. V. Vinogradov, M. G. Vorobeva, T. A. Jdanko, M. A. Itina, L. M. Levina, Yu. A. Rapoport, pp. 16-25. Moskva.

Andrianov, B. V., and E. M. Murzaev

1964   Nekotorye problemy etnografii aridnoy zony. *Sovetskaya Etnografiya*, No. 4.

Andrianov, B. V., and S. P. Tolstov

1965   *Arkheologicheskoe izuchenie zemel drevnego orosheniya v svyazi s perspektivami ikh selskokhozyaystvennogo osvoeniya. Materialy sessii, posvyashchennoy itogam arkheologicheskikh i etnograficheskikh issledovaniy 1964 g. v SSSR.* Baku.

Andrianov, B. V., and G. P. Vasileva

1957   Opyt arkheologo-etnograficheskogo izucheniya poseleniy XIX v. *Izvestiya Turkmenskoy AN SSR*, No. 2.

1958   Pokinutye turkmenskie poseleniya XIX v. v Khorezmskom oazise. *Kratkie Soobshcheniya Instituta Etnografii AN SSSR*, Vyp. 28.

Additional publications collected by Irina A. Arjantseva

1963   *Iz istorii rasseleniya karakalpakov v Nizovyakh Amu-Dari.* Karakalpaksoe gosudarstvennoe izdatelsvo. Nukus. (in Karakalpak)

1963   Opyt deshifrirovaniya aerofotosnimkov pri izuchenii drevnikh orositelnykh sistem i drugikh arkheologicheskikh pamyatnikov (na materialakh Khorezmskoy arkheologo-etnograficheskoy ekspeditsii). *Metody estestvennykh i tekhnicheskikh nauk v arkheologii*, pp. 72-74. Tezisy dokladov. Moskva.

1965   Deshifrovanie aerofotosnimkov pri izuchenii drevnikh orositelnykh system. *Materialy i Issledovaniya po Arkheologii SSSR*, No. 129:261-267 i 3 ill.

1967   Kartografirovanie drevney irrigatsii na osnove aerofotosemki. *Materialy Moskovskogo filiala geograficheskogo obshchestva SSSR. Aerometody*, 1967 g., Vyp. 1:17-18.

1969   *Drevnie orositelnye sistemy Priaralya (v svyazi s istoriey vozniknoveniya i razvitiya oroshaemogo zemledeliya).* Moskva.

1972   Proshloe I budushchee zemel drevnego orosheniya. *Priroda*, No. 9:40-47 s ill.

1973   Istoriya i sovremennaya praktika sistem zemledeliya: Konferentsiya v Moskve. *Vestnik AN SSSR*, No. 7: 91-93.

1973   K voprosu o klassifikatsii form oroshaemogo zemledeliya v Sredney Azii. *Trudy Instituta Etnografii*, Tom 98, pp. 9-15.

1973   Konferentsiya po izucheniyu sistem zemledeliya: istoriya i sovremennaya praktika. *Sovetskaya Etnografiya*, No. 4:160-162.

1973   Tysyacheletiya s vysoty. Zemlya i lyudi. *Populyarny geograficheskiy ejegodnik*, 343-346.

1974   Geograficheskaya sreda i problemy zarojdeniya zemledeliya. *Pervobytnyy chelovek: ego materialnaya kultura i prirodnaya sreda v pleystotsene i golotsene*, pp. 217-225. Moskva.

1978   *Zemledelie Nashikh Predkov.* Moskva.

1979   Irrigatsiya i ee rol v sotsialno-ekonomichekoy istorii drevney Sredney Azii. *Antichnaya kultura Sredney Azii i Kazakhstana. Tezisy dokladov Vsesoyuznogo nauchnogo soveshchaniya*, pp. 5-7. Tashkent.

1981   Arkheologicheskaya karta Khorezma. *Kultura i iskusstvo drevnego Khorezma*, pp. 60-71. Moskva.

1984   Opyt tipologizatsii oroshaemogo zemledeliya i irrigatsii v Sredney Azii i Kazakhstane (konets XIX-nachalo XX vv). *Tipologiya osnovnykh elementov traditsionnoy kultury*. Moskva.

1985   *Neosedloe naselenie mira (istoriko-etnograficheskoe issledovanie)*. Moskva.

1991   Iz istorii orosheniya v basseyne Aralskogo morya. *Aralskiy krizis (istoriko-geograficheskaya retrospektiva)*, pp. 101-122. Moskva.

1993   The History of Economic Development in the Aral Region and Its Influence on the Environment. *GeoJournal* 35(1):11-16

1998   Narodnye traditsii prirodopolzovaniya i ekologicheskie krizisy. *Priarale v drevnosti i srednevekove (K 60-letiyu Khorezmskoy arkheologo-etnograficheskoy ekspeditsii)*, pp. 60-73. Moskva.

Andrianov, B. V., and V. N. Fedchina

1980   Kompleksnoe istoriko-geograficheskoe izuchenie Sredney Azii. *Osnovnye problemy istoricheskoy geografii Rossii na sovremennom etape. Tezisy dokladov i soobshcheniy Vsesoyuznogo konferentsii po istoricheskoy geografii Rossii*, pp. 44-46. Moskva.

Andrianov, B. V., and M. A. Itina

1971   Problema kompleksnykh (geograficheskikh, arkheologicheskikh i etnograficheskikh) issledovaniy zemel drevnego orosheniya nizoviy Syr-Dari v svyazi s proektom perebroski chasti stoka vod sibirskikh rek v basseyi Aralskogo morya. *Tezisy Dokladov Sessii,posvyashchennoy itogam polevykh arkheologicheskikh issledovaniy* 1971, pp. 84-86.

Andrianov, B. V., and A. S. Kes

1967   Razvitie gidrograficheskoy seti i irrigatsii na ravninakh Sredney Azii. *Problemy preobrazovaniya prirody Sredney Azii*, pp. 24-39 i ill. Nauka, Moskva.

1974   Na zemlyakh drevnego orosheniya v nizovyakh Syr-Dari. *Arkheologicheskie otkrytiya* 1973 g.: 463-465. Nauka, Moskva.

Andrianov, B. V., and L. M. Levina

1979   Nekotorye voprosy istoricheskoy etnografii Vostochnogo Priaralya v I Tys. n. e. *Etnografiya i arkheologiya Sredney Azii*, pp. 94-100. Moskva.

Andrianov, B. V., and A. P. Mukhamedjanov

1980   Irrigatsiya i ee rol v sotsialno-ekonomicheskoy istorii drevney i srednevekovoy Sredney Azii. *Obshchestvennye Nauky v Uzbekistane*, No. 11:35-424.

Andrianov, B. V., P. I. Bazilevich, and L. E. Rodin

1957   Iz istorii zemel drevnego orosheniya Khorezma. *Izvestiya Vsesoyuznogo Geograficheskogo Obshchestva*, Tom 89, Vyp. 6:516-535.

Andrianov, B. V., M. A. Itina, A. S. Kes

1974   Zemli drevnego orosheniya Yugo-Vostochnogo Priaralya: ikh proshloe i perspektivy osvoeniya. *Sovetskaya Etnografiya*, No. 5:46-59 s ill. i kart.

1975   Zemli drevnego orosheniya v nizovyakh Syr-Dari i zadachi ikh osvoeniya. *Voprosy Geografii*, No. 99:147-156, s ill. i kart.

Andrianov, B. V., M. A. Itina, and A. S. Kes

1986   Kompleksnoe arkheologicheskoe izuchenie zemel drevnego orosheniya Yugo-Vostochnogo Priaralya i perspektiva ikh sovmestnogo orosheniya. *Arkheologicheskie issledovaniya v zonakh melioratsii. Itogi i perspektivy ikh osvoeniya*. Leningrad.

ANCIENT IRRIGATION SYSTEMS OF THE ARAL SEA AREA

Boris V. Andrianov

Foreword

N. Ya. Merpert

Responsible Redaktor

The book deals with the history of irrigation and the associated techniques in the Aral Sea area, from the Bronze Age to the Late Middle Age. In general, the book is interesting because of the huge amount of archaeological and ethnographic material, collected by the author during field expeditions.

From the Author

Agriculture represents the basis of ancient civilizations, in countries with inadequate precipitation, thus it has always been related to artificial irrigation. Water is the source of life in arid zones. K. H. Marx and F. Engels stressed many times the special importance of irrigation and its huge influence on the rise of the ancient civilizations (Marx and Engels 1957, Tom 9:132; 1961; Tom 20:152, 183-185, 188, 500; 1962; Tom 28:221).[Note 1] That is why a study of the historical evolution of irrigation elucidates different aspects of the entire economic and social history of the population of those countries.

The world literature, however, is poor in works on the history of irrigation. There is very little general research in which the entire development of irrigation is traced from the earliest period to the present. This is due, first of all, to the fact that in the modern oases the remains of ancient agriculture are poorly preserved and they occur only in the 'lands of ancient irrigation' in the shape of semi-destroyed canals, traces of small irrigation networks, etc.

Especially vast and rich, however, are ancient irrigation sites of different periods in the Aral Sea area, where, starting in 1937, the archaeological research of the Khorezm Archaeological-Ethnographic Expedition of the Academy of Sciences of the USSR, under the general direction of the Corresponding Member S. P. Tolstov, took place. In 1952, the archaeological-topographical unit, headed by the author of this work, was organized to study the ancient irrigation systems of that area. This book is a compendium of the archaeological and historical-ethnographical material accumulated on this task. We covered an area of 5 million ha with the remains of ancient irrigation canals, dams, ditches and fields surrounding the ruins of settlements and covering different periods, from the mid-2nd millennium BCE to their abandonment by the Turkmens and the Karalpaks in the 19th century. Additionally, we contributed to the study of ancient irrigation over a wide arid zone by using aerial methods; we traced the origin and development of irrigated agriculture, which made possible identifying general and specific patterns of the development of irrigation and determining their place in the Aral Sea region.

This publication is divided into two main parts. The first part concerns the use of aerial methods in archaeological and geographical fields; the desk-study of ancient irrigation systems using aerial images; the general history of irrigated agriculture and its connection with the origin and spread of plant growth; the development of working implements and irrigation skills. The second part presents the results of field archaeological-topographical research on irrigation works in the ancient lands of the Lower Amudarya, the Southern Akchadarya Delta, the Sarykamysh Delta, and the Lower Syrdarya. In conclusion, some issues of socio-economical history connected with irrigation are presented, as well as an overall view of the causes for the rise of the 'lands of ancient irrigation' and the likelihood of their evolution.

The work of the archaeological-topographical

group was carried out in accordance with the Expedition's general plan of work, and in close connection with the archaeological excavations of the largest sites located in the 'lands of ancient irrigation' of Khorezm. Many members of the Expedition participated in the archaeological-topographical unit. It is a pleasant duty to express my deep gratitude to the first chief of the Expedition, the Corresponding Member of the USSR Academy of Sciences S. P. Tolstov, the entire staff of the Khorezm Expedition as well as students and trainees of the Faculty of Geography of the Moscow State University and the Moscow Institute of Geodesy and Cartography, who participated in the survey team and were engaged in the hard work on the aerial photo interpretation and the mapping of ancient irrigation systems.

Introduction

Boris V. Andrianov

In countries with a warm climate and low precipitation, where the growth of plants was possible only thanks to artificial irrigation, the creation and maintenance of irrigation systems was an important aspect of social production. Irrigation demanded an enormous labor force for the construction and maintenance of canals as well as the construction of dams to protect the area against destructive floods. Peasants' lives were permanently devoted to the maintenance of the irrigation network. In 1874, a member of the Amudarya Expedition of the Geographic Society, the artist N. N. Karazin, described the difficulties of farmers in pre-revolutionary Central Asia: "The burning sun, the summer ten months long, the neighborhood of dead sandy deserts are strong enemies of farmers; they oppose only *chigir* against them; working hands are not enough and the sun burns everything sown. Where there is water there is life, where it is not, there is death. Both life and death border too close to each other for farmers to weaken their attention to this permanent struggle" (Karazin 1875:199).

The silent witnesses of the struggle for water - ancient irrigation works - are an important source of knowledge of the history of humanity and the history of labor and techniques. Traces of ancient irrigation are frequently found close to modern cultivated oases. The 'lands of ancient irrigation' are marked by the abundance of ruins of ancient cities, fortresses, castles, farms, dikes, dry canals, artificially planned fields no longer cultivated, vineyards and orchards. The total area of these ancient lands in the USSR is 8-10 million ha, which equals the total modern irrigated area. Almost half of these lands are in the area of the Aral Sea, the lower reaches of Amudarya and Syrdarya and their ancient deltas.

Pre-revolutionary Russian orientalists, such as Ya. V. Khanykov, N. I. Veselovskiy and V. V. Bartold, wrote many papers on the history of irrigation in Turkestan. Continuing their tradition, the Soviet researchers of Central Asian 'lands of ancient irrigation' systems began, already in the 1920-1930s, an intensive study of the history of irrigation, using both written sources and archaeological material. During those years the reconstruction of old and the construction of new systems of irrigation started in Central Asia. V. I. Lenin said that "Irrigation is needed and much more necessary to renew the land, revive it, bury its past and strengthen its transition to socialism" (Lenin, Tom 43:200).[Editor 1]

Many archaeological researchers at that time were closely connected to the practical tasks of irrigation works. In the first pages of the *Vestnik irrigatsii* (Bulletin on Irrigation), the engineer-irrigators B. N. Kastalskiy and E. M. Timofeev developed V. V. Bartold's idea of coordinating the works of irrigation specialists and archaeologists,[Note 2] and they called upon hydrologists and archaeologists to establish closer links (Kastalskiy and Timofeev 1934:53). According to their view, the revival of irrigation systems could be implemented with great consequences when the genesis and history of irrigation in this region were known. The experience of ancient irrigators over many centuries of practice in choosing the optimal canal courses should be emulated in the modern practice of irrigation construction. They rightly wrote that "it is necessary to take everything technically developed in the past and improve it using modern achievements" (Kastalskiy and Timofeev 1934:60).

While the specialists in irrigation were interested in the practical economic role of ancient irrigation systems, the historians and archaeologists paid attention to its origin and development in Central Asia. In 1924, the engineer-hydrologist and enthusiast-archaeologist D. D. Bukinich, was investigating the nature of the Transcaspian region and the ancient agricultural sites of the Anau culture. He formulated some conclusions about the first stages of irrigation in South Turkmenistan (Bukinich 1924), which were then successfully proven by archaeological research (see also Bukinich 1926, 1940, 1945).

The works by B. A. Latynin in 1930-1934 marked the beginning of the study of the history of irrigated agriculture in the Fergana Valley (Latynin 1931, 1935a, 1935b, 1956, 1957, 1959, 1962). This research indicated that the origin of irrigation canals could be dated to Prehistoric periods. They were built in connection with the construction of walls that enclosed fields to prevent floods, as in the piedmont of the Kopetdag.

Considerable documentation on the history of ancient irrigation works was accumulated by other archaeological expeditions of the 1930s. Among them the Zeravshan Expedition (directed by A. Yu. Yakubovskiy in 1934 and 1939), and the Termez Expedition (directed by M. E. Masson in 1936-1938). Also worth mentioning are: the archaeological supervision at the construction of the *Bolshoy Ferganskiy Kanal* (The Great Fergana Canal; M. E. Masson, Ya. G. Gulyamov, V. D. Jukov, T. G. Oboldueva); the archaeological supervision of the Tashkent canal construction (M. E. Voronets, A. I. Terenojkin); the archaeological supervision on the construction of the Kattakurgan reservoir (V. A. Shishkin, I. A. Sukharev); G. V. Grigorev's excavations of Kaunchi-tepe near Tashkent (Grigorev 1940a) and Tali-Barzu near Samarkand in 1934-1939 (Grigorev 1940b); the research on Varakhsha and the canals surrounding it (V. A. Shishkin 1937-1939), etc. (Yakubovskiy 1940; Tolstov and Shishkin 1942; Bernshtam 1947; Gaydukevich 1948; M. Masson 1956).

The widest territory of ancient irrigation works in the Aral Sea area corresponds to the lower reaches of the two largest Central Asian rivers, the Amudarya and the Syrdarya. They extend from the Sarykamysh depression and the Uzboy to the north of the Karakum desert, through the modern delta of the Amudarya and the dry riverbeds of the Janydarya, to the lower reaches of the Syrdarya and the Sarysu to the east. This territory is crossed by dry beds of ancient deltaic channels of the great Central Asian rivers and the remains of numerous channels branching from them. Along their banks there are ruins of settlements, fortresses and towns.

Already in the very first efforts of the archaeological expedition in Khorezm in 1934, headed by M. V. Voevodskiy, one of the task was to establish the history of the main irrigation systems (Voevodskiy 1938:235). The research was funded by the *Sredneaziatskogo gosudarstvennogo institute po proiektirovannyu vodnokhozyastvennikh i gidrotekhnicheskikh soorujeniy* (Central Asian State Institute for Projects for Water and Hydro-technical Structures). Special attention was paid to the discovery of ancient irrigation systems predating the Mongols, to establish the origin of the *chigir* irrigation in Khorezm, and to study the historical dynamics of irrigation along the Daudan and the Daryalyk. In 1934, archaeological excavations at the Zmukshir (Zamakhshar) site and the survey of sites in its surroundings were carried out. On the way back from Khorezm, members of the Expedition carried out aerial observations from their aircraft. In 1934 M. V. Voevodskiy gave special instructions to study abandoned irrigation systems (Institute of Archaeology of Moscow, Manuscripts Archive 3, N1:66-67). He recognized the need to map irrigation systems, conduct archaeological excavation of canals, and research ancient rural settlements located near these irrigation systems. Thus, M.V. Voevodskiy's work in 1934 marked the beginning of the research of ancient irrigation systems in the Aral Sea area.

In 1937, the archaeological and ethnographic works of the Khorezm Expedition, directed by S. P. Tolstov, began in the lower reaches of the Amudarya. The successful research of this Expedition in solving great historical problems, the social structure, the socio-economic, the political and cultural history of the ancient Khorezmian State are the best representations of the achievements of Soviet archaeology as a whole. For the USSR this study morphed from an auxiliary and source project into a very important and independent branch of historical science (M. Dyakonov 1949; Struve 1949).

Since the beginning, the history of irrigation in Khorezm occupied an important place among the scientific issues of the Khorezm Archaeological-Ethnographic Expedition. In prewar and immediate postwar years (until 1950), the research on the history of the irrigation network of this area provided Ya. G. Gulyamov[Note 3] with an opportunity to participate in the Khorezm Expedition in order to prepare and defend his doctoral dissertation. It was entitled *Istoriya orosheniya khorezma s drevneyshykh vremen do nashik dney* (History of Irrigation of Khorezm from the Ancient Period to Our Day), published in 1957.

In his dissertation, Ya. G. Gulyamov reviewed his archaeological supervision (1936-1950) during the construction of irrigation systems in Khorezm and as a participant of the Khorezm Archaeological-Ethnographic Expedition (1938-1946). This book summarizes his previous publications (Gulyamov 1945, 1948, 1949, 1950, 1965). In his introduction, the author especially emphasizes that he did not include any materials from the Khorezm expeditions collected on this subject after 1950.

Ya. G. Gulyamov wrote about the history of people's experience in creating artificial irrigations in Khorezm from the Bronze and Early Iron Ages to the present day. Using the abundant archaeological, historical and historical-ethnographic material, the author traced the origin of irrigation in the dry channels of the Amudarya Delta and outlined a general scheme of the development of irrigation. From the slow evolution of the use of floods of the Amudarya for fishing and hunting in Neolithic times, to the primitive agriculture and herding in the disappeared channels during the Bronze Age, and, finally, to the artificially irrigated agriculture based on the main canals in classical periods. A sharp change in development of irrigation in Khorezm took place after the victory of the Great October Socialist Revolution and the strengthening of the Soviet system in Khorezm, when the old irrigation network was rebuilt.

In his research, Ya. G. Gulyamov used numerous sources such as Oriental manuscripts, documents and literary works. His good knowledge of the archaeological sites of the Lower Amudarya enabled him to provide a general description of the main irrigation systems of Khorezm in different periods and also to draw some small scale historical maps (see Gulyamov 1957:87:fig. 6, 100:fig.7, 104:fig. 8, 118:fig.9, 133:fig.10, 161:fig.11, 177:fig. 12, 209:fig. 13, 226:fig. 14).

The author was completely right in considering the formation of class relations, the emergence of large settlements and towns, the rise of the despotic state in Khorezm in the second quarter of the 1st millennium BCE as determinant factors in the transition to artificial irrigation. In his words, the despotic state was "the force which mobilized not only tribes and clans, but also the slaves to dig the main canals, to erect dams and other structures" (Gulyamov 1957:95). The history of irrigation in the Khorezm Oasis is given by the author against a wide background of political events. For instance, the influence of the Khorezm kingdom spread south and, in Herodotus' time, the 'Akes Oasis' (located by Ya. G. Gulyamov in the basin of the Tedjen River) belonged to the people of Khorezm (Gulyamov 1957:96). After the rise of the Achaemenids in Persia and the entry of Khorezm as tributary of the Achaemenid state in the late 4th century BCE, there was a comprehensive irrigation activity by the people of Khorezm (when king Farasman was the independent ruler of Khorezm during the rule of Alexander of Macedonia).

The heyday of ancient agriculture in the oasis was dated by Ya. G. Gulyamov to the Kushan period (Gulyamov 1957:98). The reduction of the irrigation network took place in the 3rd to 4th centuries CE and its renewal in the 5th and early 6th centuries CE. The author analyzed in detail the political, and historical-geographical changes in the oasis during the Middle Ages. In order to describe the history of irrigation in Khorezm during this period, the author used numerous written sources. Ya. G. Gulyamov's deductions concerning the political, economic and cultural links between agriculture and herding of the steppe peoples and the rise and decay of irrigation systems is very important. "The agricultural population could grow and develop irrigation, organize the defense of the oasis, create cities only through close contacts between the oasis and the steppe in the mighty state system" (Gulyamov 1957:94).

Much attention was paid to irrigation techniques. The author traced the development of the main canals and the gradual shift of the head works upstream in the large flowing channels of the Amudarya. After some time, this led to a reduction in the number of single canals and the creation of wide systems based on the main riverbed (Gulyamov 1957:67). This adaptation was traced by us from archaeological material in some regions of the Aral Sea, principally in the Chermen-yab basin (Andrianov 1958b:327).

Particularly interesting in Ya. G. Gulyamov's book is the eighth chapter, dedicated to the historical and ethnographic description of the experience of Khorezmian irrigators in the construction of gravity-fed irrigation canals. These were very complicated and effective systems from a hydro-technical point of view, taking water from the river through diversion head works (saka) adapted to the river level. There was a reservoir (bedrau) to save excessive surplus water. The system consisted of large main canals (arna), distributors (yab) and narrower canals (badak), from which ditches (salma) flowed into the fields. The surplus water was directed into a terminal reservoir. The restrictions to Khorezm's irrigation system were: the unstable head water-intake due to erosion or the filling by silt sediments in the main riverbed; the labor investment required in cleaning the silt sediments, dredging the canals, etc.

This book's section describes in great detail the water-lifting system, the dams, the water exploitation and the main labor duties to maintain the irrigation systems (Gulyamov 1957:246-267).

Ya. G. Gulyamov's monograph is the most important step in studying the history of irrigation in Khorezm. However, some of his conclusions must be ameliorated and expanded on the basis of new material. The author arrived at his conclu-

sions about the beginning of irrigated agriculture in the lower reaches of the Amudarya only on historical-ethnographic analogies (Gulyamov 1957:54-65). Though he excavated in 1945-1946 the Late Bronze Age site of Djanbas 6, the archaeological proofs of the appearance of agriculture in Khorezm came solely from the find of a millstone fragment and from the similarity between some painted and burnished ceramic fragments with those of the Anau sites (Gulyamov 1957:53).

The origin of the main canals in Khorezm is dated by Ya. G. Gulyamov to the Amirabad period (i.e. 9th-8th centuries BCE) as a "continuation of the primitive methods of basin irrigation" (Gulyamov 1957:81). It is argued on the basis of finds of Amirabad pottery from the river banks of the ancient canals, which are alongside scattered streams, i.e. long uy (Gulyamov 1957:67).[Note 4] In his explanation of the origin of the main canals, the author gave great importance to floods regulation (stream overflow) of the Suyargan (on the right bank of Khorezm) and floods channels of the Daudan (on the left bank of Khorezm). These floods, which occupied the strip between 'dunes' (modern uy) "helped the early farmers with the primitive watering of sown fields" (Gulyamov 1957:62). Following the lowering of the topography, by correcting the natural thalweg, the farmers diverted the flood waters between sand hills into artificial canals. Ya. G. Gulyamov wrote that "in order to bring water to the target point, ancient farmers just retouched the natural thalweg, hence the resulting canals were small and wide" (Gulyamov 1957:89).[Note 5]

As we shall see below, a different interpretation of the origin of the main canals was given by S. P. Tolstov in his book *Drevniy Khorezm* (Ancient Khorezm). S. P. Tolstov linked the development of artificial irrigation with the observation and regulation of floods in drying deltaic channels (Tolstov 1948a:45). If Ya. G. Gulyamov supposed that ancient irrigators used the floods diverted into natural depressions, between sandy hills, as canals, S. P. Tolstov, on the other hand, connected the origin of the main canals with the regulation of the damping of deltaic riverbeds rising above the depression of the alluvial plain. Particularly clear differences in the evaluation of geomorphological conditions of that period arose in solving the problem of the Suyargan. Highlighting the origin and development of irrigation systems on the right bank of the Khorezm, Ya. G. Gulyamov gave a major role to the overflow of the Suyargan. The floods, penetrating from the Amudarya to the Djanbas-Kala *takyr*, through a chain of lakes and depressions of the Suyargan, were considered by the author as the only source of water, for all the ancient canals on the right bank of Khorezm, before the transition leading to the origin of the main river channels (Gulyamov 1957:47, 93, 97, etc.). As already mentioned in the publications of the Khorezm Expedition, archaeological and geo-morphological research indicated that the narrow width and the recent structure of the Suyargan riverbed could not have been the source of the massive ancient irrigation systems. In the Archaic period (6th–5th centuries BCE), they had their source from numerous vanished lateral channels of the Southern Akchadarya Delta (Andrianov 1959b:182; Tolstov and Kes 1960:137).

Ya. G. Gulyamov determined the origin of the large main canals in Khorezm as the time following "the culture of settlements of the 6th to 3rd centuries BCE" (Gulyamov 1957:76). This conclusion, which was very important for the entire political and economic history of Khorezm, did not fit with S. P. Tolstov's concept that the irrigation network of Khorezm's right bank and the explored part of its left bank were already built in the mid-1st millennium BCE (Tolstov 1948a:45). The discovery of the massive Archaic irrigation system (6th–5th centuries BCE) by the archaeological-topographical unit on both the left and right banks of the Amudarya confirmed, although not fully, S. P. Tolstov's thesis (Tolstov 1948a).

Despite some controversial conclusions, Ya. G. Gulyamov's work played a crucial role in our thinking about the developments of irrigation in the Amudarya. This work has not lost its great significance even though some assertions may have to be clarified and amended in light of new material.

The opinions of the head of the Khorezm Expedition, S. P. Tolstov, on the origin, development and dynamics of the ancient irrigation network of Khorezm are clearly given in his work *Drevniy Khorezm* (Ancient Khorezm) (Tolstov 1948a:37–56). The author wrote: "The analysis of the configuration of the ancient irrigation networks allows interesting considerations on the history of its origin. It entirely repeats the configuration of the ancient delta... It seems as though people reconstructed the totally vanished (by their time) ancient delta. If we consider that, long before the creation of the irrigation network, the *kair* of the ancient delta were densely populated by farming people, we can understand the meaning of this phenomenon. As people lifted the residual water from drying channels, before returning to their fields, it is highly possible that, exactly in this way, gropingly and empirically, they discovered the principle of water drainage of large canals. This is relative to the head works far upstream along the river, which could then provide water to the fields by gravity flow. It is also highly possible that, by observing the natural run of floods along the dry channels of river beds, they could define gradient techniques for the course of the canals. In any case, in the remote age when the ancient irrigation network of Khorezm was built, man did not defy nature, but just adapted his techniques, exploiting nature for his own benefit" (Tolstov 1948a:45).

Combining field surveys along canals with aerial reconnaissance and archaeological study of rural settlements of different periods, S. P. Tolstov traced the direction of the main canals of the ancient irrigation network on the right bank of Khorezm, and deduced that it "represented the modern one in a broader form". He developed methods of measurement for ancient irrigation constructions. In 1940, near Djanbas-kala, Angka-kala, and Kanga-kala, and also in the Berkut-kala Oasis, the measuring of canals revealed significant differences in the character of Antique and Medieval systems of irrigation (Tolstov 1958:102; Gulyamov 1957:90). S. P. Tolstov drew layouts of Antique and Medieval irrigation systems in Khorezm as well as archaeological-topographical plans for some of their parts (Tolstov 1948a:46-47, 134-135, 158-159). Based on above evidence, the complex reasons for the reduction and abandonment of the irrigation systems published in his book *Drevniy Khorezm* are expounded.

In the special section *Dinamika drevney irrigatsionnoy seti* (Dynamics of the ancient irrigation network), the author provided a general description of the major canals creation in the mid-1st millennium BCE on the right and left banks of the Khorezm. He also indicated the dynamics for the reduction of irrigation systems in the 4th to 6th, 8th to 9th, and 13th to 14th centuries CE. The first considerable abandonment of irrigated territories was connected by the author with the decline of the Kushan Empire. Its collapse "into separate states warring with each other and tending towards further disintegration", as well as the aggravation from social conflicts, which undermined the main pillars of the "ancient irrigation culture: communities, slave owning and centralized despotism" (Tolstov 1948a:50). The second period of decay was conditioned by social upheaval and popular riots during the Arabs' invasion. Then came the period of the Medieval Khorezmshah Empire and the beginning of an irrigation revival, which was stopped by the Mongol invasion in the early 13th century. The ensuing feudal fragmentation in Khorezm, and in many other regions of Central Asia, caused the desolation of many territories in the 13th to 15th centuries. Starting in the 16th century, and especially in the 17th century, during Abu al-Ghazi Khan's rule, and also in the early 19th century during Muhammad Rahim Khan's rule, vast irrigation works in the oasis were carried out again. These are the most fundamental milestones in the development of Khorezmian irrigation identified by S. P. Tolstov in 1948.

Widely using the data of the historical dynamics of irrigation in Khorezm and some other regions of Central Asia, S. P. Tolstov, further developing the ideas of L. S. Berg, A. I. Voeykov and others, convincingly proved that the main reason for the formation of the 'lands of ancient irrigation' was not natural catastrophic changes (progressive drying of the Asian continent, change of rivers flow, etc.), but, above all, socio-historical factors, political and economic crises. At the same time the author warned against ignoring natural processes (salting, erosion of peripheral parts of cultivated lands, moving sands), which in his view "played their role, reinforcing the effect of social and historical reasons" (Tolstov 1948a:52).

The study of the origin and development of irrigation systems allowed S. P. Tolstov to raise some important questions on the social history of ancient Khorezm: In particular he assumed that the huge irrigation network of Khorezm was realized entirely in a short time, and he deduced that "only a centralized Oriental despotism could create the great canals of Khorezm" (Tolstov 1948a:45, 49). S. P. Tolstov elucidated the historical and social conditions of the development of the class society in the lower reaches of Amudarya and characterized the ancient Khorezmian society as an "Oriental variant of the ancient slave-based social structure; we can identify it with the words 'social-slave-based structure'" (Tolstov 1948a:48; compare with Nikiforov 1968:126).

He pointed out the opposite association of the development of urban life of a "state, ruled by a mighty slave-based aristocracy, on one side, and a deeply archaic, stable and old-fashioned social system carrying diverse and strong traditions of kin system", on the other (Tolstov 1948a:124).

Considering S. P. Tolstov's more recent publications, in which his views on the history of the ancient irrigation systems of Khorezm were further developed, the essays of 1952 (Tolstov 1953a), 1951-1954 (Tolstov 1955a-b) and 1954 (Tolstov 1955a-b), should be mentioned. The general results of the whole Expedition were reported here and particularly the archaeological-topographical unit's findings (these aimed to study and draw maps of canals of the Chermen-yab basin on the left bank of the Khorezm and the area of Koy-Krylgan-kala, Bazar-kala and Djanbas-kala on the right bank). The main results of these undertakings were reported by S. P. Tolstov and B. V. Andrianov at the first conference of archaeologists and ethnographers of Central Asia in April 1955 (Tolstov and Andrianov 1957).

The chief research achievements in 1952-1954 were: 1) the discovery of the irrigation system developed at the Tazabagyab and Suyargan settlements, which changed the previous conception about the predominance of 'single' irrigation, *kair* and basin agriculture among the Prehistoric population of Khorezm; 2) the dating of artificial irrigation in the lower reaches of the Amudarya, which was anticipated more than a thousand years in comparison with the previous dating; 3) the observation of ancient irrigation works of different periods as archaeological sites with different morphometric and topographical characteristics; 4) the outline of the main differences between the Prehistoric, Antique and Medieval irrigation systems of Khorezm; 5) special table of typical examples of irrigation systems and agro-irrigation layouts (Tolstov and Andrianov 1957:8-9); 6) finally, the identification of qualitative differences in Antique and Medieval irrigation systems of Kelteminar and Chermen-yab at different periods (Tolstov and Andrianov 1957:10).[Note 6]

This new material was shown in detail by S. P. Tolstov in his report (Tolstov 1958:100-142) and also in the monographs which were the basis for the archaeological literature of the Aral Sea (Tolstov 1962a:74-77, 89-96, 246-248; 315-322). In the first of these writings, the author criticized the thesis that the main principles of irrigation technique remained unchanged. He wrote that new evidence allows us to trace the different stages of development of agriculture: starting with the nucleation process of irrigation canals in the Archaic period; through the heyday of Kangju 'classic' Khorezm; to the period of the Kushan Empire; and further to the Medieval feudal period (Tolstov 1958:102).

These 'epochal' differences in the irrigation techniques of Khorezm are illustrated by specific examples of systems from ancient Kelteminar and Chermen-yab. In the next section, S. P. Tolstov (1958:116-142) discussed the peculiar Medieval irrigation facilities of the Sarykamysh depression and the Upper Uzboy.

New data on the history of irrigation brought S. P. Tolstov, in his report of 1958, to return to the main theoretical issues of the socio-economic history of ancient Central Asia. According to him, "only the accumulation of rich archaeological materials, revealing a picture of the economic life of Khorezm and indicating a sharp qualitative changes taking place between the 5th century BCE and the 5th century CE and beyond, allow us to say that these are two different social and economic formations. In addition, the social order prevailing in Central Asia in the second half of the 1st millennium CE was undoubtedly feudal, whilst the previous one could only have been based on slave-ownership. It is impossible to deny that our argument was based mainly on indirect proof. We do not yet have direct evidence of the presence in Khorezm in that period, which we call Antique, of significantly developed slave-owning relations. Nor do we have it for Central Asia as a whole". And, at the same time, "if there were no slave-owning, then the rich irrigation culture of the Orient could not have originated" (Tolstov 1958:103-104, 106).

This point of view was developed by many other scientists based on the material of Central Asia (M. E. Masson, M. M. Dyakonov, Ya. G. Gulyamov, A. N. Bernshtam, etc.)[Note 7] and for the more ancient agricultural civilizations of the Orient (Avdiev 1934; Struve 1934, 1965a, etc.).[Note 8] In his last work, V. V. Struve wrote that "In order to construct the irrigation economy of the city-states of Sumer, or any of the nomes in the Nile valley, the labor of many slaves was required besides the labor of members of the community" (Struve 1965b:102).

The opposite point of view is also well known, according to which, in the ancient agricultural civilizations (Egypt, Assyria, Babylonia, etc.), slave labor played a minor role in agricultural production.[Note 9]

The research of the Khorezm Expedition, before and after World War II, reported in Ya. G. Gulyamov's and S. P. Tolstov's works, indicated that the 'lands of ancient irrigation' of the Aral Sea were not only a huge reserve with many hundreds of sites of different periods, but also a unique 'museum' on the history of irrigation techniques. All these factors contributed to the further development of archaeological research in ancient irrigation works in the Aral Sea area, especially after the 1950s, when the history of the hydrographic network of the deltaic areas of the Amudarya and the Syrdarya were at the center of the Khorezm Expedition work.

Chronology of Research

The task of the archaeological-topographical unit was to survey systematically the 'lands of ancient irrigation' and to map the ancient irrigation systems of the Aral Sea based on aerial methods. Moreover the task was also to reconstruct in detail the development of irrigation in this area, from its origin in the Bronze Age to the systems of modern times (settlements abandoned by the Karakalpaks and Turkmens in the 18th-19th centuries).

The Archaeological and topographical research started in 1952 (August 10-October 15) on the right bank of the Amudarya in the basin of the Chermen-yab revealed the correlations between the Antique and Medieval parts of the canals (for a total amount of 120 *poisk*) (Tolstov 1953a:179-181; 1955b:204-206; 1958:112-114; Andrianov 1958:311-317). The following year, in the spring of 1953 (April 20-May 30), the field survey in the environs of Koy-Krylgan-kala provided the possibility to trace and to draw maps with the direction of the Archaic Kangju-Kushan and Afrigid canals of the ancient Kelteminar canal (*poisk* 1-408). On the left bank of Khorezm (June 2-October 3), in the dry river-bed of the Daudan, the river-heads of the Archaic and the Kangju canals were researched, as well as the Kushan riverbed connecting them; and, the later Khorezmshah main canal of Chermen-yab. Around the ruins of Shakh-Senem, integrated soil-archaeological investigations were carried out (*poisk* 1-437) with the participation of N. I. Bazilevich and the botanist L. E. Rodin (Tolstov

1955b; Andrianov, Bazilevich and Rodin 1957; Andrianov 1958b).

In 1954 (July 15-September 1) the team continued to investigate the lower reaches of the ancient Kelteminar where, for the first time in Khorezm, Bronze Age irrigation works were discovered, and the agro-irrigation layout and farmsteads of the Antique period were also mapped (*poisk* 409-719). On the left bank of the Amudarya the canals near Kyuzeligyr were studied (*poisk* 438-448) (Tolstov 1955a:97-98, 100:fig. 10; 1955b; 1959a:10-11; 1962a:74-75, 76:fig. 34, 77; Tolstov and Andrianov 1957:5-9, Pls. *Istoricheskoe razvitie irrigatsionnikh sistem Khorezma* - Historical development of the irrigation system of Khorezm; Tolstov and Kes 1960:132-135).

In 1955 (July 16-September 28) on the right bank of the Amudarya, the ancient Kyrk-Kyz canal from the ruins of Guldursun to Big Kyrk-Kyz was researched: a wide Kangju-Kushan and a narrower Afrigid riverbeds were revealed (*poisk* 720-1012). On the left bank (September 29-October 25) the study of the abandoned Turkmen settlements and irrigation works at the Daryalyk was started (*poisk* 449-519): the Antique fortified settlements of Kaladjik, Kurgan-kala and Butentau 2 were detected (Andrianov and Vasileva 1957, 1958; Tolstov and Andrianov 1957; Tolstov 1959a:25; Andrianov 1959a:143-146; Tolstov and Kes 1960:191-192, 201-204, 146-147: *Geomorfologicheskaya i arkheologicheskaya karty prisarykamishskoy delty* - Geomorphological and archaeological map of the Sarykamysh Delta).

The team continued the mapping of the 'lands of ancient irrigation' on the right bank in 1956 (July 30-September 25). Here the Medieval system of Gavkhore was studied; the open canals and settlements of the Amirabad period (Kavat 1, 2, 3) as well as the Antique fortified settlements of Tash-Kyrman and Kazakly-Yatkan were investigated (*poisk* 1013-1195) (Tolstov 1959a:25-26; Andrianov 1959a: 146-149, and *Plan okrestnostey Kavat-kaly* - Map of Kavat-kala environs; Tolstov and Kes 1960:135). Since 1956 (August 1-September 30) work began around Barak-tama in the dry riverbed of the Northern

Akchadarya Delta and in the Lower Janydarya, where the Karakalpak irrigation systems and abandoned settlements were studied (*poisk* 1-115) (Andrianov 1960a).

In 1957 (August 1-September 30) began the research in the peripheral parts of the right bank of Khorezm, near Dingildje, Kurgashin-kala and Toprak-kala. Not far from the Afrigid castle of Yakke-Parsan, the remains of a village and canals of the Amirabad period (Yakke 2) were discovered (*poisk* 1196-1558). In the same year (October 1-26) in the lower reaches of Syrdarya the study and mapping of the Karakalpak was carried out on the Medieval irrigation systems along the riverbed of the Janydarya from Kly to Chirik-Rabat and Babish-Mulla. In that region, during the 4th to 2nd centuries BCE, irrigation was based on deltaic regulated channels, natural reservoirs and small ditches (*poisk* 116-311) (Tolstov, Vorobeva and Rapoport 1960:22-23, 40-43, fig. 30-31; Andrianov 1960a; Tolstov 1961a:123, figs. 1-2).

During the expeditions of 1958 (July 23-August 30) on the left bank of the Amudarya, the river-heads and the course of canals of the Archaic and the Kangju periods were traced on the land between the Northern (Budjunuyu Daudan) and Middle Daudan; canals of different periods of the Shamurat system were uncovered (*poisk* 520-687). The archaeological-topographical research in the lower reaches of the Syrdarya (September 5-October 20) discovered the northern border of the irrigation systems of the 4th to 2nd centuries BCE near Babish-Mulla. The zone of the Karakalpak irrigation on the Kuvandarya was mapped, as well as the southern parts of Djety-asar (*poisk* 312-536) (Tolstov 1961a:fig. 1; 1962a:fig.72; Tolstov, Jdanko and Itina 1963:18-20, 34, 83, and fig. 12: *Karta rabot Khorezmskoy ekspeditsii na drevnikh protokakh Syrdary* - Map of the work of the Khorezm Expedition in the ancient branches of the Syrdarya; Tolstov 1962a: 283, 306-311).

In the spring of 1959 (April 20-May 10), with the Dingildje team (directed by M. G. Vorobeva), I carried out a survey of the irrigation near Dingildje (*poisk* 1559-1598). In the autumn

of the same year (July 15-September 15) the archaeological-topographical unit continued to study the irrigation system on the left bank of the Upper Chermen-yab as well as the lands between the rivers Daudan and Daryalyk; study of the origin of Antique and Medieval Chermen-yab (*poisk* 688-846). An Archaeological reconnaissance was carried out by N. N. Vakturskaya on the sites of Zamakhshar, Daudan-kala and Ak-kala, and by O. A. Vishnevskaya on Medieval rural sites (Vakturskaya 1963; Vishnevskaya 1963). In 1959 (October 1-20), in the Lower Syrdarya, a large survey team (directed by S. P. Tolstov) researched the sites and the irrigation works of the Middle Inkardarya; kurgans, and sites of the '*shlakovye kurgany*' (slag-covered kurgans) culture, the group of sites of Balandy and the Medieval settlement at Uygarak were discovered. Groups of kurgans were found after the aerial surveys of tumuli groups at Tagisken and Uygarak; N. I. Igonin produced aerial photos of settlements and irrigation systems (*poisk* 1-73) (Tolstov, Jdanko and Itina 1963:20, 32-33, 47-50, 83-85, figs. 12, 32; Tolstov 1962a:170-186; Tolstov, Andrianov and Igonin 1962:9, 10:fig. 4; Igonin 1965:257).

In 1960 (August 8-October 10) a large expedition (directed by S. P. Tolstov) continued working in the Upper Inkardarya; a group of pre-Mongol Medieval fortified settlements (Zangar-kala, Sarly-tam-kala, etc.) were found; irrigation works, in particular the Asanas-Uzyak canal, were investigated; aerial surveys were carried out and photos taken (*poisk* 1-46) (Tolstov, Jdanko and Itina 1963:33, 79-82 Tolstov 1961a:144-146; Tolstov, Andrianov and Igonin 1962 Tolstov 1962a:278-281, 284-286, 291).

The survey of the archaeological-topographical unit in 1961 (April 25-May 20) in the 'lands of ancient irrigation' on the left bank of Khorezm revealed several Medieval sites, canals, dams, reservoirs and pits for *chigir* installation, west of Yarbekir-kala. N. N. Vakturskaya discovered the craft settlement of Shekhrlik, dated from 13th to 15th centuries (*poisk* 847-933) (Tolstov and Kes 1960:196-204; Tolstov, Jdanko and Itina 1963:21; Vakturskaya 1963:45-53). In the autumn of the same year (September 5-October 15) the archaeological-topographical unit, as part of a larger group (directed by S. P. Tolstov) including also a geomorphological unit from the Institute of Geography (directed by A. S. Kes), continued to collect data for the archaeological map of the lower reaches of the Syrdarya; surveys were carried out along the dry riverbeds of the Upper Inkardarya, the Janydarya and the Mayliuzyak, where Medieval (Khodja-Kazgan 1, 2, 3) and late Karakalpak fortified settlements (Khatyn 1, 2, 3) were discovered and studied. The survey of Asanas (Ashnas) followed the meandering canal of the Asanas-Uzyak. Field work was completed with an aerial reconnaissance and aerial photos of sites and canals (*poisk* 1-67) (Tolstov 1962a:278, 281-282, 311-312; 1962b:145-148; Tolstov, Jdanko and Itina 1963:79-90).

In 1962 (September 8-October 20) the survey unit (directed by S. P. Tolstov), including the geomorphological unit (directed by A. S. Kes), worked in the environs of Asanas and on the eastern sites of the Djety-asar culture, where features of Medieval irrigation, dams, reservoirs and canals were studied. In Saykuduk, artificial reservoirs were found. Aerial reconnaissance and photos were carried out at the beginning and at the end of the season (for a total number of 100 *poisk*) (Igonin 1965:258).

Field research in 1963 (September 13-October 11) also started with an aerial survey. As part of a larger team (directed by S. P. Tolstov), aerial reconnaissance and photography were carried out in the southeast border of Ustyurt, the Janydarya and the Kuvandarya basins, around Kesken-Kuyuk-kala, where settlements and irrigation were studied. Together with a geomorphological unit (directed by A. S. Kes) the team visited the southern system of the dry riverbeds northwest of Chirik-Rabat. At the end of the season an aerial survey was undertaken in the Middle Syrdarya, and Mayram-Tobe and Kyr-Uzgent (for a total number of 120 *poisk*).

In 1964 (September 4-15) the archaeological-topographical unit discovered and studied the remains of irrigation works in the area of Tazabagyab northeast of Djanbas-kala (*poisk* 1599-1640) (Itina 1967:73-79; 1968:76; Tolstov, Itina and Vinogradov 1967). During the survey on the left bank of Amudarya a kurgan and a fortress were found in Tuzgyr; the Antique and Medieval irrigation works were studied (*poisk* 934-981), N. I. Igonin and B. V. Andrianov carried out the aerial survey and photos.

The research on the Lower and Middle Syrdarya in 1966 was undertaken by the Syrdarya team of the Khorezm Expedition, together with the team from the Institute of Geography (directed by A. S. Kes). Additional material characterizing the irrigation in Djety-asar was collected. In the lower reaches of Sarysu and Chu some abandoned Kazakh settlements were discovered and, in the Middle Syrdarya, Antique and Medieval sites were investigated. The work in Oguz-sai revealed a massive system of Medieval canals (*poisk* 1-140) (Tolstov, Itina and Vinogradov 1967:307).

During 14 field seasons (1952-1964) the archaeological-topographical survey covered the 'lands of ancient irrigation' on the right bank of Khorezm (1,640 *poisk* were marked), the Sarykamysh Delta on the left bank of the Amudarya (981 *poisk*) and the huge area of the lower reaches of Syrdarya, from its source to Turkestan in the east and the Aral Sea in the west. There several areas were studied by the archaeological-topographical unit as well as the large survey team, under the direct supervision of the head of the Expedition, S. P. Tolstov (not less than 1,000 *poisk*).[Note 10] In many areas, the engineer-geodesist N. I. Igonin, provided aerial plans of different irrigation systems and ancient irrigated lands near several large settlements. The field work combined with the wide use of aerial reconnaissance and photography provided the possibility to uncover and to study in detail the areas of ancient irrigation works in the lower reaches of the Amudarya and Syrdarya, It also made possible drawing several maps and a general layout of the most typical irrigation facilities for every historical period.

Part I

Ancient Irrigation Study Methods: Origin and Development of Irrigated Agriculture

Chapter 1
Ancient Irrigation Study Methods

Boris V. Andrianov

The creation of irrigation systems is one of the ancient arts of engineering. Thus, the specialist in irrigation L. V. Dunin-Barkovskiy correctly supposed that the historical and archaeological study of ancient irrigation works and the history of the development of oases are no less important for modern irrigation planning than are the studies of natural conditions. However, for many regions of Central Asia "there are no such data yet. In this regard it is impossible to trace the development of irrigation techniques in a historical perspective" (Dunin-Barkovskiy 1960:64-65).

The long-term research of the Khorezm Archaeological-Ethnographic Expedition on the ancient irrigation of the Aral Sea area provided an opportunity to retrace the development of irrigation techniques, and thus fill a gap in the general history of irrigation development. The remains of ancient irrigation canals, the layout of fields, and the ruins of rural settlements of different periods are spread in the desert for many dozens and even hundreds of kilometers, representing remains of the material culture of ancient farmers. Throughout these remains it is possible to ascertain the main stages of development and methods of irrigation techniques, which occurred in each area with their own specific features.

In all periods, irrigation works met the same technical problem: supply water to fields and create soil moisture from the flowing water of the river. In primitive 'single' systems, the irrigation was performed by overflow banking (flood irrigation); in more complex systems (with regular water supply over time and flow rate) it was achieved through various hydro-technical facilities, such as head intake diversions, main

canals, distributors, water regulation devices, feeders, water lifting systems (e.g., *chigir*), ditch networks, etc. Labor on irrigation networks is always determined by many parameters: the nature of the water inlet, the size of the main canals, the configuration and articulation of the irrigation network, the technical facilities to regulate the water flow, slope and rate of the canals in order to combat salinization, silting and waterlogging (Glebov 1938; Kostyakov 1951).

But, in what natural and social-historical conditions the irrigation systems of various historical periods 'have worked'? What qualitative and quantitative properties they had and what was the technical progress of irrigation? It is still impossible to answer many questions and this largely depends on the poor state of preservation of ancient irrigation remains.

In most areas of ancient irrigation, each link with the irrigation systems is almost entirely destroyed. They are buried under the moving sand, having a strange appearance on the *takyr*; the massive 5-6 m high banks of canals have been eroded into small hills and riverbeds have been covered by colluvial-alluvial sediments; field layouts covered by later anthropic and cultivated layers or vanished under the inflated crust of the *solonchak*. However, the main enemy of ancient irrigation was man himself: because of land development, the ancient settlements were quickly destroyed; the ramparts and walls of dwellings were flattened to the ground, canals were reconstructed and fields were ploughed anew. The cultural oases of different periods were 'layered' on top of each other, forming a complex weave of non coeval units.

In modern oases, the ancient irrigation facilities have been almost entirely destroyed and the heads of canals, water intakes, and other hydraulic facilities rebuilt.

Many of the ancient main canals in the Lower Amudarya were located in modern contexts and cultivated areas. Thus, it is very difficult, for example, to solve the question of when changes in the water systems deriving from the main river course began. Even more complicated is to reconstruct the historical topography of the oasis as a whole. However, the good preservation of some parts of ancient (Prehistoric, Antique, Medieval) systems of irrigation might reveal many details on the historical development of irrigation techniques in the Aral Sea area.

Remains of ancient irrigation works have survived to the present because of the desert natural conditions, where, the active surface destruction and accumulation due to the aeolic processes, have created a *takyr*-shaped 'protective' crust (Gerasimov 1954) resulting in double lines of semi-destroyed embankments of big canals, slightly raised above the bottom of main water courses; lines of narrow ditches, detected only by vegetation, barely visible traces of agro-irrigation layouts on the pinkish *takyr*.

The landscape of abandoned irrigation systems strongly differs from the landscape of adjacent clay and sand deserts in its typical microrelief, the different composition of vegetation and soil cover, and the different links between many natural features. That is why it is possible to assume that the study of the 'lands of ancient irrigation' is a special case for a wider research of the so-called transformed or cultural landscape (Saushkin 1946:97ff.; Kabo 1947:5-32; Bogdanov 1951:303-305; Isachenko 1965: 207-215). The complex investigation of cultural landscapes is very complicated and it requires an interdisciplinary approach combining natural geography and human sciences. Their long separation has been detrimental to the development of general methods of natural-historical research.[Note 11] The study of areas with traces of ancient irrigation systems is impossible without both the close

contact between historical and natural sciences and the combined use of various methods (Tolstov 1947a:255ff.; Bernshtam 1949:10; Kolchin 1965:7-26; Andrianov and Tolstov 1965).

The cultural landscape is a complex natural-historical formation, in which the effects of influences of different historical periods are gradually accumulated. In every historical period the influence of society on the environment has been limited by the degree of knowledge of natural laws and the level of technological development, which in turn was determined by the laws of social development. Under the influence of different forces, continuous changes take place in the environment, the result of which have an impact on the earth's surface, space and time. Landscapes have their own history (Markov 1951a). The most striking examples of human's historical impact on the environment are: the cultural landscapes of oases in desert areas; a territory in which thousands of years of farmers' labor changed the topography; creation of a strong cultivated and irrigated soil; changing the water regime, transformation of the vegetation and even contribution to changes in climate conditions (Georgievskiy 1937:106-123; Saushkin 1946:104; Sapojnikova 1951:231). The development of cultural landscapes is closely intertwined with physical-geographical and social-historical patterns.

Peculiar formations of cultural landscapes preserve their attributes even in a desert cultural landscape, i.e., a territory where the cessation of economic activities (irrigated agriculture) led to the reclamation by the desert.

Nevertheless, there is no complete restoration to its original appearance, so long as human activities left indelible traces on its surface in the shape of dead oases with ruins of towns, fortresses, farms, orchards, fields and irrigation systems together with implements and household objects (Gerasimov 1937:51-52; Tolstov 1948a:27). These traces form a unique cultural landscape over vast territories (Letunov 1958:22-23, 175-176; Rodin 1961:120-123). They are studied by archaeologists and provide an excellent source for historical research on the economy, culture, and agriculture of ancient people, especially for the study of ancient irrigation works, with the latter forming a sort of 'topographical skeleton' of ancient and modern oases in the Aral Sea area.

Already in the early 19th century, Russian travelers drew attention to the fact that "a household in Khiva requires fields and arable lands to be very close to dwellings. The watering of fields demands close and permanent control. These led the people of Khiva to settle in individual farmsteads like our *khutor*. People settled mainly near canals and water supply facilities" (ZJ 1838, no. 6:334; Shkapskiy 1900:97). The farmers were tied to their fields and fields were located away from the river, on small canal branches, in the middle and lower parts of irrigation systems.[Note 12] These natural geographical settlement connections with irrigation (the main topographical skeleton of cultural landscapes of the ancient Aral Sea area) formed the archaeological basis for dating the existence of irrigation systems.

The problem of dating ancient irrigation is complicated. If small ditches (feeders) functioned several years without change, the main large canals did it for several centuries. Dating them on the basis of single discoveries or through a couple of excavated rural settlements is almost impossible, because the 'life' of one dwelling is significantly shorter than the 'life' of a main canal. Even more precise archaeological time markers, such as coins and so forth, are of little use here. According to S. P. Tolstov's words "the study of the ancient irrigation network of Khorezm, as well as other regions of Central Asia, is quite a complex and hard task. The preliminary condition to obtain satisfactory results in this work is a fully integrated study of archaeological sites located in these ancient irrigated lands..." (Tolstov 1948a:38).

It should be remembered that the study of irrigation by the Khorezm Expedition was preceded by a wide research of rural and urban settlements, as well as archaeological digs of the most typical sites for each historical period. During these years of research, the Expedition collected a wealth of material. The archaeological excavations of the historical sites of Khorezm and adjacent areas, combined with aerial archaeological surveys and field research of the 'lands of ancient irrigation', highlighted the history of material culture, settlements, dwellings, household items, etc. It also revealed the main stages of economic development from the Neolithic culture to the rise of feudalism in the Late Middle Ages. The study of settlement types, the history of agriculture and craft allowed to reconstruct stages of the social-economic, political, and cultural history of the Khorezm state. The Expedition collected a large array of data on the history of the people of the Khorezm Oasis and adjacent areas, covering nearly five millennia, as witnessed by the continuity of settlement patterns in the 'land of ancient irrigation' (Tolstov 1939, 1940a, 1941, 1945a, 1946a-b, 1947a, etc.).[Editor 2] As a result of postwar undertakings by the Khorezm Expedition, some corrections and additions were made to the first classification by S. P Tolstov in 11 main cultural stages and published in the 4th *Vypusk* (issue) of the Khorezm Expedition (see Tolstov and Itina 1960a), which included the study of ceramics from the Prehistoric, Antique, Early Medieval (Afrigid), Middle Ages, Late Middle Ages and pre-Mongol periods (see Tolstov 1941a:32–34; Itina 1959a; Vorobeva 1959a: Nerazik 1959; Vakturskaya 1959).

Many additions to the classification of Khorezm sites and adjacent areas were made in recent years (see A. Vinogradov 1957a, 1968; Tolstov and Itina 1960b; Tolstov 1962a-b; Levina 1966, 1967; Itina 1967; etc.).[Editor 3] The fundamental publications of the results of many years of excavations of the outstanding sites of ancient Khorezm such as Koy-Krylgan-kala, Toprak-kala, etc. are crucial. Recently, a monograph about Koy-Krylgan-kala was published (Tolstov and Vaynberg 1967). It confirmed the main work on the material culture of Khorezm in the Archaic period made previously, as well as it clarified the chronological questions of the Kangju period ceramics. It distinguishes itself as a pottery class peculiar for its light-slip vessels, which is coeval with the common pottery of the Khorezmian sites of the Kushan period. Additional work by the Expedition archaeologists on the Prehistoric (A. V. Vinogradov and M. A. Itina), Early Archaic (M. G. Vorobeva), Late Archaic and Early Middle Ages (E. E. Nerazik), Antique and Medieval sites of the Lower and Middle Syrdarya (L. M. Levina and N. N. Vakturskaya), as well as research in the fields of numismatics (B. I. Vaynberg), burial and cult structures (Yu. A. Rapoport), and trade routes (V. A. Lokhovits) will provide future details on the general chronological classification in an attempt to establish an absolute dating of sites and archaeological cultures.

In order to determine the operational periods of the large irrigation systems, the materials characterizing the main stages of long-term settlements and fortifications in the terminal parts of the main canals are important. Based on these data, S. P. Tolstov established a general picture of historical reduction rates of the irrigation network on the right bank of Khorezm. Ya. G. Gulyamov outlined a scheme of the Antique and, especially, Medieval changes in irrigation systems (Tolstov 1948a:45-47; Gulyamov 1957:99, 117-120, 162-163, etc.). The ancient canals entered the desert for many tens and even hundreds of kilometers, thus the study of fortifications and other large sites spread over a wide area is impossible without further fieldwork. The abundance of sites and the remoteness of the studied desert regions cultural oases, led the Expedition to apply new technologies, in particular aircraft and aerial photography, in the first postwar years (Tolstov and Orlov 1948).

Since 1952, the Expedition began to develop complex methods of archaeological-topographical studies of ancient irrigation based on archaeological field surveys and reconnaissance, mapping of irrigation systems from aerial photography, archaeological-topographical (instrumental), and morphometric and geomorphological studies of ancient irrigation works. Since then the dating of irrigation canals and the formation of 'lands of ancient irrigation' was based on the mapping of a large number of archaeological explorations and

massive archaeological finds. They provided a more accurate assessment of the oases main periods in ancient times and revealed a pattern of coeval sites, topographically connected to the studied irrigation systems. The Expedition improved a special method to establish the chronology of hydraulic works, closely related to archaeological excavations as well as the study of different types of material culture (ceramics, glass, jewelry, metal objects, arrows, etc.).

To determine when irrigation systems were in use and the dynamics of the 'lands of ancient irrigation', the materials from field surveys are of great importance, especially archaeological artifacts (ceramics, beads, coins, etc.), which allow to establish the life span of single areas of irrigated oases, to identify a network of contemporary agricultural settlements, topographically connected with the studied irrigation systems, and indicating when their individual parts were functioning. More intensive is the archaeological research, more accurate and detailed are the conclusions on the historical dynamics of irrigation networks. Thus, during the mapping of the right bank of the Khorezm, one of the richest area in sites around the Aral Sea, the surveyed sites (per unit area) were six times more numerous than those in the left bank of the Khorezm and twenty times those in the Lower Syrdarya.

Field research was associated with collecting and dating of artifacts, the measurement and the excavation of canals, photography, maps of both major important irrigation features and the most interesting new sites discovered. The uniform survey framework (recorded by pictures), archaeological finds, diaries, maps and photographs (methodically numbered) produced quite a precise documentation of these three main areas: right bank of Khorezm; left bank of Khorezm; Lower Syrdarya. The collection of scattered surface finds by the archaeological-topographical field unit was studied, in many cases, by other members of the Khorezm Expedition who were specialists on this subject. The finds were studied both at the Expedition base camp after the return of the team from the field and in laboratories in

Moscow, where materials (fragments of ceramics, coins, arrowheads, jewelry, etc.) were catalogued and analyzed.

Of great importance for studying ancient irrigation systems are the soil-botanical field investigations, which allow to reconstruct the geographical environment in which people lived during different periods (Andrianov, Bazilevich and Rodin 1957; Kes 1958; Zadneprovskiy and Kislyakova 1965; Lisitsyna 1963, 1964, 1965). In combination with the archaeological-topographical study of this area, the soil-botanical unit highlighted the conditions, the nature and the extent of agricultural civilizations, where visible traces of irrigation systems were not preserved on the surface. As an archaeologist looking at the rim bends of a clay bowl broken thousands of years ago reconstructs its shape and determines the age of its production, so a naturalist, studying soil layers in an excavated pit, reconstructs the anthropic traces or records its undisturbed natural structure. According to V. V. Dokuchaev, soil represents the natural-historical frame and its morphology should reflect the environment (vegetation, climate, water regime), in which it was developed. A soil, cultivated in the past, preserves forever in its structure environmental traces (Glazovskaya 1956; Gerasimov 1961). Studying the section of buried soil and analyzing its composition in the laboratory, its relative humus content (characterizing the cultivated soil), and salts, it is possible to reconstruct in great detail the geographical landscape and the conditions of agricultural activities.

The integrated archaeological and soil-botanic research was started by the Khorezm Expedition in 1953, when the soil scientist N. I. Bazilevich and the botanist L. E. Rodin (Andrianov, Bazilevich and Rodin 1957) participated in the archaeological-topographical survey of the lands of ancient irrigation on the left bank of the Amudarya. Test trenches in the region of the Shakh-Senem fortress ruins, under the ancient agro-irrigation, revealed buried *tugai* soil formed from layered alluvial deposits carrying, in the majority of cases, traces of waterlogging

and over-wetting, which took place in the most ancient Sarykamysh Delta region. These features indicated the existence in this place of active delta riverbeds during historical periods.

The existence of a long agricultural period, the presence of agro-irrigation soil layers and stages of landscape evolution due to human activities can be traced through changes in sediments' lithology; former cultivated soil emphasizes humus distribution (its highest amount is in the ancient agro-irrigation horizon) and salt.[Note 13]

The study of soil profiles can reveal the stages of evolution of the natural environment in the region and, together with archaeological data, allow the reconstruction of the ancient landscape appearance.

Mapping Irrigation Networks and Aerial Methods

The work of mapping and classifying archaeological, remains, to produce detailed maps and plans, has gained great importance in the modern progression of the archaeological science, when huge archaeological material have accumulated and archaeologists of different branches have passed from the study of single features of material culture to dealing with wide historical-geographical issues and the continuous archaeological survey of wide regions.[Note 14]

Mapping, the most effective method to study ancient irrigation systems, requires usually a large spatial extension. It is hard to chronologically differentiate irrigation systems by period, or to reconstruct the historical dynamics of ancient irrigation, without plans and maps providing in detail the configuration of irrigation systems and the location of the main waterways, distribution devices, secondary canals, etc. in relation to the sources of water supply and the large, well-studied, archaeological sites. On archaeological-topographical maps (at scales 1:25,000, 1:50,000, etc.) it is possible to assess different measurements, linear distances as well as areas of the territories occupied by irrigation systems and farms of different historical periods. Maps can reveal the natural connection of ancient hydro-technical works with relief, terrain slope, modern land cover, etc. Aerial methods play a special role in the creation of such maps. The mapping of vast territories and the extensive study of the 'lands of ancient irrigation' most ancient irrigation works, cultural landscapes and detailed studies would be impossible today without aerial mapping (see Tolstov, Andrianov and Igonin 1962; Igonin 1965; Andrianov 1965). Maps and plans drawn from aerial photographic surveys are much more accurate and detailed. They provide the possibility of obtaining a vast amount of features, to answer some questions on the history of settlement typologies, to carry on analyses on agricultural production, etc.

Aerial surveys have provided invaluable support for archaeological and topographical fieldwork. They facilitated the exploration and the 'bird's-eye' observations of invisible or barely visible sites and traces of canals, distinguishing them from the surrounding landscape according to their geometric features. Such aerial methods provide an objective and a complete documented overview of irrigation systems, which may not be clearly traced on the ground because of their state of preservation. The burrowed banks of canals were flattened and the riverbed filled by sand. According to archaeologist V. A. Shishkin, who first applied aerial methods in 1934 to study the environs of Termez and Bukhara, "the traces of irrigation canals, even the largest, gradually disappear under the accumulation of sand dunes and under the barchans and they are sometimes dispersed by wind or eroded away by water. Finding continuity of such vanished canals, even when they are one or two hundred meters long, is an extremely difficult task" (Shishkin 1957:62).

In these cases, only mapping from aerial photographs can capture the overall topography of ancient irrigation systems.

The importance of aerial methods in the archaeological study of ground surface is so great that there should be more detailed information on the history of their development (S. Pavlov 1934; V. Pavlov 1950; Crawford 1953, 1954; Chevallier 1957, 1961a; Bradford 1957) as well as their

use by the Khorezm expeditions. The history of 'aerial archaeology', i.e. archaeology using vertical aerial photography and observation in the study of archaeological sites, is inextricability linked to the origin and development of aerial photographic methods applied to the topographical study of the ground (Beazeley 1919; Daniel 1950:294-302).

The first aerial photographs for archaeological purposes were made in 1906 by Lieutenant P. Sharp who photographed Stonehenge from a balloon. During World War I, military intelligence services played an important role in the development of methods of aerial photography and their interpretation (Schuchhardt 1918).

G. A. Beazeley, an aviator of the British air detachment, during the aerial survey of Mesopotamia, discovered traces of many ancient canals, karez, dry riverbeds and remains of settlements near Baghdad (Beazeley 1919, 1920). G. A. Beazeley's observations were essentially the first attempt to use aviation in the study of ancient irrigation systems. In the same years, a German, Theodor Wiegand, conducted an aerial photographic reconnaissance around the Suez Canal and made photographs of archaeological sites (Mischrefe, Sbeita, etc.) with neighboring fields and different hydraulic devices.

Further development in 'aerial archaeology' occurred between the two World Wars associated with the names of the British O. G. S. Crawford and the French A. Poidebard. Their developments in aerial field research in the early 20th century followed their well-known passion in research issues related to geography, which is reflected in the works by J. P. Williams-Freeman, G. R. Johnson and A. H. Allcroft. In 1921, O. G. S. Crawford published a book entitled *Man and His Past*, which summarizes the possibility of a geographical approach to primitive and ancient British history.

O. G. S. Crawford's work in 'aerial archaeology' began in 1922 with the aerial photographs of Hampshire, where fields of the Roman period were revealed. In 1924, O. G. S. Crawford and A. Keiller made hundreds of vertical aerial photographs of several areas of England during the rainy season. This work represented the basis for creating and publishing in 1928 the first scientific monograph on 'aerial archaeology', *Wessex From the Air* (Crawford and Keiller 1928). Using aerial photography, O. G. S. Crawford discovered, under ploughed and sown fields, traces of ancient burials, settlements, cult buildings, Celtic and Saxon agricultural villages and fields. It is known that shades of vegetation on crops depend on the nature and the permeability of the soil, therefore, in places with buried structures, the color green is lighter and, by contrast, in the hollows of ancient canals, roads, and ditches, which accumulate more moisture, the soil color is darker and the vegetation thicker (Crawford 1923, 1924, 1929a-b).[Editor 4]

O. G. S. Crawford outlined the following classification of archaeological sites: 1) sites preserved in elevation that are clearly visible on the ground; 2) sites that are largely damaged and are recognizable on the ground only through ramparts and ditches; they also include agricultural layouts; 3) sites very badly recognizable with field survey, plowed and buried under layers of modern crops and stratifications, thus disclosed only by the color of soil and vegetation (Crawford 1953:43-50).

In 1930, pilot Major G. W. Allen, with a handheld aerial camera, made several thousand perspective aerial photographs of England and drew the attention of archaeologists for their valuable qualities of disclosing sites in a volumetric perspective. During the same year, in America, perspective aerial photographs and aerial observations were specifically used for the study of ancient irrigation systems in the valleys of the Salt and Gila rivers (see p. 105).

The studies of ancient irrigated land and ancient irrigation systems began with H.A. Beazeley in the classic ancient Orient and were successfully continued in 1929 by the above mentioned O. G. S. Crawford (1929a:342, 1929b:497-512). He made a series of vertical aerial photographs that became the basis for the photomosaics of the environs of Samarra and the Nahrwan canal at a 1:5,000 scale. In these photomosaics, the banks of a huge ancient canal, surrounding the well preserved walls of the early Medieval capital of the Abbasid caliphate, Samarra, are clearly visible (Crawford 1953:212-213). These photomosaics were successfully used by Ahmed Susa, an historian of Iraqi Medieval irrigation, who in 1948, published two volumes on the major irrigation work on the environs of Samarra during the Abbasid period. The work was illustrated by a large number of maps and plans of different hydraulic devices (Sousa 1948).

A wider scale study area of ancient irrigation in the Near East, with the help of aerial methods, was realized by A. Poidebard, whose task was to map the system of Roman military fortifications and trade routes in Syria which connected Mediterranean ports with Mesopotamia (Poidebard 1929a-b, 1932, 1934; Mouterde and Poidebard 1945). The aerial observations were fully confirmed by archaeological field surveys and excavations.

Independently from O. G. S. Crawford, A. Poidebard arrived approximately at the same conclusions about aerial photography methods for archaeological purposes. He noted that the clearest photographs of the archaeological sites were obtained in autumn, when the first rains of the season cover the steppe with new vegetation, whose shades vary depending on the nature of the underlying soil. Where hidden building stone foundations, roads, and any other structure lie buried under the ground, the vegetation is lighter and the layouts are marked by a lighter shade. Beds of ancient canals and defensive ditches differ, in their darker color, from the surrounding areas. The best time to shoot the badly preserved remains is in the morning or evening, when the oblique exposure highlights and even slightly exaggerates small ground unevenness.

A. Poidebard developed a system using different aircraft altitudes (it is believed that the most suitable height for reconnaissance is 200-600 m) to obtain pictures at different scales relative to the problem of the archaeological area of study. He also applied perspective photography from a low altitude (60-200 m) with objects taken against the sun so to provide a better elaboration of the microrelief. As we shall see below, this method was widely used in the aerial reconnaissance of largely damaged archaeological sites, especially in desert irrigation systems (Tolstov and Orlov 1948:60).

A considerable archaeological work with botanists and geologists was conducted in the desert areas of Syria, Jordan and Iraq in 1925-1950 under the direction of G. Field (1960). In the 1930s, the study of defensive walls, stone roads, aqueducts and mountain irrigation systems was started in the valley of Cuzco using aerial methods (see Johnson and Platt 1930; Shippee 1932).[Editor 5]

In the late 1930s, Americans organized a large aerial archaeological expedition in Iran, directed by archaeologist E. F. Schmidt (1940). Throughout this aerial survey (1935-1937), E. F. Schmidt covered vast areas of ancient irrigation.

An important stage in the development of aerial methods after World War II is associated with the improvement of automatic photographic equipment, the use of stereoscopic effects by overlapping images and the panoramic coverage of vast territories (Reyzer 1959).

In England, the development of 'aerial archaeology' successfully continued after O. G. S. Crawford with John Bradford, a former military decipherer, who, in 1957, published a large monograph on aerial photography, *Ancient Landscapes*, beautifully illustrated with aerial photographs. J. Bradford, continuing the tradition of classic British archaeology, considered as the main task of 'aerial archaeology' the reconstruction of the general appearance of ancient cultural landscapes, where the use of aerial photography is closely linked with archaeological, historical and natural-geographical methods (Bradford 1957:1-84; Fox 1943). According to J. Bradford, 'aerial archaeology' cannot be separated from field archaeological work and historical research. It must be combined with the study of written sources (for instance, the Roman descriptions of agriculture) dealing with modern and ancient history of settlements, and should also

be associated with geography and geology, related to the natural history of the investigated areas. Important additions to 'aerial archaeology' are the different new methods of natural and technical sciences (soil, geochemical, etc.).

The purpose of 'aerial archaeology' is to detect new, unknown sites and provide detailed plans. According to J. Bradford, it is important to combine vertical and perspective aerial photography, because oblique images give a good overview of a site general shape while vertical photography is used to draw a detailed plan. A series of stereo-paired photographs allow us the three-dimensional study of many-times resized models of a site. Aerial photographs covering a large area (depending on scale) provide an opportunity to study the site in connection with the surrounding landscape, and to reveal the historical changes of natural conditions (movement of river banks, riverbeds, sand, etc.).

J. Bradford improved the techniques proposed by O. G. S. Crawford, in the study of hidden, buried and plowed archaeological sites, which are particularly frequent in Western Europe where the 'cultural landscapes' of plowed fields, gardens, parks, etc., prevail (Figure 1). The study of 'crop-sites', i.e. archaeological sites detected by soil color and vegetation, and sites buried under layers of modern cultivation ('shadow-sites'), occupy an important place in the methodological section of J. Bradford's monograph *Ancient Landscapes* (1957). This work is very remarkable because it summarizes the experience of many countries in Western Europe that have employed aerial methods and it contains a large bibliography.

The successful investigation by A. Poidebard in Syria in the 1930s was continued in 1954-1955 by archaeologists W. J. Van Liere and J. Lauffray (Van Liere and Lauffray 1954-1955), who were commissioned by the Ministry of Agriculture to undertake aerial photographic interpretations of the Upper Jazirah (scale 1:20,000).

The work of archaeologists was devoted to reveal sites and hydraulic works of different periods. They noted that in the photographs many canals, and roads, were distinguishable by dark lines. However, the roads crossed all the uneven ground contours (sand dunes, dry wadis, etc.) whereas the canals followed, with a meandering direction, the optimal slope of the area. Irrigation basins, with main canals, distributors and irrigation ditches, form a clear system, topographically linked with fields and settlements. In the Khabur River scholars identified two types of ancient irrigation works: 1) small local systems (designed to lift water through noria); 2) large canals irrigating land above the second terraces. The length of the four major canals was from 12.5 to 35 km. They irrigated 18,400 ha.

The 1954-1955 work in Syria and, especially, the work in Iraq in the Diyala River basin in 1957-1958 (Jacobsen 1958), was connected not only with a scientific archaeological study of the area, but also with the practical economic task of rehabilitating the irrigation in these ancient irrigated lands. In the Diyala River basin, Iraqi archaeologists and scientists from the Oriental Institute of Chicago University collaborated in an integrated archaeological and soil-botanical research of ancient irrigation systems and settlements in order to identify the historical abandonment and salting of those vast areas (Jacobsen 1958; Andrianov 1960b).

The study of the historical dynamics of irrigation systems, the location of settlements in different periods, the spread and production of agricultural crops and also the identification of salted areas were based on the uninterrupted archaeological mapping survey from aerial photomosaics and from single test trenches (see also Adams 1958, 1965:119-125). Based on the recent publication of materials of R. McC. Adams' expedition, 867 large and small settlements were investigated and dated (Adams 1965). This book contains many drawings of hydraulic facilities and archaeological canal sections (Adams 1965:figs. 17-22). Particularly interesting is the description of the historical dynamics of irrigated areas and settlements of different periods from paleo-demographic material, such as the diagram of changes in population in the Diyala Basin, and the relationships between urban and rural

populations from the 4th millennium BCE to 1957 (Adams 1965:115:tab. 25).

Very particular systems of irrigation, based on seasonal mudflows, were discovered in the deserts of Southern Arabia. To these irrigation facilities, functioning from the 7th century BCE to the 1st century CE, is dedicated the thorough essay by Richard LeB. Bowen in the book *Discoveries in South Arabia* (Bowen and Albright 1958).

In the late 1950s and early 1960s, 'aerial archaeology' was largely spread in Europe and the Mediterranean Sea in connection with a special branch of Western European archaeology, the so-called 'agricultural archaeology'.

This direction is now being successfully developed in England by M. W. Beresford, M. J. Yates, J. K. St. Joseph, etc. M. W. Beresford's work concerns the geography of rural populations and the problem of 'depopulation' of villages (Beresford 1954). J. K. St. Joseph continued the research on ancient fortifications and settlements in the central regions of England started by O. G. S. Crawford (St Joseph 1945; Beresford and St. Joseph 1958).

In France, aerial methods were widely promoted by the well-known specialist in this field, R. Chevallier, who headed a commission on decoding in the International Society of Photogrammetry. She published several books about the use of aerial methods in archaeology (Chevallier 1957, 1963, 1964a) and also a bibliography on this matter (Chevallier 1957, 1963). Some of R. Chevallier's articles are devoted to the topic of agricultural archaeology in Europe and North Africa (Chevallier 1961a-b, 1964a).

The author noted that, from systematic photographs and laboratory work, it is possible to recognize agro-irrigation layouts, their size, shape and orientation in relation to other landscape elements. These data can identify types of agricultural development of different places in different periods.

In 1963, on the initiative of R. Chevallier, Paris hosted the International Conference of Aerial Archaeology (AACI), which was attended by specialists from almost all of the European countries.[Note 15]

In 1964, on the initiative of UNESCO, a conference was organized in Toulouse to discuss the use of aerial methods in the study of natural resources. At this conference, R. Chevallier provided a special report about the achievements of agricultural 'aerial archaeology' (Chevallier 1964b).

In the USSR, the beginning of aerial archaeology took place in the 1930s. In 1934, S. N. Pavlov published an article with an overview on the use of aerial photography in archaeology, reporting major information regarding the development of this subject (Gaveman 1937; Fersman 1939; Pavlov 1950). In 1934, an aerial investigation was conducted in Khorezm by the staff of the Khorezm Expedition, which carried out an aerial survey around Zmukshir, in the Upper Chermen-yab and along the Amudarya, coming back from Khorezm to Tashkent. According to the Expedition head, M. V. Voevodskiy, it "has made possible to represent, in a broader scale, the interconnection between the early settlements and their relationships with the old riverbeds and channels of the Amudarya and the preserved remains of ancient irrigation systems, clearly visible from the airplane" (Institute of Archaeology of Moscow, Manuscripts Archive 3, N1:3).

In 1934 the same aerial methods were used by archaeologist V. A. Shishkin to study the topography of Termez and Bukhara. This research was continued in 1950 and V. A. Shishkin noted that aerial methods were especially useful for studying ancient irrigation, and one of the main tasks of the archaeological study of ancient irrigated lands was their mapping using photomosaics (Shishkin 1957:62).

As noted above, the widest use of aerial methods in the USSR was with the research of the Khorezm Archaeological-Ethnographic Expedition. The new, postwar, phase of the Khorezm Expedition was characterized by a steady excavation of large sites (Toprak-kala, Koy-Krylgan-kala, etc.) with an extensive exploration of the remote outskirts of Khorezm, based on a combination of aerial reconnaissance and field surveys (detailed observation of sites, prospecting shafts, collection of ceramic material scattered on the surface) (Tolstov 1946a-b-c, 1948b:25-62).[Editor 6] At that time, the field survey reached the distant outskirts of Khorezm and the surrounding areas (Lower Syrdarya, Northern Kyzylkum and Ustyurt). During only two field seasons, in 1945 and 1946, aerial surveys and investigations by car covered more than 15,000 km, crossing the investigated territory in different directions and revealing hundreds of previously unknown sites, from the ruins of early Archaic to late Medieval Karakalpak settlements on the Janydarya and the Kuvandarya (Tolstov 1952a:10).

The work started in 1946 in the Lower Syrdarya was continued in 1948, 1949 and 1951. In 1947, aerial observation and photography were carried out on the left bank of Khorezm (Chermen-yab Basin and the northern parts of the Uzboy). In 1950 and 1951, the field survey covered the whole bed of the Uzboy up to the Caspian Sea. In 1952, the Expedition studied in detail the Medieval irrigation of the Southern Sarykamysh and this work continued (with the participation of a geomorphological unit from the Institute of Geography of the USSR Academy of Sciences directed by A. S Kes) in 1953 and 1954 (Tolstov 1958:252-253).

During these years, with aircrafts such as 'PO-2', the huge territory of the 'lands of ancient irrigation' of Khorezm and its surrounding region was surveyed. During the flight, perspective photographs from the air and observation of the territory of archaeological sites were carried out, with the purpose of finding new archaeological sites. Afterwards, detailed studies and measurements were made on sites, together with test trenches, collection of surface finds and study of ancient irrigation. In order to increase the radius of the aerial surveys, the Expedition organized some car-equipped teams to move along the accessible desert places along planned routes. The task of these car-equipped teams was to provide the members of the aircraft survey team with fuel, technical assistance, as well as water, food and scientific equipment.

The experience of the first postwar years

allowed to identify the following main aims for aerial archaeological work: 1) aerial reconnaissance of large areas observed for the first time, combined with aerial photography of single sites or site complexes; 2) aerial photography, to be carried out in places where the preliminary survey or field observations already discovered a system of sites, settlements, ancient roads or canals; 3) detailed observations and aerial photography of individual sites (Tolstov 1948b:38–39; Tolstov and Orlov 1948:60ff.).

During the aerial survey, it was possible to enhance the topographical location of sites of different sizes and to obtain photos at different scales. For the large sites identified, three main altitudes were planned: 1) from 2,000 to 1,000 m; 2) from 600 to 300 m; 3) from 300 to 130 m. Based on aerial research, S. P. Tolstov arranged and published in 1948 an archaeological map of the right bank of Khorezm, in which, together with the ruins of cities, fortresses and large farms, he also drew some large canals (Tolstov 1948b:map attached).

Since 1952, the application of aerial methods has been applied for archaeological and topographical investigation of the 'lands of ancient irrigation' of Khorezm. Aerial photography was carried out not only for individual sites and archaeological complexes, but also for the vast territories with a rich variety of archaeological sites and large irrigation systems stretched for dozens of kilometers. As a rule, most of the objects were shot at a different scale: 1:2,000, 1:4,000, and 1:6,000. The largest scale allows us to study important details of the whole layout. The small scale photography covered a larger territory: the relative position of objects is revealed better, which is very important for taking photos of irrigation systems (Tolstov, Andrianov and Igonin 1962:5–6).

Over the last years, a member of the Khorezm Expedition, engineer-geodesist N. I. Igonin, used a RMK aerial photographic camera and a specific equipped aircraft to vertically photograph single irrigation systems and neighborhoods of major sites in the 'lands of ancient irrigation' on the right bank of Khorezm and the basins of the dry

riverbed of the Syrdarya, such as the Janydarya and the Inkardarya (Igonin 1965).

Already the first years of activity of the archaeological-topographical unit in the ancient irrigated lands in the basin of Chermen-yab, on the left bank of Amudarya, and in the basin of the ancient Kelteminar, on the right bank of Amudarya, produced considerable material on the gradual development of irrigation systems in the Lower Amudarya from Prehistory, through the Archaic period up to the Early and Late Middle Ages (Tolstov and Andrianov 1957). An important achievement of the archaeological-ethnographic unit in this period was not only to further the general history of irrigation development of the Aral Sea area, but also to improve the technique based on the stereo-photogrammetric study of vertical paired aerial photographs, their decoding in the field and in the laboratory, as well as field research, identification and excavation of the canals found during the archaeological research.

The harsh climate and landscape conditions of the deserts, the survey complex organization, the nature of sources employed (aerial) and other circumstances determined the following work order: 1) preliminary knowledge with different materials and interpretation of photographs; preparation of plans for field research; 2) field observation of specific sites, collection of archaeological materials for dating canals, and chronological comparison of irrigation systems with main sites, where the Khorezm Expedition had already carried out excavations; 3) the last stage of work in the laboratory included the archaeological dating of specific collections of material objects (ceramics, coins, arrowheads, jewelry, etc.), drawing plans of irrigation systems, and producing a scientific synthesis of the results.

The effectiveness of the field survey research largely depended on the high quality and thorough planning of aerial photography, familiarity with written sources, and cartography of the region archaeological sites. Aerial surveys usually started in areas with known archaeological sites. Information from written sources and cartography 'were connected' to the photographs. On the latter, sectors of field archaeological-topographical research and field of interpretation of aerial pictures were marked. Unknown and undated sites and the main key-sectors of irrigation systems were also present on aerial photos. Afterwards, car-equipped expeditions were planned to these points. Their choice took into account the terrain conditions and the location of water sources required for many travel days, etc.

The work of the field team included the archaeological and topographical study of ancient irrigation systems, based on topographical surveys (instrumental or semi-instrumental) of individual irrigation units, the morphometric study of ancient irrigation works, and field interpretation of aerial photographs. Going beyond the well-known methods of common instrumental mapping or optical survey of ancient irrigation.[Note 16] I shall describe the application of morphometry, geomorphology and field and laboratory decoding of aerial photographs in the archaeological-topographical practice.

The methods of morphometry, widely used for the study of landforms, are very useful for measuring the width and length of ancient canals, and for revealing the size of sections, river banks, etc. However, the researcher meets many difficulties in the field: canals banks are mostly damaged and riverbeds are frequently only traceable as flat clay banks stretching on *takyr*; sometimes there are remains of canals in the form of small elongated clay remains among sand dunes and hollows, or sometimes only as a strip of vegetation or a micro-relief in a clay plain.

For an approximate reconstruction of canals sizes, transversal sections were used by both specialists of geomorphology and hydro-engineers (Glebov 1938:185-186, etc.). The preserved forms were recorded in scale with instruments and optically. Unfortunately, the state of preservation did not allow to recognize everywhere the real width between river-banks. Because of this, frequently both the remains of banks and the total width of the preserved riverbed were measured. Where possible, these measurements were performed in such a way at a determined bank height in order to give an idea of the maximal water flow. We were only able to guess the depth of ancient canals because we had too little data for identification (Figure 2).

Several difficulties arose in the study of Prehistoric irrigation works, using the aerial survey and the field work in order to recognize canals along natural landscape features. In these cases, the combined archaeological and geomorphological studies of ancient hydrography, individual relief features and a historical analysis of their formation, and the conditions of the settlement areas aided this matter (Kes 1958, 1959).

A central place in this research is occupied by field surveys, and excavations, usually dug across riverbeds and canals in order to obtain sections.[Note 17] Excavations of ancient canals had already been undertaken in the early 20th century in Mesopotamia by E. E. Herzfeld and, in 1934-1935, by the American archaeologist H. S. Gladwin in Snaketown, Arizona (Gladwin et al. 1937:106). In the Khorezm Expedition, such work was widespread in 1952-1954, when, under the leadership of A. S. Kes, the study of archaeological-geomorphological excavations and prospecting test trenches from the earliest sites near the graveyard of Kokcha 3 graveyard and Bazar-kala began (Tolstov and Kes 1960:figs. 16, 63, 71, etc.). In the same year, in 1954, the archaeological-topographical unit excavated the Archaic Kangju-Kushan and Afrigid canals near Bazar-kala and Angka-kala (see p. 162). During the combined archaeological-geo-morphological research on the 'lands of ancient irrigation', the method of approximate dating of artificial systems and natural landscapes on both geomorphological sections and natural morphological features improved.

Interpreting Ancient Irrigation with Aerial Photographs

One of the most complicated and important tasks in mapping ancient irrigation features is the development of the best method of interpretation of vertical aerial photographs. Objectively they give a picture of the soil surface sector with its natural and cultural characteristic features. All these features are represented in aerial pictures two-dimensionally and, unlike the perspective representations, they have for us an unusual layout. Thus, the ability to read and understand photos as well as identifying archaeological site characteristics required special techniques and skills. As we know, the decoding of photographs is the process of defining both the contour of cultural sites and natural features according to their photographic representation and to their quantitative and qualitative characteristics, in order to reveal the attributes of a certain place (Gospodinov 1957, 1961; Bogomolov 1963; etc.). In the study of ancient irrigation, aerial photographs helped to solve the following issues: 1) identify the key elements of the natural landscape (ancient and modern hydrography, typical contour lines, sand massifs, etc.); 2) identify remains of ancient town ruins, settlements and single dwellings in connection with the topography of ancient irrigation networks; 3) the main parts of irrigation systems, their head and their terminal sectors; 4) areas of irrigation systems over different periods, etc.

The desert landscape of ancient alluvial plains, on which are ancient irrigation systems, is characterized by the association of wide *takyr* areas with low massive sand dunes. The terrain is marked by cliffs of Tertiary plateaus, depressions of dry lakes, hollows between sand dunes, dry paleo rivers and other different forms of sand dunes. The main visible feature of the contemporary and ancient hydrographical network is the distinctive representative outline. The ancient hydrography appears in aerial pictures in the shape of outstretched light bands, delineating old rivers and riverbeds, subsequently covered by *takyr*. They both run like a meandering strip, interrupted by sand sediments, and they also appear in the form of 'shifting fans', created by the river because of its gradual spread from the main bed.

The sand ridges (barchans, chains, sand dunes

of the plain, etc.) are marked on aerial pictures by a gray shade with a specific pattern. The *takyr* differ for their light patches, sharply bordered by the dark-gray shade of the surrounding sands. Here prevails a vegetation of Artemisia and *solonchak*-related plants and rarefied black *saxaul*, which replace bushes of white *saxaul* and desert sedge on the sands, and halophytes plants on the *solonchak*. Usually *biyurgun* (*Anabasis salsa*) borders areas of the *takyr*, on which developed only films of gray-greenish algae (Miroshnichenko 1960, 1961; Vinogradov 1962).

Geobotanists and hydrogeologists, who studied the ancient irrigated lands, noted that the consistent change of vegetation groups were the result of lowering groundwater levels, which in turn depended on the duration of the process of desertification. According to L. M. Parkhomenko (1949), desertification took place in the following stages: irrigated field - fallow with weeds - camel thorn - *uldruk* - saltwort - *takyr*. S. V. Viktorov wrote about the following stages in the Kunyadarya plain: 1) a complicated combination of bushes and weeds in areas of modern irrigation; 2) in the territories abandoned 100-150 years ago (for example, the environs of At-Krylgan) *itsitek* prevails; 3) in the fallows abandoned 300-500 years ago (area of Yarbekir-kala) *itsitek* is replaced by saltwort and black *saxaul*.[Note 18]

An attempt to link geobotanical and archaeological data was undertaken by M. G. Konobeeva, a soil scientist from Tashkent (Konobeeva 1965). However, she unfortunately did not use the detailed archaeological-geomorphological map of the Sarykamysh Delta published in the book *Nizovya Amu-Dari, Sarykamysh, Uzboy* (The Lower Amudarya, Sarykamysh, Uzboy, Tolstov and Kes 1960), and also some articles (including special soil-archaeological ones) concerning this subject (Andrianov 1954, 1955; Andrianov, Bazilevich and Rodin 1957; Rodin, 1961).

Based on the Khorezm map of archaeological sites elaborated in 1948 by S. P. Tolstov, M. G. Konobeeva advanced the following conclusions in Table 1.

These data, characterizing the succession

(change) of simultaneous vegetation, can be useful for interpreting aerial photographs of separate sectors of ancient irrigated soils.

In the interpretation of archaeological sites, it should be noted that their main discovery feature, whose origin is closely connected to ancient human settlements, is, with few exceptions, their regular and geometric configuration (Mikhaylov 1959:320; Gospodinov 1957:179). The majority of archaeological sites have a shape, a size and a color, that produce quite a definite pattern on aerial photographs. On the basis of characteristics revealed by photographs, the archaeological sites of the ancient irrigated land of Khorezm belong to three main groups: 1) areal (large settlements, cities, fields, vineyards, etc.); 2) linear (canals, roads, fences); 3) point or compact (single buildings, kurgans, towers).

The photographic image of ancient settlements and fortifications, irrigation systems, fields, gardens, caravan roads, etc., shows the main, or direct features (shape and size of the object, tone of brightness, texture, structure, etc.) as well as indirect features (shadows, connection with soil and land cover, etc.). The main task of the researcher-decipherer is the rational use of all these features for archaeological and topographical study of the objects from aerial photographs.

In decoding the sites with aerial photographs, we have to consider the tone of the photographic image, i.e., the degree of emulsion darkening on the photographic paper. The tone of the photographic image delivers an achromatic range of tones and the intensity of coloration of objects, i.e., the contrast of its image on the aerial photograph, depends on the brightness of the object, the light and the color sensitivity of the photographic material. Some ancient archaeological sites are distinguished on photographs only through tonal contrasts (because they are slightly elevated on the ground), or by the color of the soil and vegetation in case of highly eroded earthen works, remains of settlements , individual buildings, agro-irrigation layouts, ancient kurgans, etc.

For those sites characterized mainly by areal

indices (cities, settlements, fields), shadows in the photographs are of great importance (emphasizing the overall configuration of the external contours), the inner layout as well as the architecture of citadel buildings. The contrast of shadows in relation to the background exceeds the contrast of objects in relation to the same background. The contrast between shadow and background is sometimes the only feature enabling the identification of objects. For instance, clay walls or ramparts on clay-sand soils, are detected almost exclusively by their shadows. Shadows on the object and the shadow of the object on the surface surrounding it, in combination with light spots on the illuminated surfaces, give a fairly clear idea of its layout.

In the decoding from shadows it must be kept in mind that, at the edges of aerial photographs, high structures produce a perspective image, resulting in their upper part being slightly tilted relative to its base.

According to their contours, ancient cities and large settlements may have different configurations, e.g., round (Turpak-kala in Tashauz Province, Turkmen SSR, quadrangular (Kunya-Uaz), complex oval-shaped or elongated (Teke-sengir, etc. The walls and ditches of these settlements are usually detected by shadows, while the inner structures by a particular 'granular' pattern (white and dark spots of dwellings, hills, lines of streets, etc.). Vertical images (stereo pairs) of major archaeological sites were widely used by the Khorezm Expedition during architectural-topographical measurements as to provide accurate plans. They represent valuable data for several instrumental measurements in the laboratory (using stereo measurement instruments, stereometers, etc.). However, these issues are not discussed here.

Among the archaeological sites in the 'lands of ancient irrigation', characterized by linear features in aerial photographs, the most interesting for us is, obviously, the ancient irrigation network (Andrianov 1965). On aerial photographs it is possible to see what cannot be seen on the field, in an attempt to recover single irrigation system links buried under sand or vegetation, in order to reconstruct the ancient hydrography, which is reflected on photographs in the shape of elongated light and dark lines.

The main feature to detect irrigation canals on photographs is a typical contour line, which depends on the scale and time of the photography, the size of the canal and its preservation on the ground (which is in turn connected with the time of function and abandonment of the system). The ancient main canals are currently slightly elevated above the surrounding terrain, with their rounded or flat tops contoured with a broken chain of remaining bank walls. For instance, the ancient Kyrk-Kyz canal, near the walls of the Great Kyrk-Kyz fortress, is 40 m wide including its banks, 20 m between the ramparts and the banks up to 6 m high (Figure 3). The banks (A) are clearly detected by shadows under lateral illumination and pale patches of light. The small irrigation network (B) is badly preserved and can be detected by the color of the soil, rare bushes of *biyurgun* and sand drifts marked on the photograph by a dark line on the lighter *takyr*. In systematic aerial photographs, the canal is distinguished by double light and dark dashed lines. The canal is easily identified by its direct, and partially indirect features. Its absolute size can be fairly well established in the laboratory using a stereometer.

In most cases, the preservation of very ancient canals is poor and it is hard to identify them on photographs. It happens that, instead of a plain largely intersected by canals and ditches, it forms a massive alluvial crust, which does not allow the identification of the object of investigation, so that only the ground color and the presence of sparse shrubs of *biyurgun* attest the presence of an ancient irrigation system. The canal banks may be completely destroyed by the erosion process and dispersed. Thus, for example, in the vertical photograph (Figure 4) of the environs of Bazar-kala, through the pattern of mobile sand dunes, a series of parallel light and dark lines become visible (A), which, during the decoding of the photograph, were interpreted as a natural

formation (broken margin of *takyr*, etc.). Surface archaeological work (sections, test trenches, etc.) demonstrated that it was a main channel of the Antique period (6th–5th centuries BCE) badly destroyed by deflation processes. Only the central part of its bed is preserved, contoured by parallel structural terraces visible in the aerial photograph as a bundle of lines. The small irrigation network is not preserved.

Canals abandoned relatively recently are easily decoded. So, in the 19th century picture of the Turkmen irrigation in the Kunyadarya basin (see Figure 11), both the main distributors and details of the network of small ditches are easily visible. This photograph allowed to draw an accurate plan of the area.

Thanks to the apparent characteristic of the soil-vegetation cover, by means of indirect signs, aerial pictures clearly show forms and directions hardly visible on the ground (and sometimes simply invisible) such as small ditches, dams, field borders and other agro-irrigation layouts (vineyards, melon fields, etc.). The use of indirect features is based on the knowledge of laws regarding the interconnection of elements of the ancient cultural landscape of deserted 'lands of ancient irrigation'. Vegetation is an important feature in aerial photographic detection. It is widely used in several complex geographical investigations in particular in geobotanical-geological (geo-indicator), hydrogeological, soil, etc. studies (Viktorov 1955; Vinogradov 1961, etc.).

The chemical and organic composition of soil under ancient canals is different from that in plots of abandoned fields under *takyr* (Andrianov, Bazilevich and Rodin 1957:518). It is more favorable to shrubs, which usually expose clearly small irrigation networks also on zenithal aerial photograph. This was the essence of this method: comparing ongoing decipherment of aerial photographs with other aerial photographs deciphered *in situ*, whose content is reliable from both the archeological-topographical and the natural-geographical side.

Through decoding of photographs and drawing plans of irrigation systems, we have to force the interpretation, even when working in the field on the key (the most typical and best-studied) place of ancient irrigated lands. In a specific paper, dealing with the questions of identification of decoded signs (markers) from well-studied key sectors on an unattended territory of other geographical regions, B. V. Vinogradov noted that the range of identification depends on geographical and technical factors (Vinogradov 1962). He identified three types of extrapolation: 1) micro-extrapolation, conducted during the field study of aerial photographs at short distances (several kilometers); 2) mid-extrapolation, or the decoding of key-parts of a geographical unit (sector, location typology, landscape), it is the most rational and effective form of extrapolation for scales from 1:10,000 to 1:50,000; 3) large extrapolation (macro-extrapolation), where the range within which the markers can move is comprised between many hundreds and even thousands kilometers (i.e., the stony deserts of the Sahara and Gobi; see also AEE 1967:81-100).

B.V. Vinogradov correctly notes that, because of a full absence of studied markers and standard signs of identification in this or that landscape, on aerial pictures it is only possible to identify a point of view to "the edges of the topographically identified layout". This simplest form of extrapolation was widely used by us during field detection of the photomosaics of ancient irrigation systems, which enabled reducing the volume of fieldwork, limiting the study of the most complex parts of the irrigation systems, to some sectors of major archaeological sites, as well as archaeologically dating the whole length of irrigation systems, whose contours were clearly seen on photographs.

During archaeological and topographical research a middle quality extrapolation was used, mainly for those regions (areas) well studied by archaeologist. During this research aerial pictures were selected of some distinctive areas for both a given period and a given cultural region. Typical photographs of the 'lands of ancient irrigation' in the Lower Amudarya are shown hereafter (Figures 5-11).

Based on the example of Khorezm, it is possible to see how the configuration and the general character of irrigation systems changed over time: in the early stages of Prehistory, irrigation systems were closely connected with the ancient hydrographical network and they followed the riverbeds, repeating their shape. The lands abandoned by farmers in Prehistory appear today like desert areas, almost without traces of human activity. (Andrianov and Kes 1967:fig. 3). This is, in particular, the area in the lower lateral channels of the Akchadarya Delta, adjacent to the Kokcha 3 graveyard, where irrigation was based on narrow lateral flow channels of the delta, from which canals were branched. Very small fields and gardens encircled by low ridges were located in the immediate proximity of the canal (see Figure 27.B-C).

In those regions where ancient irrigators' activities were permanent and prolonged, the natural landscape aspect was transformed considerably. Thus, in 1964 starting from pictures and then opened with archaeological trenches, a vast area with remains of ancient riverbeds, canals, small irrigation networks and fields was discovered north and northeast of the Djanbas-kala fortress (see p. 146). Here, there were more than a dozen major sites of the Tazabagyab culture, dated to the third quarter of the 2nd millennium BCE. In the vertical aerial photographs, there are interweaving ancient deltaic channels; dry riverbeds are often crossed by small ditches, many of which originated in riverbeds or were their continuation (see Figure 5). Field work (surveys and archaeological trenches) have proven the anthropic character of ancient structures detected on photographs (Figure 28.C-D).

In aerial photos of the ancient landscape of Prehistoric Khorezm, it is characteristic to see the predominance of very few transformed, natural environments, the shape of intercrossing rivers (part of which were deepened and collapsed) and small-sized Prehistoric irrigation systems, with their peculiar layout.

The vertical aerial photographs of irrigation systems of different historic periods and cultural areas (in particular Khorezm) differ quite clearly from each other. Parts of the ancient hydrography, the main canals, and especially the small irrigation networks provide a very characteristic structure and topography in aerial photography (Figure 6). The 'lands of ancient irrigation' in the development period of the ancient Khorezmian state (Kangju and mainly Kushan periods of the Khorezmian history), the cultural landscape is characterized by large fortifications, towns and numerous rural villages bordered by fields and vineyards, with significant lengths up to 100-120 m and widths of 60-80 m, often divided into two rectangles, each subdivided in turn into *gryad* 4.3 m wide (see p. 164). An example of a typical rural settlement of the first centuries CE is the site near Djanbas-kala (Andrianov 1965:fig. 5). As can be seen in the photo (Figure 7) a major canal was already built in the Archaic period (7th-6th centuries BCE) and later rebuilt several times. In the picture two parallel beds (A) in the shape of dark lines (due to sand sediments and vegetation) are clearly visible. A small irrigation network (B) is barely seen. Traces of vineyards (C) are clearly distinguished by a typical 'striped' pattern. Extensive well planned garden-park compounds and vineyards are also characteristic of the environs of other ancient cities of Khorezm (Ayaz-kala, Kurgashin-kala, etc.; see also Tolstov 1962a:205:fig. 118).

The early and late feudal (Afrigid) irrigation systems differ from the systems of previous periods in the greater frequency of lateral branches and branching configurations, which are clearly visible on photomosaics. An example is the Afrigid 'oasis' of Berkut-kala where, together with remains of canals (in the plots uncultivated by the farmers), more than 100 fortified farms of the late Afrigid period (7th-8th centuries CE) are preserved (Andrianov 1959a; Nerazik 1966). Even more significant is the Dingildje 'oasis', where branches of the Afrigid system intersect and cut the Antique canals (Figure 8 and Figure 35). The narrow Afrigid canals are quite clearly detected from photographs. They form a sort of 'web' around the Afrigid farms, among which Dingildje occupies the central place.

In the Middle Ages, irrigation systems were improved and, as mentioned above, acquired a branched configuration. The land inside the irrigated water basins was used more rationally, thanks to the wide use of the water lifting wheel, or *chigir*, unknown in the Antique period (see p. 208).

The Medieval cultural landscape of the right bank of Khorezm is characterized, first of all, by a large number of ruins of farms and castles, towering among the dense and intricate network of irrigation canals and fields (Figure 9). Thus, in the 12th and 13th centuries the Kavat-kala Oasis (Gavkhore Basin), over an area of 14 km^2, more than 140 farms were recorded and studied for the first time (Andrianov 1959a:fig.1) by the Khorezm Expedition in 1937-1940 (Tolstov 1948a:155).

One of the characteristic features of the Medieval landscape of irrigated lands of both the right and left banks of Khorezm is the presence of large gardens, park layouts and vineyards in the environs of major castles and cities. Very typical is, for instance, the garden-park complex near the Shakh-Senem fortress (Andrianov 1965:fig.6). In this complex (dated to the 12th-13th centuries CE), the irrigation network forms a clear and geometrically regular layout, split crosswise by lines of alleys and encircled by a quadrangular enclosure, with garden pavilions in the corners and center (A). The main canals (D) are clearly distinguished by a dark double line (due to bank shadows) against the *takyr* light background.

To a somewhat later time (14th-16th centuries) are dated the garden-park layouts in the environs of Shekhrlik (Figure 10) and of Dev-kesken, formerly Vazir (Selyuzor), which was described by A. Jenkinson in the 16th century as the most spectacular site on the left bank of Khorezm. The rectangular layout of the park (with a reservoir in the center and cultivated square gardens) is located west of the Medieval city, over which the massive Antique walls and the picturesque, fortified citadel of Dev-kesken stands on the Ustyurt cliff (Orlov 1952b:161:fig. 7). The canal reaches the settlement from southwest.

The cultural landscapes of settlements abandoned by the Turkmens at Daryalyk and in the Karakalpak territories on the Janydarya in the 18th-19th centuries, differ from Medieval agricultural settlements (compare Figure 11 with Figures 9 and 10). The Turkmen irrigation systems of the 19th century were characterized by their large variety. They ranged from massive water lifting dams (for example, Egen-klych dams near Mashryk-Sengir), to head devices with semi-dams and branching, topographically included in the general system of *sengir*, to canals with reservoirs and ending with agro-irrigation layouts of different shapes and sizes (the *gryad* of gourd plantations are especially well preserved).

A slightly different topography of sites and irrigation systems was uncovered by B. I. Vaynberg in the At-Krylgan tract, in the environs of Mangyr-Chardere and Kattakar-Chardere (see Figure 11) and also in the Uaz tract (Vaynberg 1959, 1960). There, some large abandoned Medieval irrigation systems were restored by the Karakalpaks and the Uzbeks (Bregel 1961:61,191). The lands were distributed to the Turkmen servants; the feudal lords of Khiva established there farms and gardens.

The Karakalpak lands of irrigation at the Janydarya have a completely different cultural landscape. The irrigation systems, fed from the main riverbed of the Janydarya, differ in their complex and branched layout. Very characteristic of these lands are traces of the yurts among the ploughed fields, visible on photographs in the shape of light, with middle dark circles (Andrianov 1965:fig. 7A).

These are characteristics of some typical aerial photographs of ancient irrigated lands of different periods in the lower reaches of the Amudarya. Could they be widely used in the process of mapping other areas, for instance, such as the territory of the Syrdarya?

This question must be answered negatively. Each of the major historical-cultural regions requires its own system of typical photographs. Creating such a system in a new area is impossible without archaeological field research, without surveys and archaeological excavations, without the knowledge of both the general scheme of development of irrigation works in the region over time and the chronological classification of archaeological sites. However, the method of interpreting and comparing typical photographs simplifies and accelerates the process of mapping of a single region.

Chapter 2
Origin and Development of Irrigated Agriculture

Boris V. Andrianov

Agriculture appeared in Khorezm in ancient times. Here is what the famous 10th century traveler al-Makhdisi wrote, "They say that in ancient times the king of the Orient[Note 19] was angry with 400 males of his most faithful servants of his state and ordered to eject them 100 *farsakh* away from the last settled village and that village (where now a city lies) is called Kas".[Note 20] After a long time had passed, he sent people to inform him about them. When these people came to them, they found that they were alive, built their own huts, fished and ate fish, there was a lot of firewood. When they returned to the king and told him about it, he asked, "What do they call meat?" They answered "Khor" (or Khvar). He asked: "And firewood?" They answered "Razm". He said, "So, I confirm that place for them and give it the name of Khorazm (Khvarazm)."[Note 21]

One legend says that in ancient times, when the river (Amudarya) flowed to Balkan, the king of Khorezm won the right from the king of the Orient to divert the river flow in the direction of Khorezm "for one day and night". The naughty river overflowed and "started (to flow there) till now. They derived canals from it and built cities on it" (de Goeje 1906:285; MITT 1939:185-186).

This well-known legend reminds us, not only of the controversial issues on the ancient flow of the Amudarya,[Note 22] but also that the origin and development of irrigation in the 'Land of the sun' or 'Country of good enclosure' was just one element in the general historical process of spreading and development of agriculture. In the swampy deltaic plains of the Aral Sea area there were no wild ancestors of cultivated plants and the first farmers brought grain here as well as, perhaps, irrigation skills from other more ancient centers of irrigated agriculture (see Bartold 1965:163).

The experience of Central Asian people in irrigation and agriculture was rich and various. But, on what was this diversity based? How historically developed were these or those types of local forms of irrigation or rain-fed agriculture? What was the role played in these processes by the natural characteristics of Central Asia, with its sharp contrasts? Combining vast sand and clay deserts with sparse vegetation (where, from antiquity, hunting and pastoral breeding were developed), massive mountain systems with vertical changes of natural landscape (useful for herding and agriculture) and the fertile piedmonts and alluvial river plains, where agriculture was developed since ancient times? What is the role of the Central Asian types of agriculture in the general scheme of development of agricultural systems?

The research raised some questions far beyond the limits of the lower reaches of the Amudarya and the Syrdarya. In answering them, it is clear that there are a lot of gaps in the general history of agriculture as well as no precise scheme of historical development of irrigation techniques in different geographical zones and in different historical-cultural areas (Andrianov 1968a-b). Dealing with these issues, a synthesis of natural history combined with efforts from different branches of science is required. In particular, the archaeological research provided more new information from the field of ethnobotany and genetics, from the study of archaeological vegetal remains, to ethnography, linguistics and even the introduction of compared mythology.

Recently, abroad and here in the USSR, some general archaeological works became available. They concern the origin of irrigated agriculture, the historical process of complex economy production, as well as the absolute and relative chronologies connected to this process of archaeological cultures which have been published. It is necessary to mention the work of V. G. Childe, A. L. Perkins, R. J. Braidwood, the publication of the proceedings from international conferences of ethnographers and archaeologists in 1952 in New York City (Kroeber 1953), in 1955 in Princeton (Thomas 1956), in 1960 in Austria (Braidwood and Willey 1962) and also a new edition of *Chronology in Old World Archaeology* edited by R. W. Ehrich (1965).

In the USSR, an extremely valuable summary of results of archaeological research in the Near East and in Central and South Asia was published by V. M. Masson (1964). In this book the author outlined the main areas of ancient agricultural cultures and the major stages of development from the 10th to the 2nd millennia BCE. The book on ancient irrigation works in the Eneolithic Southern Turkmenistan by G. N. Lisitsyna (1965) should also be mentioned, her work is closely linked to the studies on ancient irrigation carried out by the Khorezm Expedition.

Dealing with the origin of agriculture (especially in the paleo-geographical reconstruction of an ancient natural environment and the absolute dating of archaeological sites), great help to archaeologists is provided by the latest achievements of natural and exact sciences (Heizer 1953; Brothwell and Higgs 1963; Butzer 1964; Kolchin, 1965; Titov 1965a-b; Ehrich 1965).[Editor 7] Recent radiocarbon dating of many sites of the Mesolithic, Neolithic and Bronze Age of Southwest Asia, Europe and America allowed to clarify the absolute chronology for the origin of agriculture (Jelinek 1962; Willis 1963; Libby 1963; Clark 1965; Titov 1965a-b; Serebryannyy 1965; Ehrich 1965).[Editor 8]

The origin and development of the ancient cradles of irrigated agriculture depended on natural resources connected with the spread of these or those earliest useful plants - in brief with the history of plant-growing. These issues have been studied by the well-known botanists: C. R. Darwin, A. P. de Candolle, G. J. Mendel, V. Hehn, G.I. Tanfilev, V. A. Komarov, N. I. Vavilov, E. V. Vulf, E.

Shiman and, finally, C. D. Darlington.[Note 23] Among the recent work in this field, we should mention the fundamental research by P. M. Jukovskiy *Kulturnye rasteniya i ikh sorod- ichi* (Cultivated plants and their relatives), a book by C. O. Sauer (1952), a collection edited by I. Hutchinson (1965), and a brief overview in D. R. Harris' article (Harris 1967). If in the 19th century the answer to these questions dealt mainly with botanical, ethnographical, and historical-philological material, the 20th century began with the accumulation of archaeological data.

Following the precepts of A. P. de Candolle, who preferred 'archaeological documents', the modern paleo-botanists, relying on the progress of archaeological science, were able to approach the question of the early domestication of the main cultivated plants in the ancient cradles of Southwestern Asia and Mesoamerica. The research by H. Helbaek, K. V. Flannery, H. Kihara, K. Yamashita, M. Tanaka and J. R. Harlan in the Old World, and R. S. MacNeish, F. Engel, E. W. Haury, H. Willy and P. Armillas in the New World should be highlighted.

The origin and development of irrigation skills are closely connected with the history of tools used in irrigation. Already in 1887, H. L. Roth outlined a scheme for soil digging implements (Roth 1887:128-130, 180). Many questions on the evolution of the digging stick, shovel and hoe are given in the works of V. G. Childe, B. Klìma, H. H. Coghlan, R. Braidwood, P. I. Boriskovskiy, S. A. Semenov, Yu. F. Novikov, G. Brunton, W. H. Holmes, A. Goodwin, B. Brentjes, and H. D. Sankalia.

Archaeologists and ethnographers have recorded the wide chronological and geographical spreading of stick-hoe agriculture from the Neolithic period of ancient Egypt (G. Brunton, G. Caton-Thompson, A. J. Arkell), and Palestine (A. J. Mallon, R. Koeppel, K. M. Kenyon, R. J. Braidwood), to the Eneolithic sites of Turkmenistan (I. N. Khlopin, V. I. Sarianidi) and India (H. D. Sankalia), to modern Melanesia (H. Damm, J. Nilles, etc.) and Africa (A. J. H. Goodwin, N. I. Vavilov). E. B. Tylor, and later P. Leser and A. Steensberg, developed the idea of the gradual transformation of the primitive hoe-spade, used for furrow irrigation, into the plough. The evolution of metal digging tools in Southwest Asia and Egypt is described in the classic work of W. M. F. Petrie (1917) and in the two-volume study by J. Deshayes (1960).

Concerning the characteristics of the origin of skills in irrigated agriculture, a great place has the ethnographic material of the New and Old World, characterizing 'harvest gatherers' and primitive forms of irrigation from temporary water sources, river floods and swamps. The vast literature includes general works by J. E. Lips, A. N. Maksimov, I. N. Klingen, R. Capot-Rey; works on specific countries and people by K. Bryan, E. F. Castetter, W. H. Bell, T. R. H. Owen, R. B. Serjeant, Yu. F. Novikov, Ya. G. Gulyamov, A. P. Okladnikov, O. M. Djumaev and others. Considerably interesting are also the ethnographic descriptions of the irrigation works of the American Indians of California, in particular the Paiutes, who were not farmers and did not cultivate plants (see the work by S. W. Hopkins, J. H. Steward, C. D. Forde, A. L. Kroeber, etc.).

An attempt at a broad analysis of the development of irrigation in the main areas of the ancient world was made by R. J. Forbes in a special section of his work (Forbes 1955, Vol. II), which contains information on the history of irrigation in ancient Egypt, Mesopotamia, China, India and some other countries. There, a considerable literature is also given. However, in this work, there is no material describing the history of irrigation in Central Asia, neither the Russian researchers such as V. V. Bartold, D. D. Bukinich, etc. were mentioned. The chapter is written mainly on historical and literary sources and does not deal with archaeological work in this region. Hydro-technical issues are not sufficiently treated. It must be said that, in spite of the great progress in classical archaeological study of the ancient Orient – Egypt, Mesopotamia, India and Iran – the information on the history of irrigation techniques is poor.

Attempts for an archaeological study of the ancient canals in Mesopotamia were made in the mid-19th century by F. Jones and in the early 20th

by W. Willcocks and E. E. Herzfeld. In recent years, the subject of irrigation development was repeatedly dealt with by the famous researchers on Mesopotamia, D. Mackey, R. McC. Adams, T. Jacobsen, A. Goetze, J. W. Gruber, but, in spite of that, irrigation techniques and the dynamic of irrigation systems in ancient Mesopotamia remain still unclear. These issues are not sufficiently covered, even in the most recent work by the well-known researcher on Mesopotamia, R. McC. Adams (1965, 1966).

In the monograph *Land Behind Baghdad*, R. McC. Adams outlined three main stages in the development of irrigation in the basin of the Diyala River: first the construction of small canals and use of natural channels in Prehistory; second, from the Neo Babylonian to the Sassanid periods (when large irrigation systems were constructed); third, the heyday of irrigation during the Arabian caliphate. The book summarizes the archaeological information collected by the author (with national economic purposes and on behalf of the Iraqi government) over a long-term project in Mesopotamia, particularly in 1957-1958, during the comprehensive study of the Diyala River basin. A preliminary report of this research provides some information on the ancient irrigation works in Mesopotamia (Jacobsen 1958:58-61).

A similar work was carried out in 1954-1955 by W. J. Van Liere and J. Lauffray in the Upper Jazirah in Syria. A detailed archaeological study of ancient irrigation was done by Richard LeB. Bowen in South Arabia (Bowen 1958). Among the ethnographical works describing the irrigation techniques in this region, an excellent article by R. B. Serjeant (1964) should be mentioned. The Medieval irrigation system of Iraq was described in detail by the Arabian historian Ahmed Susa, who published two volumes on the irrigation in the environs of Samarra during the Abbasid caliphate. There are many published works, mainly in Chinese, on the history of irrigation of China (Nesteruk 1955).

Among the recent foreign works concerning the history of irrigation in Central Asia, the article by R. A. Lewis, should be mentioned. It demonstrates a good knowledge by the author of Soviet archaeological research on the history of irrigation in Central Asia; a significant place in the article is given to the works of the Khorezm Archaeological-Ethnographic Expedition (Lewis 1966:480-486).

The beginning of irrigation and ancient cradles of cultivation

From ancient times until the end of the 19th century, in the historical and geographical literature prevailed the view that the human cultural and economic development occurred through three stages: 1) gathering and hunting; 2) nomadic pastoralism; 3) agriculture. However, the studies of E. Petri, E. Hahn and others proved the theory groundless (Kramer 1967). Generalizing L. H. Morgan's periodization of ancient human history, F. Engels outlined the following milestones in the history of economy: 1) period with a predominant appropriation of ready products ('savagery' according to L. H. Morgan's scheme); 2) period starting a productive economy, herding and agriculture; 3) period of development of methods of increased production of natural products by human activities and ending with the first great social division of labor and the decay of the Prehistoric system ('barbarism' according to L. H. Morgan's scheme); 4) the beginning of civilization is connected with the separation of handicrafts from agriculture, with increasing geographical and social divisions of labor, trade, goods production in cities, emergence of class society and states.[Note 24]

In which historical stage of Prehistory did irrigation appear? In the Soviet agricultural-economic literature, the hypothesis that irrigated agriculture was just improved and later modified from agriculture based on seasonal precipitation is firmly established (Vilyams 1951:347-354). Such a view was developed by C. O. Sauer in his book *Agricultural Origins and Dispersal* (Sauer 1952:21-28). According to V. G. Childe, since the Neolithic there were two forms of agriculture: rain-fed (dry) and irrigated (wet). He wrote: "Theoretically, of course, seasonal irrigated agriculture can be also ancient" (Childe 1952:198-199, 205). In his well-known work *New Light on the Most Ancient East*, V. G. Childe stated more clearly: "... at the base of the most developed civilizations, lay primarily irrigated agriculture. This should not necessarily mean that irrigated agriculture arose later than the hoe, or horticulture, agriculture. W. J. Perry and T. Cherry, for instance, had a completely different point of view" (Childe 1956:57; see also Forde 1963:424).

It is quite clear that the origin of irrigation skills was connected to the origin of agriculture. At the dawn of human history, man acted as part of nature, adapting to natural conditions, following it and using certain laws (seasonal changes of vegetation, movement of wild animals herds, flood watering of plots). The exploitation of nature for economic purposes was broadened very slowly with the accumulation of knowledge and the first attempts to convert it (primarily the emergence of cultivated flora and fauna) must not be considered as a one-time act. Much material has been produced testifying the long duration of these processes (Forde 1963:371-377; MacNeish 1965; Flannery 1965; Harris 1967).

The new human settlements in the late Paleolithic and Mesolithic led to the formation of local cultural communities conditioned, first of all, by different sources on the territories cultivated by man (Tolstov 1960:20). It was accompanied by the formation of ancient economic-cultural types: hunters and gatherers in tropical forests and temperate zones; nomadic hunters and gatherers in mountains and arid plains, semi-settled fishers on sea coasts and deltaic areas, hunters of grazing animals in the large steppe, etc. (Levin and Cheboksarov 1955; Andrianov 1968a-b).

In Mesolithic and Neolithic periods, different degrees of progressive historical population developments in different landscape areas began to emerge. A zone in which a productive economy developed based on plant-growing and breeding of domestic animals was of special distinction. The transition from hunting, fishing and gathering of plants to the regular cultivation of useful plants and breeding of domestic animals appeared, in the history of human culture, closely linked to the greatest progress called by V. G. Childe the 'Neolithic revolution' (Childe 1953:193; 1956:55; Lorenzo 1961).

Most of the researchers, from C. R. Darwin to N. I. Vavilov and P. M. Jukovskiy, had no doubt on the thesis that agriculture originated from gathering (see also Roth 1887:102-120; Narr 1956; P. Jukovskiy 1950:9; etc.). Even now, in the 20th century, not all plants used by man can be considered cultivated. Soviet geo-botanist E. V. Vulf, one of the leading expert on cultivated plants, divided them into four groups: 1) wild growing crops used in their wild state (gathering of roots, fruits, grains, stems, etc.); 2) 'cultivated' or slightly changed species; 3) cultivated species, not found in their wild state, but whose links with wild ones can be easily traced; 4) cultivated species, which have lost their links with wild ancestors and that die out if the fields were abandoned (corn, wheat, rye, melon, flax, etc.; Vulf, 1932:195-196).

Plant resources, which are the basis for the evolution of cultivated plants, were spread unequally among continents and countries. According to N. I. Vavilov, "the initial areas of speciation of the most important cultivated plants, as you can see, are very narrowly localized" (Vavilov 1967, Tom I:39). Based on the vast floral material collected from many countries around the world, N. I. Vavilov and his students developed a theory on the main cradles (centers) of origin of, vegetables and garden plants (P. Jukovskiy 1950:5-41; Sinskaya 1966:22-31).

N. I Vavilov proposed eight important centers of origin and setting of cultivated flora (Figure 12): 1) China; 2) Indo-Malaysia, actually India and Indo-Malaysia; 3) Central Asia; 4) Near East; 5) Mediterranean; 6) Ethiopia (Abyssinia); 7) Central America and South Mexico; 8) South America: Peru, Ecuador, Bolivia, Chile and Brazil-Paraguay (Vavilov 1967, Tom I:353-393; Sinskaya 1966:22).

In recent years, archaeological and paleo-ethnobotanic works have largely proved the existence of the independent centers of domestication of the main food and economic crops, in many ways coinciding with the centers proposed by N. I. Vavilov. The conclusion that agriculture in Ethiopia was very ancient has not yet been confirmed (Helbaek 1960a:117). These studies have outlined 'the path of grain' outside its original areas, revealing different aspects of plant adaptations, hybridization with weeds and modification of species under the influence of cultivation, planting and harvesting in different geographical areas (Harris 1967:92). If we assume from the current issue of 'N. I. Vavilov's centers', it is possible to consider the following geographical centers, or regions, of ancient plant-growing.

The Southwest Asian area consists of two groups or large centers: 1) Near East (Anatolia, Syria, Palestine, Iran) and Caucasus; 2) Central Asia and Northwest India (Figure 13). Southwest Asia, the oldest area of agriculture (8th-6th millennia BCE), gave rise to many species of wheat,[Note 25] rye, small-grained flax, small-grained peas, lentils, horse beans, grass pea, chickpea, some horticultural plants and Asian cotton (Vavilov 1967; Tom I:347; Harris 1967). In the same area, were also introduced eggplant, cucumber (Northern India), yellow carrot, garlic and spinach (mountainous Central Asia). It is the homeland of almost all European fruit species (P. Jukovskiy 1950:20-21). Barley was domesticated over wide area of the Southwestern Asian countries. Its wild species are known from Cyrenaica and Cyprus to Asia Minor, Southern Turkmenistan and Pamir (Harlan and Zohary 1966:1075).[Note 26]

The beginning of agriculture in the periphery of this area is dated, in Southern Turkmenia, to the 6th millennium, in Afghanistan to the 5th millennium and in the Caucasus to the 5th millennium BCE (V. Masson 1964; Dupree 1964; Narimanov 1966).

The Caucasus is the homeland of a series of wheat species (in particular some endemic, such as tetraploid and hexaploid), rye, fruit trees and grapes (Negrul 1938; Yakubtsiner 1956; Ketskhoveli 1964). The beginning of domestication of goat, sheep, cattle, one type of pig and camel is connected with Southwestern Asia (Reed 1959, 1960; Zeuner 1963; Flannery 1965).

The Mediterranean area includes countries along the Mediterranean Sea shore, the Nile Valley in its southeastern part, the Balkan Peninsula in the north. The newest archaeological work in Anatolia (Çatal Hüyük, Hacilar) and in the Balkans (Argissa, Neo Nikomedia) suggests a close connection between these regions.[Note 27] The Mediterranean area includes a comparatively small number of autochthons plants (Vavilov 1967; Tom 1:375-379). Here are concentrated the centers of domestication of the olive, of the durum wheat (Triticum durum), oat, large-seeded flax, large-grained pea, grass pea, fava bean, sugar beet and many vegetable plants and fruit trees (P. Jukovskiy 1950:33). Agriculture began here not earlier than the 6th-5th millennia BCE. This area is known as the place of the greatest 'river' and 'sea' ancient civilizations (Egypt, Aegean). Many cultivated plants of this region were carefully selected. Species of cereals, legumes and horticulture plants differ in the large size of their fruits.

South Asia consists of three independent centers: 1) India (with the richest cultivated flora); 2) China; 3) Islands (Sunda, Philippines, New Guinea, etc.).

The South Asian area is connected with the cultivation of rice, the most important culture in the world, still feeding half of humanity. Here it is possible to find the most different species of rice, from wild weeds growing in the fields to the cultivated species of great variety (Gushchin 1938; P. Jukovskiy 1950:130-131; Ding Ying). The cultivation of rice in this area was generally preceded by a developed tropical agriculture with root-crops (taro, etc.; Sauer 1952:25-28; Harris 1967:96).[Note 28] In Southeast Asia were located the centers of formation of naked oat, naked barley and millet; the cradle of rice, barley, millet; it is the homeland of soy, many cruciferous cultivated plants and some endemic species of fruit trees (Vavilov 1967; Tom I:360-368).

Mountainous China, besides oats, barley, millet (chumiza or Siberian millet), gave the world soy-beans, buckwheat, radish, tea and mulberry (P. Jukovskiy 1950:18).

C. O. Sauer linked the domestication of pig and many types of poultry to Southeast Asia (Sauer 1952:42, 84, 86).

The Ethiopian area developed later, despite the extreme varietal diversity of many cultivated plants and the presence of endemic species (teff, barley, wheat, pea, etc.). For example, in Ethiopia there is no wild barley or wheat, and certain local plants (teff and sunfleck) did not go beyond their homeland (Schiemann 1943; P. Jukovskiy 1950:30). The coffee tree, native of the Abyssinian plateau, was widespread only in the Middle Ages. This center is the place of origin of not only teff and sunfleck, but also of some species of banana and sorghum.

To the Ethiopian cradle is linked the mountainous Arabian area, characterized by fast-ripening cereals, cereals, legumes and lucerne (Sauer 1952:76-78).

New archaeological and paleo-ethnobotanic research in Equatorial Africa allow to consider this area as an independent center of agriculture. Although the assumption of G. P. Murdock on the origin of tropical agriculture along the Niger River in the 5th millennium BCE has not been proven, a number of species, including cereals (sorghum, fonio, etc.), vegetables and roots (Guinea yam, etc.) were domesticated in the equatorial zone, probably already in the 3rd-2nd millennia BCE (Harris 1967:97-99).

Central America is divided into three centers: 1) Southern American mountains; 2) Central Mexico; 3) West Indies islands.

From this center are corn, upland-cotton and other American long-staple cotton, several species of beans, some pumpkins, cocoa, in all probability the 'sweet potato' - batata, pepper and many fruits, like guava, and several kinds of sugar-apple and avocado (Vavilov 1965:167-168).

The area of the Andes combines different centers: the proper Andes (mountainous regions of Peru, Bolivia and Ecuador), Chile (Auracanìa) and Bogotà (Eastern Columbia). This is the center of the wild and cultivated potato, and different tuberous plants (oca, ulluco, anu etc.).

N. I. Vavilov noted that some important

plants were introduced to cultures also outside the above mentioned main centers (Vavilov 1965:168). These include, for example, the date palm, which became the most important crop of the Arabian herding tribes, Southern Meso-potamia and the Sahara. In South Africa, the wild watermelon was domesticated for the first time. Within the inner and tropical South America, the cultures of cassava, pineapples, groundnuts (peanuts) and, more recently, the rubber tree were introduced. North America gave the world artichokes and sunflowers.

The history of the development of irrigation skills is tightly connected with the origin of agriculture in the most ancient centers of arid zones, characterized by large thermal resources but with lack of precipitation (200–300 mm per year), where plants cultivation is almost impossible without artificially wetting the soil.

Arid zones occupy a vast area of the Earth (more than 1/5 of the total surface). In Asia, they cover Asia Minor, the Mediterranean countries, the Arabian Peninsula, the vast Iranian plateau, Afghanistan, the plains of Central Asia (Kyzylkum, Karakum and Ustyurt, etc.), the Central Asian deserts (Taklamakan, Gobi, etc.), placed behind the massive mountain barrier of the Tian Shan stretching to the middle flow of the Hwang Ho (Yellow River). In Africa these zones are situated in the deserts and semi-deserts of North Africa; in America on the coast of Peru, the dry uplands of the Andes and Mexico and in the desert and semidesert Southern USA.

The reader is probably interested to know whether these regions were always deserts and which role climate change played in the origin of irrigated agriculture. The interrelation of small cyclical climate changes with solar activity is now established. However, the work of serious biological and geographical analysis of small and larger periodical changes in climate, based on archaeological and palynological data, have just begun (Shnitnikov 1949, 1957, 1961; Whyte 1963; Schove 1965; Butzer 1964; etc.).

The followers of geographical determinism (from E. Huntington to J. H. Steward and P.

Baker), which defined the relationships between geographical environment and society as a single-line process of mankind's adaptation to natural conditions, have repeatedly suggested that agri-culture and sedentarization appeared in the Near East as a result of progressive drying of the Asian continent (see the critiques: Markov 1951a; Lisitsyna 1965; etc.).[Editor 9] V. G. Childe, who does not belong to the followers of such a geographical deterministic theory, however, in 1952 wrote: "The conditions of increasing drought, which we will briefly mention next, should have stimulated the transition to an economy of food resources production. Greater population concentration on the banks of streams and drying sources led to a more intensive search for new means of livelihood. Animals and people reached out to the oases, which were gradually separated from each other by wide strips of deserts..." (Childe 1956:57).

The defenders of that popular-geographical hypothesis can be found, unfortunately, also in our day. For instance, the geographer A. V. Shnitnikov (1957, 1961; etc.), who widely used archaeological data, and in particular the pub-lications of the Khorezm Expedition, to prove his theory on climatic fluctuations (cycles of 1800–1900 years) also published his views. It is possible that such big fluctuations really occurred, however, in some cases the scientific arguments of the author were objectionable.[Note 29]

Concerning the historical dynamic, A. V. Shnitnikov explains the expansion and contraction, in the 'lands of ancient irrigation', of the irrigated territory through the dynamic of water resources of the Aral Sea area. If, on the basis of a long-term archaeological research, S. P. Tolstov related the first significant reduction of Khorezm irrigation to a socio-economic crisis and the barbarian invasion of the mid-1st millennium. A. V. Shnitnikov (1961:46) connected it with the raising of 'lands of ancient irrigation' caused by the water level reduction in the Lower Amudarya which occurred in that period. In his discussion, the author does not prove the fateful dependence of human activity on natural phenomena, he just describes the results concerning the history of irrigation in Khorezm. It should be said that these results do not categorically allow a basis for such anti-historical conclusions, advanced from the perspective of geographical determinism.[Note 30]

An attempt to revive the hypothesis of a continuous desiccation of Central Asia was made by the geologist V. M. Sinitsyn in his work *Tektonichesky faktor v izmenenii klimata Tsentralnoy Azii* (The tectonic factor in the change of the climate of Central Asia). He refers to the downfall of ancient agricultural oases in Bactria and Sogdiana, "whose ruins are placed deeply in the desert among its silent sands" (Sinitsyn 1949). The 'drying' Asia, according to him, caused the "migration of peoples and invading conquerors rushing into Europe from Central Asia over different times". The geomorphologist P. S. Makeev explained the desolation of 'lands of ancient irrigation' in the lower reaches of the Amudarya and the Syrdarya merely as changes due to erosion by these rivers and the shifts of the main stream (Makeev 1952:559; see also Andrianov 1954; Andrianov and Kes 1967).

A broad overview on paleogeographical changes and cyclical climate fluctuations, based on archaeological material, was recently made by K. W. Butzer, who suggests that there is no basis for recognizing sharp climatic changes over the last 10,000 years in the upland areas of Southwestern Asia, where Mesolithic and Early Neolithic sites were discovered (Butzer 1964:416-437). Paleontological research in the Zagros Mountains on the Zeribar Lake revealed the following pattern: in the Mesolithic period (12,840±300 BCE) in a cold dry climate the association with steppe vegetation prevailed; after the warming, pistachio and oak appeared, then came the warm period (3500±120 BCE) marked by the predominance of the oak (Van Zeist and Wright 1963:65-67).

In arid zones the existence of human beings was linked to water resources, therefore, as a rule, settlements were concentrated in green oases along river valleys or on lake shores. Irrigation techniques and the type of irrigation devices depended on water resources. The latter, in

different geographical landscapes, enabled the following exploitations: 1) rivers with permanent stream flow (gravity flow, flood, irrigation with artificial water lifting); 2) lakes and rivers overflow (marsh and basin irrigation); 3) surface water formation after precipitation (*sai*-brook, *kaakovoe*), irrigation, etc.); 4) underground water from springs, wells, artificial water catchment structures (wells, *karez*, irrigation, etc.) (Capot-Rey 1958:278).

The Medieval geographers al-Tabari (9th–10th centuries), and later Hamdollah Mostofi Qazvini (14th century) drew attention to the different types of irrigation in Arabian countries: brook; river (through artificial gravity-fed canals, water-lifting devices and different reservoirs); *karez* (withdrawal of groundwater in galleries); wells (with water-lifting systems); (Petrushevskiy 1960:117). These irrigation types were noted later in Iran in the 17th century by J. Chardin, who wrote: "In Persia there are four types of water (*de quatre sortes d'eaux*): two surface (*sur terre*), i.e. water of rivers (*rivière*) and water of springs (*source*), and two underground, namely water of wells (*puits*) and water of underground channels (*conduits souterrains*), which they (the Persians) call *karez* (*kerises*)" (Petrushevskiy, 1960:117).

Even now, in the 20th century, when the impact of society on the geographical environment has increased immeasurably, the techniques and character of irrigation systems are still connected to natural conditions. Soviet irrigation specialists usually divide modern systems of irrigation into mountain, piedmont, and valley (Vyzgo 1947; Legostaev and Konkov 1953; Fedorov 1953).

The irrigation specialist L. V. Dunin-Barkovskiy, who investigated the geographical locations of different irrigation systems in Central Asia and the South Caucasus, established a classification and zoning in connection with physical and geographical conditions. For example, this classification of irrigated arid zones in the USSR is shown in Table 2 (Dunin-Barkovskiy 1960:25).

The diversity of physical and geographical conditions was reflected in the nature and rate of historical development of skills in irrigation and irrigated agriculture in different regions of the world.

A special role in the origin of skillful irrigation was displayed by mountainous regions. N. I. Vavilov wrote: "The curb of the great rivers, such as the Nile, Tigris, Euphrates and others, demanded a strong despotic organization, construction of dams, control of floods, as well as it required organized massive action, which could not have been dreamt by the Prehistoric farmers of North Africa and Southwest Asia. It is highly possible that the centers of variability of species and the cradles of primitive agriculture were mountainous regions, where the use of water for irrigation demanded no such efforts (emphasis mine). Mountainous streams can be easily directed by gravity to fields. The highland regions are often suitable for non-irrigated cultures because of the large amount of precipitation in mountainous areas" (Vavilov 1967, Tom I:171).

Irrigation in mountainous valleys, where seasonal streams and mudflows were used as water sources, were probably the oldest form of artificial soil watering. This so-called sai-brook (stream overflow) irrigation, in our opinion might have occurred in Mesolithic times among the 'harvesting people'. The term sobirateli urojaya ('harvesting people' in English and ernte volker in German) was suggested in 1926-1928 by the ethnographer J. E. Lips (1953, 1954). A. N. Maksimov also mentioned the wide spreading of 'harvesting people' before the time of 'farming people'. He noticed a great similarity in the edible plants processing methods among the Australian Aborigines and the California Indians of the Great Basin. According to him, "In the history of human culture, the different labor processes appeared in a completely different sequence if compared to the sequence in which we use them today. We first plough and sow and then reap, thresh, mill and bake; our ancestors first learned to harvest, thresh, mill and only later they learned to sow and treat the fields" (Maksimov 1929:31).

The problem of harvesting was solved, as we know, by the development of the microlithic industry, the invention of reaping knives and sickles with sharp cutting obsidian or flint blades. Simple soil-digging implements, like sticks and hoes, were used for digging up roots, for planting, for digging different irrigation ditches in order to irrigate the cultivated plots, as well as for working the soil (Boriskovskiy 1961; Klìma 1955) (Figure 14). The technique of working the soil is much older than agriculture.

The study of surface implement traces from Paleolithic campsites of Eliseevichi and Pushkari I allowed S. A. Semenov the opportunity to prove that digging sticks and hoes were used as means of excavation in that period (S. Semenov 1957). The simplest tool for working the soil is a sharpened stick 1.2-1.6 m long. It was used to loosen soil which was then removed by hand. This implement and this treatment method were recorded by ethnographers among many people such as the Australians, the Semangs, the Fuegians, etc. Hands might be supplemented with a wooden trough, by animals or baskets. Using such primitive means, the Kubu tribe in Sumatra dug pits for elephants and the American Indians of California built underground huts.

The stick was also the first implement of the 'harvesting people'. They noticed that cereals, edible roots and bulbs grew especially well on naturally irrigated plots. Noting this natural pattern in the local cyclical dry spells, the 'harvesting people' could arrive at the idea of artificially watering plots of wild useful plants, which was easy to do in small mountainous valleys using small dams. Mountain brooks were locked with tree trunks, bushes and earthen banks. Small ditches dug by sticks and hoes directed water to the plots of natural cereal thickets and other edible plants. To maintain the life-giving moisture such plots were encircled by ridges.

New World

This early, if I may say, 'pre-agricultural' stage of primitive irrigation skills was noted by ethnographers among some California Indians. The American ethnographer Julian H. Steward, working on the 'harvesters' in the Owens Valley in California, reported that the Northern Paiute tribes, occupying a territory of several hundred kilometers, even in the 19th century conducted a semi-nomadic way of life (Steward 1933; Hopkins 1883; Forde 1963:32-41). In winter, the whole Paiute tribe, numbering about a hundred people, lived in one or two settlements, but in spring they dispersed over the whole territory as separate groups. The women gathered different edible plants into baskets woven of twigs and grasses. Some groups of Paiute took advantage of the artificial watering of depressions, where thickets of bulbs and grasses grew more abundantly. In spring, when the snow-melt began, small streams ran down into depressions and watered the soil. The Indians directed these small streams and mudflows into natural depressions. Before the floods, on the valley bottom, they erected artificial barrier-dams made of bushes, stones and clay (Figure 15) and, after the flood, they directed the water to the fields with ditches 1.5-2 km long. Usually, these primitive waterworks were used briefly, because they were demolished by massive floods. The dam was rebuilt the next spring in the same place. The construction of dams and ditches involved the entire community. Paradoxically, the Paiute did not yet know cultivated plants and they were not farmers in the literal sense of the word, although they already knew some primitive methods of mountain-stream–irrigation (Forde 1963:35). Different retaining embankments and irrigation ditches were recorded also among the Australian Aborigines, who did not know farming (Campbell 1965:206-207)

Among the tribes of the New World, the path from harvesting methods to plant cultivation was very long (Roth 1887; Novikov 1959; Flannery 1965). Only recently, archaeological material on the early stages of plant-growing has begun to be accumulated. The American researchers R. S. MacNeish, E. W. Haury, G. R. Willey, and others used a comprehensive methodology combining archaeological research of ancient agricultural settlements of the New World with paleo-ethnobotanic material, providing a very reliable and absolute dating from radiocarbon

analyses (Braidwood and Willey 1962:84-105, 106-131, 165-176; Armillas 1962; Heizer 1965; MacNeish 1965; Harris 1967; Gulyaev 1996a-b-c). Laboratory analysis of plants remains, in combination with very thorough documentation of the stratigraphy of cultural layers, allowed R. S. MacNeish to ascertain the gradual increase in percentage of cultivated plants feeding the population in the Sierra de Tamaulipas, boosted from 10% in the period of the irregular cultivation of legumes, pumpkins, etc. to 65-85% in the period of developed irrigated agriculture of maize (MacNeish 1958, 1965).

In the Tehuacàn Valley the most ancient cultural Ajuereado layers (10,000-6500 BCE) characterized the population of this valley as nomadic hunters and 'harvesters'. In the next stage, El Riego (6500-4900 BCE), besides hunting, the basis of economy was the gathering of grasses, wild ancestors of maize and other plants. The stone tool inventory is various: different blades, scrapers, mortars and pestles. In the settlements, together with the remains of pumpkin seeds, pepper seeds, amaranth and several balls of wild cotton were discovered. At the same time, pumpkins were probably domesticated.

The stage of Coxcatlàn (4900-3500 BCE) was characterized by an increase in planting and the beginning of cultivation of butternut squash, amaranth, pepper, beans and maize. The latter provided about 10% of the diet. Its share increased to 25% in the Abejas period (3500-2300 BCE), with settled dwellings and maize becoming the main emerging crop in irrigated agriculture.

In the latest stages of the Tehuacàn Valley historical development, in the Purron (2300-1500 BCE) and Ajalpan (1500-900 BCE) phases, the emergence of fine monochromatic ceramics, sedentary settlements and the first attempts at irrigated agriculture occurred. In the last period, the Santa Maria phase (900-200 BCE), agricultural products provided 50% of the population's diet and some developed irrigated systems based on mountain streams were recorded.

Among the reports of a symposium held in 1936 on the primitive agriculture of pre-

Columbian America, two reports drew attention to the Salt and Gila river valleys ancient irrigation works (Halseth 1936; Haury 1936). The ancient canals and the settlements, connected to them, were traced for almost 200 km along the Salt River and 96 km along the Gila River (Patrick 1903; Turney 1929; Schroeder 1943; Shetrone 1945). The largest canal reached 9.1 m in width between banks and 3 m of depth. Some of them were 16 km long. From only one of these canals the remains of 22 sites, belonging to the Hohokam culture (Gladwin 1957), were found.

The canal in Snaketown, in its last stage, reached a width of 9-10 m (in the early stage it was 2.5-3 m) and, from the description and stratigraphic sections, it had several periods of life. These periods are connected to cultural layers chronology of the main settlements of Snaketown, where seven phases and four main periods were identified: 1) pioneer (beginning of CE-500 CE); 2) colonial (500-900 CE);3) sedentary (900-1150 CE); 4) classical (1150-1450 CE); (Gladwin 1948; Schroeder 1951). According to G. Hughes' opinion, who recently renewed the excavation of irrigation systems in Snaketown, E. W. Haury was allowed to date the appearance of the first canals to the 3rd-2nd centuries BCE. In his 1967 publication E. W. Haury reported the excavations of a 7th-8th centuries CE canal, 15 m wide and another 2.5 m wide. The archaeologists ascertained that these canals were dug using wooden tools (Haury 1967:683).

P. Armillas reported the existence of such systems in Mexico during the Middle Ages (Armillas, Palerm and Wolf 1956:396). Using aerial photographs and field work in the Teotihuacan Valley, he identified canals and an old dam built on a small stream not far from its outfall into the wider part of the valley. The size of the canal was 10-12 m. Based on bronze needle finds during excavations, R. Millon dated the system to the period before the Spanish conquest (Millon 1957:163). In this place there were preserved traces of a canal head parts in the shape of pit parallel rows, which were apparently strengthened by vertical poles and limestone blocks. P. Armillas

referred to the information in historical chronicles (Velàzquez 1945:45) which confirmed the past existence of such a head work in Mexico's mountain valley rivers. He also noted that contemporary Mexican Indians often used a similar technique, strengthening the canal heads and retaining dams with the help of vertical poles and brushwood placed diagonally to the stream flow.

Very primitive forms of temporary mountain streams were noted by the ethnographer K. Bryan among the Papago Indians tribe in Arizona (Castetter and Bell 1942:168), which, still in the 19th century, placed their fields on the alluvial fans in small dry valleys (*arroyos*). Earthen dams strengthened by stakes, low walls, barriers made of bushes, and small ditches retained water and directed it to the fields. The fields were enclosed by ridges. The Papago Indians knew and improved a method, called *balsa*, for regulating temporary flows. A more complicated variant of the *balsa* system was also used by the Hopi Indians in Arizona (Figure 16) (Bryan 1929:451; Titiev 1944:182). Material from Peru (Table 3) provide a no less vivid picture of the origin of agriculture and irrigation skills, where agriculture was preceded by coastal fishermen and gatherers (Towle 1961; Engel 1963).

The material culture of the so-called 'pre-cotton' period (4000-2500 BCE) was characterized by primitive stone tools, wicker baskets, nets and wooden items. Fishermen also gathered different wild edible plants: beans, peas and pumpkins (Jennings and Norbeck 1964). In this period beans and pumpkins were domesticated.

The 'pre-cotton' period was replaced by the so-called 'Huaca Prieta' period (2500-1200 BCE), when people were already sedentary and lived in fairly large villages and the basis of economy was fishing, gathering and cultivation of beans, pumpkin, pepper, peas and cotton. They did not yet know pottery, which appeared in the layers dated to 2300 BCE.

The last period of the origin of agriculture and the cultivation of maize was between 1200-750 BCE. Later, in the second half of the 1st millennium BCE, larger settlements appeared with big religious buildings. Irrigated agriculture became widely developed, and in the mountain valleys and in the piedmonts it acquired a terraced characteristic beginning the *sai*-stream form of irrigation (Kosok 1965:169-178).[Editor 10]

In the Central Andes there are many remains of ancient artificial terraces, aqueducts and canals, most of which, until recently, were attributed to the Incas (14th-16th centuries CE). Archaeological work of recent years show a more ancient origin of these irrigation facilities. For example, in Huancaco there are canals of the 8th-9th centuries carved into the rock that reached several meters in width. Numerous branches irrigated small plots (20-30 m^2) of terraced fields and then rejoined the main canal. Most of the fields were strengthened by stone walls. Later, they developed an ingenious systems of fields within hollows. Large irrigated plots, 150 × 350 m in size, were divided into rectangular 'cells' with an average size of 2.5 × 3.5 m and separated by walls 30-50 cm thick and height (DAT 1964:517).

Some ancient structures in the mountainous regions of Peru still surprise for the acumen and perfection of their hydraulic engineering (Shippee 1932). The melted glaciers water from the mountains was skillfully collected in streams, in two rows of basins placed as steps along the slopes (Figure 17). Basins were linked between them by an underground water pipe and stone aqueducts. The development of irrigated agriculture in Peru became the economic base formation of the American Indian civilization (Figure 18) (Mason 1957; Izumi and Sono 1963) and contributed to the emergence of large permanent settlements and cities. At the time of the Spanish conquest, the cultivated plants in Peru ranged from 65% to 85% in the population diet (MacNeish 1965:90).

The main agricultural implements in America at that period were digging sticks with a weight on the top and a footrest.[Note 31] The stone weights of different sizes and shapes of the ancient settlements of California and Northern Mexico were studied (Holmes 1910; Henshaw 1887). Ethnographic materials witness the wide-spread use of stick-hoe agriculture on light alluvial soil in the arid zones of both New and Old World where agriculture first developed and also in some other regions.[Note 32]

Old World

The transition to irrigated agriculture in the Old World - Eurasia is best evidenced by the archaeological research in the Zagros Mountains in Iran and Taurus Mountains in Anatolia, Syria and Palestine, where the researchers outlined chronological milestones: sites of Late Mesolithic - 11th-9th millennia BCE (Shanidar, layers B1-2; Kebaran culture in Palestine); Mesolithic settlements of 'harvesting people' (wild barley, wheat, etc.) - 10th-8th millennia BCE (Natufian in Palestine, Zawi-Chemi-Shanidar in Zagros); Pre-Pottery Neolithic cultures, hunters-gatherers culture with elements of plant-growing in irrigated plots and possible goats breeding - 8th-7th millennia BCE (Neolithic A in Jericho; pre-pottery layers in Ali-Kosh, etc.); agricultural sites of the 'Neolithic revolution' period, the development of irrigation, the emergence of a settled, farming based economy, and cattle breeding - 7th-6th millennia BCE (Muhammad Jafar phase, Ali Kosh settlement in Iran, layer VI in Çatal Hüyük in Anatolia, etc.); Eneolithic sites of the period of the first division of labor, specialization of agricultural and herding areas, irrigation progress in Southern Mesopotamia - 6th-5th millennia BCE (Eridu XIX-X, Hassuna I-V, Tell Es-Sawwan, etc.); emergence of city-states in the late 4th millennium BCE (Eridu IX-I; Ubaid 4, Uruk, Jemdet Nasr, etc.; Porada 1965; Flannery 1965; Masson 1964, 1966a; etc.).

The development of irrigation skills in the Old World, in the most ancient centers of irrigated agriculture, took place in two main areas: the mountainous valleys and the alluvial plains of river systems. The archaeological research of the last ten years evidenced the correctness of the main findings of D. D. Bukinich, who, in the 1920s, wrote that "the estuary method of irrigation was the prototype of all modern irrigation" (Bukinich 1924:110). However, there are very few archaeological findings tracing the entire improvement process of a 'single' system of irrigation, the invention of various hydraulic devices (head water-intake structures, main canals, water-regulating and water-lifting mechanisms, etc.). In each landscape-zone and in each major historical and cultural area, irrigation methods developed according to the local topographical and hydrographic conditions, to natural soil properties and cultivated agricultural plants (see Table 12).

According to K. V. Flannery, in Southwestern Asia, the earliest cradle of plant growing, the optimal conditions for the origin of agriculture existed in the 'strip' of oak and pistachio forests of the Zagros and Taurus at an altitude of 450-900 m asl, where the annual seasonal precipitation was 400 mm. There, at an altitude of 600-1200 m asl, wild species of wheat, barley and oats grew best of all (see Figure 13). Alfalfa grasses (lucerne), species of legumes with small-sized grain, milk-vetch (astragalus), etc. were widespread. The inhabitants of these areas gathered wild rye, goatgrass, wild flax, legumes (lentil, peas, small peas) and, at higher elevations, acorns, almonds, pistachios, as well as wild grapes, apples and apricots (Flannery 1965:1247-1251; Wright 1968:338).

The archaeological site of Shanidar cave, was discovered in this area, at an altitude of 822 m asl, which gave the most significant material for an absolute chronology for the transition period to a productive economy, and a chronology of the late anthropogenesis of the whole Near East (Garrod 1930).

Man appeared in the Shanidar cave approximately 80,000-100,000 years ago (see Solecki 1955a-b, 1963, etc.).[Editor 11] According to the dating, layer C (Baradostian), lying under the Mousterian one, was accumulated in the interval from 35,080±500 to 28,700±700 years ago. The Mesolithic anthropic layer of the Zarzian B2 dated back to 12,360±412 years ago (10,400±412 BCE). The Zarzian culture was characterized by a very developed chipping blade technique. The presence in these layers of pits for storing vegetable food indicates a higher stage of development of the harvesting economy if compared with the previous Baradostian culture. Above the Zarzian layer there was, without any interruption, the proto-Neolithic horizon B1 (8640±300 BCE). It was coeval with the base level of the nearby Zawi-Chemi-Shanidar site (dated to 8910±300 BCE). The beginning of the transition to a sedentary lifestyle and the increased development of intensive gathering belong to this period. R. S. Solecki supposed that the population of this region lived in caves during winter and in open-type neighboring settlements in summer. A possible analogue of the 'proto-Neolithic' culture of Shanidar was the Natufian culture of Palestine.

'Harvesting people'

The origin of agriculture in Palestine is usually associated to the Natufian culture. This culture is now represented by a significant number of sites (Garrod 1930, 1932, 1957; Garrod and Bates 1937; Kenyon 1960).[Editor 12] In her summarizing paper, D. A. E. Garrod wrote that the late Natufian people were the first farmers (Garrod 1957:216).[Editor 13] However, R. J. Braidwood questioned the agricultural character of their economy, although it is possible to consider this period as the probable "era of incipient agriculture" (Braidwood and Howe 1960:5, 182).

Next is the stage of the 'village-farming community' (see Kempton 1938:21).[Editor 14] Very interesting observations regarding the nature of Natufian farming were expressed by R. Amiran, which dealt with the problem of the origin of ceramic production in the Near East (Amiran 1966).[Editor 15] She demonstrated the connection between the skills of cooking grain dishes with the skills of ceramic production. The secret of cooking vegetable food by the Natufian 'harvesting people' was well revealed by the materials from the Mugharet el-Wad settlement (layer B). During the excavation they found a terrace with a threshing floor, part of which still preserved some grains (Garrod and Bates 1937:11-13). The winnowing of grains occurred on this floor (Amiran 1966: 242)[Editor 16] Beside the terrace, there was a bowl-shaped hollow and mortars where the grains were milled. A big basin-shaped pit was probably used to mix the rough flour or cereals with water. The dough was baked in a hearth, which represented the 'productive' sphere of the 'harvesting people': the inhabitants of the settlements.

Wild einkorn and barley were discovered in Syria at Tell Mureybet in the cultural layers of the 9th-8th millennia BCE (Van Zeist and Casparie 1968). Material characterizing the initial stages of cereal domestication in the 'harvesting people' period were obtained during the excavations at Beidha in Jordan (Kirkbride 1966). Here, the pre-pottery Neolithic is separated by the Natufian layers by an underlying thickness of sand strata. The Neolithic layers are dated to 6800±200 BCE (see Figure 13).

The excavations revealed unique stone dwellings with painted walls and numerous stone and bone implements. A very important discovery were the remains of two-rowed barley grains, almost indistinguishable in appearance from wild barley grains (*Hordeum spontaneous*) growing in the same area. Together with barley, the wild two-grained wheat (emmer) was used in the diet. An important role in the economy of the inhabitants of this settlements was the gathering of nuts, pistachios and different legumes, as well as goat breeding, and gazelles and mountain goats hunting.

Of great importance for characterizing the period of the origin of agriculture were the American excavations, directed by F. Hole, in the Deh Luran Valley in Southwestern Iran (Hole 1962, 1964; Hole, Flannery and Neely 1965; Mortensen 1964; Flannery 1965). The valley is located at an altitude of 200 m asl and today there is a dry climate (about 300 mm per year of precipitation) and it is irrigated by small rivers.

The archaeological research of the early farming settlements in the valley, Ali Kosh and Tepe Sabz, revealed the cultural layers of the 8th-6th millennia BCE. The first settlements in this region were probably established in 8000 BCE by goat hunters and possibly cattle herders, breeding goats and using good winter pastures on the valley slopes. They regularly harvested wild

legumes and cereals and had already started to experience cereal cultivation (emmer and two-rowed wheat), which they found in the Zagros Mountains (Table 4).

The excavation of open settlements, dated by the archaeologists to the Bus Mordeh phase (7000-6500 BCE), revealed small dwellings made of mud bricks. The floor of one of the rooms showed traces of a fireplace and a layer with carbonized wheat and barley seeds, rough unfired clay female figurines, as well as millstones and flint tools. The latter resembled the flint implements from Karim Shakhir. No ceramics were found. The economic basis of the inhabitants of these dwellings was hunting and regular harvesting of wild legumes. Traces of domesticated two-grained wheat and two-rowed barley should probably be seen as the beginning of farming. Several goat bones were the sure evidence of an early domestication stage, although there is no direct evidence in the osteological material.[Note 33]

In the following period, at Ali-Kosh tepe (6500-6000 BCE), farming was already playing a major role. Dwellings were built with large adobe bricks and the walls preserved traces of coating. The number of stone millstones finds increased. Flint and obsidian implements resembled those from Jarmo. Small objects made of local copper were found. Baskets and woven mats appeared. The inhabitants of the settlement cultivated wheat and barley and also raised goats, their bone-remains bear features of domestication. Together with goats, sheep were probably domesticated. This period is characterized by an intensive hunting of wild oxen and onagers.

During the Muhammad Jafar period (6000-5700 BCE), dwellings were built with adobe bricks on stone foundations; walls plastered with clay and painted. Coarse ceramics appeared. Remarkable is the predominance of grazing over farming, which was evidenced by an increase of goat and sheep bones in comparison to the previous period, as well as the increased harvesting of cereals and beans. The cultural layers of this period have well-known analogies with the upper layers of Jarmo, Tepe Sabz and Tepe Guran.

At Tepe Sabz (5200-4000 BCE) survived the tradition of building houses on a stone foundation, with walls made of adobe bricks and their subsequent coating. This settlement was characterized by the painted ceramic 'buff-ware' with typical Susian traditions of ornamentation; there are spiral-shaped basket. The economy was based on irrigated farming and the breeding of bovines and capriovines. Among cereals, many should be mentioned: naked and six-rowed barley, one-grained wheat (einkorn), two-grained wheat (emmer), lentil, peas and flax.

According to H. Helbaek, the research in the Deh Luran valley, and also the material from other sites in the plain (Tell Es-Sawwan and Tell Mureybet), forced to reconsider the traditional opinion that agriculture emerged in the 8th-7th millennia BCE in the mountainous valleys of the Zagros and 'came down' to the alluvial plains of the Tigris and the Euphrates only in the 5th millennium BCE, and, before that time, the population had a 'Mesolithic' lifestyle based mainly on hunting and fishing (Helbaek 1964b:47).

Following N. I. Vavilov, the Danish researcher connected the centers of early plant cultivation with the areas of wild-growing species. The selection process might have already started in the period of the 'harvesting people', who cut grain ears with flint implements. The process of harvesting contributed to the gradual consolidation of plants with a crumbly stamen. The first attempts to control the growth of useful plants were followed by the selection of the most useful types. In the original centers, where there were wild relatives, the cultured plants changed little, but, when the habitat conditions changed significantly (moisture, temperature, soil-salt, etc.), the preconditions for a substantial and irreversible change in plants were created. A great role was played by irrigation (Helbaek 1959a, 1959b:193, etc.).[Editor 17]

Beginning of Irrigated Agriculture

The excavations at Ali Kosh showed that in the plain regions with little seasonal precipitation, it was possible to ascertain the presence of a mixed economy, in which gathering was combined with elements of plant growing, already in the 7th-6th centuries BCE. Agriculture was very likely irrigated.

We cannot exclude the possibility that the inhabitants of Jarmo used estuary agriculture. This site is located on one of the spurs of the Kurdish plateau. It towers above the valley of the Tauq Chai, a tributary of the Tigris. The dry bed of a mountain stream, which was periodically filled with water, is located at the foot of the hill. The thickness of the settlement cultural layers reached 7 m, which indicates a stable settlement. The finds of wheat and barley confirm the farming nature of the economy, and numerous digging sticks stone tops indicate that hoe-farming had become a permanent occupation and was carried out in the light alluvial soils. The geomorphologic conditions of the environs of Jarmo allowed the population to use flood spills into the valley for crop irrigation.

The principle of basin irrigation and flood agriculture began, at the same time, in very different regions of the Near East. It has been ascertained that the inhabitants of Çatal Hüyük (late 7th-early 6th millennia BCE) practiced irrigated agriculture. H. Helbaek revealed that the soil-salt conditions at Çatal-Hüyük in the 7th-6th millennia BCE were approximately the same as today (Helbaek 1964a).[Editor 18] The climate was dry. Streams descending from the mountains surrounding the valley served as a source of irrigation. According to H. Helbaek, farming was based on natural overflow and had a basin character; wheat (three species), barley, peas, vetch, etc. were cultivated (Table 5).

In the Hassuna culture of Northern Mesopotamia both irrigated and rain-fed farming were widely used. According to H. Helbaek's data, at the early Hassuna settlement (6th millennium BCE) of Tell es-Sawwan, located on the left bank of the Tigris alluvial plain, farming was based on seasonal stream floods, which were retained by small primitive ridges (Helbaek 1964b:47).

H. Helbaek rightly concluded that this ancient

method of basin irrigation, based on gravity-fed canals, could be seen as the beginning of irrigation (see also Jacobsen 1958:70).

The development of soil-digging implements played an important role in the evolution of irrigation skills. Their historical development in areas with light alluvial soil was in line with the improving soil-digger, which, in different parts of the world, gradually turned into a wooden spade (known, for instance, already in European Neolithic sites) and into the complicated soil-digger with footrest (Peru, Oceania), with a hook-shaped finial and a handle (Peru), with a forked working edge (*lei* in ancient China) or with two teeth (*laya* in 19th century in Spain) (Figure 19).

An important improvement of the digging stick in the Neolithic was to weigh it with a drilled stone in the shape of a carefully crafted ring or ball (Semenov 1968:62, 348; Kosambi 1968:55) (Figure 20).

The discovery of similar stone rings (with a diameter up to 9-10 cm and the hole of 2.5-3 cm in) were made in the Eneolithic settlements of Southern Turkmenistan (Khlopin 1964:98-99; Sarianidi 1965:39, Pl. XXVI.16-17,22). According to H. D. Sankalia, in India such stone rings were used as weights for digging sticks and their finds are, apparently, the evidence of the existence of a primitive agriculture; they were also used as mace tops (Sankalia 1964:85-86). A number of stone drill tools was noted by archaeologists in other areas of the New and Old World, in particular in the Palestinian Eneolithic settlement of Taleilat Ghassul in the layers of the 4th millennium BCE (Mallon, Koeppel and Neuville 1934).

While on the alluvial plains (e.g., in Egypt, Mesopotamia and China) the development of agricultural soil-digging implements were wooden digging sticks, wooden hoes, and shovels, instead, on the rocky soils of the mountainous areas of the Near East, even in the very early stages, percussive implements were developed with stone tips, conventionally called 'adze' by archaeologists (Sankalia 1964:84-86, 102).[Note 34]

The increasing size of irrigation ditches, the complexity of irrigation techniques, and the appearance of hydraulic structures, such as dams and water-distributing devices, were accompanied by the development of soil-digging tools and their progressive specialization (Table 6). The earliest agricultural soil-digging tools of percussive type are considered the large hoes from Hassuna (an early agricultural site located south of Mosul) (Lloyd, Safar and Braidwood 1945). The earliest Hassuna farmers used adzes with stone blades and also unworked stone celts as hoe working edges. There are some well-known hoes, from the lower layers Ia and II, showing traces of bitumen (Figure 21.I-III). These implements, made of schist, quartzite and sandstone, are triangular in shape and with a semi-circular working edge. They served not only to rip the areas under cultivation, but rather for excavation work in dam construction, and for digging ditches, which are essential for irrigation. Although in Northern Iran there were 'rain-fed' crops, the irrigated fields on the alluvial soils provided the inhabitants of Hassuna much higher yields.

Implements very similar to Hassuna are those from Sialk I-II, Jemdet Nasr, Ubaid and Uruk; they are very close to the stone hoes from Hassuna (Ghirshman 1938:pl. VI; Christian 1940:taf. 53.11-13, taf. 134.2, etc.) (Figure 21.IV-VII).

The stone implements used in early irrigated agriculture, apparently combined the functions of hoe and shovel. Anyway, it is quite clearly written by A. Steensberg in his very interesting article (Steensberg 1964).

The invention of metal tools increased many times the productivity of farmers and irrigators. It is known that the copper axe was three times more efficient than a stone one, the knife between eight and 10 times and the hoe probably five to seven times.

Metal appeared at different times in the vast areas of ancient irrigation, but historically it was everywhere a unique process of transition to a more effective form of labor. The beginning of metal use is now dated to the 6th millennium BCE (Kuzmina 1965).[Note 35] The oldest metallurgic centers were closely connected to the oldest centers of irrigated farming. Particularly in these areas, there appeared both metal weapons and the economic need for more efficient tools than wooden and stone ones (Figure 22).

The intensive development of lands suitable for agriculture in Southwestern Asia, Caucasus, Iran, Northern India and Central Asia contributed to the gradual spread of metalworking skills over a vast territory and the formation of independent centers, as well as metal centers working on imported raw materials in those areas without ore deposits (Forbes 1964:17:Fig.5; 21:Fig.6; Chernykh 1965:96-110; Kuzmina 1965, 1966: 86-98).

In the oldest areas such as Palestine (Jericho) and Anatolia (Hacilar VI, Çatal Hüyük), the earliest centers are dated to the 6th millennium BCE. On the Iranian plateau, in Sialk, some cold-forged items were found in layer I3 (4800 BCE) and casting in Layer III4 (3250 BCE). Copper tinned items were recorded East in Quetta and Mundigak in the 3000 BCE layers; in particular at Mundigak, axes and hoes were found in layer III6. Hoe tips with identical shape are known from Susa (Deshayes 1960, Vol.I:233, Pls. XXX.5, 8, etc.; Lamberg-Karlovsky 1967:145). The development of irrigated agriculture in Southwestern Asia and adjacent areas was accompanied by specialization of metal soil-digging and soil-cultivating implements (see Table 6, Figure 22). W. M. F. Petrie, J. Deshayes and, before them, O. Montelius outlined the evolution of percussion instruments, as adzes and hoes, from blade arrowheads, imitating stone celt, to blade with hole and shaft for handle (see Petrie 1917:16-19, Pls. XV, XVI, XVII; Deshayes 1960, Vol. I: 243; Vol. II: Pls. III, IV, VII, XV-XII, XXX, XXXLVII, XLIX; 1963) (see Table 6, Figure 22).[Note 36] Bronze winged adzes or hoes with a hole and a marked pivot are known in the layers of the 4th millennium BCE in Susa C and Sialk III (Childe 1956:218). This type of instrument with a wide working-part became, probably, the prototype of the iron *ketmen*-shaped implements which later, in the 1st millennium BCE, were widely used throughout the area of irrigated agriculture.

Together with hoes, metal shovels were used in irrigation works in Mesopotamia and the Indus Valley already in the 3rd millennium BCE. A bronze miniature shovel of the mid-3rd millennium BCE is known from Susa and it is close in form to the shovel of Chanhu-Daro (Deshayes 1960, Vol. I:374). This large-sized shovel, 42 cm long, was probably a very useful tool for earthworks in the construction of dams in the Indus Valley during the Harappan civilization. These dams protected from the devastating floods of the Indus.

The massive production of metal implements and weapons was closely associated with the separation of handicrafts as a distinct branch of social production, which was identified by F. Engels as the second large social labor division. To this period belong such archaeological complexes and cultures as Ubaid 4 and early Uruk in Mesopotamia, late Gerzean period in Egypt, Susa A-B, Sialk III and Hissar I-III on the Iranian plateau, Quetta and Mundigak I-IV in Baluchistan, Namazga IV-V in Southern Turkmenistan, and Shang (Yin) in China (Masson 1966a:160).

Irrigation and Ancient Civilizations

The spread of bronze implements in Southwestern Asia in the 4th-3rd millennia BCE increased labor productivity in irrigated agriculture, therefore a further development of irrigation techniques, and a higher agricultural production allowed the increase of an already huge population of the oases, leading to the origin and growth of cities, handicrafts, expanding of trade exchanges, the emergence of new social structures and classes that were able to arrogate for themselves part of the farmers product. In the cities there was a concentration of people (priests, state officials, traders and craftsmen) not dealing directly with agricultural production. V. G. Childe defined this process as 'urban revolution'.[Note 37] According to him, it was accompanied by an increase in the size of settlements up to 'urban' proportions, concentration of wealth in the capital city, construction of monumental buildings, invention of writing, development of a fiscal system, development and improvement of handicrafts, development of exchange, origin and spread of widely traded luxury goods, society sharp-class stratification, social organization changes based on kinship ties, state organization based on territorial principles (Childe 1950; see also Masson 1966a; Kosambi 1968).

But, what is the role of irrigation in these historical processes?

There are different theories, according to which the origin of the Old World state formations in Egypt and Mesopotamia was caused only by the need to organize large irrigation works.

Even L. I. Mechnikov, in his book *Tsivilizatsiya v velikie istoricheskie reki* (Civilization along Great Historical Rivers), divided the whole human history into three major periods: river, sea and ocean. According to him, only the need for irrigation works led to the origin of the slave-based social structure and the despotic Egyptian state. The crucial significance of river systems for the development of civilizations and states in the ancient Orient was taken up, after L. I. Mechnikov, by several Western historians.

J. H. Steward, K. A. Wittfogel, and others supported the hypothesis of the 'hydraulic' path in the formation of civilizations in the Old and New Worlds (Mechnikov 1924; Steward 1953:321; Wittfogel 1956, 1957; Gray 1963:2-8).[Note 38]

Unlike K. A. Wittfogel and the supporters of the 'hydraulic theory', and, more precisely, unlike the 'institutional' approach towards the formation of a state power, the Marxist-Leninist conception comes from the fact that the state arose from splitting the society into classes as a product of irreconcilable class contradictions (see Lenin, Tom 39:72-73).[Editor 19] The state power was only an important prerequisite (condition), and not a consequence of the successful irrigation development. In a letter to K. H. Marx on June 6 1853, F. Engels wrote about the Orient: "The first condition of farming here is irrigation, but it is a matter for the community, or for the provincial or central government. Governments in the Orient have always had only three departments: Finance (plunder at home), War (plunder at home and abroad) and Public Works (provision

for reproduction) (Marx and Engel 1962, Tom 28:221)

The development of intensive and regular irrigated agriculture in significant areas, in the fertile valleys of the 'historical rivers' (ancient Egypt and Mesopotamia, a little later the Indus Basin and, later on, the basins of the Yellow River, the Amudarya, etc.) became possible only through the stabilization of large river channels and, later, the main riverbeds, flood control of seasonal inundations using big dams, and complex and powerful hydraulic devices. The implementation of these labor-intensive projects was possible only by combining the efforts of large numbers of people with large water basins, but, in the areas with lack of hands, only by the inflow of labor from outside. Under the historical conditions of that period, the conquered neighboring peoples were forced into it.

This simple division of labor promoted the development of a slave-owning mode of production, the main features of which were: social-collective character, simple cooperation, preservation of the community-based labor and periodical, more or less permanent employment of prisoners of war in irrigation and in communal work. The scope of slave work was limited because slavery and slave relationships were not significantly developed in the early class states of the ancient Orient (Dyakonov 1963:16ff.; Adams 1966:96-98, 102-104; etc.). It is true that the organization of irrigation works has been associated with centralized management and, especially important, to a certain system of coercion of large masses of people to carry out the heavy digging projects (construction of dams, digging canals, etc.). There are still little data for an exhaustive answer to the question of correlation between community and slave labor used for irrigation. But, according to their forms of coercion, the slave character of collective irrigation work is not in doubt (see pp. 249-251).

For each historical and cultural region of the ancient world each had characteristics feature of development and productive forces, including both population and production means, consisting of items (vegetation, soil, water resources etc.) and labor, i.e., production tools, by which man acts on nature in order to produce material goods. The nature of economic development, including both population and means of production, involved labor objects (soil, water resources, etc.), and labor means by which people influenced nature in order to produce material goods. These were particular to each historical period and cultural area in the ancient world. The skills in irrigation, as a water resource for agriculture, depended on both the local geographical conditions (especially hydrographical regimes of rivers, their water rate, time and character of floods, terrain slope, soil, vegetation, etc.) and on the characteristics of technical and socio-economic development of ancient societies.

Egypt

A primitive way of retaining natural floods by embanking cultivated fields, received a classical development in ancient Egypt, in the Nile Valley, which has been recently considered as the homeland of irrigated agriculture (Gompertz 1927; Perry 1924:30). Questions about the origin of agriculture in this region is still far from being resolved (Reder 1958, 1960; Kink 1964:16-34, 140-149; Arkell and Ucko 1965:155).

The earliest Neolithic sites of the transition period from hunters-gatherers to sedentary people (with sandstones, millstones, rings, tops of scepters and ceramics) were opened by A. J. Arkell (1953) in Nubia (Shaheinab, etc.). They are dated to 5350±350 BCE. In the pre-dynastic period before dynasties (approximately 6th-5th millennia BCE) the Nile widely overflowed and left its flood for many months on the steppe border, which grew abundant swamp vegetation, and where large herds of animals came to drink (Hurst 1954:42-47; Murray 1951). Already known archaeological sites of the Neolithic and Eneolithic, Northern Egypt of the 5th-4th millennia BCE (Fayyum, Merimde Beni-Salame in the western part of the delta, El Omari and Maadi in the eastern part, etc.) contain clear evidence of agriculture (remains of wheat, flint sickles, stones for grinding grains, clay granaries, etc.; Hayes 1964; Arkell and Ucko, 1965; Baumgartel 1955, 1960, 1965; Ehrich1965:1-46).[Editor 20]

In the Fayyum Oasis, the Neolithic sites were discovered by G. Caton-Thompson on the upland obliquely located over the lake (Caton-Thompson and Gardner 1934). The place was useful for both *kair* crops on the shore moistened by the lake overflow and fishing. The inhabitants of this area cultivated emmer and barley, which, according to H. Helbaek (1955), differed from their wild-growing ancestors (Braidwood and Howe 1960:111). In Kh. A. Kink's opinion (Kink 1964:144), barley was the predominant crop. The crops were reaped by composite tools with flint inserts and perhaps digging sticks were used, as there are disk-shaped stone tops and rounded pebbles with grooves to attach to a stick (Brunton 1937:XLII.16-21; Caton-Thompson and Gardner 1934, Vol. I:33; Vol. II:pls. XII.26, XXX.2-3).[Editor 21] Harvested grains were stored in special pits, lined with straw mats.

The inhabitants of the Fayyum Oasis were engaged in fishing (they wove cords of cultivated flax for fishing nets), hunting and possibly herding (Reed 1959:1637; Kink 1964:149-155).

A similar pattern of sedentary life with elements of farming, fishing, herding and hunting, can also be seen in another Neolithic site: Merimde Beni-Salame. The site is located in the Nile Delta, on the sandy strip about 2 km west of the Rosetta deltaic branch (Junker 1945). At Merimde, during the Neolithic, pear-shaped pommel stones were found,[Note 39] one of them, as Egyptian hieroglyph (Gardiner 1957; Sign-list:510 T3). D. G. Reder rightly connected the beginning of human settlements in the delta with the development of a primitive integrated agro-pastoral and fishing economy. He wrote that conditions for the emergence of irrigated agriculture were "initially more favorable in Lower Egypt than in Upper. Using water from individual branches was easier than adapting the large river flood along its entire length, from its first edge to the southern mouth of the delta. Irrigation works are likely to have begun in the north" (Reder 1960:177).

According to a legend, the beginning of artificial irrigation dates back to the mythical god of fertility, Osiris, who seemed to have settled in the Nile dams in the mountains of Ethiopia and prevented soil swamping in the valley (Diodorus I, 19, 5).[Editor 22] The most remarkable achievement of early Neolithic Egypt inhabitants was the discovery and development of skills in irrigated agriculture with river floods (Forbes 1955, Vol. II:22-30). According to Russian agronomist I. N. Klingen (1898:5), who studied Egyptian farming in the last century, "the progress of Egyptian agriculture still describes the progress of regulating the Nile water regime". I. N. Klingen outlined four stages of skill development of irrigation in the Nile valley: 1) marsh; 2) regulation of alluvial sediments (i.e. estuary 'single' irrigation); 3) irrigation (basin irrigation); 4) mountain and steppe-plain (with the development of water-lifting devices) (Klingen 1898:301).

Herodotus rightly called Egypt "the gift of the river" (Herodotus II, 5). During the annual overflow of the Nile, extended from the 20th of June to October, the soil of the valley and delta was not only watered but also fertilized by the new sediments of river silt (Barois:13-15). It has been calculated that the riverbed, together with the cultivated lands of Egypt, raised an average of 1 m over thousands of years (0.9-1.2 m) (Hurst 1954:43).

Flood water did not cover the whole valley and delta of the Nile, although overflow from the first cataract is about 15 m high and in the delta it is 6-8 m high. Thus the Egyptians distinguished the natural irrigation based on freshet water (single) from the artificial, regulated by hydraulic devices. Concerning the farmers of ancient Memphis, Herodotus said: "Now they really gather the fruits of the earth with less labor than other people or the rest of the Egyptians; they do not plow furrows, loosen soil with picks, or perform any other work on the arable fields, binding for every other people. The river floods and irrigates the fields by itself and, after watering, it returns into its banks; then everyone sows his field and drives pigs to it, they press seeds into the soil; then they

wait for harvest time, thresh seeds using pigs and thus receive bread" (Herodotus II, 14).

J. C. D. Clark dated the beginning of simple irrigation devices to the first pre-dynastic period of Egypt (Amratian), which, according to him, could be dated back to the period between the 5th-6th millennia BCE (Clark 1965), while E. J. Baumgartel, on the contrary, dated the spread of irrigation only to the second pre-dynastic period (Gerzean, 4th millennium BCE) (Baumgartel 1947:46) It is hard to say in which period emerged drainage canals, embankments and dams to regulate and protect the fields against floods, and irrigation canals, which gradually formed the basin irrigation system, which became the basis of agriculture and the development of urban civilization in Egypt (Kink 1964:142; Saveleva 1962:26-86).

The ancient Egyptian basin irrigation was linked to the seasonal rise of Nile water and the gravity flooding of irrigated fields (Barois 1904; OITD 1940:141-160). The main scheme of basin irrigation was to retain the Nile overflow through a system of longitudinal and transverse dikes (Klingen 1960,:208-218; Hurst 1954:47-54; Forbes 1955, Vol. II:3, 22-30). As the river flowed downstream its own sediments formed riverine levees. However, huge dams were necessary to regulate the water flow. The land along the river was divided into basins, with embankments or dams built along or close to the river, and the transverse levees stretched from the embankments to the desert margin. In ancient times, the area of a basin probably did not exceed 2,000 ha. The basins, in turn, were divided into small squares surrounded by low embankments.

During the rising level of the Nile, water was allowed to flood the land an average of 1-2 m, which ranged from 7,500 to 15,000 m³ of water. Water remained in the area from 40 to 60 days, till the silt was deposited and the soil was impregnated with moisture. The surplus of water was then discharged back. After the water receded, wheat, barley and other grains were sown into the liquid mud which then matured during winter.

After the harvest in March and April, the flooded soil lay fallow, warmed by the hot rays of the sun; it cracked and was ventilated. Salts on the surface were later washed away by the Nile floods and carried out to sea. All these elements ensured the exploitation of the same land for thousands of years.

In the period of lowest water (May and June) the population cleaned the canals, restored the dams and, to irrigate the upper fields, some strips of basins were built. Lateral canals brought water over long distances at higher levels. The chain of basins and lateral canals with branches formed a 'water province' (Barois 1904:11; Audebeau 1932).

Agriculture based on the basin system, gave the most benefit when the use of water across the river valley was merged into a single well-organized system (Avdiev 1934:70–83; Sholpo 1941:96). Thus economic interests contributed to the formation of a centralized despotic state with a developed apparatus of governance, which carried out not only the process, but also the control of irrigation works (Struve 1934, 1965a; Forbes 1955, Vol. II:23). Gravity-basin irrigated agriculture required large and heavy labor to build dams on the river and embankments, and for the water to reach the fields. Therefore, the main function of the state apparatus of ancient Egypt was to organize the forced labor (for irrigating and enhancing the agricultural expansion) in order to increase the share of the surplus product for the ruling elite.

The organization, on a large scale, of basin irrigation in Egypt is attributed to the legendary pharaoh of the first dynasty Menes (Meni), who built dams and a series of basins along the west bank of the Nile in Lower Egypt. But, apparently, it happened somewhat earlier. According to T. N. Saveleva, the appearance of pictograms in sites of the early kingdoms depicting the land divided into quadrangular plots (basins) proves that the basin system already existed in some parts of Egypt (Saveleva 1962:34). A basin (?) and the scene of the opening of a canal are depicted on the famous pear-shaped 'mace' of Scorpion, king of Hierakonpolis

(Forbes 1955, Vol. II:21:fig. 4; Saveleva 1962:32, fig. 2).[40] It is possible that its prototype in the pre-dynastic period was not only a fighting 'mace', a ritual object and a symbol of power, but also the stone top of a diggings stick.[41]

Africa is known for a number of petroglyphs, which depict women with diggings sticks topped with balls, several scenes of hunting and military battles featuring a man with a 'mace' and, finally, the ritual scenes where an important role is played by tops and the leader's rod with a top (Stow 1930:pls. 29, 35, 70; Aliman 1960:140) (Figure 23).

The 'mace' of the Scorpion king depicts the king himself opening a canal with a wooden hoe. [42] It depicted part of a basin, a river dike (that the king is going to open) and a palm-tree. Three men apparently strengthen the canal banks. According to T. N. Saveleva, the representation on the 'mace-head' of the king's figure and people on the canal bank resembles an archaic pictogram representing a human walking or swimming along a canal (Saveleva 1962:32). This sign is found in various forms in the titles of officials and rulers of nomes in sites of the Early Dynastic Period.

The entire collection of such pictograms was recently published by P. Kaplony (1963:N. 2, 24, 75, 92, 319). According to T. N. Saveleva, the majority are representations of officials connected with the organization of irrigation works. The various Egyptian titles such as 'chief of irrigation', 'watchers for the inundation administration', 'chief of the canal workmen', 'who opens the dams', 'inspector of the inundation', 'watchers of Nilometers', testify a great complexity, both of the whole irrigation system and its management (see Forbes 1955, Vol. II:24, 63). T. N. Saveleva provides data on 'a minister of the canal', 'ministers of the two canals' (of the two shrines) during the 1st–4th dynasties (Saveleva 1962:47).

V. V. Struve, V. I. Avdiev and N. A. Sholpo assumed that the position of the ruler of the nomes was originally connected with irrigation works and fishing (Struve 1934:37; Avdiev 1934:80; Sholpo 1941:83). One of the highest positions, 'chief of all the royal works', known from titles of the Old Kingdom, was also likely to be linked with irrigation management (Saveleva 1962:48), which was headed by the highest official (*vezir*). The most important task of this manager was monitoring the Nile water level. Diodorus (I, 36, 11) wrote that "worried about the level of the Nile, the king built a nilometer in Memphis" (see Breasted 1906, Vol. 1:90-167; OIT 1940:258-259; Popper 1951;Drioton 1953; Borchardt 1906, 1934).

The harvest in ancient Egypt was directly dependent on the level of flood waters, followed by the priests (measuring and observing the level of the Nile, linking it with the seasons changes and the calculations of time). The priests also predicted the time of river floods by considering the appearance of bright stars (helical rise of Sirius). In ancient times the Nile (Hapi) was called 'great stream', 'great river'; as well as the main and large canals were also called 'stream', and 'river' (*itrw*). The distributors, branching from the 'river', were called 'dug stream' (see OIT 1940:255). Once the flood waters reached the needed level, it passed through a 'gate' on the banks and remained on the fields at a height of 0.5-2m for 6-9 weeks (Figure 24). With the lowering of the river, the dams were closed. The protective dams (not less than 4 m high), on which roads connecting settlements were laid, were strengthened by vegetation, poles and reed mats. In May-June the cleaning of the large 'royal' canal and small irrigation systems took place. The basins were also cleaned from excess sediments, sometimes reaching 5,000-10,000 m³.

The construction of reservoirs was already mentioned in the annals of the 'Palermo Stone' of the 5th Dynasty (Old Kingdom). The inscriptions of the rulers of the nomes at the end of the Old Kingdom and the transition to the Middle Kingdom reported dug canals in Upper and Lower Egypt, the restoration of an abandoned network, etc. (Saveleva 1962:45). To that time, must probably be ascribed the invention of hydraulic head works - sluices, which were later described by Strabo (XVIII, I, 37).[Note 43] Strabo noted that sluices were similar to the Babylonian ones, but had wooden shutters.

The construction of the 'gate' to the mouth in the nome called Siut is described in the inscription of the ruler Kheti II (2125 BCE): "I brought a gift (of water) for this city ... I substituted a canal of ten cubits... I excavated for it upon the arable land. I equipped a gate for its (mouth ?) ... I supplied water in the highland district, I made a water supply for the city of Middle Egypt in the highlands which had not seen water... I made the elevated land a swamp. I caused the water of the Nile to flood over the ancient landmarks. Every neighbor was supplied with water and every citizen had Nile water to his heart's desire" (Breasted, Vol. 1:407 quoted in Forbes 1955, Vol. II:25).

The success of hydraulic engineering in the Old Kingdom can also be seen by the construction of navigation canals on the Nile at the first cataract. They were built during Mernere (2400 BCE) (Breasted 1906, Vol. 1:324; Saveleva 1962:45). In the Middle Kingdom, during Amenemhat III, the lands close to the Fayyum Lake were cultivated, a dam was erected and water directed into reservoirs for irrigating the environs of Memphis (Forbes 1955, Vol. II:26).

The Egyptian system of basin irrigation was highly efficient and productive in comparison with the older estuary ('marshy') one and it provided regular crops, as never before. It allowed taking a big step forward in all spheres of life, even using very primitive and mainly wooden tools (wooden composite hoe). This system promoted the formation and strengthening of the ancient Egyptian state in the 3rd millennium BCE. However, it required enormous labor costs from many millions of peasants, a vast army of workers. Still, Yu. P. Frantsev linked the term 'hoe' with 'servants', according to Yu. Ya. Perepelkin. V. V. Struve, E. V. Cherezov. A. M. Bakir and some others have considered those people as slaves. W. Helck and E. Edel supposed that they were serfs and I. A. Stuchevskiy supposed independent farm producers (Stuchevskiy 1966:63-75; Saveleva 1962:184). Convincingly, Yu. Ya. Perepelkin drew a picture of exploitation in the agricultural economic production of that time and the work of the employers who were controlled by overseers.

But, especially important for us, are the features of those ancient Egyptian people who suffered the duties of hydraulic work.

According to O. D. Berlev, the heavy construction and excavation work, in particular the huge irrigation works ('the king's works') for dam building and canal cleaning, were carried out by special groups of men called (*hesbu/hesebu*) in the Middle Kingdom sources. They were collected in special camps, (*stany*), from which the exit was not free. The majority of the (*hesbu/hesebu*) were 'slaves of the king' and they could work (free from digging and other work) for the rich officials and as slaves for part of the year (Berlev 1965:16).

In Egypt, the basin system was improved rather slowly. Apparently, there was a process to increase the area with regular artificial watering of the fields (especially in the delta and Fayyum), in order to preserve the area irrigated by natural flood waters. A noticeable progress of hydraulic engineering, especially water-lifting and water regulation, was observed in the Hellenistic period, which may be connected to the fourth period (according to I. N. Klingen) in the history of Egyptian irrigation.

By that time, the irrigation system became more complex. The sources referred to the different categories of canals: 'main canals' or 'canals', 'royal canals' (with the same function as the main canals), the lateral distribution canals (1st order), distributors (2nd order), irrigation canals and drainage canals to collect the excess or waste water. Along with the canals, a particular important role was played by the dams ('embankments', 'royal embankments') as well as the hub head works. To build dams, reed mats and woven bundles were used and some dams were strengthened with stones (Forbes 1955, Vol. II:28; Zelin 1960:213, 244, 382; etc.). The embankments were closed by various small dams and they had water distribution devices, whose sluices were made of wood. In this age of irrigation, the shaduf became widely used, it is a water-lifting wheel, 'Archimedes' screw', etc. (Diodorus I, 34, 2; Kryuger 1935; Boak 1926a-b; Yeivin 1930; Tenney 1936:7-25; etc.).[Editor 23]

While in the Middle Kingdom the irrigation duties for 'royal canals' and 'royal dams' were carried out forcibly by the adult male population of the oasis and largely by special working 'teams' (*hesbu/hesebu* for O. D. Berlev), in the Hellenistic period, there was a special labor conscription for each member of a rural community (adult male) who was obliged to work five days a year. A series of papyri containing data on the seasonal irrigation during low water were analyzed by scientists as W. L. Westermann, A. E. Boak and, later, O. M. Pearl (see Boak 1926b; Pearl 1950; Taubenschlag 1955). O.M. Pearl reports about the distribution of farmers for work on the 'six-gate sluice': the main water intake of the Fayyum canal (Pearl 1950:226-229).

The major reason for the evolution from the hardest 'royal works' of antiquity to the Hellenistic five-day duties seems to be found in the significantly increasing Egyptian population. The inhabitants of the Nile Valley in the period before agriculture were only a few tens of thousands (Ohlin 1965:9). During the Old Kingdom there were 3-6 million people; by the beginning of our era 5-9 million and, according to J. C. Russell, in the 4th-5th centuries it was reduced to 3 million people (Russell 1958:7-8, 78-80, 90-91; Beloch 1886:501-507; Ohlin 1965:9, 21-22). Only at the end of the 19th century the level of the Ptolemaic age was exceeded and the population reached 9.7 million people (in the same irrigated areas, approximately 2.4 million ha). The labor duties for cleaning canals and repairing dams was 40-60 days, without counting the time for road work (Barois 1904:108-112; Hurst 1954:58). At the beginning of the 19th century, when Muhammad Ali attempted to widely extend the areas with controlled irrigation, canal building employed 400,000 people permanently. W. Willcocks (1889:274) reported that, in some regions, the irrigation labor conscription reached 180 days per year.

Thus in Egypt the transition from primitive-single estuary irrigation to a regular flood basin irrigation, hanged not only the character of the economy but also the character of socio-economic relations, embodied in the ancient Egyptian state as an 'apparatus of violence'.[Note 44] The sharp increase in work volume and the intensification and labor seasonality in the period of low water contributed to change from a democratic character of irrigation duties (resulting from the community or tribal traditions of mutual aid and collective works); violent forms of coercion appeared.[Note 45] As mentioned above, later, in the Hellenistic period, the nature of irrigation duties changed, and some population groups were completely exempt from duties or had the opportunity to engage, in their stead, a stand-in from the population poorest (Taubenschlag 1955:618).[Note 46] Coercive irrigation duty (called by some 'corvèe') was abolished in Egypt in 1890s and replaced by the labor of salaried workers (Hurst 1954:58-59)

Mesopotamia

In ancient Mesopotamia, irrigation also began with an estuary or marsh single-stage. In the 5th-4th millennia BCE, in the valleys of the Tigris and Euphrates there was the second most important center of development of irrigated agriculture and urban civilization, which had a particularly great influence on the political, economic and cultural history of the ancient world. This country, called by the Greeks Mesopotamia ('between the rivers'), in the 6th millennium BCE had similar geographical conditions to the Nile and to the Lower Amudarya of the 3rd-2nd millennia BCE; it was primarily marshy, rich in lakes and alluvial plains (Lees and Falcon 1952; Willcocks 1903:20).

The Tigris and Euphrates rivers, like the Amudarya, carried a large amount of silt and their deltas grew at a rate of 2.5 km every 100 years (Table 7). The regime of the rivers favored agriculture, because the biggest water flow took place in summer, when farmers needed water most. The rivers often broke their banks, changing their beds and forming lateral channels.

In contrast to the gradual rising level of the Nile floods, sometimes the Tigris and the Euphrates (see Table 5) took the form of catastrophic and devastating floods and therefore, already in the early stages, the construction of protective dams and drainage systems was a priority (Willcocks 1903:20-22; Gruber 1948:72; Tyumenev 1956: 29-31; Wittfogel 1957:25). It is not by chance that in ancient Mesopotamia the myth of the flood killing mankind existed. Archaeologists discovered traces of large catastrophic floods that occurred in different periods and covered different parts of Mesopotamia.[Note 47]

River sediments, the processes of water-logging and salinization and, most important, the continuous agricultural activities over many millennia have hidden the early stages of irrigation in the Mesopotamian plain to the modern archaeologists. The first attempts for an archaeological study of the irrigation facilities in that area were made in the early 20th century by W. Willcocks and, later, by E. E. Herzfeld (Willcocks 1903; Herzfeld 1909:345). The excavations of the embankment of one of the canals showed that it existed for a thousand years. In recent years, the issue of irrigation development have repeatedly been faced by the well-known scholars of Mesopotamia: D. Mackay, R. McC. Adams, T. Jacobsen and A. Goetze (Mackay 1945; Jacobsen and Adams; Adams 1965, 1958; Jacobsen 1960; Goetze 1955).[Editor 24] However, the process of development techniques, the periods of irrigation technology formation and the formation time of ancient Mesopotamia irrigation systems are still largely unclear (Gruber 1948:72). The problem of irrigation in ancient Mesopotamia was doubly complicated. It was necessary to subject the fields to controlled flood irrigation and, at the same time, to protect the crops from devastating floods.

Strabo, referring to Polycletus of Larissa, reported that the beginning of irrigated agriculture in this region was connected with soil drainage. He wrote: "The fact is that the Euphrates, in early summer, overflows, and water increases in the spring during the snow melt in Armenia, for this reason the river inevitably forms marshes and floods the arable lands, overflowing from the river banks, if it is not drained on the surface using ditches and canals, just like they do with the waters of the Nile in Egypt. That is why the canals appeared" (Strabo XVI, I, 9).

The farmers who settled on the marshy river banks already had some experience in cereal cultivation and especially in irrigation. These skills could have been brought from eastern or northern regions, where mountain-brook farming (cultivation of wheat and barley) and estuary farming from river spills were already known, as mentioned above, at the contemporaries Hassuna and Tell es-Sawwan (Helbaek 1964b:47).

Between the 6th–5th millennia BCE, near the valleys of the Tigris and Euphrates rivers, there developed a peculiar culture of sedentary farmers (pastoralist and fishers) called by archaeologists Ubaid. Centers such as Eridu, Ur, Uruk and Telloh (Girsu), which at the time were villages, later (3rd millennium BCE) became major cities. The Ubaid culture, according to recent publications, chronologically continued over a long period of time, from 5300 to 3500 BCE (Porada 1965:149-152) (Table 8). At the end of the 4th millennium BCE this culture in Southern Mesopotamia followed the urban civilization of Uruk. Rivalry began between city-states, for example, the king of Uruk claimed to rule over many cities.

Many aspects of the material culture of the Ubaid period were studied in great detail, but despite this, we still know very little of the nature of irrigation works of this time.

The greatest achievement of the farmers of the Ubaid period was the establishment of a system of irrigated agriculture, adapted to the specific conditions of the delta. At the beginning there was, apparently, diking of small dumping channels and building of miniature 'basin' systems, of artificial estuary type, then the construction of primitive control flood areas and different water-regulating head units (Forbes 1955, Vol. II:16-22). The soft alluvial soil allowed the use of simple tools such as wooden hoes and shovels. However, at the basis of it all was the continuous labor of creating fields from the erratic marshes and wetlands.

The early stage of irrigated agriculture in Mesopotamia, before proper irrigation, i.e. the stage of the estuary-single irrigation, could not be far prolonged (see Table 8). Where did the first farming settlements appear? They were located at the confluence of channels into overflows, along the banks of riverbeds and lakes (Jacobsen 1958:60; Adams 1965:33-36) as in the outskirts of marshes and floods, where primitive measures were more easily implemented to retain flood water in plots diked by levees. These places were favorable for both agriculture and livestock breeding, fishing and hunting. The gradual movement of sedentary settlements and an increase in their number, in naturally irrigated spaces, forced the inhabitants to drain their lands (Goetze 1955).[Editor 25] It is possible to assume that, in a deltaic landscape, the development of irrigation techniques had to pass through the stage of controlling riverbeds, i.e. lateral river channels adapted with dikes, as in other deltaic regions such as the Amudarya and the Zeravshan (see pp. 131-133). This is proven by some indirect data. Thus, the ancient Sumerians natural streams and artificial canals had the same name *íd* (Jacobsen 1958:60).

T. Jacobsen and R. McC. Adams, who provided a great job on the study of the Sumerian and Babylonian written sources, studying the irrigation in the Diyala River basin, reported that originally the natural channels served as irrigation (Jacobsen 1958:60), reaching the width of a hundred meters or more. The lateral branches and ditches supplying water directly to fields were, in contrast, small, and did not exceed a width of 1-1.5 m between banks and a depth of 0.5-1 m. The combination of wide natural deltaic channels with small irrigation ditches is one of the characteristic features of the early stage of irrigation development in deltaic regions.[Note 48]

According to T. Jacobsen and R. McC. Adams, the main principles of gravity irrigation were discovered long before the Early Dynasty period, i.e. in the second half of the 4th millennium BCE (Jacobsen 1958:60). I. M. Dyakonov also dated a 'regulation' of the river to the end of 4th millennium BCE (Dyakonov 1959:157). But it is unlikely that the process of creating complex flood dikes along the rivers was a one-time act.

The Tigris and the Euphrates rivers had different flood water rises; so, the level of the Tigris raised before the Euphrates and lasted much longer, but the water intake from the Tigris was less suitable (Willcocks 1917)[Editor 26] and therefore the Euphrates served as the main water source. The Tigris water was used mainly on its left bank. The 'regulation' of the river apparently began in the lower regions, in the deltaic branches of these rivers, where the archaeologists found the most ancient centers of Sumerian civilization.

The continuous hydrographic changes demanded intensive labor to build protective dams and canals (Jacobsen and Adams 1958). The process of irrigation development in ancient Mesopotamia was impossible without the organized collaboration of a large number of people and without the association of individual communities into a kind of 'water alliances' (Avdiev 1934:73; Fish 1935:98; Ionides 1938; Tyumenev 1956:200; Dyakonov 1959:130). On the basis of these unions, the first small city-states of Sumer,[Note 49] such as Kish, Lagash, Uruk, etc., appeared. Their formation coincided with the beginning of dams and irrigation systems built in large deltaic channels. R. J. Braidwood (1952:39) dated the establishment of the Sumerian irrigation at the time of Late Uruk (3300-3100 BCE) and Jemdet Nasr (3100-2900 BCE) periods (Braidwood 1952:39).

D. Mackay assumed that most of the major Sumerian settlements were placed in the meandering channels (Mackay 1945), which were used for irrigation as well as transportation routes. Reclamation and drainage work played an important role. The experience in making drainage ditches, apparently, formed the basis for the establishment of irrigation systems (Gruber 1948:71). A number of drainage canals was created parallel to the river and served to prevent flooding (Willcocks 1903:20; Contenau 1954:41). Especially labor-intensive was the work of building dams and weirs, using primitive hoes, wooden shovels and baskets to take away the soil. The fertile silt was not a suitable material, so the large dikes and the banks of ditches were strengthened with layers of reed mats (Forbes 1955, Vol. II:18). Wooden hoes with a few teeth were used in irrigation work. This is recorded in the remarkable Sumerian literary sources of the 3rd millennium BCE, for example in the following verses of the *Disputes between Hoe and Plow* (Kramer 1965:95-96):

> Look here. Hoe, Hoe bearing knot, Hoe (from) a mulberry tree, whose teeth of cornel,
> Hoe (from) a tamarisk, whose teeth of the 'sea' tree,
> Hoe (with) two teeth, (with) four teeth, Hoe, son of a poor man, support of a man in rags ...
> Hoe challenges Plow.

Somewhat later Hoe lists what It can do and Plow cannot:

> I multiply, (but) what do you multiply? I extend, (but) what do you extend?
> When the water overflows (through eroded dam), you do not contain it,
> You do not fill baskets with mud, You do not fill them with clay, you do not make bricks...

In turn Plow says to Hoe:

> I am the Plow made by a mighty hand, assembled by a mighty hand.
> ...
> The king holds me by handle, Harnesses my bulls in the yoke, All know is next to me,
> All countries adore me,
> All the people are happy to look upon me, My presence among furrows is the decoration of the fields,
> Before the ears, which I cultivate in the fields.

This highly figurative 'debate' between the two most important agricultural implements suggests that during this period the plow only began to enter in the prosperous economy of most of the population of Sumer. From the text we have also seen that the hoe served as a universal implement. The hoe was both a proper hoe and plow for the 'man in rags'.

A. Steensberg proved that the majority of the large triangle-shaped basalt blades from the layer J6 of Hama in Syria (2300 BCE), should be seen as hoes-shovels for irrigating furrows (Figure 25.I-V). He sees in such a hoe (traction-spades) a famous prototype of plow implements, managed by two people, one of which pulls the tool by a rope. According to the data of A. Steensberg, until recently near Hama shovel-hoes were used with a thrust to lift the earth on the bank of irrigation canals during their cleaning.[Note 50] Such a shovel was called *mishiyya* by the local population.

In the 3rd-2nd millennia BCE, metal began to enter widely within the sphere of economic activities of ancient farmers. The capacity of metal to take any shape and to become hardened provided the opportunity to produce various metal implements. By this time, in irrigated agriculture, besides the wooden hoe and shovel and the stone hoe-shovel, the metal shovel existed for irrigating furrows.

The shovel played an important role in the economy of ancient Mesopotamia. Thus, A. Steensberg interpreted the 'divine rod', the arrow-shaped sign *marru*, as a picture of a triangle-shaped shovel. This sign was found on one of the triangular hoes from Hama dated to 2150 BCE (Steensberg 1964:116-119). The purpose of this hoe-shovel was to make furrows. In modern Iraq the spade is called *marr* and the digger *marrâr*. The Soviet Sumerologist A. A. Vayman considered this term Akkadian. According to oral information, the Sumerians called the spade *mar*. The corresponding pictograph character recalls a shovel with a rectangular rather than triangular working part. These images are very interesting as they give reason to assume that the older Sumerian shovel, with a rectangular blade, was replaced by the shovel with a triangular blade in the Akkadian period, when irrigation furrows appeared. Marduk, the Babylonian God of farming, "who always takes care of the furrows, keeps plowed fields, dams and canals in order" (Avdiev 1953:107), was depicted in the 13th-10th centuries BCE with a rod in the shape of a triangular shovel with a forked stick.[Note 51]

One of the Old Babylonian texts, dated to the first half of the 2nd millennium BCE, referred the promise to reimburse borrowed 'copper' and it added: "As soon as we make (of this bronze) hoes and shovels (spades), which are needed to work in the fields, we shall give back (refund) to you, what we owe you" (DAT 1963:139-140).

The ancient people of Mesopotamia, the Sumerians and Akkadians in the 4th-3rd millennia BCE, created a high culture which formed the basis for the later Babylonian culture. By the end of the Uruk period, writing appears on clay tablets and now these written sources enable us to gather some information on the development of irrigation techniques. A. Falkenstein states that a series of pictograms of the first half of the 3rd millennium BCE are connected with irrigation (Falkenstein 1936:34, N. 115, Signs 850 and 851). There are special signs for 'gardens' or rather canals. The channels and canals served as main roads in the delta. It is possible that, at the turn of the 4th and the 3rd millennia BCE, the methods of basin irrigation received a certain development. Thus, A. A. Vayman, a scholar of pictographic texts of the Uruk and Jemdet Nasr periods, reports about huge 'fields' that extend in the form of long strips a few tens of kilometers and 0.5-2 km wide (Vayman 1963:14-18). Perhaps, it is about this method of irrigation the later text of Hammurabi (1792-1750 BCE) refers to: "The flood is coming to us and there will be a lot of water. So, open the sluices in direction of the 'marsh' in order to fill up with water the field near Larsa" (DAT 1964:520).

However, the main direction of development of Mesopotamian irrigation was towards improving systems based on the main canals, which "became larger and larger till the entire surface of the country was covered with huge canals" (Willcocks 1903:20). From the archaic texts of the Uruk period, there are terms that indicate canals (*palgu, hiritu, pattu, atappu*, etc.) and it implies that "they have wide (channels), which flowed directly from the river and also this river has the same names as the channels; there were also narrower ones connected to the first: from the shipping canal to canals, to row furrows or simple irrigation ditches" (DAT 1963:216). In the Uruk period complex irrigation systems had already been created with regular water supplies at fixed rate, through a variety of devices, main canals, distributors and irrigators (Delattre 1888). There were also head-water regulating devices. A. I. Tyumenev, who studied the household economic documents of Uruk, Ur, Shuruppak (Tell Fara) and the temple of Bau in Lagash, wrote that at this time there already existed main canals 10 to 30 m wide between banks, which served both for irrigation and as waterways (Tyumenev 1956:195-196). The literature contains many examples of how the rulers of Sumerian city-states gave themselves the special credit for work on new irrigation systems and reservoirs (Tyumenev 1956:195-196; Barton 1929:14; Dyakonov 1959:79; DAT 1963:215-216).

I. M. Dyakonov rightly divided all the Sumerian lands into those naturally irrigated from rivers or canals, and those 'high', with artificial irrigation. He wrote: "And until now, the artificially irrigated lands are most exclusively on the middle course of the Euphrates and the Tigris, i.e. where only city-colonies (Mari, Ashur) emerged on the trade roads. In contrast, the lands in the lower reaches of the Euphrates, where there were the cities of Sippar (Tell Abu Habbah in the modern Babil Governorate, Iraq), Kish, Dilbat (modern Al-Qadisiyyah), Nippur, Adab, Shuruppak, Uruk, Ur, Larsa, in the lower Diyala Valley, where Eshnunna is placed, and in the very lower part of the Tigris old riverbed (Shatt al Hai, Al Muntafiq Province, Iraq), where Umma and Lagash are located, represent a continuous territory of natural irrigation" (Dyakonov 1959:86).

In this case, natural irrigation should be interpreted as gravity irrigation by flood canals, which was later called in Babylonian as *bīt mē* (Laessøe 1953:7). In the Old Babylonian texts there was a term for irrigation based on water drawing *eqel dilūtim*. In the Neo Babylonian texts, it is identical to the term *bīt dalū* (Laessøe, 1953:7). The Sumerians differentiated between the lands along the river, or *GAN-ID*, and the lands beyond the river, or *GAN-GA* (Forbes 1955, Vol. II:62: note 5).[Editor 27]

Part of the lands of Mesopotamia were situated above the river level of stream overflow already in the 3rd millennium BCE, therefore spouted vessels were developed as well as primitive manual water-lifting equipment for irrigating gardens, such as shaduf, which were frequently depicted on cylindrical seals.[Note 52] R. J. Forbes published a picture of Sargon's seal (2371-2316 BCE) showing several shaduf (Forbes 1955, Vol. II:10; DAT 1964:521).

The improvement and spread of metal and bronze implements in the 3rd millennium BCE, in particular shovels, in Mesopotamia allowed to accelerate the construction process of complicated systems of parallel levees, main and secondary canals, dams and reservoirs, where the water was collected during floods and used as needed throughout the season (Forbes 1955, Vol. II:17).[Editor 28] During the cleaning and digging of canals and irrigation furrows, combined tools (hoe-shovel) with a thrust for lifting began to be used. According to A. Steensberg and P. Leser, these implements became the prototype of the handheld wooden plough.

In a small essay on the history of irrigation of ancient Mesopotamia, R. J. Forbes outlined four main stages, considered as a unique period in relation to Hammurabi's administration, the third of these stages is connected with the political direction of Hammurabi (1792-1750 BCE; Forbes 1955, Vol. II:16-22).[Editor 29] The work of construction of irrigation systems continued without interruption for over nine years. In the famous Code of Hammurabi, the laws 53-56 impose a punishment for the negligent use of the irrigation networks, demanded the strengthening of dikes and dams and made obligatory the periodical cleaning of canals (KhDM 1950:156; Dyakonov 1952).

There were very interesting teachings and recommendations for a farmer in the cuneiform text, *Farmer's Calendar*, dated to the early 2nd millennium BCE (Kramer 1965:78-85), which suggests that irrigated agriculture had already turned into a very complex branch of the economy. Interesting remarks are from the Sumerian teacher of irrigation: "When you start to deal with your field (to proceed with its cultivation), be alert to sluices of dams, ditches and dikes (so that), when you water the field, the water level has not risen too high. When you drain the water from the fields, make sure that the fertility of the wet soil has been retained as you desire" (Kramer 1965:82).

From this it may be concluded that in Sumer and Babylonia in the 3rd and 2nd millennia BCE, the irrigation of fields with complex watering (flooding) prevailed. A cuneiform text at the Louvre Museum said: "the field is covered by water and I no longer see its surface" (DAT 1964:521).[Editor 30] There was also irrigation with furrows (Kramer 1965:96, 98).

The development of irrigation techniques in ancient Mesopotamia, the vast land management work, the construction of dams, a complex network of flood canals, aqueducts (Jacobsen and Lloyd 1936) and reservoirs to regulate the water flow contributed to the prosperity of handicrafts and the emergence of sciences (Gadd 1962; Contenau 1954; Kramer 1965:78-79). There were many cuneiform Sumerian-Babylonian texts, where in school tasks there were references to the data of canal sizes, the amount of land for constructing dams, etc. (Schneider 1931:n. 33; Unger 1935; Vayman 1961a:38, 1961b:240-241; etc.). Particularly interesting are the early data on the special rules for excavation work, which show a very rigid system of daily 'lessons' for working on irrigation canals and dams. Thanks to the courtesy of A. A. Vayman, who is engaged in the translation of mathematical texts, I am able to give some examples. Thus, for the construction of a canal 300 *GAR* long (1,800 m) and 3 *kùš* wide (3 cubits = 1.5 m), the standard of earthwork at the depth of the first cubit was 1/3 *sar* (6 m³) while the depth of the next two cubits was 1/6 *sar* (3 m³). In other examples, the 'task' for a day is again 3 m³. These were the standards to fill baskets with earth, etc.

According to I. M. Dyakonov, the period of Jemdet Nasr, on the eve of the formation of the Sumer statehood (end of the 4th millennium BCE), shows a picture of the rapid development in productive forces: 'mastering' the rivers, emergence of well-organized irrigation, use of wooden sickles and ploughs by using copper axes and shovels, etc. In this period a fundamental

social reorganization took place: the wealth of society increased and, at the same time, began the division of property and the exploitation of slave labor (Dyakonov 1959:157).

Since the Mesopotamian state was formed in 3000-2500 BCE, the system of forced irrigation work was also developed. Already in 1934, V. V. Struve wrote that in ancient Sumer the union of farming communities had stimulated interest in irrigation economy, imperatively demanding "the union of all members of the community for the successful management of the river. Together with slaves, the members of the community themselves took part in the work of digging canals. Also in the period of Urukagina, in the list of communities working on digging canals, warriors, scribes and priests were also included with the indication of their quota required" (Struve 1934:31).

Thus, during the reign of Urukagina (from 2319 BCE) and the ruler of Lagash, the irrigation economy demanded the full effort in intensive labor by the entire male population. Coercions and cruel 'lessons' were necessary. According to I. M. Dyakonov (1968:6, 18), in the 2nd millennium BCE the whole population of Babylonia was subjected to irrigation duties and "this conscription was communal and royal, as the water was communal and royal. Each family community and each adult male had to work up to two or more months a year. Irrigation activities such as the curb of the 'Biblical Flood' and the construction of massive embankment-dams, etc. were possible only through a large irrigation operation, achievable only by violent means, specifically because just the existence of a separate-property society demanded state violence" (Dyakonov 1963:30).

China

There are many similarities in the development of irrigation skills in Mesopotamia and China. The history of mastering water resources of huge river systems of China is the millennial struggle with devastation, entailing the hard work of millions of peasants to build massive protective dams and an extensive system of drainage and irrigation canals along rivers and lakes. However, irrigated agriculture along the banks of the great river systems of China developed between the 2nd-1st millennia BCE, later than in Mesopotamia and Egypt (late Bronze Age).

As in Southwestern Asia, the early Neolithic farmers of the great Chinese rivers basins were naturally preceded by Mesolithic hunters, fishers and gatherers. However, the archaeological sites of that time have not been discovered in the river valleys. Mesolithic and early Neolithic sites are known from more northern areas (Okladnikov 1966a) and from the Japanese islands.

But, where are the origins of the ancient farming culture of China?

On this matter, there are two points of view in the archaeological literature (Kryukov 1964). One is given by the first researcher of the Yangshao culture, J. G. Andersson (1923, 1943), and supported by some scientists who connect the origin of farming in China with the newcomers from the Caspian area: the bearers of the painted ware culture (see also L. S. Vasilev 1962; Ehrich 1965:507-509); the other view rejects the direct diffusion from the West, although it admits the possibility that some cereals appeared in China due to cultural contacts (Kryukov 1964; Laufer 1919). In his book *The Archaeology of Ancient China*, Kwang-Chih Chang rightly pointed out the complexity of the issue, the lack of data about the absolute age of the Neolithic cultures of China, the undoubted autochthonous origin of three species of millet, buckwheat, soybean, azuki beans, as well as domestic animals such as dog and pig (Chang 1963:53-57; Kryukov 1964; Vavilov 1967:353-359). In the Neolithic the Siberian millet characterized the cultivation in the fertile loess lands of the Huang He Valley, while rice was the basis of farming in the Yangtze Valley (Table 9).

Rice was cultivated before the appearance of irrigation skills in Southwestern Asia (Bakhteev 1960:43; Gushchin 1938). Its homeland are the mountainous slopes of the Himalayas, featuring a large amount of precipitation. Chinese researchers (Ding Ying 1958) supposed that cultivated rice appeared in Southern China, while Indian archaeologists concluded that rice replaced millet and became an important crop in the middle of the 2nd millennium BCE in Bengal and already penetrated from there into China (Das Gupta 1964:14). However, because of the questionable original cradle of rice domestication (there were probably a few) the question is still unsolved; the wide use of artificial irrigation for this crop belongs to a later period.

In the Neolithic period rain-fed and slash-and-burn agriculture prevailed (see Licent and Teilhard de Chardin 1925). In the Shang (Yin) period (Early Bronze Age) fertilizers begin to be applied together with weeding, tilling and some other agronomic activities that increased crop yields (Vasilev 1962:80-85). Worries about rain, judging by 'fortune-telling inscriptions', occupied an important place in the social life of that time. To bring rain, prayers were said and even human sacrifice practiced (Kryukov 1960). A warm climate with enough precipitation allowed the loess of fertile soil a relatively high yield even using primitive agricultural implements such as: digging sticks (*lei*); wooden ploughs (*sy*); reaping stone knives, etc. (Kryukov 1964:97-98).

In the Shang period urban civilizations arose, and the foundation of the state and the class society of ancient China were laid out. However, the role of irrigation works in these processes was not significant (Chang 1963:307). In the process of community unification and the formation of the first centers of urban civilization, a major role was played by the organization of collective work in the struggle against floods. Still in the 19th century, for example, the Huang Ho broke through a dam and flooded across the plain for thousands of square kilometers and killed about one million people (Popov 1925:31).

M. V. Kryukov rightly noted that "like a system of artificial irrigation in the Nile Valley, the struggle with the consequence of floods in the basin of the Hwang Ho was an essential and permanent conditions of social production" (Kryukov 1960:53).

A real opportunity for controlling the overflow of the massive river systems of China appeared only with the spread of bronze and especially iron tools (Chang 1963:198-202, 289). F. Ya. Nesteruk, referring to Chinese chronicles, wrote that, at the end of the 2nd millennium BCE, in Central China there were "virgin forests and marshes fed by rivers which widely overflowed during high water, forming vast lakes and locally also salty marshes, and only on elevated plateaus and mountainous south-facing slopes, meadows and steppes" (Nesteruk 1955:8).

The Chinese written sources (Shu Jing, Mencius, etc.) attributed the beginning of dam construction along the Hwang Ho and the drainage of wetlands to the legendary Yu (Nesteruk 1955), who supposedly said these words: "I opened the way for the rivers across the nine provinces and directed them to the sea and I also deepened the canals and sent them to rivers" (Giles 1927).[Editor 31]

The basic principles of hydraulic activity in the Shang period was limited to attempts at regulating the flow of the major rivers by draining canals and streams. Man learned from nature and followed it. Mencius said: "To regulate the river by the method of Yu, it means to use the river in accordance with its features; thus it gives the impression that Yu did not deal with the rivers but only provided an opportunity to properly develop the peculiar forces of channels" (quoted in Nesteruk 1955:12).

Significant advances in the development of hydraulic engineering took place in the West Zhou dynasty (11th-8th centuries BCE). From that time the *Zhouli* has survived, which describes all the provinces with their rivers and reservoirs (Nesteruk 1955:17). A special part of this book (*Kaogongji*) includes advice for the construction of dams, drainage of canals, distributing facilities, etc. Thus, for instance, "the thickness (width) of a dam on the bottom should be equal to its assigned height, but at its top (crest), it decreases to one-third. The slope of a dam should be designed away from the water (dry slope)" (Nesteruk 1955:18).

In the construction of canals, the Chinese hydraulic engineers skillfully employed the massive river flows (water washout technique) and the river sediments formed by the presence of protective dikes. They strengthened them with stones, trees and bundles of bamboo and brushwood (Von Li 1931; Lowdermilk and Wickes 1942). Rivers embanking was absolutely necessary for many rivers in China. During rainfall and snow melt at river head, water rose 10-15 m or more (on the Yangtze a flood was recorded up to 28 m) (Chu Shao-Tan 1953). In the lower reaches of the river, stream water could reach 5-10 m above the surrounding areas. The estuaries frequently moved. Thus, over two thousand years, the mouth of the Huang Ho at sea, has shifted almost 800 km from north to south.

Along with the creation of protective dams, which stretched hundreds of kilometers along river banks and reached 20 m in height, drainage hydraulic systems were constructed in the basins of rivers (Huang Ho, Yangtze, etc.). During the excavations of the Shang capital in Xiaotun, the drainage system of north-south canals was discovered, it was 40-70 cm in width, 126 cm in depth and a maximum length of 60 m. The main canals had smaller branches. Similar systems were also found in the early Shang settlement of Zhengzhou (Kryukov 1960:49-51; Chang 1963:172).

In the middle of the 1st millennium BCE, during the Eastern Zhou dynasty (722-481 BCE), lighter and more economic iron tools replaced bronze and wooden ones for farming (Chang 1963:197-198). The assortment of agricultural crops changed and broadened. Irrigated rice expanded and its yield was twice that of rain-fed rice. The gradual worsening of climate conditions, due to deforestation, contributed to the development of irrigation. Droughts became a common phenomenon.

The most ancient (mentioned in the chronicles) irrigation system was established in the Wei Bei area in 246 BCE (Nesteruk 1955:51). Using the main Zhengguo Canal, 173 km long (with many lateral branches), a territory of 162,000 ha (40,000 *qing*) was irrigated. A retaining dam 300 m long and 30 m high, was built for lifting water from the Jing River to the canal. The water intake facilities were rebuilt several times. In 211 BCE, in the basin of the Yangtze River, a fan-shaped irrigation system was completed, fed by water of the Min River. The basic idea was to accumulate the stream water overflow, using a series of distributing dikes, and to divert it into fan shaped canals and thence across the plain. The system irrigated about 200,000 ha (Nesteruk 1955:59-63). The biggest canal in China was the Dà Yùnhé, or Grand Canal, with a total length of 1,782 km and 60 to 300 m wide, which had mainly a transportation function. Its construction was completed in 1289.

On the loess plateau of China a terraced agricultural system was developed long ago. The fertile loess soil and the hot summers with a long growing season promoted the wide spread of many important crops (King 1911). Irrigation appeared there not earlier than the 3rd century BCE (Forbes 1955, Vol. II:5); this contributed to the invention of a number of water-lifting systems. The construction of the first hydraulic chain pump with square blades (fan che) was attributed to Bi Lan who died in 186 CE (Petek 1965:28). This pump consisted of a closed chain with fastened blades, which, passing along a vertical trough, lifted the water up to 5 m, using human or animal power. According to other sources, the invention of a primitive pump (ta che) was made in the Zhou period (Nesteruk 1955:13). However, gear wheels were mentioned for the first time in the Han dynasty sources (206-220 CE) (Petek 1965:28). Together with devices such as the norias, in the 1st millennium CE, large water-drawing wheels were used, made with bamboo and powered by water flow (Nesteruk 1955:14-16).

In some mountain regions of Xinjiang, the system of irrigation using karez (underground tunnels) raised underground water to the surface (Cressey 1958). F. Ya. Nesteruk cited data from Chinese sources about the appearance of 'the canal with wells' between the 3rd and the 2nd centuries BCE. In Xinjiang, all the oases (Handu, Karakhodja, Karys, etc.) still exist thanks to the ancient karez system. They were described in the late 80s of the 19th century by the Russian traveller G. E. Grum-Grjimaylo (1896).

Central Asia

Among the many historical and cultural regions of Eurasia, Central Asia is notable for the large variety and sharp contrasts of natural conditions. It combines vast sand and clay deserts with sparse vegetation (where pastoralism has long been extant); high mountains with a vertical change in natural landscape (used for both herding and farming); with fertile plains and river valleys which were the ancient centers of agriculture. The most favorable regions to develop skills in irrigated agriculture were: the foothills and isolated mountain valleys with small river alluvial terraces covered with green meadows, bordered by fading mudflows and by upper deltaic brooks; plains, especially deltaic areas with swamps, a labyrinth of lakes and streams that carried water to wet shores thickly overgrown with vegetation (Bukinich 1924:121; Gulyamov 1957:54; Andrianov 1961:141; Latynin 1962:23-26).

The investigations of S. P. Tolstov, M. E. Masson, A. P. Okladnikov, Ya. G. Gulyamov, V. M. Masson and other archaeologists[Note 53] proved that, since the Mesolithic, Central Asia was the area of contact of three main important economic and cultural zones: the zone of Neolithic settled farmers and pastoralist of Southwestern Asia (the northern periphery, which, in the 6th-3rd millennia BCE, was settled by the early farming cultures of Southern Turkmenia); the steppe zone of hunters, fishers and gatherers; the mountains area (Okladnikov 1966a:215-221; Coon 1957). Like other historical and cultural areas, in Central Asia Neolithic farmers and herders were preceded by Mesolithic hunters-gatherers (Okladnikov 1966b:73; Masson 1966b: 76).[Note 54]

The excavation of S. P. Tolstov in the Djanbas 4 encampment in the Aral Sea area became a classic site of the Neolithic Kelteminar culture, showing that the Kelteminar bearers, hunters and fishers, settled the shores of water bodies of the Akchadarya Delta, overgrown with reeds and tugai. In his detailed review of Neolithic sites of the Akchadarya Delta, A. V. Vinogradov (1968a:152) divided them into three groups: early (second half of the 4th millennium BCE); middle (first half-mid of the 3rd millennium BCE); late (second half of the 3rd-beginning of the 2nd millennia BCE). He distinguished the proper Kelteminar culture of the Akchadarya from the closely related Neolithic culture of the Lower Zeravshan and Lyavlyakan, and also from other local Neolithic cultures of the Caspian Sea area, the Lower Uzboy and Western Kazakhstan (Vinogradov 1968a:153-158; Vinogradov 1968b:12-13).

In the 6th-5th millennia BCE, in the mountainous regions of Southeastern Central Asia,[Note 55] the Neolithic period was characterized by the possible origin of a productive economy, which could be seen by examples of late Neolithic settlements of the Hissar culture (Tutkaul, Kuy-Bulen, Tepe-Gazien), discovered in the Hissar Valley on the hill slopes near mountainous brooks (Okladnikov 1959:176-184; Korobkova and Ranov 1968:18-21). The horizon of the Tutkaul II settlement was dated by 14C to 5150±140 BCE. Basically, it is possible that this culture economy was already herding. According to the level of technological development, A. P. Okladnikov and V.A. Ranov compared the inhabitants of Kuy-Bulen with the early farmers of Jarmo, and they presumed that there was a beginning of plant-growing (ITN 1963:91).

In the 6th-5th millennia BCE, in the piedmont zone of Southwestern Central Asia,[Note 56] Mesolithic hunters and gatherers were replaced by the so-called Djeytun culture (Kuftin 1956; Masson 1957a, 1961b, 1964:18-38; Ershov 1956; Berdyev 1966). This culture combined traditions of Mesolithic hunters and gatherers and new advanced features, reflecting a transition to settled agriculture (Masson 1962a:159).

The sites of the Djeytun culture occupied the northeastern border of the Midlle East vast territory, where the transition to agriculture and herding took place (Masson 1964:37-38). D. D. Bukinich, studying the piedmont of the Kopetdag in the 1920s, drew attention to the favorable conditions for the development of irrigated farming. He outlined the successive stages in the development of irrigation skills, which were connected to different natural regions (see also

Gulyamov 1957:54-65; Latynin 1962:23-26). These consist of mountain highlands with centers of wild growing grasses (barley), the edges of dampened mud flows and mountainous streams, the heads of deltaic brooks and, finally, the valleys and deltas of large rivers (Bukinich 1924:121; see also Sarianidi 1965:42).

The first crop in these regions were grown in piedmont strips, on the flat areas of flood dampened silt. D. D. Bukinich wrote: "The microrelief of these areas was so useful for making fields that a farmer did not need any kind of leveling or construction of any water facilities. It was very important to build a small dike along the margins of a field to retain water for a certain period" (Bukinich 1924:110). This very primitive method of irrigation, which consists in the artificial retention of flood water into diked areas, was needed for growing crops and it started the development of irrigation. The ancient farmers "could reduce all operations for cultivating useful plants to throwing seeds into the wet silt, without any plowing" (Bukinich 1924:113). D. D. Bukinich observed the Ustyurt Kazakhs who, in order to press the thrown wheat grains into the wet silt, drove herds of sheep in the fields several times.

The observation of D. D. Bukinich was largely confirmed by the archaeological study of the Djeytun culture sites as Djeytun, Chopan-Depe, Bami, Chagylly-Depe, etc. Numerous finds of typical harvesting knife blades (in Chopan-Depe a bone base was found), mortars and millstone and also seeds of dwarf beans and common wheat (Chagylly-Depe), prints of stems and barley seeds preserved in the clay coating floors (Djeytun, Chopan-Depe) indicate that the inhabitants of these settlements already knew farming very well. Most of the settlements were located on the delta plains of piedmont streams and *sai*, where rain-fed farming would have been impossible. Canals and fields in sites of the Djeytun culture have not yet been discovered.

In 1956 V. M. Masson and geologist L. G. Dobrin surveyed the environs of the Djeytun, where water of the Kara-su stream crossed a large sand dune, then watered a depression near the settlement, where irrigation was based on floods. The prospecting trench made by G. N. Lisitsyna in the area of a supposed ancient field disclosed buried soil apparently dated to the time of the Djeytun culture (Lisitsyna 1965:25). According to A. A. Marushchenko, the irrigation was not based on estuary farming with floods, as V. M. Masson and G. N. Lisitsyna thought, but on dammed valley-brooks (Masson 1964:20; Lisitsyna 1965:24). There were no clearly marked floods in the brooks of this area, and the farmers, locking the course of the brooks with dams, provided artificial overflows, which were probably encircled by embankments.

Developing the idea of D. D. Bukinich, V. M. Masson wrote that "in the primitive irrigation of the estuary type, it was sufficient to enclose certain plots with ridges, regulating the flow of water and to throw grass seeds into the wet mud" (Masson 1964:20; Lewis 1966:472). According to the paleogeographical research of G. N. Lisitsyna, in the piedmont zone of the Kopet Dag, there existed favorable conditions for the transition to a sedentary lifestyle and irrigated agriculture in the mountain streams and watercourses; on the mountain slopes grew juniper forests and green pastures (Lisitsyna 1965:89-90).

Another site of the Djeytun culture, Chopan-Depe, was located in the lower reaches of the Altyyab River, where traces of a former small riverbed were seen at the base of the site (Ershov 1956:13). Perhaps, there was a dam in the riverbed in order to water the sown plots (Masson 1964:21). In recent years, sites of the Djeytun Neolithic culture have been discovered in Southeastern Turkmenistan and also in deltaic sediments of mountainous brooks (Chagylly-Depe, Chakmakly-Depe) (Berdyev 1966:3).

The estuary and mountain-brook character of irrigated farming was typical not only of Neolithic Southern Turkmenistan, but also of the later Anau culture farming, which developed in small oases of the piedmont (Table 10). However, at the time of Namazga I (Anau I B), settlements were already located in the middle flow of streams

and small rivers, and the ancient inhabitants apparently knew well methods of flood control through retaining dams and small ditches (as was done under similar environmental conditions in Northern Mexico and in early farming settlements of the Hohokam Indians) (see p. 105).

In the period of Namazga I, the major farming-herding settlements developed over an area of more than 10 ha in size (Kara-Depe, Namazga-Depe, Ulug-Depe, etc.) (Khlopin 1964:97-98). To the material culture of that time, belong some well-known and preserved flint blades from reaping knives, millstones, mortars, pestles, grinding stones, as well as stone pommels (from Anau North), which served as weights or digging sticks. The main crops of that period were two-rowed barley and common wheat (Pumpelly 1908, Vol. II:471-473; Khlopin 1964:93; Sarianidi 1965:41).

In the late 5th-early 4th millennia BCE the zone of productive economy moved northeast (Adykov and Masson 1962:61; Khlopin 1968:34). The limited natural resources of the piedmont zone of Turkmenistan forced farmers to develop their economy on mountain streams and brooks and the already exploited principles of gravity irrigation (see Table 8), to seek out greater and more permanent irrigation sources (Sarianidi 1965:47). By that time, metal implements appeared in the Anau farming tribes, including axe-adzes that may have been used by the ancient irrigators in the late stages of Namazga I. At the beginning of the 4th millennium BCE, the farmers moved to the shores of the Tedjen River deltaic channels, where a group of sites was conventionally known as the Geoksyur Eneolithic Oasis. It was studied in 1956-1963 by the combined forces of 14th unit of the YuTAKE and the Karakum team of the Institute of Archaeology of the Academy of Sciences of the USSR. As a result of research and excavations, V. M. Masson, V. I. Sarianidi and I. N. Khlopin found the cultural and historical features of Eneolithic sites and developed their precise chronology. G. N. Lisitsyna did a paleogeographical study of the oasis and, through aerial photography,[Note 57] she discovered some unique sites of ancient irrigation works of the late 4th-early 3rd millennia BCE.

At present, all the 'oasis' is a *takyr* plain covered with dunes and sand ridges, cut across by traces of numerous shifting deltaic channels, the main one flowed from southeast to northwest, following the general terrain slope. The archaeologists were able to identify three groups of buried riverbeds corresponding to the main stages of the oasis existence (Yalangach, Geoksyur I and Geoksyur II stages) (Lisitsyna 1965:52). In the 4th-3rd millennia BCE this deltaic area was a green oasis with *tugai* forests, cane thickets along the riverbeds and overflows. Canes were widely used by ancient farmers for a variety of household needs, in particular for the manufacture of baskets. The latter were discovered during archaeological digs (Sarianidi 1961). According to G. N. Lisitsyna, the baskets could have been used to remove soil during the construction of canals (Lisitsyna 1965:101-106).

A small irrigation system, now only detected at low elevation, was discovered by G. N. Lisitsyna near the settlement of Geoksyur I. It consisted of two canals branching from the riverbed of the Geoksyur I phase almost at right angles. This system was traced on the terrain for more than 2.5 km. According to G. N. Lisitsyna, the oldest canal associated with the Yalangach phase of riverbed flooding (late 4th millennium BCE) was covered by some later canals. The trenches dug through the canals revealed a clear lens of aeolian sand deposit. Canal 1 was 3.47 m wide between ridges and 1.2 m deep. Canal 2 was 5.05 m wide and had a depth of 1.24 m below the ancient soil level (Lisitsyna 1965:116-118).

In 1964, a female statue of the goddess of fertility was found in one canal, not far from its head (Lisitsyna 1966:99-100). The slope of the system revealed a reduction of the canal bed with an average of 20-50 cm per 1 km. Lisitsyna successfully compared the results of her archaeological research on canals with Sumerian and Babylonian written sources. It appeared that the size of canals near Geoksyur I were very close to the irrigation canal described by A. A. Vayman

(1961a-b) based on cuneiform texts from Ur of the mid-18th century BCE.[Editor 32] This canal was fed by the Euphrates and was 4-6 m wide (two narrow and two broad parts) and 1 m deep.

According to the evaluation of G. N. Lisitsyna, the construction of a canal in the Geoksyur Oasis with a cross-section of 2.5 m^2 required 2,500 men, i.e. 100 men might have dug it in 25 days (Lisitsyna 1965:128-129). In 1964 the soil scientist N.G. Minashina, researching the canals of the Geoksyur Oasis, concluded that the functioning of these canals was very short. The canals were made after the delta dried up because of the shifting of the Tedjen channels. In her view, the process of salinization was probably aggravated by the general xerophytization of Southern Central Asia and the neighboring eastern countries (Whyte 1961). In the Yalangach period (end of the 4th millennium BCE), watering the territory was still significant, but by the beginning of the 3rd millennium BCE many eastern channels of the Tedjen Delta were filled with alluvial deposits. Several settlements were abandoned. Water was not sufficient and farmers began to build artificial reservoirs together with irrigation canals. One of the reservoirs studied by G. N. Lisitsyna, at the western edge of Mullali-Depe, was round in shape, 35 m in diameter and connected with a 'sleeve' to the riverbed of the Geoksyur II phase. According to G. N. Lisitsyna (1965:111), this reservoir was reused as early as the Yalangach period.

Of great interest in the work by G. N. Lisitsyna (1965:41-74), in addition to the descriptions of canals and reservoirs, is the material concerning small buried river beds (8 to 20 m wide), covered by later sediments and crossed by trenches. The riverbed of the Geoksyur I site had a width of more than 19 m and a depth of -2.65 m in its central part. In section an older (Yalangach phase) but bigger canal (Lisitsyna 1965:61-62) was clearly evident. The published profiles suggest frequent hydrographic channel changes, which gradually filled with silt and then faded out. However, the archaeological excavations revealed clear traces of their artificial embankment.

From the material published by G. N. Lisitsyna,

we concluded that the people of Geoksyur quite successfully applied the mountain-brook irrigation experience in the piedmont zone of the deltaic plain area, but failed to realize further possibilities for water regulation with diked channels. Without any embankments, the river moved northwest. The channels became silted and dried up and, in the mid-3rd millennium BCE, the Geoksyur Oasis was abandoned (Lisitsyna 1963; Sarianidi 1965:44).[Note 58]

Between the 3rd and the 2nd millennia BCE, bronze metallurgy skills started entering the steppes of Central Asia and Kazakhstan and independent centers appeared (Kiselev 1965:102; Kuzmina 1966:88-90). During this period, the Central Asian plains completed the process of internal development of the Neolithic tribes economy, which had borrowed herding skills (cattle and horse breeding) and farming, as well as the skills in metal working through cultural contacts with the agricultural civilizations of the Caspian Sea area, the Iranian plateau and Northern India (Tolstov and Itina 1960b:17-23; Kiselev 1965:53-57; Kuzmina 1966:89-90; Itina 1967:79). These processes were accompanied by migration of the steppe tribes. They contributed to the emergence of skills in irrigation on the Murgab, Zeravshan, Fergana and Amudarya Valley, where, in general, the favorable conditions of the moist delta developed preconditions for *kair* and estuary agriculture, followed by artificial irrigation based on dampened regulated riverbeds and derived from smaller canals.

The primitive rain-fed agriculture (the main economic activity together with herding) appeared at this time also in neighboring steppe areas of Northern Kazakhstan and the Ural region. In 1951 A. A. Formozov, and later V. M. Masson, suggested that wheat and domestic animals appeared in these northern regions under the influence of ancient farming tribes, bearers of the painted ceramic culture (Formozov 1951:5; Masson 1959:114).

According to S. V. Kiselev (1951:99-102), farming played an important role in the economy of the Andronovo pastoral tribes and he linked their origin with the transition to farming.[Note 59] In his article published posthumously on the cultures of the Bronze Age, S. V. Kiselev (1965:34) noted the farming character of the early Srubnaya culture (late 3rd-early 2nd millennia BCE), whose economic activities were linked to the river, rich with floodplains, good wetlands, lush meadows and forests with abundant wild animals and birds. According to K. V. Salnikov, the Andronovo tribes of the Southern Urals in the mid-2nd millennium BCE already represented a population completely devoted to herding and farming (Salnikov 1951a, 1951b:124-126, 1965:23). Their settlements were usually placed on the margin of a river terrace or in the modern floodplain and their economy was based on herding and floodplain hoe farming.

Between the cultures of the Eurasian steppes, and the ancient centers of settled farming in the South, there were several connections. Even S. P. Tolstov, in publishing the material of Djanbas 4 Neolithic site, noted the strong influence of the southern cultures of the early Anau of the Kelteminar culture (Tolstov 1948a:66; Vinogradov 1968a:170-171). This was confirmed by further studies on the Kelteminar culture provided by A. V. Vinogradov (1957a, 1968:135-152; Marushchenko 1957:7; Masson 1964:178).

The connection of the steppe cultures with the southern tribes of settled farmers of later periods (2nd-early 3rd millennia BCE) was traced in Turkmenistan,[Note 60] where the steppe type pottery of the Bronze Age was found not only in the vast areas of the Karakum and the Uzboy (Tolstov 1952b:4-6, 1958:52-57; Itina 1959c:259), but also in the layers of the 2nd millennium BCE in sedentary agricultural sites of the piedmont zone (in Anau, Tekkem-Depe, Namazga-Depe and sites in the Murgab Delta) (Pumpelly 1908: Vol. I:49, 149; Ganyalin 1956:86; Masson 1959:117).

In the East, in the Zeravshan Valley, where in 1950 Gulyamov found the Zaman-Baba graveyard, contacts of southern and northern cultures were also traced (Gulyamov 1956; V. Masson 1956:305, 1957b; Kuzmina 1958; Askarov 1962a-b, 1963). [Editor 33] V. I. Sarianidi traced the analogies of flint items, copper and stone ornaments, as well as the construction of ceramic kilns between Zaman-Baba and Khapuz-Depe. He even considered possible the infiltration of part of the population from the piedmont plain to the northeast (Sarianidi 1964:65). Whatever the nature of cultural contacts between the population of the southwestern farming oases of Turkmenistan and the population of large river valleys and deltas of Central Asia, these contacts contributed to the formation of unique farming cultures.

Judging from the Zaman-Baba settlement digs, located 0.5 km east of the Zaman-Baba graveyard on the dry riverbed of the Bujayli, the economic basis was represented by *kair* agriculture and herding (Askarov 1962b, 1963). Impressions of straw, wheat and barley were found in the settlement of Zaman-Baba. By definition of the botanist F. Kh. Bakhteev, in one of the clay pieces there was a well-preserved imprint of barley film (Bakhteev 1962). The discovery of broken parts of millstones, pestles and sickle blades from the settlement and the graveyard also confirmed the important role played by agriculture in the economy of the Zaman-Baba bearers.

The research material on the ancient irrigation of the Lower Zeravshan was recently published by Ya. G. Gulyamov (Gulyamov, Islamov and Askarov 1966). The authors outlined the main stages of development of irrigated farming, elucidated the different aspects of culture and economy, and hypothesized that the absence of large diverting systems and the preservation of traces of only a small network were evidence of the use of fading riverbeds, the creation of artificial overflows and their regulation in natural and artificial estuaries (Gulyamov, Islamov and Askarov 1966:16-17). According to Ya. G. Gulyamov, the beginning of irrigation in the Zeravshan was connected with damming the ancient deltaic channel of the Gurdush and the creation of small irrigation canals. Judging by the report of A. R. Mukhamedjanov around Zaman-Baba, a small canal 3 km long and 2-2.5 m wide was found, from which narrow irrigation canals 0.5 m wide were diverted. No pottery was found, but it is possible that it belonged to the Bronze Age. It should be noted that, near Zaman-Baba, sites of the mid-2nd millennium BCE were also known and that the majority of the researchers considered the sites very close to the Tazabagyab culture of Khorezm (Gulyamov 1956:149; Askarov 1962a:17). As in Khorezm, the inhabitants of the deltaic areas practiced irrigated agriculture along with pastoralism.

The settlement excavated by B. A. Litvinskiy in 1955-1956 at Kayrakkumakh provided some evidence of late Bronze Age irrigated agriculture in the Upper Syrdarya (Litvinskiy 1959:191-196). However, the main occupation of the people of Kayrakkumakh was herding.

Sites of the settled-farming culture of the late Bronze Age were discovered in several places in the Fergana Valley near Eylatan, Kuga, Tyuyachi, Chakana and especially near Chust (Voronets 1951; Sprishevskiy 1957; etc.) and Dalverzin (Zadneprovskiy 1962:11-37). The so-called Chust culture of painted pottery had some similarities with the culture of Anau III, Namazga VI and Murgab (Takhirbay, Auchin-Depe) (see Zadneprovskiy 1966).

The agricultural sites of the Fergana Valley in the Bronze Age were mainly located on the second terrace above the floodplain of the Syrdarya and its main channels (Zadneprovskiy and Kislyakova 1965:237). The natural conditions of late 2nd-early 1st millennia BCE differ little from today and agriculture in that period was irrigated. Yu. A. Zadneprovskiy suggested that agricultural development in the Fergana Valley started under the influence of the more developed southwestern regions of Central Asia (Zadneprovskiy 1962:200). At the Chust settlement, sickles, millstones, pestles as well as grains of common wheat, barley and an unknown plant similar to millet (Khudayberdyev 1962) were discovered. Yu. A. Zadneprovskiy published the data from a geomorphologic study of the *sai* in the Eastern Fergana Valley, such as Andijan-sai and Shaarikhan-sai, which were, according to the geologists, originally derived from the Karadarya (Ryjkov 1957:72).

The scheme of artificial irrigation development

in the Fergana Valley and its presumed historical stages are outlined in the work of B. A. Latynin. He identified three main stages in the development of irrigation techniques in Central Asia (Figure 26). The first stage, covering a considerable period of time (from Neolithic to late Bronze Age and early Iron Age), was the period originating irrigation skills and the development of naturally irrigated estuary-farming, the use of estuary brook floods fading onto the plain, and the use of temporary streams as well as the lower parts of river floodplains in the foothills of the Fergana Valley. From the simple embankments around plots, farmers gradually progressed to drain the excess water into neighboring lowlands and cleaning silted deltaic channels. The oldest system with lateral canals, presumably appeared as a consequence of land reclamation work, regulating overflows and riverbeds.[Note 61]

The second stage was characterized by the transition to artificial irrigation methods and the creation of small irrigation systems with a water distribution network on the slopes of foothills, and devices such as dikes and reservoirs at the mouth of *sai* ravines.[Note 62] At that time simple methods of water management were invented, in the form of water barrages (wooden tripods or *sepaya*, stone embankments, etc.) and head works. The territories of the upper terraces were connected to swampy lowlands and the lands flooded during freshets. In Fergana, this second period lasted from the 3rd century BCE to the 4th century CE, which coincided with the period of intensive development of class relations and the formation of an ancient state: the Davan kingdom (Latynin 1962:28).

The third stage began in the piedmont regions with the appearance of more complex fan-shaped systems with special head works and significant main canals, demanding large earthworks and constant sediment removal. Above the main upper distributing nodes of irrigation system there were fortified settlements: tepe (Sarykurgan on the Sokh, Shosh-tepe and Sharikhan-sai, Kala-i-bolo in the Isfayram Valley, etc.).[Note 63] That period, coincided with the beginning of the Fergana feudalism (5th-7th centuries CE), iron

implements were improved (heavy iron *ketmen*), the area of irrigated lands discernibly increased and the efficiency of irrigation and water-lifting facilities widespread (Latynin 1962; Gaydukevich 1947, 1948).

Considerable work on the study of irrigation in the late Bronze and early Iron Ages was provided by the Southern Turkmenistan Complex Expedition (YuTAKE), in connection with the study of the ancient agricultural settlements of the Meshed-Misrian plain and the Murgab Delta (M. Masson 1955; V. Masson 1954a:5-7; 1954b, 1956a-b).[Editor 34]

V. M. Masson spent 1951-1953 in a complex archaeological survey of the Misrian plain and its ancient irrigation structures. He traced the riverbed of one of the main canals of the settlement of Chat (at the confluence of Sumbar into the Atrek) prior to its outflow into the plain. It dug 'mysterious' ridges, which, according to some authors (S. A. Ershov, S. A. Shuvalov), were water distribution facilities of estuary irrigation based on local waters. The ridges appeared as ancient canals calculated on the basis of regular water flow. V. M. Masson dated the creation of this developed canal system to the late 2nd-early 1st millennia BCE (the culture of Archaic Dahistan).

S. P. Tolstov and M. A. Itina connected the beginning of agriculture in the lower reaches of the Amudarya with the early Suyargan-Kamyshli culture, when "perhaps, there were the rudiments of farming, but hunting, fishing and, likely, herding played a large role" (Tolstov and Itina 1960:28; Tolstov 1948a:348, 1948b:77-78; Tolstov and Kes 1960:82-89).

A. V. Vinogradov, based on the materials of the excavated late Kelteminar encampment of Kavat 7, deduced a direct genetic relationship between the late Kelteminar and the early Suyargan people (see also Gulyamov 1957:51-54; Masson 1964:184:note 103). A. V. Vinogradov wrote, in a rather valid comment, that the Kamyshli culture population "marked a qualitatively new stage in the history of the population of Khorezm, a stage which we connect with the appearance of the first metal implements, farming and herding.

According to the archaeological periodization this one is already Eneolithic. From here, there were a lot of features in the material culture, unusual for Kelteminar, in particular flat-based pottery, impoverishment of ornamentation, etc." (Vinogradov 1968a:176).

By 1948, S. P. Tolstov had noted the close relationships between the Suyargan culture with the farming cultures of the painted pottery of Anau, Transcaucasus, more ancient sites of Northern Mesopotamia and Northwestern Iran (Tolstov 1948b:78). V. M. Masson rightly pointed out that in the marshy lowlands of the Amudarya Delta wild grasses were absent and the first wheat in the Khorezmian fields was received only from the south (Masson 1964:184). This is probably confirmed by the fact that the contemporary traditional cereals of Khorezm, the endemic club wheat, has a typical mountain origin (Vavilov 1967, Tom 1:105; Jukovskiy 1950:87).Note 64 It was spread throughout Southern Turkmenia in the 4th-3rd millennia BCE (Anau I, Namazga I, Namazga IV-V) (Yakubtsiner 1956:108). According to V. M. Masson, from there the wheat and the barley were brought to the Lower Zeravshan, to the population of Zaman-Baba, and then to the lower reaches of the Amudarya "together with vessels with certain shapes" (Masson 1964:184).

In the mid-2nd millennium BCE (and possibly somewhat earlier),Note 65 Suyargan tribes of farmers and herders, together with tribes of another culture, called by S. P. Tolstov Tazabagyab, settled in the lower Amudarya reaches (Tolstov 1939:174-176; 1948a:66-67; 1948b: 76-77; 1957a:36-42; Itina, 1959a-b, 1961, 1962, 1967).Editor 35 According to M. A. Itina (1962), the Tazabagyab culture was formed in Khorezm as a result of the merging of newcomers from the contact area between the Srubnaya and the Andronovo cultures in the Ural region with the autochthonous people, bearers of the Suyargan culture.

The Tazabagyab tribes, bearing the skills of primitive floodplain agriculture of the Srubnaya-Andronovo tribes, created in Khorezm the particular type of integrated economy of semi-settled farmers (in delta drying channels) and herders (Itina 1967:75). Apparently, M. A. Itina (1967:73), was right in considering the irrigation farming of the Tazabagyab tribes as the main evidence for distinguishing the Tazabagyab as a special culture of the Steppe Bronze Age (compare with Masson 1957:53). As we shall see, the original Bronze Age irrigation was an early step in furthering improvements of irrigation facilities in the Aral Sea area.

* * *

An overview of the origin and development of irrigated agriculture in different areas of the Old and New Worlds allows to highlight some of the most ancient, geographically isolated and chronologically different areas of irrigated farming: the Near East and East Mediterranean (8th-6th millennia BCE); Iran-Central Asia (6th-5th millennia BCE); Nile (5th-4th millennia BCE); India (4th-2nd millennia); India-Malaysia (3rd-1st millennia); China (3rd-1st millennia BCE); Mesoamerica and Peru (3rd millennium BCE-1st millennium CE) (see Figure 14). Natural resources, which form the basis for the development of irrigated agriculture, were distributed unevenly among the continents. Even F. Engels noted that the Old World "had almost all the animals suitable for domestication and all kinds of grasses useful for cultivation, with the only exception of one. While in the western lands, America, among all the mammals suitable for domestication, only the llama was present and only in the southern part of the continent, and among all the grasses suitable for cultivation, only one was present but it was the best of all: maize. Because of this gap in natural conditions, the human settlement of each hemisphere developed its own path and the peculiarities of every phase of sociocultural development were different for each of the hemispheres" (Marx and Engels 1961; Tom 21:30).

In the Old World, the distribution pattern of the earliest settled farming settlements, such as Jericho in Palestine, Jarmo, Sarab and Tepe Guran in Iraqi Kurdistan, Ali Kosh and Tepe

Sabz in Southwestern Iran, and Çatal Hüyük and Hacilar in Anatolia, suggests that in the most ancient Southwestern Asian area, the center of origin of irrigated agriculture, the transition from the stage of 'harvesting people' to regular farming and cultivation of cereals, in some parts of this area took place almost simultaneously and independently.[Note 66] There were some initial centers, which apparently developed separately before 6500-6000 BCE (Masson 1964:39-40). Their inhabitants used local natural resources, and wild species of useful plants, adapting them to local peculiarities of soil, topography and climate (precipitation or river flooding, streams and lakes, etc.). Already in this very early stage, there existed sowing in natural overflows and artificially irrigated lands.

Irrigation skills were highly dependent on local conditions. Their transition from the ancient original areas was a complicated historical and cultural process, which cannot be considered as a process of linear diffusion from a single center. Each independent area of irrigated farming had its own individual, and chronologically different, local centers, where consistently and gradually it developed the complex adaptation process of local vegetation and of water resources to the needs of farmers. But it would be a mistake to assume that, in any local center of ancient irrigation culture (in particular, we investigated the Aral Sea area), the problem of plant domestication and the invention of irrigation methods were solved in complete isolation. For instance, in Mesopotamia there was a well-known cultural continuity between farming on the mountain streams in Hassuna in the 7th-6th millennia BCE, the estuary agriculture of Tell es-Sawwan on the banks of the middle Tigris in the 6th millennium BCE, and the Ubaid farmers in the 5th millenium BCE on the deltaic channels of Southern Mesopotamia. This cultural continuity existed also between farming cultures very far from each other even from a chronological point of view. Thus, we can trace a number of links between: irrigated estuary agriculture of the 6th-5th millennia BCE in the piedmonts of Kopet Dag (the Djeytun culture); deltaic farming and the canals of the 4th-early 3rd millennia BCE in Geoksyur (with typical species of barley, common and club wheat); agriculture and irrigation of the Bronze Age on the Makhandarya in the 2nd millennium BCE (with barley and wheat); and the peculiar deltaic irrigation of the Tazabagyab tribes of Khorezm in the third quarter of the 2nd millennium BCE.

Part II

The Lands of Ancient Irrigation of the Aral Sea Area

Chapter 3
The Southern Delta of the Akchadarya

Boris V. Andrianov

The territory of Khorezm has been inhabited since ancient times. The great Khorezmian scientist al-Biruni (973-1048), in his book *Vestiges of the Past*, considered the beginning of farming in Khorezm, and the settlement of the country, 980 years before Alexander the Great (1292 BCE), and about the arrival of "Siyavush, son of Kaikaus, and about the accession to the throne of Kai Khusrau and his descendants, who moved to Khorezm" (BiruNI 1957:47).[Note 67] If V. V. Bartold believed that the information in this book to be just a legend, S. P. Tolstov considered it as the echo of a real migration of Iranian-speaking farmers and herders of the Bronze Age to the lower reaches of the Amudarya (Bartold 1965:26, 545; Tolstov 1957b:XVIII-XIX; see also Sachau 1873).[Editor 36] The myths, legends and religious rules of Iranian tribes of the 2nd-early 1st millennia BCE were reflected in the oldest parts of the Zoroastrian canon (*Avesta*): *Gathas* and *Yashts* (Sokolov 1961:13, 21; Livshits 1963:137).[Note 68] In the *Yashts*, the myths about kings–snake-fighters were replaced by the legends on the Kayanids kings (*Kavi*), whose initial period of rule is probably to be placed in the 10th-8th centuries BCE (Klyashtornyy 1964:169).

According to most of the scientists, the Amudarya basin, together with the neighboring areas in the south (from Herat and the Paropamisus Mountains to the Merv Oasis), including the banks of the Middle and Lower Amudarya up to the Aral Sea, and the areas along the Syrdarya, was the area of the Iranian tribes and the area of spread of Zarathustra's teachings. In the tenth *Yashta*, dedicated to Mithra (Mihr Yasht), describing this territory, Khorezm was first mentioned. The main river of the Arian land was the Dāitya, which is identified with the Amudarya. It is located near the original area of the preacher Zarathustra,

Airyana Vaējah (later Ērān-wēz). This region is described in the *Bundahishn* as the country of countless rivers and streams, branching off from the Daraga River,[Note 69] on the bank of which also Zarathustra was born (Müller 1880, Bundahishn:ch. XX:32; Bundahishn:ch. XXIV:15). Adjacent to Airyana Vaējah, the non-Arian country of Gōbad was also located on the Dāitya River, but to its north. Here lived the Saka (Turanian) ruler Gōbadshāh (brother or nephew of the severe king of Turan: Afrasiab). Gōbadshāh which means 'the king of cattle', 'the owner of bulls and cows' (Inostrantsev 1911:315, note 1; Müller 1882, Dâdistân-î Dînîk:ch. XC:4; Müller 1880, Bundahishn:ch. XIX:13).

As mentioned above (see page 132), the archaeological research revealed that the settlements of the first half of the 2nd millennium BCE belonged to pastoralist tribes of the Kamyshli culture of Khorezm in the Northern Akchadarya, which carried the waters of the Amudarya (Dāitya River) directly north. It is highly possible that these places (very convenient for year round grazing), during the early years of *Avesta*, were the countries of the Saka, or Turan, 'the owners of oxen and cows'. From that period there is no certain evidence of the existence of irrigated farming or any other irrigation skills in Khorezm. The beginning of a productive economy in the lower reaches of the Amudarya was apparently connected to tribes of herders-farmers of the Bronze Age. Irrigated agriculture, i.e. the base of the economy in Southern Central Asia, Tedjen, Murgab and the Zeravshan (Zaman-Baba culture) between the 3rd and 2nd millennia BCE, in the Lower Amudarya was apparently accompanied by cattle breeding, fishing and hunting.[Note 70]

M. A. Itina rightly noted the complexity of the issue about the origins of agricultural civilization of Khorezm. She wrote: "Even if we assume that the ancient Khorezm received grains of wheat and barley from outside, nevertheless the development of forms of irrigated farming in Khorezm took place through the accumulation of experience under certain conditions and this process cannot be merely explained as a southern adoption" (Itina 1968:83).

It is possible to see from above examples describing the origin and development of irrigation in different countries of the Old and New Worlds, that every major historic and cultural region developed its own method of irrigation, according to local geological, hydrological and climatic conditions and, finally, the features of the population historical development.

From the review of the history of agriculture, the development and distribution of productive skills (irrigation implements) and irrigation skills, it can be seen how the latter is most dependent on specific local conditions. Hence, there is reason to believe that V. V. Bartold was wrong when, regarding the irrigation in Khorezm, he wrote that "in spite of the perfection of this system, it would be difficult to expect that in an isolated country, with very poor construction materials, sophisticated techniques to control a river such as the Amudarya could be devised" (Bartold 1965:163).

The independence of the Khorezmian irrigation tradition, improved over several millennia, is confirmed not only by the remains of ancient irrigation works, whose description we are going to see, but also by the brilliant work on astronomy, mathematics, geology and hydrography representative of the Khorezmian Shool of scientists, headed by the Medieval encyclopedic, Khorezmian al-Biruni.[Note 71]

Natural Conditions of the Lower Amudarya

The 'lands of ancient irrigation' researched by the Khorezm Archaeological-Ethnographic Expedition are located both on the right bank of the Amudarya (the territory of the Southern Akchadarya Delta in the Karakalpak Autonomous SSR) and on the left bank, within the limits of the Sarykamysh Delta (the territory of Tashauz province in Turkmenistan SSR). The territory of the whole ancient alluvial plain of the lower reaches of the Amudarya, including the modern delta, reaches 4.5 million ha (Akulov 1957). On a territory of almost 2 million ha, archaeologists have found the remains of an ancient agricultural civilization in the form of settlement ruins, abandoned canals and fields. These lands on the right and left banks of the river, together with contemporary cultural oases, are included in a single historical and cultural area, called since ancient times 'Khorezm' as confirmed by the *Avesta* (see p. 95). They include the oases of the lower reaches of the Amudarya from the Dargan-Ata to the Aral Sea, and from the dry riverbeds of the Akchadarya on the east to the Sarykamysh depression and the Upper Uzboy on the west. Forming a single historical, cultural, and geographical area, as we shall see later, the right bank and the left bank of Khorezm differ in both archaeological and natural respects. On the right bank, the sites, starting from the Neolithic (4th–3rd millennia BCE) to the Khorezmshah settlements of the early 13th century, are well preserved. On the left bank, the remains of the Prehistoric encampments were covered by later layers (Tolstov 1955c:192) and only the Antique, Medieval, and contemporary settlements up to the abandoned sites of the 19th century are preserved.

The lower reaches of the Amudarya are surrounded by deserts and have a sharply continental dry climate with insignificant precipitation (less than 800 mm per year), dry air, strong winds and very high evaporation, which excludes the possibility of farming without artificial irrigation. The arid climate of the Aral Sea area dates back to the late Tertiary period. I. P. Gerasimov outlined the following stages of climate change for the Turan: glacial pluvial, postglacial xerothermic and modern arid (Gerasimov 1937:26). This is confirmed by K. W. Butzer's scheme for the lower latitudes (see Table 4). However, the cyclical climate change did not critically change the arid and semiarid landscapes.

The ancient deltaic plains of the Amudarya

were formed in the wide Aral-Sarykamysh depression in the late Tertiary period (Yamnov and Kunin 1953; Eberzin 1952; Kes 1957). In the Middle and Lower Pleistocene, the river wandered around the Karakum and flowed into the Caspian Sea (Gerasimov 1937:63; Fedorovich 1946:152-173, 182-184; 1950a:204-213; Tolstov and Kes 1960:16-17, 18:fig. 2). From the early Khvalinsk period, the Amudarya turned north and began to fill the depressions with sediments, resulting in three deltas formed in distinct periods: Akchadarya (South and North); Sarykamysh; modern Aral (Tolstov and Kes 1956:327-336; Tolstov and Kes 1960:17-21).

Modern geological, geomorphological, and archaeological studies have confirmed the hypothesis, advanced in the 10th century by al-Biruni, about the gradual migration of the river: first west into the Caspian (Khazarian) Sea; then north, to "the margins of the land of the Oghuz', through the Fam al-Asad ('lion's mouth') gorge";[Note 72] then "to the right toward Farab, along the river called today al-Fahmi (dry riverbeds of the Akchadarya and Janydarya) (Tolstov and Kes 1960:9, 35, 66; Tolstov 1962a:21-26, 274); and then left, to the land of Pechenegs, along the riverbed known as Wadi Mazdubast",[Note 73] in the Sarykamysh depression (Khiz Tanqizi) (Gulyamov 1957:25; Tolstov 1962a:21; Tolstov and Kes 1960:8); and, finally, north into the Aral Sea (Biruni 1966:95-96; Gulyamov, 1950:85-92; Tolstov and Kes 1960:8-25).

The riverbed of the lower Amudarya, wondering across the plain, has repeatedly changed its direction and, depositing the alluvium, formed several tiers of deltaic sediments (Skvortsov 1959; Bogdanovich 1955). In the Late Khvalinsk period, the Akchadarya channels began to decay. But later, between the 3rd and 2nd millennia BCE, when there was a new shift of the Amudarya River eastward, the channels of the Akchadarya Delta supplied water again as proven by loamy sand sediments (up to 1.5 m) covering the Neolithic encampments (Itina 1968:76-78; Vinogradov, 1968:32). The widespread Bronze Age sites of the 2nd millennium on the banks of the channels

indicate the duration of the new watering period (Tolstov and Kes 1956, Tolstov and Kes 1960:24, 35-133).

With the appearance of the first herders and farmers, the 'Neolithic' landscape of Khorezm (lakes, swamps and forests) gradually began to change. Human labor transformed the marshy 'kair' lands into fertile fields and cane thickets into pastures. The forests were cut down and artificial plantations began to appear. In turn, some physical-geographical processes, in particular the drainage process (early 2nd millennium BCE) and later the gradual fading of the southern delta tributaries, were reflected in the area historical development of population and economic activity. The damping and sealing of the deltaic channels facilitated flood control and contributed to the advancement of skills in irrigation (Tolstov and Kes 1960:132; Andrianov and Kes 1967:29).

In the 2nd and 1st millennia BCE, the Amudarya had a huge double delta consisting of numerous channels (the river Darga of the Airyana Vaējah - Aryan expanse?) some flowing into the Sarykamysh depression and others into the Aral Sea. The natural tendency led to the fading of branches in the upper part of the delta and the displacement of a major mass of water and sediments towards the Aral Sea.[Note 74] These processes initially favored the regulation of lateral channels and then the creation of many kilometers of artificial canals on the Akchadarya and Sarykamysh deltas, fed by the flood water. Under the influence of human activities, especially after the formation of a large Khorezm state in the 6th-5th centuries BCE, intensive changes in the geographical environment occurred (Saushkin 1947:276; Tolstov 1848a:45; Andrianov 1951:323). Among numerous factors, closely linked to each other, causing during the centuries successive changes of 'the main direction' of the lower course of the river, the most important were hydrologic and climatic factors, causing uneven alluvial deposits (effective factor), and economic-transformation activities of the population (see Andrianov 1951:326-327; 1966:148-149; Andrianov and Kes 1967:38).

Almost the whole alluvial plain of the

Amudarya is an ancient terrace of the river, 1-1.5 m higher than the floodplain terrace (*kair* lands). It has an even flat surface with a slight slope from the river and large channels mostly in the northwest direction (an average of 18 cm per km); to the south the plain is limited by the Karakum sands, where the waste water from irrigation canals is collected, forming a chain of lakes surrounding the oasis. The prevailing slopes of the place favored the development of gravity irrigation and water supply to the farthest distance from the rivers. In the area immediately adjacent to the river, the plain is mainly covered by contemporary irrigated and cultivated lands, between which there are large tracts of salt marshes, swampy depressions and small lakes that serve to discharge irrigation water. A system of protective dikes was built along the banks of the Amudarya, which extends, in several rows, for many hundreds of kilometers. In territories of anthropogenic irrigated deposits, the walled dikes grew significantly slower than those in the floodplains located between the river and the oasis. This led to the formation of differences in height, representing a serious threat to the cultural development of these territories, and in case of breakthrough of the dikes, it often represented the cause of repeated devastating floods.

The hydrographic regime of the Amudarya is largely determined by the supply of its snow and ice (see Table 7). The water flow of the river differs significantly from year to year, and that is why the Amudarya is favorable if compared with the Tigris and the Euphrates (with their winter and spring floods), the Nile (where the highest flood falls in August and September), and the Indus and the Hwang Ho with their harsh and frequent catastrophic fluctuations of water levels. Al-Biruni observed that "the Djeykhun rises when the water level of the Tigris and the Euphrates decreases" (Biruni 1957:286). He also revealed the physical causes for the freshet overflow of the rivers and he highlighted the long-time observations of the Khorezmians thanks to 'Moon stations', which created, long before the Arabs, a particular Khorezmian school of astronomical and calendar observations for the purpose of irrigated

agriculture (Biruni 1957:259; Tolstov 1957b:X; Tolstov and Vaynberg 1967:251-264).

The river regime is extremely favorable for irrigated agriculture, since the period of maximum water level coincides completely with the period of sprouting vegetation. The beginning of the vegetation growing period is in March, around the twentieth, when the first flood, called by people 'the flood of the green canes', begins (Gulyamov 1957:237). The second flood ('vimba flood') falls in the middle of April, the third one in the middle of May. This high water period is connected to the appearance of the Pleiades in the sky. The longest flood period is the fourth; it begins in the second half of June and ends in early August. The irrigation season ends in October. The water flow rate in autumn is minimum. The average annual water flow rate at the exit of the plain is approximately 2000 m³/sec; at the beginning of the delta, near the city of Nukus, it is 1,500 m³/sec.

The studies of V. L. Shults indicated that the delta annually received an average of 48,124 million m³ of water (from 37,000 to 65,000 m³) (Shults, 1948:64).[Editor 37] Of these, the average evaporation is about 11 million m³ per year, about 22.5% of the water entering into the bed of the river. The remaining 77.5% (37,732 million m³ per year) flowed into the Aral Sea. According to V. L. Shults, only 37.1% of the annual run-off flowed into the sea. During the summer floods, when much water was used for irrigation, this percentage decreased to 25%. In winter it was over 50%.

According to S. G. Altunin's calculations, the river carried annually up to 1,200 million m³ of sand and its water was very muddy (Altunin 1951; Altunin and Buzukov 1950:184-185). Of this amount, less than half (43%) is deposited on the floodplain, riverbeds, and irrigation systems of Khorezm. Of the remaining 57% of sediments, 28-30% is deposited within the 'living' delta.

At that time, V. V. Tsinzerling drew attention to the great similarity between the deltaic areas of the Amudarya and the Nile valley and delta (Tsinzerling 1927:193). The soil of the Amudarya Delta is not less fertile than the soil of the Nile and the conditions for land use also are not less favorable (see Table 7). The development of soil cover in the Amudarya Delta was meadow, meadow-swamp, and riparian vegetation at the first stage of the alluvial sediments, as already noted by many authors (N. A. Dimo, S. S. Neustruevym, I. P. Gerasimov, V. A. Kovda, etc.).

In order to better understand the following description, it is necessary to focus on the soils. In the modern and ancient deltas, the soils tend to have a complex evolution, depending on the water-salt regime: from the alluvium to meadow turf soil with high water table; then to salt and *solonchak* soil with different degrees of salinity to salt marshes and, further, with low water table, the desert-*takyr* type of soil (Bogdanovich 1955). The rise of groundwater and the increase of evaporation in the vast irrigated areas have contributed to a more intense flow of salts in the soil and caused a gradual salinization of the cultivated soil. In the past, during low levels of water use and agricultural techniques (abandonment of irrigated land becoming fallow) the primitive character of irrigation systems aggravated the process of salt accumulation and, thus, reduced the soil area suitable for farming. According to the calculation of P. A. Letunov, across 2,500 years of Khorezm's agriculture, every hectare of irrigated land had received up to 250 tons of water-soluble salt (Letunov 1958:50).

As the hydrological regime changed toward drying, and the water level sufficiently lowered, the reverse process began with the formation of desert and *takyr*-like soil, *takyr* and loose sand desert soils (Kovda 1947:23–24; Letunov 1958:115). Since the structure of soil and ground was extremely colorful and their exploitation in the past was notable for its diversity, thus the whole territory of both modern and ancient oases are extremely complex and characterized by a mosaic of soil covers (Favorin, Ostrovnaya and Timoshkina 1956).

A large variety of water-salt regime determined the mixed character of vegetation in the 'lands of ancient irrigation', from the almost empty *takyr* and *solonchak* soil with rare bushes of *biyurgun*, to dense thickets of *yantak* licorice and even reeds along the shores of floods on the border of the modern cultivated areas. The modern geographical distribution of land cover largely reflects the history of agriculture in the lower reaches of the Amudarya, the repeated changes in the irrigated areas, the direction and the magnitude of the irrigation systems, etc. (Shuvalov 1950:37–38).

The Southern Delta of the Akchadarya

The whole territory of ancient irrigation in the Southern Akchadarya Delta is 160,000 ha and the area has a triangular configuration with its vertex at Turtkul. To the west the river is 150 km long; to the north there are the rocky mountains of the Sultanuizdag; to the east there are massive reddish-yellow sand dunes of the Kyzylkum. Because of the exceptional abundance and diversity of archaeological sites (ruins of cities, fortified and unfortified rural settlements, surrounded by ancient canals), these lands have been named as the great 'museum' of sites of the ancient Khorezmian technique of irrigation culture (Tolstov, 1958:100). To the west, the 'lands of ancient irrigation' are adjacent to a contemporary oasis, before which are located three narrow green wedges of verdant fields, farms and canals, which correspond to the three modern irrigation systems of Kelteminar, Kyrk-Kyz (whose former name was Tazabagyab) and Amirabad (Tolstov 1948a:46).

The river 'front' of the modern oasis is more than 150 km with a downstream slope (20 cm by 1 km); the slopes of the dry riverbeds of the Akchadarya is 4–25 by 1 km. The system of dikes protecting the oasis against flood water existed from ancient times. These dikes were more than 2 m high, 5–10 m wide and, sometimes, disposed in 2 or 3 rows. Their overall cubage was estimated by specialists in the beginning of 20th century at around 450,000 m^3 (EOZU 1914:331).

The modern canals, as well as the ancient irrigation systems, followed the direction of the three major deltaic riverbeds of the Akchadarya,[Note 75] and their lines can be traced with great difficulty within the 'lands of ancient irrigation'. The riverbeds are dispersed and filled up with moving sand with very rare lines of preserved clay-*takyr* parts (Tolstov and Kes 1960:34-35: *Geomorfologicheskaya i arkheologicheskaya karty akcha-darinskikh delt* - Geomorphological and archaeological map of the Sarykamysh Delta). The entire alluvial plain is dissected by the system of dry riverbeds, differing in size and degree of preservation, with separate massifs of moving sand and narrow (2-3 km) lines of *takyr* bands along the ancient channels stretching north and northeast (Tolstov and Kes 1960:35-45). The plain is dominated by combinations of *takyr* soil, *takyr* and *solonchak* (the latter mainly in the western outskirts around Toprak-kala), as well as territories half-covered by gray and yellow-gray sand dunes and fixed sands with numerous depressions, fractures and remains of erosions. In many upper soil layers, the ancient layers were exposed as a result of the wind deflation process. The northern and northwestern parts of the plain are formed of multilayered alluvium (powder clay and loams); loams and silty clay with alternating layers of sand prevail southeast (Fersman 1934, Tom I:75-100; Tom II:7-34, 77-88).[Editor 38]

The system of riverbeds dismamtling the southern delta starts above the city of Turtkul and extends north, northeast, till the latitude of Sultanuizdag, where the majority of channels sharply turn east along the innermost Kyzylkum hills, and then north forming the single valley of the Akchadarya (Figure 27.A). The most western riverbed, as a continuation of the direction of the Amirabad system, sharply turns northwest up to the neighborhoods of Toprak-kala, then it gets lost in the contemporary cultivated area. This part of the large ancient riverbed is still well pronounced in relief; it is filled with water.

Next, to the east, is another large riverbed of the Akchadarya (Kokcha), up to 4-5 km wide, covered with sand and separating the modern Kyrk-Kyz Oasis from the ancient Kelteminar Oasis (Tolstov and Kes 1960:39-40). North of Koy-Krylgan-kala, the depression becomes a typical riverbed. Some smaller branches depart from it. In the ruins of the Big Kyrk-Kyz, the main riverbed (200-300 m wide and 10 m deep in some places) sharply turns east and then, skirting the Kokcha Mountain from west and north, it enters into the Akchadarya corridor, where it merges with the two riverbeds approaching from the south. These riverbeds are marked by an accumulations of sand, *takyr* bands, vegetation and, in general, they have a north-easterly direction. Their inception can be traced in the contemporary cultivated area near the ruins of Eres-kala. In the vicinity of Bazar-kala the riverbed divides into two: the right branch goes around the plateau of Djanbas-kala from south and east, the left from northwest.

From the Djanbas-kala plateau, to the south towards the contemporary Amudarya, among the inner Kyzylkum sands, a north-south chain of small lakes and the meandering channel of the Suyargan, often watered by discharging water, stretch along a depression (Tolstov and Kes 1960:41, 138). This north-south depression, apparently, is of very ancient origin. Near the Amudarya it has a wide clay bottom, through which passes a narrow riverbed 10 m wide, which serves to discharge the high flood water from the Suyargan lakes connected to each other by meandering streams. The northernmost of them brings water of the Suyargan into a wide valley of the Akchadarya riverbed southwest of Djanbas-kala. On this riverbed, sometimes the contemporary Amudarya waters flow far north, aside the Akchadarya corridor (Tolstov and Kes 1960:138).[Note 76]

During the archaeological and topographic research and mapping (based on the vertical aerial survey) remains of many dry riverbeds in the southern delta of the Akchadarya, 10 to 150 m wide and 1.5-2 m deep, were identified in addition to major channels. In some places, the branching of these riverbeds formed small internal 'deltas' (Andrianov and Kes 1967:Fig. 3). At present, only the larger riverbeds (more than 50 m wide) are marked by the lowest part of the relief; the small riverbeds are silted and filled up with sand. Frequently such riverbeds, completely covered by river sediments, are preserved in the shape of meandering *takyr* lanes, raised on the surrounding ground and contoured by vegetation above them. As a rule, they do not exceed an average width of 10-30 m and are distinguished by characteristic structured terraces 5-10, or even 30 cm high on the flat *takyr* surface (Tolstov and Kes 1960:42-44, figs. 14-15). The majority of Prehistoric sites and irrigation works of the Bronze Age discovered in the course of field research are connected to these '*takyr*' riverbeds.

Irrigation Works of the Bronze Age
New geomorphological materials, and the remains of ancient irrigation systems of the third quarter of the 2nd millennium BCE (which will be discussed below) characterize the southern delta of the Akchadarya in the Bronze Age as a territory with continuous shifting channels, lakes and inner small 'deltas'. All this allows to update the portrayals of skills for irrigated agriculture and natural conditions of the Bronze Age Prehistoric irrigation. These were considered at his time by Ya. G Gulyamov, though the main hypothesis (which he developed following D. D. Bukinich) on the initial estuary irrigation and farming on *kair* lands, should be accepted (Gulyamov 1949:11, 1956:157, 1957:49-65; Gulyamov, Islamov and Askarov 1966:16; Itina 1968:83-84). This most ancient stage of irrigation on alluvial plains, when farmers used the areas watered by overflows for their crops, is not yet exposed by archaeologists in the Aral Sea area. In the future the study should not be focused on to the identification of ancient agricultural-irrigation layouts (which could not have been preserved), but to search for slag remains in the cultural layers of the first half of the 2nd millennium BCE sites. Anyway, we can share the certainty of M. I. Itina that the Khorezm people of the Bronze Age were familiar with irrigated agriculture (Itina 1968:84). The ethnographic materials can serve as proof of that.

Even recently, the inhabitants of the lower and marshy sectors of the valleys of the major Central Asian rivers, i.e. Karakalpaks, Uzbeks of the Aral Sea, and some groups of Kazakhs and Turkmens,

combine primitive agriculture with primitive farming with cattle-breeding and fishing, and grow crops extensively (melon, pumpkin, millet) on flood overflows and *kair* lands near the rivers. (see Kaulbars 1881:552, 562; Levshin 1832:200; Georgievskiy 1937:99; Bregel 1961:50-55; Andrianov 1958a:58, 1961:139, 1963: *Karta khozyaystvennykj typov Sredney Azii i Kazakhstana* - Map of the economic types of Central Asia and Kazakhstan). As pointed out by the research of S. P. Tolstov and T. A. Jdanko, the inhabitants of desert areas near the sea, lakes and banks of the steppe rivers, and less populated vast deltaic regions of Central Asia, inherited the archaic traditions of integrated farming - herding and fishing economy dating back in time (Tolstov 1947b:71-90; 1948b:99; 1962a:308-309; Jdanko 1952:163; 1958a:633-634; 1961, 1964:17-19).[Editor 39] The essence of this primitive economy is figuratively given in the folk tradition saying: "Three months melon, three months milk, three months pumpkin, three months fish" (Gulyamov 1957:64). *Kair* farming refers to the most primitive forms. According to Ya. G. Gulyamov, in early summer the inhabitants of Khorezm sowed melon seeds into the drying *kair* soil and returned at the end of the summer. Here they arranged a small hut for themselves, fished in the channels and gathered melons until late autumn (Gulyamov 1957:64-68).

Ya. G. Gulyamov had the very interesting idea about the name of the modern Khorezmian seed plot units (*kulcha*), which refers to a diked area filled with irrigated water. He compared this name with *kul* (lake, estuary)[Note 77] and rightly saw in the term *kulcha* the ancient tradition of estuary irrigation (Gulyamov 1957:63). In an attempt to reconstruct the daily life of Prehistoric farmers of the Bronze Age, Ya. G. Gulyamov used historical and ethnographic material and reported that, in the 19th century, small groups of Karakalpaks and Uzbeks families migrated in early summer from their winter places, and settled near channels which had overflowed during summer's high water, or at the ends of the major channels, where the lowlands were seeded with millet, rice and melons and the cattle grazed in the reeds. In autumn, after the harvest, the same groups migrated deep into the cane thickets which, in late autumn and winter, sheltered the cattle against the cold winds and also provided people with fuel. These pastoralist-farmers were distinguished by their mobile agriculture and settlements (Gulyamov 1957:63).

There are many examples from different areas of arid zones (see Wittfogel 1957:24–25; Gray 1963; Owen 1937; etc.). V. G. Childe already referred to the Beja-Hadendoa way of life and other Nilotic tribes, occupying an intermediate position between the hunter-gatherers and settled farmers in Egypt (Childe 1956:68).[Note 78]

The process of gradual transformation of *kair* farming into estuary irrigation was discovered by L. S. Tolstova in the highly swampy soils of the territory along the Syrdarya exploited by the Fergana Karakalpaks. She cites memories of the elders: "Before, irrigation ditches and canals were not there when floods happened, it flooded the shore for 2-3 km and, after the water retired, these lands were sown" (Tolstova 1959:32). The drainage of some wetlands allowed to use the vacated lands for farming. First they began to divert canals from lakes and flood areas near which they were settled. Then the larger canals were dug (Tolstova 1959:31–35). A similar process of distribution of *kair* and estuary irrigation along rivers and mastering the management of wetlands in the territory of the Amudarya Delta was researched by us in the Northern Khorezm, where the Karakalpaks of the Janydarya moved from the Lower Syrdarya in the 16-18th centuries (Andrianov 1952a, 1958a, 1966).[Editor 40] The Karakalpaks were familiar with the use of crops in *kair* wetland islands and flood plains of river deltas, the bottom of the drained or dried-up lakes, former riverbeds, temporary flood rivers. They were able to drain swamps and to adapt to the changing water and soil conditions (Andrianov 1958a:117; Jdanko 1964:17).

Ya. G. Gulyamov reported the use of lake floods for estuary crops in peripheral areas of the Amudarya Delta. According to him, since the 16th century, the population of the Aral Sea area began to artificially flood the 'Kuygun' depressions under the Ustyurt plateau (Bartold 1965:91, 177). Until September, they were shallow lakes where a large quantity of fertile silt accumulated and the soil was saturated with moisture. "The sedentary and nomadic people of the Aral Sea area began to gradually arrive there as the lake dried in order to sow seeds directly into the slurry of land designated for each of them; the yield was harvested the following summer" (Gulyamov 1957:60).

In Southern Turkmenia, on the slopes of the Kopetdag and the basin of the Atrek (in particular on the Sumbar River), the system of estuary irrigation existed among the Turkmen even recently (Vasileva 1954:107). With this method (*loya-sepma*), the plots of land silted during freshets were sown. The seeds were planted directly into the fissures formed by the cracking surface after the water recession (Grigorev 1932:6). In the more advanced stage (*suomi*), transverse small ridges were built to retain the moisture. The method *darava* consisted in the whole system of ditches draining rainwater to the estuary, where the wheat or other crops (most often melon) were sown. These sown areas were confined to the margins of the *takyr* depressions where, during the rainy season, the water runoff was stored (Djumaev 1951). A similar irrigation method with river overflows (*suolna*) were spread among the Kazakhs of the Syrdarya (Gulyamov 1957:61).

Remains of the most ancient irrigation works preserved in the Aral Sea area were found in the southern delta of the Akchadarya, both in the Tazabagyab sites (third quarter of the 2nd millennium BCE) and in the coeval Suyargan sites (the so-called Bazar-kala or Tazabagyab-Suyargan stage of development of this culture) (Tolstov and Andrianov 1957:5–6).[Note 79] A significant number of Bronze Age sites in the form of accumulations of pottery, scattered settlements, some dwellings, and traces of primitive small irrigation works were identified on the banks of ancient riverbeds north of the ruins of Bazar-kala.[Note 80] Here, a lateral channel, 60-70 m wide with a valley clearly visible on the ground, branched from the major bed of the Akchadarya, with a decrease of 1.5-2 m in depth. In some places the valley was weakly indicated and in other places it was completely buried under the sand. However, it was traceable on the ground as well as on aerial photographs for 20 km from the head (west of Bazar-kala) to the lower highly-branched ancient riverbeds southwest of the Kokcha plateau. The majority of Bronze Age ancient irrigation works were located along the banks of this large riverbed and on small (from 10 to 30 m) *takyr*, slightly elevated above the riverbeds, which sometimes formed a strange intertwining (see Andrianov and Kes 1967:fig. 3). The trenches indicated that *takyr* riverbeds were formed by a layer of gray sand of the Amudarya, hard and light loam, and clay. These channels, in antiquity, were similar to narrow riverbeds, with slowly flowing water abounding in fish, with banks covered with reeds and *tugai*; but they gradually became silted over with river sediments (Tolstov and Kes 1960:42-45).

Primitive agro-irrigation layouts and irrigation works of the Bronze Age (tiny 'fields') were discovered by S. P. Tolstov in 1954 during the Akchadarya survey, southeast of the site and the graveyard of Kokcha 3 (dated to the third quarter of the 2nd millennium BCE), as well as in the later site of Kokcha 1 (Tolstov 1955a:102-104; Tolstov and Andrianov 1957:5-6). The archaeological-topographic unit spent 1954 carrying out the topographic survey of these lands and studying structures. The graveyard of Kokcha 3 (Tolstov 1955a-b, 1957a, 1958; Itina 1961) was located on the bank of a small Akchadarya riverbed, on an upper terrace, behind a somewhat horseshoe-shaped, almost enclosed sand dune. 'Fields', or rather the tiny gardens plots,[Note 81] were located 150 km southeast of the *takyr* with burials and south of a settlement on the edge of a narrow meandering *takyr* strip reminding a riverbed, but not detectable in relief (Figure 27.B).

Somewhat to the south, the remains of a ridge-diked channel parallel to this *takyr* strip was discovered.[Note 82] It was preserved for a height of

5-10 cm and a width of 1 m, and appeared as a dark-brown line (Tolstov and Kes 1960:95, fig. 50). The banking was traced for a few tens of meters on the west and on the east banks of the riverbed, where it was better preserved. There 'fields' were discovered, irregular quadrangular diked areas measuring 2.6-3 m by 3.5-4.8 m (Figure 27.C). The 'fields' were grouped in several rows. Their character was similar to the layouts in the environs of the later Suyargan site of Kokcha 1, situated southwest of Kokcha 3. The layouts were rhomboid-shaped with the small sides 2.2-2.9 m and the large sides 3.4-4.7 m (Figure 28.D). Slightly lowered areas were located in a row between two flat banks and rising 3-5 cm above the *takyr* were, apparently, the remains of irrigation ditches 50-70 cm wide (see Tolstov and Andrianov 1957:6:fig. 1).

Trenches were made in the small riverbed and the 'fields' near Kokcha 3 (Tolstov and Kes 1960:figs. 16 and 50). Trench 35 extended through part of the field and part of the banked riverbed, its length was more than 10 m, the width of the riverbed lens was approximately 9 m and the depth was 1.2 m (Figure 28.A). Here is a description of the trench.

Layers: 1) *takyr*-shaped thin crust, gray, clay loam, fine porous, loose, of 1-3 cm; 2) yellow-grayish clay loam, dusty, light, partially turning into sandy loam, with marked horizontal stratification, heterogeneous, dug, with unclear lower border, of 20-25 cm; 3) gray-brownish clay loam, of medium density, partially alternated with more compact interlayers of clay, and partially with a high presence of sand, up to 25 cm; 4) light gray-yellow clay loam, thin, dusty, partially with sandy loam, with a weak marked horizontal stratification, up to 25 cm; 5) brownish clay loam, heavy, compact with salt efflorescence, up to 10-60 cm; 6) light yellow-gray sandy loam, thin, dusty with rusty spots and lines, compacted at the bottom, of 20 cm; 7) gray sand, micaceous, thin layered, with admixture of clay loams and iron spots, up to 20 cm; 8) gray sand with inclusion of uneven interlayers of loams and small pieces of clay, gradually turning into a layer of

10-25 cm; 9) yellow-gray sand, fine-grained in the shape of inclined lens, turning into sandy loam at the bottom, up to 15 cm; 10) sandy-clay loam heterogeneous sediments.

According to A. S. Kes, the section of this trench, as well as other neighboring trenches, indicates that there was a narrow deltaic channel, which turned into a former riverbed with a lake, and then again water when heterogeneous alluvial-colluvial sediments were deposited there. The functioning of irrigation facilities was connected to this period (Tolstov and Kes 1960:95). In addition, it is possible that the revival of the flows is likely to have been linked to the activities of farmers, who regulated the flow of flood water by artificial means in the former riverbed (as we shall see below, such a primitive method of flood water regulation was the base of Antique and Medieval irrigation in the Lower Syrdarya and the Amudarya Delta).[Note 83] Otherwise it is difficult to explain the scarce importance of the protective embankment (up to 1 m) separating the riverbed from fields. Therefore, it would be necessary to point out that S. P. Tolstov and B. V. Andrianov wrongly compared Kokcha's 'fields' with the ancient Egyptian basins (Tolstov and Andrianov 1957:6). There the flood water was regulated by the scheme: river - dam - field (later: river - dam - distributor - field); here by the scheme: former riverbed - field (see Figure 27.C).

In other sites of the Bronze Age, irrigation ditches leading water into larger irrigated plots were also recorded. Such systems resembled the Eneolithic canals of Geoksyur described by G. N. Lisitsyna (1965:115:fig. 30) (see p. 129). Their principle is similar: river - distributor - field.

Small ditches were studied by us, in particular, in the Suyargan encampment of Bazar 2 (see also Tolstov and Itina 1960b:22; Tolstov and Kes 1960:119-123). The encampment was located at the former riverbed on the right bank of the lateral channel of the Akchadarya, here 150 m wide. From the former lake derived a small irrigation ditch and an accumulation of ceramics and traces of burnt dwellings were found next to it. A small trench 4 m in size was dug through the irrigation ditch (Tolstov and Kes 1960:120-121, fig. 63) (Figure 28.B). The existence of an ancient irrigation ditch is well supported by the peculiar stratification of sandy loam and clay loam with typical characteristic layering on the bottom of the ditch.

More complicated irrigation facilities, connected with sites of the Tazabagyab culture, were discovered and investigated by us in 1964, southeast of the Kokcha 3 burial on the right bank of the Akchadarya channel. This plot of ancient irrigated land was not covered with later agro-irrigation cultural layers. It was located north of the Djanbas plateau between two large dry riverbeds of the Akchadarya Delta, and stretched for 20 km from south to north and 15 km from west to east. The whole space between the Djanbas-kala plateau, Kokcha Mountain and the riverbeds was filled with moving barchans sand, *takyr* surfaces of different levels, and several meandering, small, deltaic dried '*takyr*' riverbeds, crossing each other (5 to 25 m wide) (see also Tolstov and Kes 1960:42-44). In some places they were scattered and covered by moving barchan sands, in some places preserved only in the form of characteristic bands of vegetation (Figure 29).

Moving away from the ruins of the Djanbas-kala site (4th century BCE to 3rd century CE), the ceramic production finds of that period became rare on the *takyr* and, in the zones of narrow *takyr* lines alternating with low sand dunes, there were, some small and some large (50 × 100 m), accumulations of Bronze Age dark ceramics ornamented with incised and stamped patterns. In some places unclear traces of dwellings and layouts of narrow short ditches, revealed by the soil color and the microrelief, were visible (Itina 1967:figs. 4, 5; 1968:figs. 1, 2).

More than a dozen major settlements-villages were discovered in this territory in 1954 and 1964,[Note 84] consisting of semi-earthen houses widespread along the riverbeds, some of which, however, were clearly revealed on the light *takyr* surface because their darker soil color. Judging by the outer contours, the prevailing size of the dwellings was 6-8 m × 8-10 m (Itina

1968:78). They were dotted with fragments of pottery with characteristic geometric patterns, typical of Tazabagyab ceramics. From many sites, rich scatterings of surface material were collected, their features completely identical to ceramic complexes from well-known excavations and publications of the Tazabagyab culture of Khorezm (Tolstov 1939; Itina 1959a-b, 1960, 1961, 1963, 1967, 1968).[Editor 41]

In one of the largest settlements (*poisk* 1611 or Kokcha 16), the archaeological-topographic unit found out flint knife-shaped blades, a bronze knife with blade and a large millstone. The prospecting trench in the dwelling of that settlement carried out in 1964 by M. I. Itina, and the archaeological work of the group directed by her, which studied the sites of the Bronze Age in 1965-1966, dated this village and other neighboring sites to the third quarter of the 2nd millennium BCE. In 1964, unique sites of former riverbeds were discovered and surveyed. They were adapted for irrigation purposes, artificial head structures and minor ation canals. Later (1965-1966), during rchaeological work of the team studying ronze Age sites, several field layouts were vered.[Note 85]

rom the historical perspective of irrigation nology, of great interest are the preserved ains of ancient irrigation systems, i.e. traces former riverbeds adapted for irrigation, and remains belonging to distributor canals and ditches derived from them (see Figure 27.F-G).

Trenches were dug through some riverbeds, as in the environs of Kokcha 3. In the site of Kokcha 16, a trench 5 m long and 1 m deep was dug through the '*takyr*' riverbed (see Figure 28.C). Here is the description: 1) *takyr* thin crust; 2a) light-gray clay loam, dusty, stratified as thin plates, its lower border, of 10-15 cm is unclear; 2b) the same light-gray clay loam with different composition, with clear traces of digging, its lower border of 20 cm is sharp and uneven; 3) ashy-yellow clay loam, heavy compact, loam, with typical vertical pole-shaped single parts, numerous small pores with traces of roots and spots of salt accumulations, traces of digging

decreasing in the direction of the riverbed, of 20-50 cm; 4) dark-gray clay loam, compact, badly stratified into separate parts, sloping to the riverbed, with visible buried silt; 5) brownish clay loam, heavy, compact, horizontal stratification and vertical cracks with ochre-colored spots and bluish-gray pipe-shaped holes, lying slanting and in the shape of lens, of 10-30 cm; 6a) ashy-yellow sand, fine dusty, with traces of alluvial stratification, layers inclined towards the riverbed, the rare clay loam layers are characteristic, up to 60 cm; 6b) the same sand, with more clearly expressed stratification of different compositions, lens-shaped in the center of the riverbed; 7) gray sand, unstratified, partially oxidized, lens-shaped of aeolian origin, of 15 cm.

The description of the section (see also Figure 28.C)[Note 86] indicates that alluvial riverbed sediments lie in the lowest layer (6a); under the field these sediments are covered by loam with traces of agro-irrigation activity (layer 2b); on the edge of the riverbed, a layer (3) of compact clay loam with traces of digging infiltrated into sand. The upper border of this layer has a particular raised shape and an uneven wedge-shaped border with fluvial sands; this, according to the measured accumulation of the river stratification, highlights the fact that the ridge was rebuilt several times in the same place. Beside the riverbed, on alluvial sand there are heavy and compact loams (layer 5) formed from slowly flowing water and, when the channel began to fade, farmers, apparently, appeared. The buried silt (layer 4) and the sand of aeolian origin indicate that the water flow stopped; then, when the channel was supplied again with water for a while, alluvial sand (layer 6b) was deposited.

Thus, judging from the archaeological-geomorphological research, the construction of irrigation systems began on regressing lateral riverbeds, and former riverbeds of the small internal 'deltas' of the Akchadarya. By their nature, these channels are very similar to the deltaic riverbeds of Mesopotamia and with those of the Tedjen described by G. N. Lisitsyna (1965:42-74, Figures 4-11). The only difference being that the

Tazabagyab channels are better preserved on the surface (they are 1,500 years 'younger' than those of Geoksyur). In several places of their banks there were small, apparently artificial ridges (see Figure 27.F). So a plan was made of a part of the riverbed near the settlement of Kokcha 16 (Figure 30). On these places we identified both different sectors of low embankments (30-50 cm to 1 m wide), weakly expressed in microrelief and located along the north-south riverbed, and accumulations of ceramics, layout of dwellings and numerous diversion ditches, 0.5 to 1.5 m wide and several tens of meters long. Many distributors branched from the riverbed at right angles and some with sharp angles.

At the settlement of Kokcha 15, M. I. Itina traced the changes in topography of a small irrigation network: first, the system of small ditches departed at a sharp angle from the distributors to take water from the riverbed; later, when the former riverbed dried up, the canals were moved upstream and were parallel to the riverbed, repeating its meanderings (Tolstov, Itina and Vinogradov 1967:304; Itina 1968:81). The numerous intersections of canals in this region and in other places of the Tazabagyab 'oasis' show a very fast sedimentation and a short functioning of the irrigation systems. Canals and fields, and dwellings, were often moved from place to place following changes of the watered lands of the deltaic region which, as a matter of fact, became good archaeological evidence (Itina 1968:78-82).

The frequent shifting of irrigated plots indicates the instability of deltaic forms of the Tazabagyab's primitive irrigation. According to its schematic diagram (river - former channel distributor - field), it can be compared with irrigation of Southern Mesopotamia in the 6th-5th millennia BCE (where, initially, the role of the main irrigation works was attained with natural channels, whence the water fed the fields with small canals) (see p. 199). Similarities were also found with the irrigation of the Sakas, the Medieval tribes of the Lower Syrdarya, and even with the delta irrigation of the Karakalpaks in the 18th-19th centuries.

The continuous hydrographic changes in the deltaic channels, and the fluctuations of their level, forced the ancient irrigators to seek ways to artificially control the flood water level in the former riverbeds adapted for irrigation. The next extremely important step in the development of irrigation technology - the invention of the head works - was made in Tazabagyab.

In the Tazabagyab 'oasis' we surveyed some irrigation systems with head structures consisting of watering canals. The origins of one of the largest systems were preserved in the shape of two flat walls 8 and 13 m wide, 30-50 cm high above the *takyr*, scattered with the typical Tazabagyab ceramics (see Figure 27.G). These structures were first found by the archaeological-topographic unit in 1954 (*poisk* 662). They began in the now vanished riverbed. The origin of the system was covered by a small sandy ridge, behind which a scattered site was found due to the accumulation of ceramics over an area of 30 × 50 m. On the *takyr* around the canal, which crossed this accumulation, there were many traces of small canals 0.5 m wide.

During the repeated investigation in 1964, we followed the irrigation system further northeast, where two canals 'merged' into a former riverbed (*poisk* 1605). It was meandering and clearly expressed by the microrelief of the *takyr* surface and also by vegetation; it was 11-13 m wide between the two banks. This riverbed was traced further for 2.5 km. In the lower riverbed the small natural or artificial (?) ridges disappeared and the general direction of the riverbed continued with a small (2-3 m wide) artificial canal 2.5 km long. The total area covered by this system in the environs of the settlement was 70-90 ha. Almost throughout the whole length of the canal, small canals (0.5-0.7 m wide) branched from it at right angles and they were scattered around the remains of broadly dispersed dwellings, with pottery and other cultural remains all around. The settlement stretched for 1,200 m along the canal (see Figure 27.G) and the canal itself could be traced for further 1,300 m. Its slopes were similar to those of the former riverbed (0.2-0.3 m/

km) and quite enough for the water to flow from the former riverbed along the canal and continue in its direction. This system of irrigation of the third quarter of the 2nd millennium BCE well illustrates the thesis that the observation of the natural movement of flood water along the bed of drying channels gave rise to the development of the technique of leveling the canals path (Tolstov 1948a:45).

The trench made through the canal revealed at least two periods of life of this irrigation system. Here is its description (see also Figure 28.D): 1) *takyr* thin crust of 5 cm; 2) ashy-yellowish clay loam, loose, granular, partially with big lumps, stratified in the center, with many rusty ochre-colored spots and traces of digging, up to 20 cm; 3) dark-gray clay loam, dusty, compact and stratified, with weak ochre-colored spots and traces of gleysol, traces of roots and layers of sands at the bottom, of 30 cm; 4a) dark-gray clay loam, slightly sandy, stratified, with pieces of charcoal and other cultural remains, of 30 cm; 4b) the same clay loam, but without a marked stratification and with traces of digging, 50 cm; 5a) massive brownish stratified silt of heavy loam composition, with rusty spots and traces of roots; 5b) gray-yellow clay loam with pale ochre-colored spots; 6) yellow-gray sand.

As we can see from the description and the drawing (see Figure 28.D), there are clearly two main phases of this irrigation system, between which a break occurred. After the first phase the canal was cleared up to its previous depth and this was evident in the irregular border between layers 3 and 4b. It is possible that the second period of the system was connected to the construction of a new head structure found 1 km southwest of the old one. The canal function could hardly have been continuous for a long time, and, judging by the fact that the whole system does not intersect other canals but, on the contrary, it cuts across older ones (for example, the former riverbed and canals near the site of Kokcha 16), therefore it should be considered as the latest one in this place.

Of the total area of 70–90 ha covered by the settlement and the network of small ditches, the fields occupied no more than 20-30 ha. They were approximately the same size as the fields recorded by M. I. Itina's team on site of Kokcha 15 (7 × 7, 10 × 10, 10 × 12 m) (Itina 1967:74; 1968:81). We know nothing about the crops cultivated in these Tazabagyab fields. It might have been millet (typical of the Eurasian steppes in the Bronze Age) or barley, wheat and rye (recorded in Bronze Age sites in Southern Turkmenia and the Zeravshan Basin) (see pp. 130-131). It is very difficult to determine also the total number of dwellings and, consequently, the population of the area controlling this system, forming a *rustak* (*rotastak* in Middle Persian) or 'cultivating land, plowing field' (*shoytra* in *Avesta*) (ITN 1963:144, 507:note 49). If we proceed from the Zoroastrian tradition of kinship in the ancient Iranian villages (*vis'gm*), the number of families there were at least 15 and the tribes no less than 30 (ITN 1963:144, 507:note 60). The most recent purely hypothetical number can be compared with data taken by A. G. Perikhanyan from the Sassanid code of laws (6th-7th centuries CE), according to which the traditional (and going back to ancient times) average number of people was 100 people - males (Perikhanyan 1968:37). Perhaps in the observed Tazabagyab site there also lived at least 100 males, sufficient (according to earthworks calculations) to carry out and exploit the communal irrigation system.

Similar, but not identical, irrigation systems consisting of adapted former riverbeds and small distributors were found by the unit in the environs of Bazar-kala, in particular in the late Suyargan encampments of Bazar 1 and Bazar 3, dated by M. I. Itina to the 11-10 centuries BCE (Tolstov and Kes 1960:123-124, 132-133).

The encampment of Bazar 3 is located 9 km north-northwest of Bazar-kala on the dispersed *takyr* with an uneven loose surface. The *takyr* was surrounded by sand barchans covering the riverbed located 50 m west (see Figure 27.H). According to the evidence, a badly preserved dwelling with an oval-shaped plan (12 × 9 m in size), was discovered on the bank of a small '*takyr*' channel, and marked on the ground by a series of characteristic structural ledges. The overall width of the channel was 13 m. At the encampment a trench was made on the channel (Trench 44; see Tolstov and Kes 1960:132-133). The irrigation facilities consisted of several, almost parallel, meandering ditches preserved in the shape of gullies. They were revealed during topographical surveys on both the southern and northern sides of the encampment. North some canals were traced for 116-120 m; they were close to the transverse canal, through which the water discharged into the riverbed. The whole irrigation system area was less than 1 ha. The visible width of the canal, where another trench was opened (Trench 45), reached 1.7 m (see Figure 27.H and also Tolstov and Kes 1960:133:fig. 71).[Editor 42] The surface of the canal was uneven up to 1-2 cm, and there were clearly expressed steep ledges up to 3-5 cm high. Some places provided surface finds of late Suyargan ceramics.

Trench 45 concerned both a buried ditch and part of the field (see Tolstov and Kes 1960:133:fig.71). In the stratification of the ditches it was detected a lens-shaped deposit in the compact grayish-yellow clay loam and in the gray fine-grained, thin horizontal stratified sand, with layers of clay loams and clay lumps in the lower parts. Two thin layers of sandy loam, lying under the sand, repeated the contours of sand lenses (see Figure 27.I). Traces of digging were found at the lens side of the ditch. Somewhat different in character was Trench 44 dug through the 'takyr' riverbed which reached, as mentioned above, a width of 13 m near the site of Bazar 3.

The compact sand lens (Layer 4) lies in a dark-brown loam of marshy-cultivated origin, as indicated by the layers of buried soil with black humus spots and charred plant roots. Above the loam there were strata of thin layered sandy loam, which show a gradual change in the water regime and the appearance of flowing water. The sandy loam was covered by a massive lens of gray, fine-grained and stratified sand with inclusions of sandy loam and clay lumps in the lower part, a very typical feature of fast flowing river water. Such sand was also recorded in the trench, dug

through the ditch, indicating the simultaneity of their exploitation. It is very important to note that lateral areas of the channel, especially from the Bazar 3 encampment and fields, preserved traces of digging. They can be attributed to the clearing and deepening of the riverbed, and perhaps the strengthening of the channel banks, protecting dwellings and fields against the rising flood water.

Further improvement of irrigation systems, based on delta channels and former riverbeds, was traced by the team studying the environs of the Amirabad sites (see Tolstov l946c:66-67; 1948a:68-70; l948b:89-90; 1962a:68-74; Tolstov and Kes 1960:133-136; Itina 1963; Yagodin 1963) of Bazar 8, Bazar 10 and Yakke-Parsan 2 dated to the 9th-8th centuries BCE (see Figure 27.K-L). They are located in an area where the preserved sites of Bronze Age irrigation were connected and covered by later and more massive Archaic systems of the 6th-5th centuries BCE. Here, the surface was heavily eroded in the past by irrigation waters and then dispelled. Some areas have become a fanciful conglomeration of residual outcrops, towering among the vast sandy hollows, marked by sandy loam and widespread dunes.

The encampment of Bazar 8 was located 11 km northeast of Bazar-kala (see also *poisk* 571-576, 1954). The riverbed is expressed here by a bank 20 m in size with clear ledges of destroyed *takyr* surface. In the lower reaches, where the bank was close to the encampment, the width was reduced to 10-12 m and the height above the *takyr* level was from 1.5 m to 30 cm. From the riverbed, long (up to 1 km) and narrow (0.5 m to 1 m) distributing ditches departed at right angles, and are clearly marked on the ground as small narrow hollows with *biyurgun* and *saxaul* bushes (see Figure 27.K). In contrast to the very short feeders, in the encampment of Bazar 3 these long (up to 1 km) canals have, in the lower part, ramifications, which can be considered as a further step in the increase and complexity of irrigation systems, as well as the appearance of a new element: the distributors. The irrigation system in the environs of Bazar 8 has a more extensive control area (200 ha), which is five times more than the Tazabagyab systems at *poisk* 1609-1610, and almost 200 times larger than the system of Bazar 3, which is 1 ha.

The majority of Amirabad sites, and the remains of irrigation were discovered west of the Southern Akchadarya Delta, located along banks of the meridian riverbed, which along its whole length, beginning from Little Kavat-kala to Yakke-Parsan, was affiliated with sites of the Bronze and early Iron Ages (Tolstov 1962:fig. 29).[Note 87] From this channel, 3.8 km southwest of Yakke-Parsan, a small, rather meandering, dry 'takyr' riverbed, rising above the surrounding *takyr*, branched off at a right angle. It was preserved in the shape of a flat bank 15 m wide. The riverbed was 1.8 km long. It ended in a branching of delta channels with clear traces of irrigation ditches (2-3 m wide). Right here we discovered a whole Amirabad village consisting of about twenty semi-earthen houses; it was called 'Yakke-Parsan 2' (Tolstov 1962a:68-74; Itina 1963:fig. 1).

The study of the test trench dug on the mainstream south of this site by M. I. Itina's archaeological unit was particularly interesting (Trench 5). Here is its description (Figure 31). The trench was 18 m long and 1.4 m deep. The layers are: 1) *takyr* thin crust preserved only on the left bank of the riverbed; 2) light-gray clay loam, ochre-colored, compact, up to 40 cm; 3) light-brown clay loam, thin dusty, compact, forming in the riverbed three mixed layers divided by sands (layer 4); 4) dark-brown clay loam, compact, with salt efflorescence, the upper border is unclear, the layer is inclined toward the left bank of the riverbed, in the center, a lens-shaped layer (the lens length is approximately 4 m long), of 40 cm; 4a) light sandy loam, grayish-yellow, dusty, compacted, with rare layers of loam and traces of digging near the riverbed, of 30 cm; 4b) the same sandy loam with stratification of clay loam and of digging, inclined and wedged towards the bottom of the river, up to 80 cm; 4c) dark-gray sandy loam, loose, with lumps, inclined, wedged towards the bottom of the riverbed; 4d) light ochre-colored

sandy loam up to 100 cm; 5) gray sand, with high inclusions of mica and traces of alluvial stratification forming a layer intermixed with loam and filling the riverbed; 6) swampy layer.

In this trench successive stages can be traced of the gradual silting and the consequent turning of a wider (15 to 20 m), fading natural riverbed into an artificially deepened, regulated riverbed of smaller size (8-10 m wide). The alternating layers of light-brown loam, with lenses of gray alluvial sand, illustrate well stages of the riverbed covered by sediments and the sharp changes in water regime: fast flowing waters depositing sands and slow water depositing silt particles. Traces of digging in the clay and sandy loams sediments along the banks indicate that this regime change was caused by activities of the inhabitants of the Amirabad village, who regularly cleared the sources and the bottom of the riverbed (see also Itina 1963:128). As a result, the riverbed gradually assumed the character of an artificial main canal and its size decreased from 8-10 m to 3.5-4 m, and 2.5 m between banks. The last stage of life is connected with a layer of dark-brown heavy loam formed under silting conditions and fading of the system. According to the remains of buried swamp sediments (layer 7 at a depth of 35-40 cm), high floods occasionally broke through the banks and flooded the surrounding lands; the flooding of dwellings (after which they were abandoned) was recorded by M. I. Itina during the archaeological excavations of Yakke 2 (Itina 1963:111:fig. 3).

Within the limits of the village several trenches were also dug, in particular a trench (Trench 1) through the narrow 2 m canal, which was a continuation of the main canal mentioned above. On the surface there were remains and traces of parallel banks of a narrow canal, which was dug in the sediments of a canal more than 10 m wide. M. I. Itina found Amirabad ceramics in the sediments of this more ancient canal (Itina 1963:110).[Editor 43] The total trench width in the later canal was approximately 4 m and 1.8 m between the banks. The layers are: 1) light-gray *takyr* thin crust, of 5 cm; 2) gray *takyr* clay loam, with inclusions of sand and traces of digging, up to 20 cm; 3) loose clay loam, of 40 cm; 4) light-brown clay loam, compact, with rusty spots; 5) gray-yellow sandy loam, inclined, of 10-15 cm; 6) gray sand, fine-grained, thin stratified, lens-shaped, on the bottom of which a pottery fragment was found, thickness of 40 cm; 7) light-gray sand, thin stratified sand, with high quantity of mica, inclined along the bottom of the canal, of 10-20 cm; 8) layer of buried swampy soil at a depth of 35-40 cm.

The active process of river sediments was very characteristic for the irrigation system of Yakke-Parsan 2, so it is interesting to try to estimate the amount of labor expended for cleaning the irrigation works. Considering the reduction of cross-sections (from 8-10 m to 3.5-4 m and 2.5 m), in Trench 5 cleaning was only partial and an estimate of its total amount is rather difficult. In the last stage, when the section was no more than 3 m^3, 3,000-4,000 m^3 of soil had to be removed for cleaning the system.

How many people lived in the village of Yakke-Parsan 2? If we accept that the settlement had about 20 dwellings (Tolstov 1962a:69), and every house was inhabited by an expected average of no more than 20-25 people,[Note 88] the total population was likely to be 400-500 people, including 100-150 working men. They spent from 10 to 15 days cleaning the irrigation system (with an average daily rate of no more than 2 m^3).[Note 89]

Typical of Amirabad irrigation was the combination of former channels or regulated riverbeds, transformed over time in main canals, and small, short distribution and irrigation ditches were also recognized in many other ancient settlements of that time, in particular in the encampment of Kavat 2 (9th-early 7th centuries BCE) (see Yagodin 1963). The encampment of Kavat 2, located on the bank of a small dry channel, was marked by clusters of pottery, fired stones and bones traced along the canal. The latter was preserved as a dark meandering line 2-2.5 m wide. With the silting process and further artificial cleaning, it gradually assumed the character of a main canal.

The main canal of the Amirabad period, 3 km long and a completely artificial irrigation construction, was found and studied in 1954 by the archaeological-topographical unit north of Bazar-kala near the end of the Archaic irrigation system. In one of these areas, in close proximity to a massive Archaic canal, which survived as a *takyr* strip 40-50 m wide and some remains of natural hills, traces of a more ancient, heavily damaged, canal were found, with a total width of 11-18 m, beginning in the riverbed northeast of Bazar-kala (*poisk* 504, etc.). In the vicinity of its head Antique pottery was absent, but clusters of late Suyargan and Amirabad ceramics (*poisk* 503-510) were found. The middle part of the canal intersected with the later Archaic one covering it. Here, on the *takyr* surface, the Amirabad ceramics mixed with Archaic ones. Three kilometer from the head, the canal intersects the vast Amirabad sites (500 m long and 100 m wide) of Bazar 10 and 11 (see Figures 27.L and 29).

The development of irrigation technology in the Lower Amudarya during the Bronze and early Iron Ages was a complex process of research and accumulation of experience by the ancient irrigators, grasping the knowledge of natural laws which rule water movement in fading delta channels and former riverbeds, and beginning to use them to regulate the water supply to fields. That was the way in different deltaic areas of the ancient world, from a very primitive single and temporary estuary form of irrigation to the establishment of irrigation systems in small fields placed along the banks of former channels or fading riverbeds adapted for irrigation (see pp. 109-110, 132-133).

In the Tazabagyab sites of the third quarter of the 2nd millennium BCE irrigation systems different for their complexity were recorded: vegetable-fields, small irrigation canals on former riverbeds, as well as more developed systems, consisting of head canals, former channels or regulated (re-deepened and embanked), and the small distributors derived from them.

People learned from nature to adapt the fading (silting) deltaic channels for gravity irrigation. However the progress of irrigation techniques in

Prehistory was very slow. Only in the early 1st millennium BCE there was a gradual development of skills in irrigation. The short irrigation canals of the early Bronze Age, i.e. lateral canals (not exceeding a few tens or hundreds meters), in the Amirabad period (9th-8th centuries BCE) were replaced by longer ones (up to 1 km) with branched distributors. Thus new features appeared in the irrigation system from riverbed to field. The regulated former riverbeds were gradually changed into small main canals.[Note 90]

Along with the changing character of irrigation facilities in the Amirabad period there was a noticeable increase of irrigated areas. Thus, in the primitive irrigation of Khorezm we have both qualitative (complex water supply systems to the fields) and quantitative changes. They characterize the development of productive forces and the emergence, already in Prehistoric communities, of the main irrigation devices that represent the following historical period, i.e. the Ancient Khorezmian state (Tolstov and Andrianov 1957:6).

Archaeological excavations of Khorezm sites of the Bronze and early Iron Ages indicated that farmers lived in small villages consisting of several dozens semi-earthen houses. These settlements were placed along the banks of small irrigation systems, and the sphere of their agricultural production activity was limited to the basin of the settlements. Thus, in the *Gathas*, not accidentally the same term refers to 'tribe' (*shoytra*) and 'cultivated land', 'field', 'district' (Livshits 1963:147). The scale of irrigation systems did not exceed the physical capabilities of the family communities that had no more than a hundred men's hands.

The material describing the irrigation facilities of the Southern Akchadarya Delta in Prehistory helps to understand the overall process of gradual transformation of former riverbeds, through silting and re-deepening, into artificial canals, as well as the origin of irrigation systems with multi-headed hydraulic structures ensuring a more stable and permanent diversion of water from the lateral deltaic channels. However, during Prehistory, when the Amudarya, in the words of V. V.

Bartold, "subordinated the people of Khorezm to itself", the channels had not yet been brought under farmers' control, for this reason the heads of canals were often blurred, and the canals and fields were subsequently moved from one place to another (in areas with a more favorable regime of flood water). The farmers depended mostly on the continuous changes of delta channels, which they could overcome only by regulating flood waters of the Amudarya large channels, using dams and extensive long-term irrigation systems. This required the organization of large-scale irrigation farming.

Origin of the Ancient Khorezm State and the Development of Irrigation

The deltaic plains of the Amudarya were already densely populated by farming peoples as early as the Amirabad period (Tolstov 1948a:45). The character of irrigation was broadly correspondent with a society that still did not know the state, preserved a tribal organization, and stood on the threshold of a new socio-economic formation (Tolstov 1939). At that time, in Khorezm as in many other agricultural regions of Central Asia located near the world of pastoral steppes and supporting them with close economic ties, the relationships between the different populations intensified, antagonisms between economic structures arose, and the growth of property differentiation between and inside tribes began.

In the first quarter of the 1st millennium BCE, the Eastern Iranian tribes experienced a period of expansion of primitive communal societies due to social and property differentiation and the formation of a class society (Tolstov 1940b; 1948b:95; Dyakonov 1954; V. Masson 1959:122-135; Livshits 1963:149). These transitional societal features, which stand at the beginning of class formation but have the characteristics of a military democratic society, were reflected in the ancient sections of the *Avesta* texts (Tolstov 1935a, 1940b; Artamonov 1947:83; Kosven 1960; M. Dyakonov 1961:58-65; Livshits 1963; Masson 1968).

S. P. Tolstov, connecting the homeland of

Zarathustra's doctrine with Khorezm, wrote that the *Avesta* describes the society of "settled herders and farmers breeding cattle, horses and camels. All property interests rotated around the livestock" (Tolstov 1948b:96). Cattle breeding suggests the traditions of the complex deltaic economy of the semi-settled herder-farmers, where a major role in the food supply was played by naturally irrigated pastures and reed thickets, so typical of the landscape of the lower reaches of the major Central Asian rivers. Along cattle-breeding the population was engaged in agriculture. One of the parts of the *Avesta*, the *Videvdat*, preaches: "The root of faith is to revere Mazda ... tillage; who cultivates cereal, also cultivates the Truth". The *Avesta* reported "on the vast land that produces a bright abundance", in which "the water flows through canals, the plants grow to feed cattle and people, to feed the Aryan countries" (Yasht 13:9-10). The texts refers to channels or canals, for instance (Yashts 8, 29, 3; Herzfeld 1947:539):

Good fortune to your springs and your fields,
Good fortune to you and (your) lands,
Your channels (canals?) will be watered,
Let full-weight yield of grains pour on the fields
And yield of grass pour on pastures.

Interestingly, to note that a community farmer was called in the *Avesta* a 'supplier of grass to cattle' (*vāstrya-fśuyant*), which, according to E. Benveniste, reflects the relatively recent transition of population engaged in farming (see also Dyakonov 1961:363).[Note 91]

The rich pastures of the deltaic regions of the major Central Asian rivers, which served as the basis for development of cattle-breeding to the present day, in that remote period could not have been the cause of violent tribal conflict and wars. In his very interesting article on the economic and socio-historical characteristics of the Eastern Iranian tribes of the 8th-6th centuries BCE, based on the information from the *Avesta*, V.I. Abaev noted that the ideology of early Zoroastrianism emerged mainly from the antagonisms and struggles of settled herders,

the Aryas, with the nomadic herders, the Tura (certainly, Iranian speaking), related to the Aryas. According to V. I. Abaev, this struggle with the 'Scythian way of life', which became a social program of the preacher Zoroaster (Zarathustra) was not only a struggle with neighboring Saka (Scythian) nomads, but the struggle with those groups within the same people who have sought to follow a 'Scythian way of life'.

The struggle for unification, defense against forays (*Aēshma*), establishment of strong power (*Khshathra*) and a common world (*Aramati*) on the entire Aryan land, stretching from Herat, the Paropamisus Mountains to the Merv Oasis, the banks of the Amudarya and the Aral Sea: all this is consistent with the ideology of the dominant social groups to create a strong irrigation-based agricultural economy and a strong government.

The historical-social interpretation of the *Avesta* texts made by S. P. Tolstov (1940b, 1948a-b), V. V. Struve (1948, 1949), M. M. Dyakonov (1954, 1961), I. M. Dyakonov (1956) and V. I. Abaev (1956) reveals a picture of the struggle between the settled farmers of Khorezm, Margiana, Sogdiana and Bactria and the pastoral tribes in the valleys of the great Central Asian rivers.[Editor 44] This is a picture of the increasing crisis between the primitive communal relations and the establishment of a strong power unifying different tribal groups, without which it would have been impossible to organize a large-scale irrigated economy. This process of forming the first state organizations on the territory of Central Asia took place amid violent political and social conflicts. Ideologically, this process was reflected in a religious worldview, as in the passionate sermon where Zoroaster proclaims the farming as the first virtue of the true followers of Ahura Mazda (see Struve 1949:149; Bartholomae 1905:127, 133).

According to the archaeological material from the Murgab Oasis, the process of organization of a large-scale irrigation economy (based on the canals Gati-Akar and Guni-yab),[Note 92] and of a local state power, was established there earlier than in Khorezm (V. Masson 1956a:22-23; 1958:59-61;

1959:90-91; ITN 1963:149). Thus, in the Murgab Oasis of the 9th-7th centuries BCE, there are small holder centers (settlements with citadels of the Yaz 1 culture) and the major main canals have a length of a few tens of kilometers.

How can the earlier development of the Murgab Oasis be explained? As in many countries of the ancient world, in the territory of Central Asia the borders of many early state unions coincided "with the borders of irrigation systems and their irrigated lands" (Dyakonov 1954:178). In the fading channels of the lower reaches of the Northern Murgab Delta, the regulation of riverbeds was less complicated and less difficult than the regulation of the full-flowing channels of the Amudarya River, therefore, earlier (in favorable historical and cultural conditions), the local state of Margiana formed here also. It is in this area (on the Murgab or the Tedjen) that many researchers have localized what Herodotus described as complex regulated hydraulic works on the Ak River (Akes), flowing on the junction of the "land of Khorezmians themselves, Hyrcanians, Parthians, Drangians and Thamanaei" (see Herodotus III, 117; Markwart 1938:9ff.; Henning 1951:42; Gulyamov 1957:96; Dyakonov 1956:357-358; Masson 1959:125).

However, in the period before the Achaemenids, in the Khorezmian Oasis with its vast natural and human resources, the mighty Saka-Massaget military-democratic confederation of tribes was formed and gradually became a unified state headed by the Siyavushids' kin (Tolstov 1948a:341-342). The formation of such a union was the historical prerequisite for further progressive development of the economy of the Khorezm Oasis. Initially, when the tradition of military democracy still predominated and wars were, according to F. Engels, 'a permanent job', the mighty tribal alliance of the Saka-Tura, or Saka-Khorasmian, directed all their forces to conquer the neighboring, more developed, southern countries (see Pyankov 1963).

According to Herodotus (III, 117), before the establishment of the Persian rule, the Ak Valley belonged to the Khorezmians (Bartold 1965: 544).[Note 93] On the basis of this report, J. Markwart and S. P. Tolstov concluded that before the Achaemenids there was a powerful state union headed by Khorezm, the 'Great Khorezm', (see Markwart 1938; Tolstov 1948a). They identified Khorezm itself as the legendary homeland and the first area of the spreading of Zarathustra's teachings, the Airyana Vaējah. (Benveniste 1934; Marquart 1901:118, 155; Markwart 1938:10-11; Tolstov 1948a:19ff., 286ff., 341; 1948b:103ff.; Henning 1951:42-43; Gershevitsh 1959:14-21; 269-299; Livshits 1963:151).

In the Greek sources, Khorezm ('city of Khorasmiya' and country of 'Khorasmians') was first mentioned by Hecateus of Miletus, predecessor of Herodotus, as a populated, partly flat and partly hilly, country east of Parthia (Bartold 1965:23-24, 544). This clearly described the area where the Khorezmians ruled politically (ITN 1963:152; Dyakonov 1961:64). In particular, on the traditional 'Great Khorezm' political confederation of the pre-Achaemenid period (according to S. P. Tolstov), was also based the inclusion of the Khorezmians during the Achaemenid Empire in one taxed province (16th Satrapy), together with Sogdians, Hariana and Parthians (Tolstov 1948a:341). M. Dyakonov rejected the legitimacy of this comparison (Dyakonov 1956:348). However, as rightly pointed out by V. A. Livshits, "It is largely possible to assume that the Khorezmians controlled vast territories in Central Asia over a long period of time" (Livshits 1963:153).

However, these very difficult problems of Khorezm political history and its neighboring countries in the 8th-7th centuries BCE were solved, one thing is completely clear: the acquaintance of the Khorezmians with the southern, more ancient agricultural, oases of Margiana and Bactria contributed to the rapid progress in irrigation skills in Khorezm. Perhaps, the influence of the experience of the southern farmer-irrigators played an important role in this process. The Archaic Khorezmian culture of the 6th-5th centuries BCE, which replaced the Amirabad culture, is characterized by an extremely close relationship with the cultures of Southern Central Asia, especially Margiana (Tolstov 1962a:108). They are clearly manifested in the similarities of the ceramic complexes (Vorobeva 1958:344).

Irrigation Works of Antiquity

Archaic Period (6th-5th centuries BCE)

This is the time of construction of massive irrigation systems both on the right and left banks of the Amudarya, the formation of a class state in ancient Khorezm, the wide agricultural development on vast deltaic plains, and the development of handicrafts, especially pottery (Tolstov 1948b:101-103; Vorobeva 1959a:66-84).

In the second half of this period, Khorezm was included in the Achaemenid Empire (see Struve 1946; Dyakonov 1961:77; Masson 1959:139-145) and, apparently, it contributed to the overall development of agriculture and handicrafts, and also the broadening of the range of crops (at that time the grape appeared in Khorezm). According to V. V. Bartold and S. P. Tolstov, by the end of the Achaemenid Empire, the population of Khorezm already formed an independent kingdom (Bartold 1965:544; Tolstov, 1948a:17; 1948b:103-109).

In her study of ceramics of the Antique period of Khorezm, M. G. Vorobeva led many arguments for dating the pottery of Archaic Khorezm to the 6th-5th centuries BCE (Vorobeva 1959a:65-84; see also Tolstov 1941:178; 1948a:77). This pottery characterized the numerous plowed and dispersed settlements on the banks of great canals, differing sharply in size from the previous Amirabad system of irrigation.[Note 94]

The archaic canals were dug parallel to the deltaic channels, as if repeating the hydrography contours. The strong pattern was recognized in the near Bazar-Kala, whose shape, perhaps, reflects most aptly the ancient Iranian name of the country 'Khorezm' as 'A country with a good enclosure' (Bogolyubov 1962:370).

Weak traces of canals tracked by us on the west and east of Bazar-kala were, however, dated to the more recent Kangju or even Kushan periods

(Tolstov, 1962a:104). The banks of the eastern canal were completely destroyed by erosion and preserved only in the bottom central part, bordered by a series of parallel ledges, detected on photographs by light and dark lines visible through pattern of shifting sands (see Figure 5). The impressive walls of Bazar-kala, towering above this sea of sand, and the hardly detectable traces of ancient activities of farmers, according to the abundant dispersions of typical pottery, are here attributable to the Archaic period (see also Gulyamov 1957:77). The Archaic ware accompanies this canal along its entire length: from the beginning (3.5 km south of Bazar-kala, where it branched out from a 60-70 m lateral channel of the Akchadarya) to its lower parts near the encampments of Bazar 1. There it was preserved slightly better and was expressed on the flat terrain by a 40 m wall rising 1-1.5 m above the *takyr* (Figure 32). The canal seemed to be a continuation of the channel and resembled the supply scheme (river - former riverbed - main canal - feeders - field) of the previous Amirabad systems but sharply different in size: the length is 8-9 km long and the total width between the banks is about 40 m.

The western canal was also preserved in some parts as a bank of 40 m. Like the eastern one, it meandered and continued in the direction of another small dry channel, expressed by a depression 60-80 m wide filled with sand. In its head section (*poisk* 359) we found ceramic production remains of the Archaic period (see also Vorobeva 1961:167). In the area of the canal we identified remains of small irrigation ditches, branching only on one side. At one of the feeders (*poisk* 701), with a total visible width of 2.7 m, we dug a trench revealing the lens of a ditch. The banks were not preserved and the canal was marked on the surface as a weak depression of 2-3 cm, covered in some parts with sand, *Salsola*, and small bushes of *saxaul*. Archaic ceramics were collected inside the canal and the fragment of an Archaic vessel was found in the trench at a depth of 10 cm.

A diked field, $88 \times 90 \times 92 \times 112$ m, was found beside the ditch. The field was covered with some fragments of Archaic and Kangju ceramics, based on that we dated the period of agro-irrigation activity to the inhabitants of this region, which coincided, probably, with the main period of life of Bazar-kala, i.e. with the late Kangju and even Kushan periods (Tolstov 1962a:104). The latest sites of this region had remains of ceramic production (both on the mound and outside it) belonging mainly to the Kushan period (poisk 320, 616, etc.) (Vorobeva 1961:166-167).

Regarding the history of irrigation works in the environs of Bazar-kala, the third Archaic canal, constructed later than the two systems described above and crossing them in its lower reaches, is the most interesting. This larger canal (more than 40 m wide between banks) had its source not in the lateral channel or former riverbed, but already in the main riverbed of the Akchadarya. Moving the head of the main canal to a major river channel represents the next important step in the development of irrigation skills in Archaic Khorezm.

The canal was dug strictly parallel to the lateral 'Kokcha' channel, already well known to us from the graveyard of Kokcha 3 in its lower reaches, for the numerous Suyargan and Amirabad encampments, and for the remains of small irrigation systems, the latest of which, apparently, immediately preceded the Archaic systems. This was proven by the fact that on the banks of the ancient channel (which had a maximum width of about 100 m), where it branched off from the Akchadarya riverbed, we found the Amirabad period intake as well as Archaic systems (poisk 454, 463, 555, 556). Some parts on the left bank of the channel show traces of several canal reconstructions. Their beds were largely dispersed, detection was only possible by soil and vegetation, and covered with Amirabad and Archaic ceramics.

The massive Archaic canal was dug along the right bank of the channel. It was heavily eroded and destroyed. Not far from the head its total width was 42-45 m. In the central part, the microrelief and the strip of vegetation exposed a more ancient and narrow (15 m wide) Amirabad

bed, well traceable downstream (2-3 km north of the ruins of Bazar-kala). There, both banks (1 m high) were well preserved and their width was 40 m (Figure 34.A). These sizes remained the same over the whole length of the canal, which was very straight and had few branches, mainly in one direction. The canal, apparently, did not have a significant depth compared to its wide width (compare with Gulyamov 1957:90). It was traced for 11 km, starting from the head (Gulyamov 1957:77). According to our modest calculation, the construction of such a canal would have required 500 workers for 35-45 days in order to extract 50,000-65,000 m^3 of soil (at a rate of 3 m^3 per day).

In the Archaic period when the intensive drainage of the Southern Akchadarya Delta channels started, the population constructed not only artificial 'rivers' (as exemplified by the above mentioned Archaic canals), but also tried to create small artificial 'deltas'. The ancient Kelteminar is an example of such kind of hydraulic solution suggested by nature. The irrigation network of the Archaic Kelteminar reproduced in plan a small river delta layout (Figure 34.A) (see also Tolstov 1958:112-113: *Karta Drevnyaya irrigatsionnaya sistema kelteminara* - Map of Kelteminar ancient irrigation system).

Archaic Kelteminar

A complex network of Archaic main canals, together with the remains of dwellings spread an accumulations of Archaic ceramics, along them, starting 10 km south of the Koy-Krylgan-kala ruins in the environs of Eres-kala.[Note 95] Here, from the large north-south Akchadarya channel, another, even smaller riverbed departed northeast to the right towards Djanbas-kala. On the right bank of the lateral channel, our studies revealed a complex knot of head canals from mutiple periods feeding the ancient Dingildje Oasis and, on the heavy sands on the left, the origins of the Archaic Kelteminar system. This place, near the riverbed bifurcation was well chosen. It met the necessary requirements described by Ya. G. Gulyamov (1957:239).

The main canal was made along the Akchad-arya riverbed on the edge of the *takyr* shield, which here was 1–2 km wide and stretched for 15 km in different directions across large and small canals, remains of fields, gardens and vineyards, badly damaged by erosion. Among them, sometimes there were eroded mounds of ancient dwellings with material culture remains of the ancient inhabitants. All these were covered by shifting sand, forming entire dunes, or opening up vast clay areas of *takyr*, with rare sand dunes almost without vegetation.

Judging from the numerous pottery finds of the Archaic period, the largest and most ancient canal bordered the left area of *takyr* loams located between two large Akchadarya dry riverbeds. The canal left the ruins of Karga-kala, Adamli-kala and Koy-Krylgan-kala (Figure 34) right, skirting the latter site from north, in the shape of branching *takyr* bands and heavily destroyed *takyr* remains, and continued flowing northeast in the direction of Bazar-kala. Apparently, it was rebuilt several times, as proven by the remains of at least four parallel canal beds, situated at a distance of 25 m to 100 m from each other. This indicates the long-term activity of the whole system. It lasted with reconstructions, from the 6th–5th centuries BCE to the 4th century CE (in the environs of Koy-Krylgan-kala) and up to the 7th–8th centuries CE (near Adamli-kala). The canal silted quickly and, when the filling reached the canal banks, it was more economic, in terms of labor costs, to dig a new main canal, whose head was built upstream.

The biggest canal (the western) of the Kelteminar system had a total width of 50 m, including bank ridges. From the upper part of this canal, only *takyr* remains (*poisk* 142) (Figure 33.11) were preserved in deep sand. The total width of the banks was 52 m with a relative height of 2–2.5 m and a 12 m distance between banks. The head parts were washed away. One kilometer downstream, the canal narrowed to 45 m wide (*poisk* 42) and the width between the banks was 9 m. Near the ruins of Koy-Krylgan-kala, the canal was largely destroyed (*poisk* 113)

and it was difficult to establish its width but, apparently, it did not exceed 45 m. North of Koy-Krylgan-kala the canal was preserved without levees in the form of a flat bank (45 m wide) buried by sand dunes (*poisk* 367) (Figure 33b:6). From there, and further northeast, the main canal branched into a number of very meandering, narrower distributors, between which were the remains of dwellings in the shape of eroded hills, and different accumulations of Archaic pottery, as well as fields (Tolstov and Vaynberg 1967:7–10, fig. 3).

The most significant branch of the main Archaic canal began in the area between the Afrigid ruins of Karga-kala and Adamli-kala, from where the canal ran along the other large branch of the Akchadarya to Djanbas-kala. Here the canal was preserved without levees in the shape of a massive flat dike about 40 m wide (*poisk* 126). The majority of branches of the ancient Kelteminar system watering the environs of Koy-Krylgan-kala were derived from the main canal at right angles, when they did not follow the branches of the ancient riverbed (see also Lisitsyna 1965:127). The rare, very badly preserved small irrigation networks, frequently were based directly on the main canal and departed at a right angle. As the main canal ran along the ancient riverbed, as a rule the branches were located only on its opposite side. The whole irrigation network covered an area of approximately 2,000 ha and, knowing the dimensions of the canals, it is possible to calculate the labor costs required for its construction and exploitation. On the main canal, 15 km long (and 25 m² in the middle cross-section), at least 350,000 m³ of soil were removed, and 50,000 m³ from the lateral branches. For these works, not less than 5,000 diggers were engaged for 30 days (at a rate of 3 m³ per day). After the construction, the canal had to be cleaned from the annual silt sediment (which amount was not less than 30–40%) (Smirnov 1933:9). Two thousand people were required for canal cleaning every year at the same time period. The organization of such work was possible only with state power.

Dingildje

The Archaic irrigation systems of the ancient oasis of Dingildje are similar in size and layout to those of Kelteminar (Figure 35). This system also had bunched branches in its lower reaches. Most of the distributors and irrigation ditches branched right. Among ancient fields, the remains of many Archaic settlements and particularly the remains of many Archaic houses[Note 96] were excavated by archaeologists under the direction of M. G. Vorobeva in 1956 and 1959-1960, and dated by them to the mid-5th century BCE (Vorobeva 1959b:78)

The ancient agricultural oasis of Dingildje stretched 5-6 km from south to north and 2-2.5 km from west to east in a flat, slightly south declining desert plain, entering like a sharp wedge into the innermost sand of the Kyzylkum desert. The channel of the Akchadarya southern delta flowed, describing an arc, toward it from southeast. It bordered the plain on the north and flowed northeast in the direction of Djanbas-kala. The southern main canal, up to 40 m wide including banks which were not preserved, was traced along the former channel on the right bank. It was traced also in the environs of Eres-kala. The head parts of this canal were placed, possibly, south of the sources of the Kelteminar, in the main channel of the Akchadarya. Here, the remains of banks and an elevated central part of the bed, largely eroded and covered with fine fragments of ceramics, were preserved only in some places. About 500-600 m west of the ruins of the Afrigid castle of Dingildje, the canal gradually turned north and it branched. Its left branch took the shape of an arc following the curve of the former riverbed. North of the Archaic Dingildje house, the canal was marked only by a series of *takyr* bands 35 m wide. Four lateral distributing canals branched from the main canal, they were preserved as weak traces in the shape of narrow (1.5 m) depressions covered with small Archaic ceramics in the environs of the Archaic Dingildje house. Another large branch on the right side flowed first eastward then, perhaps following a small branching channel, turned sharply south along high sand ridges. The remains of a large

Archaic site, in the form of ceramic accumulations with an unclear layout, were preserved on its banks among the moving sands. Although the *takyr* surface was largely damaged, it was possible to trace the width of the canal in some places. In profile it looked like a very flat wall, decreasing 1-1.2 m in the central part, and up to 40 m in width. On the surface it showed traces of later small irrigation ditches.

According to the ceramic deposits, six or seven rather large settlements existed in the basin of the Archaic Dingildje system. Traces of a rectangular layout, with lines belonging to the remains of a vineyard 50 × 35 m in size, were preserved near the Archaic Dingildje house excavated by M. G. Vorobeva.[Note 97] The vineyard was cut in the center by irrigation ditches. The width of a *gryad*, known in modern Khorezm as *karyk*, had a traditional size of approximately 4 m. The vineyard agro-irrigation layout was connected to the house and the irrigation system in a single topographic complex.[Note 98]

The Archaic fields (diked plots), recorded by us north of the main canal and the Archaic Dingildje house, were used for different cereals.[Note 99] Many fields had irregular quadrangular layouts; one of the fields measured 30 × 17 × 25 × 38 m (*poisk* 155). They were three to four times bigger than the Bronze Age fields.

Along with the well-known neighborhoods of Dingildje, another area northeast of Dingildje, near an Archaic two-roomed dwelling, was interesting (*poisk* 1204). The remains of a small irrigation network were preserved around the dwelling. The ditches were indicated on the *takyr* surface by depressions 2-3 m wide and the accumulations of pottery found in some places of the bed. The main distributor had a preserved bed 5 m wide. There were remains of diked fields on both sides. The shape was mainly sub-rectangular with the prevailing sizes of 23 × 22 × 21 × 25 m, 25 × 38 × 17 × 30 m, etc. (see Figure 33.5).

Archaic irrigation systems (individual parts of canals and scattered sites) were noted in many places of the Southern Akchadarya Delta and everywhere they followed the direction of the natural channels, which indicated the continuation of the Prehistoric irrigation tradition. However, the huge difference in the water supply pattern (presence of massive main canals and branching distributors) should be emphasized and especially the sharp increase in canal sizes. It is necessary to note that it is hard to establish the real size of canals, as the remains of the banks were preserved only in few places. Therefore it was impossible to detect also the reduction of their cross-sections from the heads to the lower reaches. Very little can be said about the character of the head works of that period (the latter is much better preserved on the left bank of the 'lands of ancient irrigation'). According to the Bazar-kala system, it is possible to assume that, through the transfer of head structures to large channels, the systems lasted longer and functioned better.

The Archaic canals of the right bank of Khorezm repeated the configuration of the Akchadarya ancient delta and followed the direction of the major channels. This general pattern was characteristic for many deltaic areas of the ancient world (see also Forbes 1955:17; Willcocks 1903:20). V. A. Shishkin, for example, came to similar conclusions for the Bukhara Oasis, where deltaic channels of the Zeravshan defined the most appropriate direction of the ancient canals (Shishkin 1963:228). Compared to the Amirabad canals, the Archaic canals were characterized by a huge size (up to several meters between banks), but were, apparently, not deep. Ya. G. Gulyamov explained this circumstance by the fact that the creators of the major canals followed the overflow and just adapted "the natural thalweg, with the result of obtaining small and wide canals. They compensated the width with a shallower riverbed depth, otherwise the canal could not have taken all the freshet water" (Gulyamov 1957:79).

The transition to wide main canals in the Archaic period was likely dictated by the ancient irrigators's desire to reduce silting in the irrigation systems head parts. Modern specialists in irrigation believe that wide and shallow canals become silted less fast than narrow and deep canals (see

Glebov 1938:130; Tsinzerling 1927:399-406; Dunin-Barkovskiy 1960:54).

The Archaic irrigation system was very uneconomical in terms of water consumption, not only for the canals width, but also because of the peculiarities of the lateral branches and distributors. The branches departed from the main canal at right angles, whenever they did not follow branches of the ancient riverbeds. The rare distribution network supplying water to fields, located near the main canal, branched at a right angle. This 'sub-rectangular' irrigation was typical of the Tazabagyab systems of Khorezm and the earlier stages of irrigation skill development in other areas of the ancient Orient. In particular, these systems belong to the Eneolithic oasis of Geoksyur described by G. N. Lisitsyna (1965:124). She compared the Geoksyur network with the Sumerian one, which was depicted on a clay tablet from Lagash (Lisitsyna 1965:126). On this tablet, the canals and fields had a typical 'sub-rectangular' configuration. Another striking example, from Southern Arabia, is most of the agro-layouts of ancient Qataban in the 1st millennium BCE, which had, according to plans and aerial photography, a 'sub-rectangular' configuration (Bowen 1958:90, 106-111).

The 'sub-rectangular' character of the Archaic irrigation systems of Khorezm led to an excessive water consumption and a quick silting of canals, which increased the volume of work needed to clean the distributing network.

The lands watered by the Archaic canals were considerable and the size of cultivated plots between huge deserts was small. There is reason to assume that what was sown was no more than 5-10% of the 'whole irrigated' land. The coefficient efficiency of these systems was still relatively small and the water expenditure very high and uneconomical. The construction and maintenance (seasonal sediments cleaning, strengthening of levees, etc.) of such irrigation systems (artificial 'rivers' and 'river deltas') demanded a huge labor cost, work input, many times the irrigation cost in Prehistoric Khorezm. The whole character of irrigation economy and farming production changed. The changes in production were accompanied by drastic transformations in the entire economic structure of society, such as the formation of classes, a strong state power and a strong coercive apparatus for the farmers to organize, and maintain the large-scale irrigation systems.

In the history of Khorezm, the Kangju and Kushan periods, from the 4th century BCE to the 4th century CE,[Note 100] characterized the heyday of the Khorezmian state (especially in the second period). It was represented by the construction of large fortifications and towns in the central part of the Khorezm Oasis and its outskirts, the further development of irrigation techniques, handicraft production and the wide diffusion of one type of high quality pottery (Tolstov 1948a:32, 1948b:113ff.; Vorobeva 1959a:84-124).

At the time of Alexander, the Khorezmians were no longer subjected to the Achaemenids. In the spring of 328 BCE, Alexander received the Khorezmian king Farasman, who claimed that his kingdom extended west to the Colchis (Bartold 1965:544). Since that time, Khorezm, and all the Aral Sea area (together with joint deltas of the Amudarya and Syrdarya), started playing an active cultural and political role in the northern part of the Central Asian *mesopotamia*, where the massive union of steppe tribes, the Kangju (Kangkha, Kangdiz), was formed.[Note 101] By the 1st century BCE they controlled the steppes from Eastern Turkestan to the Caspian Sea and the Black Sea areas (McGovern 1939:241). In the 1st-3rd centuries CE, the Kangju lost control of Khorezm and Chach and, according to S. P. Tolstov and Ya. G. Gulyamov, became part of the Kushan state (see Tolstov 1948a:180-181; 1948b:151; Gulyamov 1968:9). M. E. Masson disputed S.P. Tolstov's conclusion, drawn from numismatic data, and believed that Khorezm and Kangju did not enter in the Kushan Empire (Masson 1968). Until now, the borders and the absolute chronology of the rulers of this state, and thus the whole system of archaeological dating based on numismatic material, are debatable. Some authors have assumed 78 CE as the start

date of the reign of Kanishka I (Fergusson 1870; Oldenberg 1881; Tolstov 1963; see Tolstov 1961), for others 144 CE, and even 278 CE (Bhandarkar 1902a-b; Zeymal 1965:4-6; 1968:132-133).

The Kushan period brought within the countries of Central Asia a rapid rise in economic life, the revival of trade from south to north (India-Aral Sea area) and from east to west ('the Great Silk Road') and the emergence of the Buddhist world religion.

In Khorezm new cities and villages appeared with a high level of handicrafts production, irrigation systems were created over a broad expanse and large tracts of land were cultivated anew. The majority of irrigation systems created in the Kangju period continued to function (often in an extended form) in the Kushan period, thus their mapping created many difficulties.

In many parts of the Akchadarya south delta the archaeological-topographic research revealed Kangju and Kushan settlements and their associated irrigation ditches, which frequently coincided or simply overlapped. This indicates a great stability of the main contours of the irrigation systems and the main canals continuous functioning throughout the entire period. The creation of long-term irrigation systems require high skills and knowledge of hydraulic engineering.

Since then, the Khorezmian began to form specialist schools in irrigation and scientist-priests, which probably lasted until the time of the campaign of Qutayba in Khorezm. These schools combined the knowledge of mathematics, hydraulic engineering, cartography, astronomy and calendar observations, that are important for a large-scale irrigation economy (Tolstov 1957b:X).

In the Kangju and, especially, in the Kushan periods they created massive main canals with large water intake devices along the banks of large channels and main riverbeds, as well as a whole series of fortress-cities were built both in the upper and end parts of systems. Koy-Krylgan-kala and Bazar-kala were built in the basin of Kelteminar; Djanbas-kala in the Djanbas-kala branch of the system; Guldursun, Big Kyrk-Kyz-kala and Kurgashin-kala on the Kyrk-Kyz canal; Ayaz-kala I and III in the lower reaches of the Yakke-Parsan canal; Kyzyl-kala and Toprak-kala in the system of the ancient Gavkhore (Gulyamov 1957:99; Tolstov 1962a:91).[Editor 45]

Ancient Kelteminar

During the Kangju period, a basic reconstruction of the whole irrigation system took place in Kelteminar.[Note 102] A new canal, with a total width of 40 m and 6-8 m between banks, was constructed parallel to an abandoned Archaic canal. South of the Afrigid ruins of Adamli-kala, the canal turned north in the direction of Bazar-kala (see also Gulyamov 1957:77). In contrast to the Archaic canals, the main Kangju canal was built along the middle line of the *takyr* clay loam crusts with distributors branching, as a rule at right angles, from its right and left sides (see Figure 34.C). To the side of Koy-Krylgan-kala, built up between the 5th and the 4th centuries BCE (see Tolstov 1939:181; 1948a:99-100; 1948b:112, 119-120; 1953b:317-318; 1954:255-258; 1958:168-192; 1962a:117-135; Tolstov and Vaynberg 1967)[Editor 46] only narrow branches and small canals 1-1.5 m wide supplied the fields adjacent the settlement (see Tolstov and Vaynberg 1967:10, fig. 3).

The main bed directed toward Bazar-kala was, however, smaller in size than the Archaic systems of this region.[Note 103] It was 30-35 m wide in the environs of Angka-kala and 6-8 m wide between banks, which were 10 m wide and 4 m high. As it approached Bazar-kala, the total width of the canal bed increased from 30-35 m to 39 m (*poisk* 282), and to 40 m and 43 m (*poisk* 282). When a cross section was cut in the lower part of the canal, a later (Kushan) bed, 10 m wide, could clearly be distinguished.

In the Kushan period, the sources of the Kelteminar moved further south, to the main riverbed of the Amudarya according to S. P. Tolstov. The upper parts of the Kushan bed of the Kelteminar were traced by us near Eres-kala, where the canal was overall 28-32 m wide, the banks were 12-14 m wide and 3-5 m between banks. This canal was connected with the Bazar-

kala canal through a depression (the former riverbed of the Akchadarya).

We investigated the Kelteminar system with four trenches (Trenches 45, 46, 47 and 48) in the Kushan fields and canals surrounding Angka-kala (1.5 km south), near the ruins of a Kushan dwelling of the late 3rd century CE. It showed that the beginning of farming in this area belonged to the Archaic period and the thickness of the agro-irrigation sediments exceeded 125 cm in height. The size of the small, diked field was 13 × 19 m. There were some sand accumulations and many surface fragments of Kushan pottery scattered on the *takyr*. The layers were: 1) *takyr* thin crust clay loam, compact, fine porous, of 7 cm; 2) gray clay loam, compact, with lumps and slightly sandy, of 10 cm; 3) brown-gray clay loamy, heavy, with lumps, remixed with cracks and small sand particles, the visible thickness was up to 75 cm. Fragments of less distinctive ceramics were found at a depth of 60-70 cm.

A trench (Trench 48) was dug in the irrigation ditch which provided water to the field (Figure 36). On the surface, the ditch was quite clearly marked by banks 10-20 cm high, 1.2 m wide and 1.5 m in-between them. The canal had a bed depression covered with sand. The trench was 4.3 m long and 1.5 m deep. The layer were: 1) gray *takyr* thin crust (preserved at the canal's edge), loose, up to 7 cm; 2a) yellow-gray clay loam, compact with traces of remixing, with small particles of sandy clay loam, up to 70 cm, at a depth of 37, 48, 54, 57, and 69 cm, fragments of Archaic pottery were found (mainly rim fragments with typical Archaic rolled decorations); 2b) gray clay loam, compact, heterogeneous, forming ridges, of 20 cm; 3) light, gray-yellow clay loam, compact, more homogeneous with sand particles; the transition toward the lower layer was very gradual, up to 125 cm; 4) gray clay loam, with lumps and highly cracked, ditch lens-shaped, of 20 cm; 5) light brown sandy loam, micaceous, partially turning into fine grained sand; traces of agro-irrigation activity were missing.

Archaic pottery was found in the trench at a depth of 50-70 cm and this suggested that the overlying strata were formed later, in the Kangju-Kushan period. The nearby canal was about 2 m between banks, which were 1.2-1.3 m wide and preserved 15-20 cm in height. The trench (Trench 47) was 5 m long and 1.8 m deep. The layers were: 1) light gray *takyr* thin crust, porous, of 5 cm; 2a) gray clay loam, compact heterogeneous, with lumps, cracked and split by holes, of 45-50 cm; 2b) white-gray clay loam, compact, with high cracks mostly vertical; 3) gray clay loam, heavy, with lumps, heterogeneous, with vertical cracks and particles of clay loam, lens-shaped; at a depth of 72, 95 and 97 cm fragments of Kangju ceramics were found, the thickness of the layers was up to 100 cm; 4) yellow-gray clay loam, compact, heavy, at a depth of 110-130 cm with abundance of loam particles; at a depth of 1 m there were fragments of less distinctive ceramics; 5) brown-gray sandy loam, weakly compacted, micaceous; from 1.4 to 1.8 m. The general character of borders between layers was very unclear, which can probably be attributed to the infiltration of ditch water.

In the system of the Kelteminar, the right Djanbas-kala branch, built, as seen above, already in the Archaic period, still functioned in the Kangju and Kushan periods. The preserved parts of this main canal provided the following sizes: not far from the ruins of Angka-kala the bank was 35 m wide, including banks, and in some areas, the remains of hills, 3-4 m high, were preserved. In the center there were visible remains of a later canal bed 10 m wide. In the environs of Djanbas-kala the overall size of the preserved banks increased up to 42 m. Traces of several canal reconstructions were visible (see Figure 7). In the area of the Djanbas-kala canal and for a distance of 25 km, the survey revealed a series of secondary branches of the main canal, small ditches, fields, vineyards, and numerous ruins of single buildings and more complex living compounds with well-preserved layout and remains of basement walls. A large number of scattered pottery enabled us to date most of the ruins to the Kangju and Kushan periods.[Note 104] The best preserved farmsteads were those of the first centuries CE (*poisk* 610, 611, etc.). One of these

farmsteads was studied by E. E. Nerazik, who in 1964 excavated a five-room dwelling (Nerazik 1966:11,120:note104).

Of considerable interest for the history of irrigation in the environs of Djanbas-kala are the remains of head hydraulic structures on the riverbed of the Akchadarya (*poisk* 603-608). Considering the fact that the canal derived here from the riverbed, it flowed under the main canal of Djanbas-kala (*poisk* 622), and it was perhaps more ancient. The head works consisted of two rows of dams (5 m wide), whose remains were preserved in some places. The canal was cut through them and its bed preserved as a *takyr* strip 6 m wide. The construction of the canal was likely to have been connected not to the functioning of the Akchadarya riverbed but rather to the water supply from the Suyargan. Breakthrough floods occurred in Antiquity and the Middle Ages, particularly in the 9th-10th centuries, when they used to irrigate melons and vineyards (*poisk* 323, 324) located at the foot of the ruins of Djanbas-kala (see more details in Tolstov and Kes 1960:138).

According to S. P. Tolstov, in the *rustak* of Djanbas-kala, along the 25 km long main canal, no more than 4,000-5,000 people could live there,[Note 105] including the inhabitants of the city (Tolstov 1958:115). Of course, it is very difficult to evaluate the number of people and dwellings existing at the same time. The detailed archaeological research of the Kangju and Kushan rural settlements will help in the future to refine the data of S. P. Tolstov. Based on the available archaeological-topographic materials, it is possible that this figure is overestimated, but, even if we use it as a base for the evaluation of the labor necessary to clean the canal, it appears that the number of adult males (from $1/4$ to $1/5$ of the total population) was not able to fulfill these tasks by themselves.[Note 106]

During the Kangju and, especially, the Kushan periods, the further development of agricultural techniques (see more details in Tolstov 1948a:55-56) was connected with the appearance of large fields and increasing crop varieties. Besides the finds from archaeological excavations,[Note 107] it was also supported by the different agro-irrigation layouts of vineyards and melons identified in the ancient Kelteminar area. The layout of a farm in the lower part of the canal north of Djanbas-kala can be considered to be the earliest one (*poisk* 631). In the ruins of a farm we found remains of walls, traces of room layouts covered with typical ceramics of the Kangju period, similar to pottery of the lower layers of Koy-Krylgan-kala. The field layout (53 × 45 m) had very weak traces of narrow strips and was bordered by a fence 1 m high (Figure 37). Such field layout may have been reflected in one section of the *Avesta*, *Videvdat* (much of the texts of which belongs to the Parthian period). They said that the farmer must take the water flow (with a width and depth equal to a 'dog'!?) to a piece of land which has to be "so great that it could be watered through this flow from both sides" (Müller 1879:170).[Editor 47]

A very interesting agro-irrigation layout of vineyards and melons were uncovered in some farms west of Djanbas-kala (*poisk* 610, 611). They were marked by the alternation of narrow (1.2-1.8 m) and wide (3.3-4.4 m) lines of different soil color and, in some places, by their microrelief (see Figure 7). On the side of a vineyard there were traces of a narrow rectangular room with nine huge Kushan *khum* in a row, dug into the ground. In the ruins of the building there was an abundance of Kushan ceramics of the 1st centuries CE, as well as Kangju ceramics indicating a long settlement life. During our archaeological-topographic work, in one of the rooms we found a clay figure of a man with grapes in his hand (the ancient Dionysus, God of the grape harvest), which, along with other evidence, suggested that in such 'striped' fields grapes were cultivated in the past (see Tolstov 1962a:126, figs. 66-66a). This was also confirmed by ethnographic parallels (see below). A whole series of such agro-irrigation layouts was studied in the environs of Koy-Krylgan-kala.

The agro-irrigation layout south of Koy-Krylgan-kala (*poisk* 67) had a size of 107 × 85 m (see Figure 34) and it was divided by 5 m lines into two rectangles, which in turn were divided into irrigated plots-*gryad* 3-4 m wide. Another layout with a rhomboid configuration (*poisk* 70), with side dimensions of 90 × 120 m, was close to it. Over the long side it was adjacent to the canal, and its *gryad* was 3-4 m wide. Near the Afrigid castle of Karga-kala, the layout had an overall size of 60 × 72 m. It was divided by strips of 4 m into two equal parts, each of them including 9-11 *gryad*. The size of these *gryad* were similar to the ones described above (3.7 m) and the distance between them was 1.2 m (Figure 38).

Vineyards similar in size were recorded north of Koy-Krylgan-kala. Large layouts, with sides 100 × 120 m in size, adjoined the site from the southwest and touched the outer ring with its own corner (see Tolstov and Vaynberg 1967:9:fig. 3). In the lower layers of Koy-Krylgan-kala (4th-3rd centuries BCE), many grape seeds and a representation of grape gatherers were found (see Tolstov and Vaynberg 1967:206). According to Prof. A. M. Negrul, the grape seeds belonged to different species, for wine making and, the large berries, for the table.

The traditional sizes of a *gryad*, 3.2-3.6 m for vineyards and narrower (up to 2.2 m) for melons, are preserved, as mentioned above, to this day. In the Turtkul region a gardener, Sultan Ismagilov (80 years old), said that gourd production needs a low place, while for vineyards and orchards a higher one. In spring the vineyard was entirely filled with water three times and then *salma* (50 cm wide and 40 cm deep) were made.[Note 108] After watering in July, they were leveled again in August. The dimensions of the *gryad* (*karyk*), with bushes planted along the edges, were 3-3.2 m. Their partition was made by a rope. The gourd *karyk* was 2.3 m wide. It is necessarily shorter than that of a vineyard, because watering is faster and, after irrigation, the water does not touch the melon roots. The plants are irrigated with a full supply of water like cereals, so they do not have *salma* and *gryad*, defining the quadrangular diked areas. This simple layout for irrigated plots was widespread in the 'lands of ancient irrigation'. In the basin of the ancient Kelteminar fields have been found in the shape of quadrangular diked plots surrounded by ridges 20-30 m long and 15-25 m wide, and sometimes more (see Figure 34). These fields were apparently used to cultivate cereals, because their sizes differed little from the modern Khorezmian *kulcha* of 500-700 m² (see Saushkin 1949:290) [Editor 48].

In the Kangju and, especially, in the Kushan periods there were major renovations not only of the Kelteminar but also of the Archaic system of the Dingildje 'oasis' irrigation.[Note 109] At that time, the heads of the main canal of Dingildje were moved far upstream of the Akchadarya channel. The massive lateral branch of the Archaic canal was turned into the main canal (see Figure 35). Narrow distributors branched from it in different directions, covering almost the entire oasis area. If in the Archaic period the irrigated territory did not exceed 60-100 ha, in the Kangju and, especially, in the Kushan periods it increased to 200-250 ha.

Well preserved traces of quadrangular agro-irrigation layouts were found southeast of the ruins of the early Afrigid castle of Dingildje. Nowadays, in many places, there are traces of ancient melons and vineyards. In the Kushan period, in the same 'oasis', a narrower canal was detached from the old main canal toward the central site, stretching like a sharp arc, previously already included by an Archaic canal (*poisk* 1559-1565). This canal, 5 m in size between banks, supplied water to the central part, now preserved in the shape of mounds 2 m high, and from there the water flowed east to a large Kushan farmstead (see Figure 35.B). Based on the variety of field layouts, gardens and vineyards, the farming here, in the Kushan period, acquired a more intensive character; vineyards and orchards occupied a large area.

Ancient Kyrk-Kyz

A recently developed territory of irrigation along the ancient Kyrk-Kyz canal, from which the modern main canal was made, is located parallel to the Kelteminar system behind the heavy sand of the Akchadarya dispersed riverbed (see Jdanko 1958:705-730). Already in 1913, D. D. Bukinich and V. V. Tsinzerling traced this canal "from the fortress of Guldursun, passing by the fortresses of Berkut, Uy, Big Kyrk-Kyz and Kempyr for a distance of more than 40 *verst* till the point placed along the axe of the canal, at a distance of 70-80 *verst* from the Amudarya" (EOZU 1914:324). In 1937-1940, when the Khorezm Expedition started the exploration of this territory (see Terenojkin 1940a-b; Tolstov 1948a:28-29, 128-153, fig. 95; Nerazik 1959b, 1963, 1966), the cultivated irrigated lands ended near the ruins of Guldursun. Further north lay the desert: barchans sands and *takyr* covered with pottery fragments, on the clay surface. There were large fortified settlements (Guldursun, Kum-Baskan-kala, Berkut-kala, Uy-kala, Big Kyrk-Kyz-kala), and the preserved remains of 140 large and small agricultural farms belonging mainly to the 7th-8th centuries, many of which were remains of more ancient dwellings. Now the Kyrk-Kyz canal runs between Guldursun and Uy-kala, in a contemporary very narrow cultivated zone, and its banks are sometimes largely damaged. Several parts along the canal were paved for a road.

In the environs of Big Guldursun, a massive Kushan fortress reconstructed in the Khorezmshah period, and the bed of the canal were preserved in the shape of several bank remains 4.5-5 m high above cotton fields (see the schematic map in Nerazik 1966:fig. 1). The overall width of the canal exceeded 40 m and 7.5 m between banks. In the north, separate canal parts were preserved in the shape of very flat clay banks 30-50 m wide and 1-1.5 m high, and partially in the shape of two banks. The remains of a Kushan ceramic production, studied by M. G. Vorobeva in 1956, were found in the environs of Uy-kala near the main canal on the bank of a pond (Vorobeva 1961:150-151). The northern canal sharply turned east toward Atsyz-kala and, along the left bank of the Akchadarya riverbed, it approached, with a smooth arc, the ruins of the Kushan site of Big Kyrk-Kyz, which existed till the 6th century CE. Here the canal branched in the direction of the Kangju fortress of the Small Kyrk-Kyz and then, south of the Big Kyrk-Kyz fortress, sharply changed direction from north to east toward Kurgashin-kala, whose upper layers were also dated to the Kushan period. Here are the well-preserved huge levees, 6-7 m high and more than 20 m wide between banks.

As shown by the topographic survey and profiles (17 transverse profiles were made every 2 km), the flat bank of the Kyrk-Kyz canal had a complicated structure (Figure 39). It was possible to trace two beds of a canal almost along its entire length from Kum-Baskan-kala to Big Kyrk-Kyz (Figure 39.7-12). A bigger one was completely blocked or broken up by the other, narrower, Afrigid bed, which was 30-50 m from it, had an average size of 40-50 m, 10-11 m between banks and in some places more than 20 m. According to the numerous archaeological finds on the banks, accumulations of ceramics and remains of settlements on the preserved lateral branches, this canal was dated to the Kangju and Kushan periods.[Note 110]

The remains of ancient irrigation systems, dated by Kushan and Kangju ceramics, were preserved along the margins of Afrigid fields, in areas not affected by modern irrigation: north of Kum-Baskan-kala (an Afrigid settlement with a Kushan layer); west of the Afrigid castle of Teshik-kala; and in the environs of the Afrigid castles located on ruins of such Kushan fortresses as Berkut-kala, Uy-kala, Atsyz-kala, and also near Big Kyrk-Kyz, which existed for a long time and ended in the 6th century (Tolstov 1948a:128-153; Nerazik 1966:11).

The most ancient parts of the oasis were preserved best of all east of Uy-kala, where the main canal turned slightly northeast 3 km from the castle. The overall width of the Kangju-Kushan bed was 50 m and 10-11 m between banks. The banks were largely destroyed, but the bed was made quite visible by two earthen banks. The canals and ditches branching right ran towards the ancient sites. Over a dozen clusters of Kangju and Kushan ceramics were recorded on the narrow *takyr* strip between the canal and the Akchadarya riverbed. The largest of them was located in the lower reaches of the lateral canal. It was well

marked in relief and it was 15-19 m wide and 7-8 m between banks. The site covered an area of 50 × 100 m (*poisk* 975). The fragments of many Kangju *khum*, small cup-shaped vessels and other pottery, as well as a millstone, were scattered throughout the area (see also Nerazik 1966:9).

In the neighborhood of the Big Kyrk-Kyz, the main canal preserved traces of several reconstructions, apparently connected to the main phases of its history. The most right branches, located on the bank of a huge bend of the Akchadarya riverbed, were dated by Kangju potteries. The overall width of one of these branches was more than 40 m at *poisk* 833. Although the canal banks were eroded and dispersed, they rose 2.5 m above the *takyr*. The upper part of the largest left branch was almost entirely destroyed during a later period (lines and low banks were preserved on the *takyr*). It appeared again in the shape of a flat bank 30 m wide with a latitudinal direction, 1 km west of Big Kyrk-Kyz. This canal irrigated the environs of the Little Kyrk-Kyz in the Kangju period (see also Gulyamov 1957:99, 101) and was affiliated with Kangju pottery along its whole length. Near the Little Kyrk-Kyz, the Kangju fortress, built on a plateau, the canal was 9 m wide between banks and the banks were preserved 3 m in height (*poisk* 854).

In the environs of the Big Kyrk-Kyz, just north of the fortress, our team explored a large layout of vineyards, with irregular rhomboid shapes and sides of 330 × 520 × 400 × 550 m, over a total area of 20 ha (see Figure 4). The territory was divided into *gryad*, as highlighted by the soil color. The darker strips were a little higher and had a width of 2.3, 2.8, 2.6, 2.3, 2.7 m, etc. In contrast, the darker lines were higher and wider (3.0, 3.3, 3.6 m, etc.). The borders between strips were very unclear. The entire field territory was divided into strips with a north-south direction, except for the southwestern part.

East of the ruins of the Big Kyrk-Kyz, the canal branched into two beds, along which we recorded Kangju and Kushan settlements.[Note 111] A more ancient northern bed was preserved in the shape of a flat dike 23 m wide with slightly raised banks 0.5-1.0 m above the *takyr*. On one section of the southern branch we recorded traces of canal reconstruction. On the left bank, a later narrow canal (3.5 m wide), traced the ruins of Kurgashin-kala, could be dated by S. P. Tolstov to the Kushan period (see Tolstov 1948a:111). At the foot of Kurgashin-kala, which was built on a plateau representing the output of local Tertiary rock, the remains of a large park ensemble were preserved, with gardens and vineyards, as well as a square area (90 × 80 m), covered with pottery sherds, which probably served as a market square. The park ensemble was characterized by *gryad* (*karyk*), typical of ancient vineyards, with 3-4 m dimensions between irrigation ditches, whose width varied from 1.3 to 1.6 m.

The preserved part of the Kyrk-Kyz canal, from the ruins of Guldursun to Kurgashin-kala, had a total length of 60 km. It may be assumed that, in ancient times, its length from the head in the Amudarya to the lower reaches, was approximately 90 km. Considering a cross-section dimension of 25 m² as the most typical of the Kangju period, it is easy to calculate that the total amount of soil dug was more than 2.2 million m³. It means 15,000 workers for two months and 6,000-7,000 people in order to maintain the system.

What might have been the population of this place? The answer to this question is difficult because the archaeological data are scarce and require special research on rural settlements. However, by analogy with the later Afrigid oasis of Berkut-kala (where, according to E. E. Nerazik, no more than 4,000-5,000 people lived there), we can conclude that here the ancient population density was no higher and, consequently, the inhabitants no more than in the 7th-8th centuries, when the local forces were not sufficient to clean the main canal (see p. 173). It is true that the Antique Kyrk-Kyz canal cross section was twice that of the Afrigid bed and thus demanded a much larger volume of earthwork (for example, the annual cleaning of the main canal in the Kangju period alone required the labor of 6,000-7,000 diggers). With the likely average size reduction of the main canal in the Kushan period, the total

annual labor costs for cleaning decreased at least by twice the amount. To carry out this work, 3,000-4,000 diggers would have been needed. Consequently, not less than 12,000-16,000 people could live in the oasis at that time, which is unlikely.

Yakke-Parsan canal

The large Kyrk-Kyz canal in the environs of Guldursun branched only in a north–south direction towards the ruins of Ayaz-kala. The large Guldursun and Ayaz-kala fortifications were built probably in the Kushan period, at the branching or at the end of the canal (Tolstov 1948a:103, 170; Gulyamov 1957:99-101). The latter consisted of a massive rectangular fortress at the edge of an elevation (Ayaz 1), a round castle on a cone-shaped rock (Ayaz 2), and a vast fortified farm placed among vineyards (Ayaz 3) (see Tolstov 1948a:102-108; 1962a:205: fig. 118). The Yakke-Parsan canal was traced over its whole length from the ruins in the area of the Early Medieval city of Narindjan to Yakke-Parsan and Ayaz-kala, during the aerial survey of 1946-1948 (see Tolstov, 1948b:41-45 and attached map). The field survey provided remains of Archaic and more recent Kangju, Kushan, Kushan-Afrigid and Afrigid settlements.[Note 112]

The main bed was apparently already dug in the Kangju period. The preserved bank of a destroyed bed was between 28 to 50 m wide (Figure 39.13-25). The prevailing overall width of 25-32 m and 9-10 m between banks refers to the later Kushan-Afrigid and Afrigid periods. In places where the total width was more than 40 m, parts of the more ancient Kangju-Kushan canals were clearly preserved as indicated by pottery finds.

North of the ruins of the large Afrigid castle of Kum-kala, near Yakke-Parsan, the total width of the main canal was more than 40 m and largely destroyed canals branched at right angles from it. Many of them were recorded in the Kangju and Kushan settlements plowed in the Afrigid period. Single clusters of pottery covered an area of 100 × 200 m and 200 × 200 m (*poisk* 752, 754, etc.). A lot of surface material collected belonged to the Archaic and Kushan-Afrigid periods. At *poisk* 1063, the main Yakke-Parsan canal was approximately 50 m wide and 9 m in the central part between banks. The bank remains rose 3-3.5 m above the *takyr*. At Ayaz-kala, this canal had a total width of 28 m and 6 m between banks. The bottom was 1.6-2 m above the surrounding *takyr*. Another canal, supplying water to Ayaz-kala, was somewhat smaller, with an overall width of about 25 m and 6.5 m between banks at *poisk* 1066. This canal branched near Kavat-kala from the ancient Gavkhore system (see also Gulyamov 1957:110). Its total width was 30 m and 6 m between banks, which were almost completely destroyed.

Ancient Gavkhore

As noted above, the system of the Gavkhore laid on the right bank of the western massive riverbed of the Akchadarya. It consisted of a series of main canals of different periods, sometimes flowing parallel, sometimes crossing each other and forming complex bundles. The remains of cultivated fields and highly damaged ancient dwellings were found in several places.[Note 113] Unlike in the the Later Medieval complex, branched layouts of the 12th-early 13th centuries, the Gavkhore ancient system configuration differed in its 'sub-rectangular' and angular shapes, perhaps dating back to the Archaic period. The majority of branches were directed to one side (Figure 40).

In the vicinity of the ruins of Duman-kala (whose upper layers dated to the Kushan-Afrigid period), we found the Archaic main canal, functioning in the Kangju-Kushan period and later. Three km east, the remains of its bed, in the form of a bank, was gradually resized and its overall width reduced from 70 to 50 m and even 40 m. Less than 5 km before the ruins of Djildyk, this Kangju-Kushan system was interrupted, replaced by the Medieval system of Gavkhore.

In the environs of Djildyk and Kavat-kala, within the limits of the densely crossed small ditch network of the Medieval oasis of the 11th-beginning of the 13th centuries, the remains of the Kangju-Kushan Gavkhore canal were detected only in a few places (see Figure 40).

East of Djildyk, the ancient bed in the shape of a very flat and low bank was traced for 1 km (*poisk* 1132). The remains of massive banks 4.5 m high were preserved at its end. The overall width of the bed here was more than 60 m and 20 m between banks, both of which were 20 m thick. On the hills and around there were rare fragments of Kushan and Afrigid ceramics. North of the Khorezmshah Castle IV the canal was entirely destroyed and plowed under.

Northwest of the Khorezmshah fortified farms of Kavat-kala, the remains of an Antique canal, directed toward Toprak-kala, were preserved in the form of huge hills 5-6 m high and 10-11 m wide between banks (*poisk* 1017, 1145). The left branch of the ancient Gavkhore canal, not far from this place, was approximately 40 m wide and 11 m between banks (*poisk* 1166). About 2.5 km south of Toprak-kala (*poisk* 1378) the ancient Gavkhore bed was traced between sand hills in the shape of a flat bank 30 m long. Two kilometers southwest of Toprak-kala the bed of the main canal had a total width of more than 35 m. Weak traces of a narrow (5-6 m) canal were noted in its central part.

The general tenor of life in the environs of the capital of Khorezm, Toprak-kala, between the 3-4th centuries (Tolstov 1946b:84-85; 1946c:69-72; 1948a:119-124; 1948b:164-190; 1952a:31-44; 1958:195-216; 1962a:204-226) was revealed through ceramics finds[Note 114] and coins (studied by B. I. Vaynberg).[Note 115] The massive discovery of many Kushan coins of Vazamar type dated to the late 3rd century (*poisk* 1414, 1478, etc.), as well as later coins of the 4th-5th and 7th-8th centuries, testify to the vibrant city and the region trading life which lasted, without interruption, through the Kushan, Kushan-Afrigid and Afrigid periods.

Northwest of Toprak-kala, the main Gavkhore canal flowed close to Kzyl-kala, which was built in the Kushan period and reconstructed twice in the 6th-8th and 12th-13th centuries (Tolstov 1948a:123; 1948b:168). The canal had a total width of 34 m and 8 m between banks. Northwest of Kzyl-kala the terminal parts of the ancient Gavkhore were traced near the Sultanuizdag upland. Here the canal was preserved in the shape of a flat bank, covered by sandstone gravel, 32 m wide and 8 m between banks.

The ancient Gavkhore branched north–south towards the Kangju fortress of Burly-kala and northeast toward Ayaz-kala. The Burly-kala canal was largely eroded and mainly Kushan-Afrigid sites were located along it.[Note 116] The canal was distinguished only by the ground color. An embankment 45 m wide was preserved, north of the ruins of Kosh-Parsan, and it branched right 5 km from this site. Further along the canal was well marked on the ground by a flat bank 45 m long and 1.5 m high above the *takyr*. In the central part we measured a 2 m narrow canal in the form of a puffy dark line covered with small fragments of pottery. This ditch should apparently be attributed to the last (Afrigid) stage of the Burly-kala main canal.

In these areas there are numerous interesting remains of large Kushan-Afrigid pottery which were preserved in small hills of destroyed kilns from 10 to 50 m long and 1.5–2 m high (*poisk* 1118, 1120). The northeastern branch of the ancient Gavkhore was marked north of Kavat-kala as a flat bank more than 30 m wide. The banks were almost entirely destroyed. The central part was preserved as a slightly lowered line 6 m wide. On *poisk* 1066, its overall width was 25 m and 6.5 m between banks. The banks were preserved to a height of 50-60 cm. This canal, like the Yakke-Parsan, irrigated the environs of the fortified enclosure of Ayaz-kala 3, around which small unfortified houses of the 1st–3rd centuries were located (Tolstov 1940a:74–75; 1948a:108).

The massive fortress of Ayaz-kala 1 was located on the picturesque spur of the Sultanuizdag, above a Kushan farm and the Kushan-Afrigid castle on the Ayaz-kala 2 rock. At the foot of the ruins, the canal ended in a vast agro-irrigation layout, already familiar to us, with vineyards and gourd plantations. (Tolstov 1962a:205:fig. 118). The dimensions of this layout was up to 700 m in length and 300 m in width. Layouts smaller in size were also discovered in other places, in particular the garden-park layouts and vineyards

east of the Kushan farm were 100 × 100 m and 200 × 200 m in size. They were irrigated by the Yakke-Parsan canal water.

Tash-Kyrman Canal
During the survey of 1956 west of the Kavat-kala *takyr*, in an area inaccessible to motor vehicles, plots of 'lands of ancient irrigation' with Archaic, Kangju and Kushan settlements were discovered. Remains of massive ancient irrigation networks were preserved on the *takyr* and they extended in a north–south direction. The main canal had an overall width of 30 m and 17 m between banks (*poisk* 1107). Kangju and Kushan ceramics were found in the canal and around it.[Note 117] In the central part of the territory rise the ruins of Tash-Kyrman where Archaic and early Kangju ceramics prevail. At the end of the Kushan period this canal had ceased to function. West of Tash-Kyrman, at the edge of the modern oasis, a vast, strongly fortified settlement, now buried under sand, was found. According to the layout, the defensive structure and huge size resembles Bazar-kala. This settlement, Kazakly-Yatkan, was dated by S.P. Tolstov to the Kushan-Afrigid period, although Kushan remains were also found.

* * *

Summarizing the surveys of ancient irrigation works of the Khorezm right bank, it should be noted, first of all, that the technical level of ancient irrigation sharply differed from the previous pre-historic one (construction of local and large main canals with various distributors, head units, etc.) and in the irrigated areas scale.

In the Archaic period the main canals originated from the lateral channels of the Akchadarya and followed the natural riverbed directions. Some systems (Kelteminar, Dingildje, etc.) had an extremely characteristic cluster-shaped branching of distributors and feeders, representing something similar to artificial deltas. The small irrigation network topographic patterns showed differences in their 'sub-rectangular' shapes. The irrigation systems did not exceed the length of 10-15 km and the main canals were mostly very wide, up to 20-40 m, between banks. In the second half of the 1st millennium BCE the main directions of the Amudarya deltaic channels were enclosed by protective dikes and floods were controlled by people. The beginning of dam construction coincided with the formation of the ancient Khorezmian state (see Tolstov 1948a) (see also p. 153). Such a process, as noted above (p. 112), occurred in ancient Mesopotamia at the end of the 4th millennium BCE and in China in the 1st millennium BCE.

The Archaic canals were recorded throughout almost the entire territory of the 'lands of ancient irrigation' on the right bank of Khorezm. According to approximate calculations (based on cartometric measurements and large-scale maps), from 120 to 150 km of main canals were constructed in the Archaic period. In the next Kangju period a radical reconstruction of almost the entire system of the right bank of Khorezm took place. The total length of the main canals increased at least two or three times for an overall length of 250-300 km. The head works of many systems (in particular Djanbas-kala, Bazar-kala and Gavkhore) were moved far upstream of the riverbeds.[Note 118] The topography changed: during the Kangju and Kushan periods the main canals were dug along the central strip of a *takyr* between two rivers and this greatly expanded the irrigated area.

The artificial 'rivers' and 'deltas' of the Archaic period were replaced in this time by more rational irrigation systems. Particularly evident are the changes in the Kelteminar basin, where the massive Archaic canals were abandoned in the Kangju and, especially, in the Kushan period. They were replaced by small canals, but towards Bazar-kala, a straight water course ran for a length of 20 km long and a width of 10-12 m.

The Kushan period was characterized by the maximum spread of irrigation canals. However, in the eastern side of the Khorezm right bank (in the systems of Kelteminar, Kyrk-Kyz, etc.) a sort of reduction of the lateral branches and of terminal parts took place at that time. Thus the lower part of the Djanbas-kala canal (near Farm 631) was abandoned, and part of the Kyrk-Kyz canal watering the environs of Small Kyrk-Kyz ceased to function. In the lower part of the Kyrk-Kyz canal several reconstructions were undertaken. By contrast, a significant expansion of irrigated areas took place in the first centuries CE in the western regions, especially in the ancient Gavkhore system. Traces of farming culture were especially extensive in the environs of Toprak-kala, the major urban center of Khorezm and the residence of its rulers before they moved to Kyat on the bank of the Amudarya in 305 CE. The city of Toprak-kala existed until the 6th century and agricultural activities continued in its surroundings. (Tolstov 1948b:187).

In the first centuries CE, the Archaic and Kangju canals in the environs of Ayaz-kala 1 and 3 were revived and rebuilt, towards which the large canal from the Gavkhore system was brought. In this period the Kangju bed of the Yakke-Parsan canal was also rebuilt.

In the Antique period progress in irrigation techniques included developing water supply schemes (appearance of new units, improvement of water distribution facilities, etc.), and also changes in the cross-sections of main canals. The wide and shallow Archaic canals (replacing the fading delta channels of the Akchadarya) were changed in the Kangju and, especially, in the Kushan periods, with a reduced cross-section but with more depth (see Gulyamov 1957:89-90; Tolstov and Andrianov 1957:7-8). However, when comparing cross-sections, it should be taken into consideration that on the right bank of the Khorezm the upper parts of the Kushan and Medieval irrigation systems were not preserved, since they were located in the contemporary cultivated area. For instance, a comparison of the Kangiu canals sections heads with those of Kushan (recorded in the same basin), misrepresents the situation. Only if we accept an average decrease in cross-section of ¼ (from the heads to the lower reaches of the major main canals), the general trend of cross-sections reduction in the Kushan period can be ascertained.

For example, the Archaic canals around Bazar-kala were 8-10 km long, but their upper parts were 40 m wide between banks. In the Kelteminar basin, where the Archaic canals were largely destroyed, these cross-sections were smaller. On the Kyrk-Kyz canal, over 60 to 90 km long, the preserved parts of the Kangju period had sections of 20 m between banks and 10-11 m only in the Kushan period. In the Kushan period all the canals were based on the main riverbed and the presence of dams guaranteed a long-term function of the irrigation systems. However, the replacement of canal heads in that period led to the lengthening of idle parts of the canal (from *saka*, i.e. the head structures to the distributor releasing the first gravity-flow),[Note 119] which increased the amount of work required for cleaning the canals. The major canals undoubtedly retreated with the fading heads of the Akchadarya channels, where the water of the Amudarya was taken by *saka* in a larger area as it occurred later on the left bank of Khorezm (see Gulyamov 1957:243; VIR 1927, no. 3:41; etc.).

In Khorezm, during the Kangju and, especially, the Kushan periods, farming techniques and methods of soil fertility restoration were improved (to enhance the mechanical and chemical composition) with 'wall fertilizer': lands with damaged buildings and earthen banks of abandoned channel were rich in potassium salts, fertilized with silt sediments of irrigation water, ameliorated with applied sand, and the removal of *takyr* crusts, etc. (Tolstov 1948a:55-56; Fersman 1934, Tom II:82; etc.).[Editor 49] The assortment of cereals increased sharply.

In the right bank of Khorezm, during antiquity, there was the process of improving water diversion and the system of distributors (which first branched from the main canal at right angles then at a sharp angle), and the decrease of the main canals cross sections and a general increase of their length.

The development of more effective methods of water supply to the fields contributed to the expansion of sown areas and irrigated plots and an overall reduction of areas occupied by irrigating systems. Perhaps the general reduction of labor costs was due to the progress in irrigation techniques and to the activity of irrigation experts such as the Zoroastrian priests, or magi. The priests not only performed calendar and astronomical observations (in the 4th-3rd centuries BCE in Koy-Krylgan-kala), but they also improved the mathematical and physical basis of irrigation farming in the Khorezm oasis (Tolstov 1957b:X; Tolstov and Vaynberg 1967:251-264).

Medieval Irrigation Works

Kushan-Afrigid period (4th-6th centuries)

At that time there was a change in the whole material culture of Khorezm. There are signs of decay of the traditional forms of Khorezm cultural and economic life, its handicrafts and agriculture, together with the desolation of a number of cities and some rural settlements, especially in the eastern part of the right bank of the Khorezm. (Tolstov 1948a:50; Gulyamov 1957:110). In the basin of the ancient Kelteminar the branches of the Djanbas-kala and Bazar-kala systems collapsed; the bed of the Antique Dingildje Oasis was abandoned, though new narrow canals were dug around the castle in the 6th century; the terminal parts of the Kyrk-Kyz and Yakke-Parsan canals and some branches of Gavkhore were abandoned. In the west the Tash-Kyrman canal was completely abandoned. However, in the 5th and 6th centuries, the ancient Gavkhore continued to function without major reconstruction, irrigating the environs of Kosh-Parsan and the major urban center of Khorezm, i. e. Toprak-kala. The main canal of Yakke-Parsan apparently continued to function, possibly with interruption, and in its lower part the fortified farmstead of Yakke-Parsan was constructed in the 4th-5th centuries. According to the excavation of E. E. Nerazik, this settlement experienced three major reconstructions and existed from the 4th-5th centuries to the beginning of the 8th century (Nerazik 1963:37).

The typical features of the Kushan-Afrigid period irrigation were the abandonment of a large part of territory both in the east (Kelteminar, etc.) and in the west (Tash-Kyrman canal); the dying end parts and large branches; the gradual disappearance of ancient systems (especially in the eastern part of the delta) and, at the same time, the beginning of a radical reconstruction of irrigation systems based on new, already Medieval, hydraulic solutions. This corresponds to a further progress in the technological process (in the processing of agricultural products): the transition from the grinding stone to the millstone (see Gulyamov 1957:121) suggests the Medieval introduction of the principles of rotation in the construction of water-lifting mechanisms of the 9th-10th centuries.

Al-Biruni reported the construction by Afrigids of a fortress on the edge of Khorezm city in the year 616 of the Seleucid era (304 CE) and its ruin in the turbulent water of the Djeykhun River in in the year 994 of the Seleucid era (Biruni 1957:I:48). Khorezm city, or Kat (Kas), was watered by a canal which Ya. G. Gulyamov connected with the late Kushan and Kushan-Afrigid sites of that region (Sarkop-kala, Sim-ata, Karakol-kala, Pil-kala) (Gulyamov 1957:110). The Kyat canal (according to Ya. G. Gulyamov constructed in the 3rd century) perhaps played an important role in the Medieval irrigation of the Khorezm right bank, since the transfer of the head upstream of the channel, gradually merged the sources of all previous systems. The written sources reported that it turned into the Medieval Gavkhore system by the 9th-10th centuries (Bartold 1963, Tom I:199; 1965:72:note 40) (see also p. 176).

In the history of Central Asia, the 4th-6th centuries were a period of huge upheavals and the decline of many urban centers in Parthia, Bactria, Sogdiana, Margiana and Khorezm (see Masson 1949:52-53; Dyakonov 1953:292; Shishkin 1940: 44-45; 1963:199; Mandelshtam, 1964:26-28; Tolstov 1948a:119; etc.); the collapse of slave-owning states and the emergence of new feudal ones; people riots, movement of the steppe pastoral tribes and assimilation, the ethnical composition and renovation of the settled agricultural oases (Tolstov 1948a:50; Nerazik 1959b:224; Masson 1968:100).

The desolation of some settlements on the right bank of Khorezm at the end of the 6th century, such as Toprak-kala, Eres-kala and Duman-kala (whose upper layers were dated to this period), might be connected with the devastating campaigns of the Turkic nomads (the Tyurkyuts). In the middle of the 6th century, they created a vast domain from China to the shores of Djeykhun and the Aral Sea, including also Khorezm (Gumilev 1967:35).

The significant reconstruction of irrigation systems on the right bank of the Khorezm was dated to the Late Afrigid period, in the 7th-8th centuries. The main features of economic and cultural life of this period were described by E. E. Nerazik, based on the Berkut-kala Oasis. She developed S. P. Tolstov's thesis about the formation of early feudal relations in that period and traced the differences between the topography of the environs of Berkut-kala and the lower part of the Kyrk-Kyz canal, where the relationships between sites and canals remained unchanged (Nerazik 1966:48-49). The author identified economic inequality and societal stratification, as reflected in the size of single farmsteads, the dimensions of their adjacent vineyards and melon plantations (Nerazik 1966:110). The distinctive features of many settlements and of all oases, were a combination of farming and trade, the relative slow barter development and the internal market narrowness. The population ethnic composition also changed under the influx of incoming barbarian tribes. This process began at the end of the 1st millennium BCE and was reflected in the appearance of the peculiar handmade light-slipped ceramics of the middle and upper horizons of Koy-Krylgan-kala. In the 4th-5th centuries this was reflected in the appearance of Chionite elements in the Kushan-Afrigid culture. In the 7th-8th centuries it was reflected in the strong influence of Syrdarya herding-farming tribes in the formation of the Afrigid culture (Tolstov 1962a:252; Nerazik 1966:122-129; Tolstov and Vaynberg 1967:20:130-131).

Great changes took place in the field of irrigation. The Afrigid ruins of Karga-kala, Adamli-kala and small buildings in the neighborhoods of Angka-kala in the ancient irrigation basin of the Kelteminar are dated to this period.[Note 120] The irrigation canals of the Afrigid period were characterized by their small size; the largest one, in the area of Adamli-kala, was no more than 5-6 m between banks. However, the small irrigation network was different in comparison with the Antique one for the great number, regularity and branched configuration.

The changes in the overall 'picture' of irrigation systems should be considered, perhaps, as further progress in irrigation techniques. The main goal pursued by the Medieval irrigators, who changed the 'sub-rectangular' ancient irrigation systems into branched ones, was to reduce the canal silting and the work required for their upkeep.

In the Afrigid period, the irrigation system of the Dingildje 'oasis' was entirely rebuilt (see Figure 35). The oasis water supply diminished, and people started using water more economically. Instead of large and massive main canals, numerous narrow canals were constructed to form a complex web, which either intersected the ancient Kangju-Kushan canals or repeated their configuration.[Note 121] The main source, however, preserved an older one: the main canal flowing from Eres-kala.

After a sharp irrigation network decline along the Big Kyrk-Kyz canal in the 4th-6th centuries, later on, a second, narrower (6-9 m) and deeper bed was dug in the 7th-8th centuries, which supplied water to many Afrigid settlements, castles and farms of the Berkut-kala Afrigid Oasis (see also Nerazik 1966:8-91). The Afrigid canal, in its middle reaches, continued east the Kangju-Kushan canal and it ran very far from it and then rejoined it again in the same flat embankment. To the right of the canal there was a vast space, which stretched for 5 km from south to north, narrow, almost free of sand *takyr*, bordered by a series of fluvial terraces and *takyr* remains with sand from the dry riverbed of the Akchadarya. In some places the *takyr* was covered by few barchans and in other places by more frequent barchans,

including the towering walls of the large Afrigid castle (Walls 9, 10, 11, 58, 60) and smaller ruins (remains of donjon and square fences). Traces of canals, gardens layouts, accumulations of Afrigid and Kushan-Afrigid ceramics were found on the surface.[Note 122]

The basic main canal to the north of Uy-kala had a double bed with a general width of approximately 80 m (7-8 m between banks of the Afrigid bed). Three kilometers from Uy-kala the canal turned northeast. The Afrigid bed was again situated on the left. Here are traces of the irrigation canals and ditches feeding the castle neighbors with water, the origins of which led toward the Afrigid main canal. The prevailing size of this canal was 7-9 m between banks (see Figure 34.D).

For digging the Afrigid bed of the Kyrk-Kyz canal it was necessary to remove 600,000 m^3 of soil, which means 4,000 diggers over 50 days at a rate of 3 m^3 a day. In order to maintain the canal, the labor of 2,500 workers was required each year. According to the calculation of E. E. Nerazik, no more than 4-5000 people could live in the oasis, including about a thousand working men. This calculation shows the local force was not sufficient to carry out all irrigation work, but the difference between the required number of diggers and the number of working men living on this irrigation basin was not as huge as in previous historical periods.

The configuration of the agro-irrigation layouts in the Afrigid period became increasingly diverse and complex. It was noted that, in the Afrigid fields of the Dingildje 'oasis', the more complex and varied layouts of the 7th-8th centuries were the evidence of a new approach to field cultivation techniques for this period. In the Kyrk-Kyz Oasis of the Afrigid period, different types of fields were also recorded. It should be noted, for instance, the remains of a vineyard near Afrigid Farm 66. It was rectangular, the layout oriented north with the bigger side 90 m and the smaller one 29 m. The rectangle was divided into wider *gryad* of 3.3-3.4 m and narrower *gryad* of 1.4 m (in the middle they were almost washed away). They appear to be the remains of small ditches that were preserved in the shape of narrow flat banks, rising 15-29 cm above the broad *gryad*. E. E. Nerazik discovered similar rectangular field layouts for the cultivation of cereals (with sides of 9 × 34 m, 20 × 20 m, and 27 × 28 m, etc.) (Nerazik 1966:93).

During excavations at the Afrigid castle of Teshik-kala, S. P. Tolstov discovered remains of millet, barley, wheat, beans, mung beans, grapes, peaches, apricots, plums, cotton, melons, pumpkins and cucumbers, as well as the remains of domestic animals (Tolstov 1948a:142). The research by E. E. Nerazik on the rural sites of the Afrigid oasis increased this collection. She noticed that the most widespread (apparently, the prevailing) was millet (which was typical for Antique sites according to Koy-Krylgan-kala). During the excavations of farms, seeds of sorghum, wheat, barley, melons, watermelons, grapes, cotton, plums, cherries, apples, apricots and peaches were also found (Nerazik 1966:92).

After the reduction of irrigation on the right bank of Khorezm in the Kushan-Afrigid period, and the later revival of the irrigated farming during the Afrigid period in the 7th-mid-8th centuries (until the mid-8th century),[Note 123] a new abandonment of the irrigated land occurred in the second half of the 8th and 9th centuries, when Khorezm became the arena of feudal enemy factions, dramatic political events and popular riots.[Note 124] All this led to the weakening of the central power (formation of two states in the oasis) and, consequently, the reduction of irrigated lands (Tolstov 1948a:46; Gulyamov 1957:123). The lower reaches of Kelteminar (environs of Adamli-kala and Karga-kala) were deserted, and the whole Kyrk-Kyz canal, Gavkhore, and other irrigation systems fell out of use (Tolstov 1948b:231). Only near the city of Narindjan the upper section of the Yakke-Parsan canal continued to function.

In spite of the Arab conquest of Khorezm in 712 CE, and the reprisal of Qutayba ibn Muslim of the Khorezmian magicians (magi) - i.e. the guardians of writing, astronomical calendar, mathematical and hydraulic knowledge so necessary for large-scale irrigation farming - the

irrigation technical tradition survived. Although, according to al-Biruni, "the Khorezmians became illiterate and if they needed something they relied upon memory" (Biruni 1957:63).

The end of the 8th-beginning of the 9th centuries dates the life and activity of Muhammad ibn Musa al-Khwarizmi, the founder of 'Arabian' mathematics connecting Greek geometry and Indian algebra (the name 'algebra' comes from the name of his treatise *al-jabr*). Generally accepted for his success, to a large extent, he was indebted to the secular tradition of natural and exact sciences, which originated on the basis of irrigation practical needs in Khorezm and in remote travel exchanges (Tolstov, 1948a:267-268). He is widely known for his dictionary of irrigation terms in the Merv Oasis. Among the words that are worth of mention: *mufriga*, an artificial discharge for excess water (*bedrau* in the 19th century); *kuvalidja*, a channel made above the distribution facility; *tiraz* (or *taran*), a canal water separator (*daraka* and *mazraka* in Maverranahr); *musanna*, a dam; *azala*, the amount of land of 100 cubic ells on which are employed the diggers; *sakiy*, irrigated crops; *kazaim*, the *karez*, i.e. underground water gallery; *gil*, a sort of swamp, i.e. the discharging flood waters used for irrigation; *garb*, that which is irrigated by buckets (water-lifting devices); the higher lands were irrigated through different water-lifting devices like *dulayb*, *dalia*, *garrafa*, *zurnuk*, *nasura*, *mandjanun* (MITT 1939:217-218; Bartold 1965:142-143; Gulyamov 1957:10, 237-259).

The high level of hydraulic skill of Khorezmian irrigators of the 9th-11th centuries are testified by the work of the prominent scientist al-Biruni, who was a native of Kyat suburbs. His work refers to the instruments for leveling the slopes of a canal, "which is measured (at ground level), level the ground, dig and build canals" (Biruni 1957:288).

A great revival of economic relations with the Arabian Caliphate promoted the development of geographical knowledge and the emergence of a whole group of geographers and historians in the 9th-11th centuries (Ibn-Rust, al-Istakhri, al-Makhdisi, Yakut, etc.), whose work describe the life of the Khorezmian Oasis, the condition of irrigation, canals, cities and settlements. A detailed study of these written sources was provided by V. V. Bartold, who dealt with the history of irrigation of Turkestan. The material describing the irrigation of the Khorezmian city were later studied by S. P. Tolstov and Ya. G. Gulyamov (see for example Gulyamov 1957:125-163; etc.).

Returning to the description of the irrigation works on the right bank of Khorezm in the Afrigid-Samanid period (9th-11th centuries), it should be noted that in the 'lands of ancient irrigation', the irrigated areas of these periods were preserved only in a few places, mainly on the borders of the contemporary cultivated area. The archaeological-topographical survey, for example, revealed, near Narindjan, an Afrigid-Samanid network of small canals and ditches irrigating in the past the area surrounding the city. Now they are densely covered by fragments of brown, light ochre-colored and gray Early Medieval vessels, and sherds of glazed pottery with red-brownish painting with geometric patterns, Kufic Arabic inscriptions, etc.[Note 125]

The settlement of Narindjan covered the bed of an Early Medieval canal near the southern wall, which was marked by a flat 13 m depression (Gulyamov 1957:139). The bottom of the canal was 1-1.5 m lower than the Early Medieval *takyr* and its total width, together with banks, was 25-27 m. It was covered with small pottery fragments of the 7th-8th and 9th-11th centuries. The canal was dug north of the settlement. Its visible width (the width of the bottom between banks) was 5.1 m. Near the ruins of an Early Medieval brick kiln, the canal was marked by a small hill and insignificant embankments. The flat bottom was covered with small dark-brown pottery fragments, which occupied a 9-10 m wide strip. Its overall width was probably 23-28 m. Five kilometers north of Narindjan the canal was preserved in the shape of a flat bank with an overall width of 15 m. In the canal Early Medieval glazed pottery of the 9-11th centuries was found and, on its left side, there were the ruins (*poisk* 746) of a small Medieval fortification

which preserved the southern earthen wall, 5 m in height and with narrow outlets.

A number of gourd plantations, vineyards, small canals and single traces of destroyed dwellings of the Afrigid-Samanid period were found near the ruins of Buran 1 (10th-11th centuries) and Buran 2 (9th-10th centuries).[Note 126]

Near Kum-kala we also discovered Afrigid-Samanid canals, fields, vineyards and gourd fields of different shapes. Thus, in Kum-kala the layout of the vineyards had sides of 10 × 20 m. The width of a narrow line was 0.8-0.7 m. It was expressed by a slight ground lowering. The *gryad* were 3.6, 3.8, 3.5 m etc. wide.

On the *takyr* west of Kum Kala, interspersed with small sand barchans, the layouts of large quadrangular fields with sides of 22 × 20 and 7 × 28 m were preserved. In the lower parts of the main canal we found individual field layouts that were adjacent to the main bed. So, at *poisk* 739, we surveyed the layout of a vineyard consisting of two uneven parts. Between them was a 6 m wide strip. The length of one side was 38 × 44 m, the other side 31 × 44 m. The sizes of its narrow strips were 1.5, 1.3 m, etc. The wider strips were 3.6, 3.5 m, etc. wide. These layouts remind us that, in the 9th-11th centuries, grapes and especially gourds from the Khorezm Oasis were highly appreciated in the Islamic Orient. Al-Tha'alibi reported that watermelons from Khorezm were brought to Baghdad to the courtyard of the caliphs al-Mamun (813-833 CE) and al-Wathiq (842-847 CE) in lead containers covered with snow (Bartold 1963, Tom I:297).

Khorezmshah Period
In the history of Khorezm this is the time (12th-early 13th centuries) of a radical reconstruction of the old irrigation systems and further development of Medieval irrigation techniques. It was the heyday of the feudal monarchy in Khorezm and, according to S. P. Tolstov, "before us there were huge irrigation works, revival of hundreds and thousands of hectares of fertile land, construction of border fortresses on the edge of the desert, strengthening of strategic pathways to Khorasan,

Maverranahr, deep into the steppes of Dasht-i Kipchak, a new flowering of urban life, handicrafts and trade" (Tolstov 1948b:276). This was the time of the rise of the mighty state of Khorezmshah, broadening their borders, at the beginning of the 13th century, from the Aral Sea and the Syrdarya to the shores of the Indus on the south, and from the Tian Shan on the east to the Azerbaijan on the west. Within the limits of the right bank of Khorezm, in that period the canals in the basin of Gavkhore were reconstructed.

Medieval Gavkhore

S. P. Tolstov provided enough proof to identify the Medieval systems in the Lower Amirabad with the *rustak*[Note 127] of Gavkhore described by the authors of the 10-13th centuries (Tolstov 1948a:46). Al-Istakhri, who in turn used an essay by Abu Zayd al-Balkhi (919 or 920 CE), described this canal: "Six *farsakh* from Garabkhashny, the canal branched from the Djeykhun feeding the rural area to the city (the capital Kyat). This canal is called Gaukhore, which means "food of the cows", the width of this canal is approximately five ells,[Note 128] the depth was approximately the height of a two men, (according to him) floating ship. From the (canal) Gaukhore, when it held five *farsakh*, departed a canal called Karikh and it irrigates a part of the *rustak*" (MITT 1939:179). Ya. G. Gulyamov supposed that its sources were in the Akkamysh. He studied in detail the etymology of the name 'Gavkhore' and he linked it with the name of the channel ('Gau') in the province of 'Kh(v) or' (Gulyamov, 1957: 93, 138).[Note 129] He rightly pointed out that the canal branching from Gavkhore, the Karikh (Gire according to V. V. Bartold), should be connected with the canal irrigating the environs of Guldursun, built in the Middle Ages, and the Medieval city of Narindjan.

The dead 'oasis' of Gavkhore stretched along the old riverbeds and contemporary lakes, nearly 30 km north of the modern cultivated area, in the shape of a narrow strip of *takyr* entirely scattered with several ruins of single farmsteads and castles of the 12th-early 13th centuries, towering above the dense and intricate network of canals, ditches

and fields (see Andrianov 1959:146-149, fig.1). On the west this oasis was bordered with sand, dry riverbeds and contemporary lakes, on the east it bordered with sand and the deflation basin formed in water discharge places of the early irrigation systems. The Gavkhore system, within the contemporary cultivated area, branched in the direction of Duman-kala. This canal irrigated the environs of the Naib-kala 1 castle (10th-11th centuries) (Tolstov 1948a:115). Moving from the cultivated area, Gavkhore, in the area of a small lake, it had a width of 6-5.5 m between banks and the banks were 2 m high. Three kilometers from the cultivated area the canal split in two. The total riverbed width was more than 40 m. The width of the left bed, was apparently preserved for approximately 20 m and 6 m between banks. Slightly below the Medieval system, close to southwest, there was an ancient system, forming an almost continuous strip of damaged irrigation canals and ditches of different periods, more than 200 m wide (*poisk* 1171). To the east a small Early Medieval canal was 3 m wide between banks and its bed was clearly marked in relief with a width of 6-6.5 m and 11-12 m between banks. The banks covered with sand and bushes were 1.6-2 m high above the surrounding *takyr*. The complex structure of the Gavkhore system was preserved also in its lower part with the only difference that the Medieval main canals were in some places preserved 350-400 m away from the Antique ones, and even their number increased.

Eight kilometers from the cultivated area (3 km from the lake), small distributors branched right from the Medieval Gavkhore system, forming a continuous network of small ditches and fields, among which, here and there between sand dunes, Medieval farms and castles lay. Here, over an area of 8 km², almost a hundred ruins were recorded. As the main Gavkhore system approached Djildyk, running along the bank of another lake (the former riverbeds of the Akchadarya), it again decreased to 100-120 m. Its structure was similar to the above described one. In the west there was a massive bank 40-50 m in size covered with Kangju and Kushan pottery, in the east there was an assortment of small Medieval canals highly damaged by erosion. The largest one was 56 m wide between banks (*poisk* 1145). In the environs of Castle 5, the Medieval system of Gavkhore greatly branched out (*poisk* 1123). Its total width exceeded 220 m. In spite of heavy destruction, in the transverse section it was possible to recognize 11 Medieval canals, differing from each other in size and forming an almost continuous series of small hills and terraces: the remains of river banks did not exceed the height of 1 m. Fragments of Medieval unglazed and glazed pottery of the 12th-early 13th centuries prevailed among the scattered surface materials.[Note 130]

Between Castles IV and V, the Gavkhore system highly branched: the distributing canals were both on the left and on the right. Cluster of Medieval canals deviated slightly west and approached Castle IV in the shape of a 150 m area of destroyed irrigation ditches and canals, whose central part did not exceed the 3-3.5 m.

The Antique and Medieval systems ran in parallel to Castle IV, but, between Castles III and IV the pattern became complicated. In some places the entire range of parallel and intersecting irrigation ditches reached 400 m in size. In order to reveal the system structure, we carried out two perpendicular profiles. The transverse latitudinal profile between Castles III and IV revealed 5 small Medieval canals (*poisk* 1155), of which the largest and most important was situated on the western border. Its width was 15 m and 4.5 m between banks. It was directed toward Castle III and then to Kavat-kala, where a large number of small irrigation ditches departed from it.

Already in the 1940's, during the photographic mapping of the Kavat-kala Oasis, S. P. Tolstov noted a massive group of canals located between Castles III and IV and turning eastward (Tolstov 1948a:158:fig. 95). We dug a north-south trench with a total length about 400 m, which revealed a very complex web of narrow Medieval ditches branching from the main Gavkhore canal, and adjoining the ancient, completely destroyed, canals.

Northwest of Kavat-kala, the Medieval Gavkhore was 12 m wide and 4 m between banks. The canal turned northwest along ancient systems in the direction of Toprak-kala. The Medieval canals entered the older lands from the southeast and formed a wedge, whose top approached Toprak-kala. At 1.5 km southeast of the ancient capital of Khorezm was located the main Medieval center of this region, Toprak-kala 2. In some places between the Medieval fields and ditches, almost along a straight line between Kavat-kala and Toprak-kala 2, rose the ruins of rural farmsteads of the Khorezmian period, with typical remains of *kaptarkhana*: the high ceremonial-room, like the Uzbek *mekhman-khana* (see Tolstov 1948a:159-162; 1948b:280-282).

The Medieval 'lands of ancient irrigation' of the 12th-early 13th centuries were revealed by the team northwest of Toprak-kala, along the main Gavkhore canal in the environs of Kzyl-kala. Here, at 1 km northeast of Kzyl-kala, in a *solonchak* area, a quadrangular (40 × 40 m) highly damaged fortification was found, only a low wall was preserved. On the surface of this *solonchak* crust with white salt efflorescence spots, the finds of ceramics of the 12th-early 13th centuries were very rare. It should be noted that the latest network (the lower part of Gavkhore), in the form of narrow (3-4 m) irrigation canals, crossed the Antique distributors, maintaining a rectangular and 'sub-rectangular' configuration of agro-irrigation layouts (fields and irrigation ditches), inherited from the previous periods.

The archaeological-topographic research in the environs of Kavat-kala identified a group of peasant farmsteads along single canals of the Gavkhore system. Clusters of 10-20 farms were located along the canals, as a rule lower than the castles around which they gravitated. Thus, Castle I was adjacent to Farms 1-10, 59, 60, etc., Castle I and Castle II, controlled the canal where Castles 19, 22-26, etc. were placed. Farms 36, 45, 46, 63-70, etc. were placed below Castle III (see Andrianov 1959: fig. 1).

In the Gavkhore, the irrigation system plan with its complicated branching shapes and numerous narrow branches, along which farms were placed in connection with individual castles, it is possible to see, according to S. P. Tolstov, the graphic materialization of the feudal relations existing at that time. The peculiar topography of these dwellings gave S. P. Tolstov the reason for concluding that the *rustak* of Gavkhore was an example of a completely feudal type of settlement, with the residence of the major feudal lord (the ruler), in Kavat-kala, the castles of his vassals and numerous unfortified farms (Tolstov 1948a:150; 1948b:282).

The Medieval oasis of Gavkhore gave us an idea about an extremely high level of agricultural production in the Khorezmshah period. The rural population density in the irrigated farming zone was high at that time. The total area covered by the Medieval fields in the basin of Gavkhore reached 35 km². In the central, densely populated part adjoining the small Early Medieval settlement of Kavat-kala, more than 140 farms were found over an area of 14 km², while, in the Afrigid Oasis of the 7th-8th centuries, only 100 farms were located in an area of 35 km². According to S. P. Tolstov, the population density increased fourfold (Tolstov 1948b:280).

The smallest farmsteads in the environs of Kavat-kala reached 150-300 m² and, according to S. P. Tolstov, a family of the Khorezmshah period still retained the large family tradition (Tolstov 1948a:160-164; see also Nerazik 1966:116-120). Excavated Farm 1 had five rooms, including room for guests, or *kaptarkhana*. If we assume that at least 10-20 people lived in every farmstead, it appears that the population living around the Kavat-kala Oasis was from 1,200 to 2,500. The population density in the irrigated area reached 80-150 persons per 1 km². It was a rather large number for that time.[Note 131] Not by chance, the Arab geographer and traveller Yakut (1179-1229), who visited Khorezm in 1219 CE, wrote that he "had never seen a place more populated than it (Khorezm)... Continuous population, villages close to each other, many individual houses and castles in its steppes, and rarely your eyes see an uncultivated place in its district (*rustak*). There are a lot of trees. Most are

mulberry trees and poplars. They need them for buildings and feeding silkworms. And there is no difference (in population), when you go over all of its districts and when you walk through all the bazaars. And it may be assumed that in the world there are provinces excelling Khorezm in welfare and more populated than it is... The majority of villages of Khorezm are cities with markets, much material goods and shops. It is rare a village without a market. All this with the total security and full serenity" (quoted in Tolstov 1948a:156-158).

The Medieval 'oasis' of Gavkhore gives us an idea of an extremely high level of agricultural production and irrigation technique. Apparently, by that time, the irrigation systems had already all the main elements recorded at a later time: protective dams on the Amudarya; system of *saka*, head constructions; idle part of main canal (frequently connected to *saka*); reserve drainage canal (*bedrau* or *mufriga*); main canal (*arna*); major distributors (*yab* or *yap*); small distributors (*badak*) with water regulating devices; feeders (*salma*), from which excessive water flowed to end-water escape ponds. In order to avoid erosion, the head of the *badak* was equipped with sluices (*doldarga*). Often some water-regulating wooden devices (*tokurtka*) were put on the *salma*. The control of water using regulators at the heads of canals was not practiced. The regulators were used only in the distributing network (see Gulyamov 1957:243-244).

On the right bank of Khorezm in the Khorezmshah period, irrigation was only by gravity. On the left bank, as we shall see below, to this scheme should be added supply ditches to *chigir* (water-lifting devices) and the *chigir* themselves, which received a widespread use in Medieval Khorezm, between the 9th-11th centuries, (see Vakturskaya 1959:269).

The irrigation systems of the 12th-early 13th centuries were quite perfect and close, to some extent, to the Late Medieval systems. The main canals were narrower but deeper than in ancient time.[Note 132] The layout of the system was already characterized by a complex branched configuration (see Figure 34.J). This made possible the irrigation of more land within the same irrigation system. The coefficient of the irrigated area, i.e. the percentage of irrigated land within a general controll area, in the Middle Ages probably rose by a high 30-40%.[Note 133]

The branching parts of irrigation canals were not covered with sediments as fast as the 'sub-rectangular' systems of Antiquity. The volume of earthwork decreased. The reduction in transverse sections and the overall labor costs for the main canals also decreased. The main Gavkhore canal extended for 60 km with an average cross-section of 5 m², thus the volume of the soil removed was probably about 300,000 m³. This could have been done by 2,000 diggers in 50 days (with an average rate of 3 m³ per day) and, in order to maintain the system functioning, 800-1,000 men were required annually. As we mentioned above, only 1,200 to 2,500 people lived in an area of 14 km² around Kavat-kala. In the whole Gavkhore basin (including the zone of the main canal within limits of the contemporary oasis), there apparently lived from 4,000 to 8,000 people, among whom there were 1,000-2,000 working men. Based on these calculations, some important conclusions can be drawn (for the history of the Khorezm irrigation) that in the Middle Ages the local labor resources were sufficient for the implementation of the entire irrigation cycle, i.e. digging and cleaning.

Labor conscription to clean up the canals was administrated by the central authority and, according to the sources, in the Middle Ages it was characterized by the well-known feudal corvèe, *begar* or *bigar*. This corvèe was wide-spread in countries of the Near East and Central Asia (Minorsky 1939:947, 950; Petrushevskiy 1960:394-396; Ali-Zade 1956:228-230). The Khorezmshah Sultan Tekesh (1172-1200 CE) mentioned the natural conscription *khashar va begar* in a letter to his son Malik Shah, ruler of Djend, in connection with very hard and constrained work and recruitment of peasants for military home guard (Horst 1964:59, 71, 121). I. P. Petrushevskiy states that, during the Mongol

rule in the 14th century, the rulers of Herat gathered people for *bigar* as repression of 'breach'.

V. V. Bartold noted that, in the Early Middle Ages, new canals construction with peasants forced labor was considered a popular calamity. He mentioned al-Tabari's account about a message of the Caliph Yazid III (who came to the throne in the 744 CE), promising not to erect buildings and not to dig canals (Bartold 1965:115).

The term *begar* is ancient, perhaps of Pahlavi origin (Petrushevskiy 1960:394-396) and means natural conscription of rural population.[Note 134]

Returning to the Gavkhore basin people, it should be noted that, in the Middle Ages, in this region the population density rose sharply and areas of irrigation increased with a general reduction of the territory occupied. A similar process was also noted by R. McC. Adams (1965:115) in the environs of Samarra and the basin of the Diyala River, where the total population reached its maximum (more than 800,000 people) during the rule of Harun al-Rashid.

The process of economic development in the right bank of the Khorezm was interrupted in the early 13th century (1220 CE) by the catastrophic Mongol invasion. After weak attempts to revive the economy in the early 14th century, life here was frozen till the 19th century (see Tolstov 1948a:51; Masson 1940:114; Shishkin 1963:243-244; Petrushevskiy 1960:36-46,67-83; Vakturskaya and Vishnevskaya 1959:161).[Note 135]

The small Medieval distributing and irrigation network differed from the Antique one in its great frequency and branch configuration. The origin of complex branched layouts changed the 'sub-rectangular' and angular ancient systems of irrigation and promoted the reduction of canal silting, thus preserving for a greater length (up to hundreds of km), the Medieval canals which were several times narrower than the ancient ones. This was especially clear when comparing the Gavkhore Kangju-Kushan bed, which was 20 m wide, with the Khorezmshah period canals, 3-5 m wide.

In the Middle Ages, as in ancient time, there were width differences in the canals upper and lower parts. However, the upper part of Gavkhore lie in the contemporary cultivated zone and it is difficult to ascertain the precise size of the main canal in the 12th–early 13th centuries. It should be recalled that the insignificant width of the canal irrigating the "*rustak* to the capital of Khorezm" was reported by Medieval geographers (Bartold 1965:165). It has also been noted by W. Willcocks in Mesopotamia (the reduction of cross-sections of ancient canals from 50 to 5-10 m in width) (Willcocks 1903:1-13). The general decrease of canal cross-sections led to a reduction of work required in their maintenance.

Apparently, the irrigation systems of the right bank of the Khorezm in the Khorezmshah period already featured all the main elements of the complex Khorezm irrigation of modern times: few heads (*saka*) - drainage canal - distributor canals (1st and 2nd order) - feeders - fields. The longitudinal profile of ancient and Medieval canals indicated that they were designed for a river water level very similar to the modern one.

Chapter 4
The Sarykamysh Delta

Boris V. Andrianov

In the 'lands of ancient irrigation' the territory of the Sarykamysh Delta consists of a vast flat plain with a general slope west and northwest (with an average of 0.2-0.4 m/km; see details in Tolstov and Kes 1960:147-174; Doskach 1940). The plain decreases from the highest elevation of 80 m asl east near the Amudarya to 50 m west of the Sarykamysh depression.

The 'lands of ancient irrigation' are bordered in the east by fields of a contemporary oasis, by the steep Ustyurt plateau in the north, the Sarykamysh in the west, and the high sand ridges of the Transunguz Karakum south (Tolstov 1962a: *Karta Zemli drevnego orosheniya i nizovyakh Amudari i Syrdari* – Map of the 'Lands of ancient irrigation' in the Lower Amudarya and Syrdarya). The area of this territory is approximately 1 million ha. It is crossed by dry riverbeds (Kangadarya, Tunydarya, Daudan, Daryalyk or Kunyadarya) and the land in-between is rich with barely visible traces, reliefs, with many small dry channels marked by hollows, narrow sand strips or flat *takyr* lines (see Tolstov and Kes 1960:146: 146-147: *Geomorfologicheskaya i arkheologicheskaya karty prisarykamishkoy delty* – Geomorphological and archeological map of the Sarykamysh Delta).

The ancient hydrography is visible on the ground and aerial photographs in the shape of elongated light lines broken by sand sediments, forming meanders or in the shape of 'wandering fans' left by the river after its gradual progress to the flood-land (see Figure 6). The dense ancient delta network of channels begins within the limits of the contemporary cultivated area between Khiva and Urgench (Georgievskiy 1937:65). The central and southern parts of the delta are characterized by the highly branched Daudan channel system (Glukhovskoy 1893:183). The most northern branch of the Daudan, sometimes

called 'Budjunuyu Daudan', passes by the foot of the southern slopes of the Mangyr small elevation. As a network of channels it flows into the Sarykamysh depression between the elevations of Tarymkaya and Buten-tau (Tolstov and Kes 1960:159-162; Georgievskiy 1937:70-76).

The Daryalyk riverbed can be traced north of the city of Tashauz, whence it is directed to Kunya-Urgench and the narrow promontory of the Ustyurt, where the ruins of Shiran-kala and Dev-kesken (Vazira) tower above a 200-250 m wide well preserved valley. The Daryalyk then runs along the northern foot of the Buten-tau, and narrowing it takes the form of a canyon, finally flowing into the Sarykamysh depression (Tolstov and Kes 1960:153-159; Georgievskiy 1937:77-81). The relief shapes here are very different: from active moving sand barchans and less mobile to stable, fixed sandy hills and plains (Gelman 1891; Doskach 1948). The alternation of sands and *takyr* clearly limited the depression, covered with thin clay sediments, which filled over a more or less extended time period during spring rains and rare downpours. The *takyr* is mainly distributed along former riverbeds in the southern and, especially, in the northern parts of the ancient deltaic plain, where surface sediments have clay with a more organic composition (Rodin 1961:13-14; Konobeeva 1965:129).

In the northern half of the delta, traces of farming activity are expressed more clearly. The ruins of abandoned Turkmen settlements of the 19th century and rare sand accumulations rise above the clay *takyr* surface, crossed by canals and field ridges. In recent years, this area was actively developed by cotton-growers of the Kunya-Urgench region, who ploughed the abandoned fields and used many Medieval and Turkmen canals.

In some places the remains of a Tertiary

plateau with limestone, clay, gypsum and marl (elevations of Kuyusaygyr, Tarymkaya, Gyaurgyr, Buten-tau, Tuzgyr, etc.) rise 20-60 m high above the deltaic plain. Besides large elevation remains, there are also smaller remains of Transunguz pliocenic stratification: ridges stretching from north to south (Tolstov and Kes 1960:162-172). There are ruins of ancient fortresses (such as Kyuzeligyr-kala, Kalalygyr-kala I and II, etc.) which lie on the main ridge areas.

The delta of the Sarykamysh is formed by stratifications of sandy clay loam and alluvial deltaic sediments several meters high. This sequence was generated in the Upper Khvalinsk and Neo-Caspian period, when the Amudarya formed the Akchadarya Delta, then curved west and started to supply water to the Sarykamysh depression (Tolstov and Kes 1960:16-23, fig. 4, 173-174). The Amudarya waters filled the Sarykamysh and the adjacent Assake-Audan depressions and formed a huge fresh water lake. Once the lake water level reached 53-60 m in height, the water flowed and thoroughly washed away the Uzboy riverbed, through which the Amudarya had reached the Caspian Sea (Kes 1939, 1954; Tolstov, Kes and Jdanko 1954, 1955; Tolstov and Kes 1960:21).

The Uzboy ceased to flow continuously in the early 1st millennium BCE, which was proven by geomorphologic and archaeological research, although through a breach in the Amudarya water also flowed at a later time (Tolstov and Kes 1960:23-25, 29-31, 343; Bartold 1965:173-184, 321-325; see also Kaulbars, 1887; Konshin, 1897; Bartold 1965:15-94, 99-120; Tolstov and Kes 1960:5-13; Tolstov 1962a:17-26). The end of the Uzboy permanent water flow was connected to a 50 m lowering in the level of the Sarykamysh Lake. Most of the Amudarya water started flowing into the Aral Sea, and a significant water quantity began to be used for irrigated agriculture in the deltaic channels of the Sarykamysh as far as the 1st millennium BCE. Active deltas and riverbeds of the Amudarya tended to move downstream, i.e. in the direction of the Aral Sea. Farming activities of the inhabitants of ancient Khorezm accelerated

this natural process of moving the active deltas further downstream (Andrianov 1955:356).

The thick layers of alluvial sediments in its upper part were subjected to significantly lithological changes. The process of mud deposits and the accumulations of sand and soil by irrigation waters led to the formation of irrigated loamy sediments, coating the land (Gerasimov, Ivanova and Tarasov 1935:24; Georgievskiy 1937:98; Letunov 1958:43).

In the regions of ancient regularly-irrigated agriculture in the Southern Khorezm, cultural sediments reached a thickness of 2 to 5 m along the irrigation canals (Georgievskiy 1937:112). In the basin of the Daryalyk and the Daudan cultural sediments were 1.3-1.5 m thick. In the lower delta the sediments were less thick, just 25-50 cm thick. Sediments associated with a population cultural-economic activities are differentiated by their origin. Thus, the geologist B. M. Georgievskiy identified five provenance types on the left bank of the Khorezm: 1) cultivated-irrigation sediments; 2) irrigation sediments; 3) cultural-urban sediments; 4) sediments of old cemeteries; 5) artificial earthen mounds, such as dams, dikes, etc. (Georgievskiy 1937:98-118). Considering that irrigation sediments increase the surface of irrigated fields at 1 mm each year, and the compost introduced in the soil adds another 0.5 mm (i.e. 1.5 mm per year), B. M. Georgievskiy tried to establish the antiquity of agriculture in different parts of the Southern Khorezm.

According to his calculations, farming activities were: 1) on the left bank of the Daudan Basin 2,000 years old; 2) between the ancient floodplains of the Daudan and Daryalyk former riverbeds 1,750 years; between the ancient floodplains of the former riverbeds of the Daryalyk and the *kair* lands of the Amudarya floodplain approximately 900-1,000 years; 4) in the Daudan floodplain 800-1,000 years; 5) in the Daryalyk floodplain 400-500 years; 6) in the *kair* lands of the Amudarya floodplain in the area furthest from the river 250-300 years, and in the area closest to the river no more than 50 years (Georgievskiy 1937:108; see also Gulyamov 1957:89-92).

I. P. Gerasimov, E. N. Ivanova and D. I. Tarasov supposed that the annual layer of irrigation sediments did not exceeded 0.6 mm, thus it took approximately 1,500 years to reach a layer of irrigation sediment (without fertilizers) 1 m thick (Gerasimov, Ivanova and Tarasov 1935:52; Saushkin 1947:294; Letunov 1958:43). As for sand and manure sediments or fertilizers (which appeared in Khorezm, in S. P. Tolstov's opinion, in the 5th-6th centuries and possibly sometime earlier), according to data of the early 20th century the field soil level could have raised annually by 6.4 mm (Gerasimov, Ivanova and Tarasov 1935:33). According to above authors, it is impossible to evaluate the annual increase of a cultivated layer, taking into account, that not all areas were treated equally. The co-efficient of land use was not huge, only 0.2-0.3-0.35% (Letunov 1958:43). The rate and thickness of agro-irrigation sediments on fields varied, depending on the nature of irrigation and the exploitation of each field in different historical periods, which leads us to take B. M. Georgievskiy's approach with great caution. However, he was completely right when he connected the process of rising cultural-irrigation sediments in the Southern Khorezm with: the gradual deepening of ditches; lowering the horizons of irrigation water; and changing from gravity irrigation to the *chigir*, i.e. artificial water lifting to irrigate fields, which was, as we shall see below, widespread in Khorezm starting from the 9th-11th centuries (Georgievskiy 1937: 109) (see also p. 208).

The first to formulate a thesis that *kair* agriculture represented the beginning of farming in the Lower Amudarya was B. M. Georgievskiy. Later it was developed by S. P. Tolstov and especially Ya. G. Gulyamov (Tolstov 1948a:41, 67; 1948b:76-77; Gulyamov 1957:59-60, 74, 89; etc.). According to B. M. Georgievskiy, "Originating on the alluvial plain of the delta, first on both sides of the ancient flood-land on the active channel of the Daudan, irrigated agriculture, then developed on both sides of the ancient Daryalyk channel flood-land, and after the channels dried, it also spread to the ancient flood-lands of the

Daudan, to the Daryalyk, and finally into the modern flood-land of the Amudarya forming 'kair lands'" (Georgievskiy 1937:106).

If on the left bank it was possible to identify the undeniable evidence of prehistoric irrigated agriculture (in particular, the Tazabagyab and Suyargan sites of the third quarter of the 2nd millennium BCE), in the 'lands of ancient irrigation' on the right bank of the Khorezm traces of such early irrigation were not yet detected. The Bronze Age finds in the Sarykamysh Delta are rare (Tolstov and Kes 1960:179).[Note 136] According to S. P. Tolstov, here these "sites were washed up or fully covered by additional later sediments" (Tolstov 1955c:192; Tolstov and Kes 1960:180).

Ya. G. Gulyamov supposed that the wandering channels of the Daudan became the habitat for ancient herders and farmers, apparently, almost simultaneously with the right bank (Gulyamov 1957:74). However, according to a different point of view, irrigated farming and irrigation facilities appeared on the left bank of the Amudarya somewhat later. The Sarykamysh Delta (the river Mazdubast according to al-Biruni) is younger than the Akchadarya Delta (the riverbed al-Fahmi).

The history of irrigation on the right bank of the Khorezm gives us an example of the very close connection of ancient irrigation works with the oldest deltaic channels. Based on geological data, B. M. Georgievskiy supposed that "the spreading of irrigated agriculture in Southern Khorezm took place simultaneously with the history of the ancient delta last period of formation and proceeded in parallel with it" (Georgievskiy 1937:106). According to that scheme, B. M. Georgievskiy suggested that the left bank most ancient irrigation should be found in the upper reaches of the Daudan and Zeykash. That is to say, within the limits of the contemporary cultivated area of Southern Khorezm, where the irrigation network functioned continuously, forming new cultural layers thus making it difficult to identify prehistoric sites, hidden under the soil (Gulyamov 1957:92).

The research of Bronze Age sites in the Upper Uzboy, i.e. in the region adjoining the Sarykamysh

Delta from southwest, enabled M. A. Itina some conclusions on the appearance of cattle breeding in the economy of the inhabitants of this area (Itina 1959c:308).

How was the Sarykamysh Delta in that period? Herodotus' writing about the Araxes River (Amudarya?), which flows into the Caspian Sea from the east, fits very well with it. At its mouth the river was divided into 40 branches, one of which "flows through an open area into the Caspian Sea from the east" while others "flow into swamps and lagoons" (Herodotus I, 202; Bartold 1965:24; Fedchina 1967:5-6).

In the earliest period of farming culture, swamps and waterlogging in this region were proven by soil research in the environs of the Shakh-Senem elevation and ruins on the Daudan southern channels where several stages of natural history were identified. In the pre-farming period, delta freshets predominated when the riverbed sandy environment was changed by the appearance of the marsh-*tugai* soil formation. This landscape is very close to that of the modern Amudarya Delta, with its lakes, cane and *tugai* thickets (Andrianov, Bazilevich and Rodin 1957:525). The test trench at *poisk* 13 (Section 151) is very characteristic (Figure 41), showing the prevailing Archaic pottery and providing many Kushan and Medieval ceramic finds.

Here is a description of the prospecting test trench. Layers: 1) *takyr* crust; 2) ashy-gray, in some places dark-gray, medium dense loamy clay with lumps; sand in the upper part, affected by traces of modern soil formation process (agro-irrigation sediments), from 2 to 14 cm thick; 3) ashy-gray loamy clay with sand and traces of digging, rusty veins and swamping, organic remains (roots) and salt deposits (agro-irrigation sediments), from 14-27 cm thick; 4) light gray loamy clay with rusty spots and in some places sand, salt; handmade 'barbaric' and badly burnt pottery fragments were discovered at a depth of 40 cm (agro-irrigation sediments), the middle layer was 27-40 cm thick; 5) brown-ashy fine porous loamy clay with lumps, rusty spots and remains of gleysol, deposited on the lower layer with a break (upper part of low

agro-irrigation layer), 45-68 cm thick; 6) loamy clay with sand, traces of digging and a horizon of swampy soil in the low part (low agro-irrigation layer), 68-90 cm thick; 7) loamy, swampy and riparian soil with no traces of agriculture, 90-95 cm thick.

From this section it is clear that the irregular agricultural activities began in the area of wetland *kair* lands apparently already before the Archaic period. A picture of the life and economy of recent inhabitants of these horizon were vividly depicted by A. V. Kaulbars (1881:531, 543 and others).

In the 19th century, in regions with changing river channels, ancient traditions dating back to the Bronze Age were preserved with a very typical primitive integrated economy: vast cane thickets served as cattle pastures; the inhabitants fished in channels and floods and cultivated millet, melons and pumpkin in the *kair* (Andrianov, 1958a:113; see also Gulyamov, 1957:63). In order to reconstruct the economy of inhabitants of the Lower Amudarya in the Bronze Age, Ya. G. Gulyamov reviewed historical-ethnographic materials and concluded that, from the outside, the way of life of some Karakalpak and Uzbek tribes partially represented, to some extent, the Bronze Age farmers' life (Gulyamov 1957:63).

The cattle occupied a central place in the economy of inhabitants of the deltaic areas rich in channels, lakes and cane thickets. This was widely reflected in traditions, religious concepts and cult complexes of the local Iranian-speaking population of Central Asia, in which the bull is identified with water and vegetation (Trever 1940:71-86; Tolstov 1935b:16; 1948a:295; 1948b:87; Snesarev 1960:199).[Note 137] Much attention was given to the bull in religious and Iranian epic traditions where the feasts of the first Kayanids' feats took place on the banks of the Dātya River, in a country rich in lakes, where the first people came on a sacred ox named Srīsōk and where the hero Yima-Jamsh (Jamshid), according to Biruni's words, "ordered to dig canals" (Tolstov 1948b:87; Biruni 1957:228).[Note 138] According to S. P. Tolstov, the mythological image of the Turks' ancestor, 'Oghuz', in the Turkish language means 'bull' and 'river', and specially (in the form okuz), the Amudarya (Tolstov 1935b:16; 1948a:295).

The mythological images of ancient Iranian legends and the *Avesta* religious texts, describe the struggle of the Aryans as settled farmers and pastoralists of Margiana, Sogdiana, Bactria and Khorezm (Airyana Vaējah) with the Tura as their northern nomadic neighbors (Saka-Massagets tribes), reflected the real historical process of the spreading of developed agriculture (Bartholomae 1905:127, 133; Marquart 1901:156; Struve 1949:149; Abaev 1956:41).

Perhaps a few archaeological sites of the Sarykamysh Delta of the early 1st millennium BCE are connected to this historical developments. In particular, 3 km from Kuyusaygyr, remains of a large site of semi-settled herders were found, represented by five clusters of rough ceramic of 'barbaric type'[Note 139] over a total area of 250 × 100 m (*poisk* 82). On the site there were many large and small cattle bones, as well as a three-winged bronze arrowhead with a long shaft of the 7th-5th centuries BCE (Tolstov and Kes 1960:179). B. A. Litvinskiy (1968:88),[Editor 50] referring to K. F. Smirnov, dated such arrowheads to the 7th-5th centuries BCE. The ceramics, according to E. E. Kuzmina, showed great similarity with the molded pottery of Southern Turkmenia. These temporary herder settlements were widely spread across the Sarykamysh Delta, which, in the 7th-6th centuries BCE, was a highly watered deltaic area, intersected by channels and rich in meadows and lakes.

Irrigation Works of Antiquity

Archaic period (6th-5th centuries BCE)
In that period a strong state evolution developed in Khorezm, the construction of irrigation systems began in the Sarykamysh Delta on the Daudan lateral channels (Gulyamov 1957:74; Tolstov 1958:112; Tolstov and Andrianov 1957:9; Tolstov and Kes 1960:182-185). The remains of these systems were best preserved in the ancient Chermen-yab basin on the left bank of the Daudan

near the ruins of Kyuzeligyr-kala (see Andrianov 1958b:312-313: *Karta Drevnyaya irrigatsionnaya sistema Chermen-yaba* - Map of Chermen-yab ancient irrigation system). Kyuzeligyr-kala (lit. 'ceramic hill') was already discovered in 1939 and investigated again by the Khorezm Expedition in 1950. In 1953-1954 Kyuzeligyr-kala was the subject of a long-term excavation (see 1947a:3-8; Tolstov 1948a:77-82; 1948b:93-109; 1953b:318; 1955b:176-177; 1955c:193-197; 1958:143-153; Vorobeva 1955:73-74; 1958:329-346; 1959a:66-84).[Editor 51] Based on finds such as Scythian type bronze arrowheads, beads, rich ceramic material similar to Afrasiab I, Kobadian I, Bactria and Margiana, the site of Kyuzeligyr-kala was dated to the 6th-5th centuries BCE (Tolstov 1958:146-149, fig. 56; Vorobeva 1958:344).

The economic activity of the inhabitants of Kyuzeligyr-kala was characterized by a combination of regularly irrigated (and perhaps irregular *kair*) farming and cattle-breeding. This was proven by the remains of a massive irrigation work of that period in the site environs, the discovery of an iron sickle and stone millstones during archaeological excavations, and numerous bones of domestic animals. A special place still belonged to bovines. It was evidenced by the results of V. I. Tsalkin, who identified bones of domestic animals found in the ruins of Kyuzeligyr-kala, where they represented 52% of the total livestock (including cattle, 40.9%; horses, 8.9%; camels, 2.2%; Tolstov 1958:150; Tsalkin 1966:150). According to Ya. G. Gulyamov, the inhabitants of this areas grazed animals on the plain, were engaged in agriculture, and in case of approaching enemies they drove the cattle into the fortress (Gulyamov 1957:74).

The site excavations of 1953-1954 led to a correction of S. P. Tolstov's early assumption on Kyuzeligyr-kala as a "city with residential walls". However, confirmation of the fact that there were no buildings in the inner part main space made a valid comparison of this site with the Avestan enclosure already advanced in the book *Drevniy Khorezm* (Ancient Khorezm) (see Tolstov 1948a:79-82, fig. 20, 1958:143-153, fig. 58). The linguist M. N. Bogolyubov suggested a new reading of the name 'Khorezm' as 'Country with good cattle fortifications' or 'Country with good enclosures', supporting the archaeologists' conclusions.

The fortification of Kyuzeligyr-kala covered a significant agricultural area of the Archaic period (6th-5th centuries BCE), extending along the Daudan southern channels west and east of the settlement.[Note 140] In this region the Archaic main canals were identified (Figure 42.A). The westernmost of them, found in 1952, started in the area of Kuyusaygyr (*poisk* 31, 35, 89) and was traced along a narrow channel (50 m wide) which had here an east-west direction. The canal was the artificial extension of this small channel, which was a well-known irrigation method in Bronze Age and in some Archaic canals of the right bank of the Khorezm (see p. 156).

A large meander of the channel straightened the canal in its central part. Almost along its whole length (6 km), small mounds (0.5-1.2 m high) were preserved on the left bank. At the point, where the right bank was preserved, the width between banks was 16 m and the overall canal width was 25-30 m (Figure 43.1). A few narrow ditches departed from this canal, irrigating the regions and stretching to the Taygyr, where the main riverbed of the Southern Daudan had been lain. A test trench was dug in one of the lateral branches (*poisk* 33). Here is the description of the horizons: 1) *takyr* crust; 2) ashy-gray, heavy loamy clay with traces of digging, sand in the upper part and an uneven border in the lower part (agro-irrigation horizon), 0-12 cm deep; 3) light-gray, heavy and thin-stratified clay loam with rusty spots and weak traces of agriculture (lower agro-irrigation horizon), 12-30 cm deep; 4) alluvial sand with traces of swamping, 30-70 cm deep. The irrigation layer was 30 cm deep. According to finds of Archaic ceramics (without any later settlements and irrigation systems) these sediments should be dated to the Archaic period.

Different clusters of Archaic ceramics were found in many places of this region. Directly near the above mentioned canal, which at this point ran along the river channel, we found traces of an Archaic settlement, with scattered ceramics clusters typical of the 6th-5th centuries BCE (*poisk* 31,32, 35, 89, etc.). The collected pottery allowed us to date to the same period a wide but short (up to 10 km) canal west of Kyuzeligyr-kala, and several canals to the east. The canals branched from the Daudan riverbed almost perpendicularly, in a north-south direction, and ended after 5-6 km in small branches of the same riverbed. The last one likely closed from the south a group of narrow deltaic islands, with remains of scattered hills ('rocks') widespread between the riverbeds, topped by ruins of ancient fortresses (Kyuzeligyr, Kalalygyr I and II).

The most massive north-south canal, very clearly seen from aerial photography (see Figure 6), had an overall width of 70 m and 40 m between banks (Figure 43.2). In some places, the banks were preserved to a height of 1-.2 m. In the central part there were visible traces of a narrow bed 12-15 m wide, now expressed only by a strand of *biyurgun* bushes. The small irrigation canals branched from the main riverbed at right angles, which gave the whole system a primitive, angular and 'sub-rectangular' character. In its lower part the canal was cut across by Medieval systems with an east-west direction. It flowed into a small east-west channel, which is preserved and very visible on vertical aerial photographs thanks to the typical 'wandering fan' pattern marked by modern soil and vegetation (see 'B' in Figure 6).

Not far from the canal head there was a dispersed site with Archaic ceramics of the 6th-5th centuries BCE (*poisk* 130). The canal started from a lateral Daudan channel at the foot of a small sandy hill and then veered south between two elevations. The location of the canal heads in the hard ground between hills, apparently, was not causal. It was easy to build stable head works there, which would be preserved for a long time and would prevent the destructive effect of flood water from natural channels. In his research on the irrigation of Khorezm, Ya. G. Gulyamov noted the importance of selecting the right location for head constructions. The head of flood canals always had to be constructed in places with stable ground in order to prevent the rapid erosion and the head and the system drainage (Gulyamov 1957:239).

As indicated by our archaeological-topographical study, the head works of the north–south canals consisted of several head canals passing directly over the edge of a fan elevation in the hard ground (Figure 42.A). The supplying canal, 15 m wide (probably the riverbed latest functioning stage) was divided into three branches: the western, 35 m wide (15 m between banks); the central, also 35 m wide (13 m between banks); the eastern, 18 m wide (8 m between banks). If the first two canals supplied the main canal bed, the eastern one was intended to drain excess water away from settlements and fields. This method, called *bedrau* (*mufriga* according to al-Khwarizmi), recorded by Ya. G. Gulyamov in the late Medieval irrigation of Khorezm, was apparently known to irrigators of the Archaic period, indicating a high level of engineering and hydraulic skills.

Thus, the relatively well-preserved canal of the 6th–5th centuries BCE had a complicated supply scheme: riverbed – head works (two) – drainage (*bedrau*) – main canal – feeder – field. The principal part of this scheme is to some extent close to the Bronze Age irrigation systems, but different for its huge size: the main canal 40 m wide was an artificial construction. The peculiarities of its topography and the sharp difference between width and length are explained by the fact that the canal connected two east–west channels (a method widespread in deltaic irrigation of the lower reaches of the Syrdarya) (see pp. 218–219).

Quite unusual was also the other, neighboring Archaic canal (*poisk* 186-191). Its overall width was 35 m, 14 m between banks and 2.5 km long (see Figure 6). The canal was at the foot of another sandy hill. Its central part was 1.5-1.8 m high above the *takyr* in the shape of a flat bank. In the lower part there was a fan-shaped branch with a whole series of narrow canals (see 'C' in Figure 6). One of them had well-preserved banks under sand deposits (*poisk* 187). Its total width was 8 m and approximately 2 m between banks (1 m high). Numerous fragments of Archaic ceramics were found in these banks and around them.

The bed of the Archaic Chermen-yab, took water in the proximity of Kyuzeligyr, where the canal was greatly eroded and only one earthen wall was preserved on the riverbed bank. It should be attributed to the ancient period of regular irrigated agriculture. The leveling of the canal head revealed that, over a length of 900 m, its level lowered in the south up to 93 cm. Traces of a complex interweaving of bank lines, floodwaters, and artificial head works indicated its long functioning. The oldest probably even dates back to the Kyuzeligyr-kala settlement (6th–5th centuries BCE). The abundance of Kangju ceramics at the head canals indicated that, at a later period, these were the irrigation system sources of the Chermen-yab, which watered the environs of the Shakh-Senem plateau. The cultural layers in the environs of Kyuzeligyr-kala, dated to Antiquity and the Middle Ages, covered almost entirely the central part of this irrigation system. Only the margins of irrigation systems were preserved, little affected by later cultures.

The Archaic canals of the Chermen-yab basin greatly resembled the former riverbeds of the Bronze Age, adapted for irrigation and similar to the Archaic canals in the environs of Bazar-kala (see p. 157).

Quite interesting is the evaluation of labor cost required to construct these Archaic canals in the Chermen-yab basin. The above described north–south canal was 2.6 km long and 50-70 m^2 in cross-section. In order to dig such a massive bed, build the head works and dig the canals, it was necessary to remove about 130,000-180,000 m^3 of soil. This work could have been carried out by 1,500 people in 30-40 days. To build the neighboring, smaller sized canal (*poisk* 185-190), 200 men were required for 50 days. In the environs of the canals clusters of pottery of the 6th–5th centuries BCE were recorded, indicating remains of Archaic sedentary agricultural settlements, in which no more than one or two hundred people could have lived.

Remains of irrigation systems and sites of the Archaic and early Kangju period were found in the area between the Southern and Northern Daudan rivers. In particular, a north–south Archaic canal (20 m wide), and remains of a fortified settlement, with Archaic and early Kangju material were discovered 10 km east of Mangyr-kala (*poisk* 650, 653, 655).

More sophisticated and vaster in size are the early Antique systems in the environs of Kunya-Uaz, amongst which a huge Archaic canal which also functioned in the later Kangju period, 50 km long and 70-75 m wide (Figure 42.B). The sources of the ancient canal were located on a lateral small channel of the Daudan. The canal was directed towards the ruins of Turpak-kala and then it continued to Tuzgyr along a small channel. Its continuation north of Tuzgyr was a natural riverbed. Traces of sites and clusters of ceramics of the Early Kangju period were found in the canal and in the small irrigation network.[Note 141]

In its upper part the canal was largely destroyed and covered by later systems. In the area of the ruins of the Kunya-Uaz fortress (whose lower layers seem to be dated to the Kangju period) the canal flowed along the banks of the Southern Daudan. Its width was more than 40 m, 70 m including its banks (Figure 43.3). Here these banks were preserved for a height of 0.75-1 m. In many places the canal was cut across by narrow Medieval ditches. Next to the canal, traces of an Archaic settlement with a rich accumulations of Archaic ceramics were found (*poisk* 258, etc.). South of the ruins of Turpak-kala, this canal was preserved with the same general width, i.e. about 70 m (Figure 43.4). The better preserved banks are 2.5-3.0 m high above the *takyr* and 45 m wide between each other. The narrow Medieval ditch was visible in the central canal area (*poisk* 420).

In the lower part, below a large branch in the direction of the ruins of Turpak-kala, the canal sharply reduced its size. It was approximately 40 m wide and 20 m between banks (*poisk* 289) (Figure 43.6); in its lowest part, it was 32 m wide and 12 m between banks (*poisk* 572). Here, next to the canal, was found a small fortified square settlement 80 × 80 m. Its corner walls were barely discernible, traces of rounded towers along the

walls and visible remains of a circular corridor were perceived. The walls were made of adobe bricks 45 × 42 cm in size. Ceramics were spread along the walls, and the room layouts were visible in places. After studying the collected surface material in 1953, M. G. Vorobeva dated the site to the late Archaic period.

In 1965, B.I. Vaynberg found at this site early Kangju material, proving that the surrounding area irrigating system functioned in the 5th-early 4th centuries BCE.[Note 142] As we know, the lower layers of Kunya-Uaz were dated to the Kangju period (Nerazik 1958:371). The settlement was located in the upper canal area, and perhaps at that time it was already a large settlement. The large Archaic, Kangju and Kushan canals were found in its surroundings (see p. 196).

In its middle and lower parts, the Archaic Kunya-Uaz canal ran for a long distance very close to the Southern Daudan (see Figure 42.B). On one side of the river the irrigated area was controlled by a dam, some sections of which are preserved with earthen walls 3-5 m high and a peculiar horseshoe-shaped construction (60 × 30 m), its function unclear. The chronological correlation between canal and dike is unclear, although fragments of archaic vessels (*poisk* 416) were found on the latter. On a small section of the canal (between *poisk* 412 and 414), the dike was very close to the canal and it was located along its destroyed southern edge. This creates the impression that the dam was more recent than the canal itself.[Note 143]

At *poisk* 415 the dam was cut across by a small canal covered with Early Medieval ceramics, thus the dam was probably older than it. It is possible to assume that it was built in the late Kushan period, when channels of the Sarykamysh Delta were highly watered and when the level of the Sarykamysh Lake rose. Water was drained into the Uzboy, then the Igdy-kala was built on an irrigation riverbed in the 4th-5th centuries (see Tolstov 1962a:235).

The dam was constructed to protect fields and settlements against destructive floods. Traces of floods are still visible south of the bank along the Daudan in the shape of fan-shaped lines of vegetation, and sand spits very clearly seen in vertical aerial photographs. These lines are very close to the canal and the protective dike. The dam had two banks in the environs of the ruins of Turpak-kala (poisk 410) and the second, less massive bank surrounded a large oval shaped hollow depression which was, without any doubt, an artificial basin 600 × 300 m in size. Its bed is now 1-0.8 m lower than the surrounding takyr. On the banks there were traces of standing water. In the basin there were no traces of Medieval pottery but, instead, very abundant Archaic and Kangju-Kushan ceramics. There is reason to believe that this basin was directly connected with the whole protective system complex of Antique irrigation and it is a very interesting prototype of modern protective lakes placed between rows of dams on the Amudarya.

The dams in the region of Kunya-Uaz are impressive ancient monuments of Khorezmian farmers' struggles against river floods natural forces. Similar dams were also described by R. J. Forbes regarding Sumerian irrigation (Forbes 1955, Vol. II:17). He also mentioned Polycletus' story about the activities of Babylonian rulers, under whose guidance protective dams were constructed in Mesopotamia. I.E., "in order to maintain high water and, on the contrary, to prevent the filling of canals, which is caused by the accumulation of silt, by cleaning the canals and opening the mouths. Although the cleaning of the canals is not difficult, the construction of dams demanded a lot of manual labor" (Strabo XVI, 1, 10).

The Russian travelers of the early 19th century reported on the protective dams in the Khiva Oasis. They wrote about banks "as high as a person and more", built around lakes and along channels "near where they located the arable land". With the destruction of the dams "all the neighboring inhabitants immediately came running to close up the breach" (ZJ 1838, N. 6:330-332).

Returning to the description of the Archaic and early Kangju periods canal of Kunya-Uaz, it should be noted that the capacity of its transverse flow in the upper part (up to 45-40 m) and the narrowing of its branched large distributors, provided the canal water with the necessary speed for gravity irrigation. The system is characterized by angular 'sub-rectangular' layouts. Minor small networks of ditches were rare and insignificant. Knowing the cross-section of the canal in its single parts, it was possible to calculate the amount of labor for its construction and exploitation. The volume of removed soil was, perhaps, 1,600,000-1,750,000 m^3. In order to accomplish such work, an army of 12,000-14,000 diggers for 50 days were needed. Cleaning up the canal could have demanded 5,000-6,000 diggers, which would have required centralized action over the whole oasis. The organization of such large-scale irrigated farming was possible only through the exercise of state power.

The creation of such a huge irrigation system in the late Archaic and early Kangju periods on the right bank of the Daudan, is evidence of a significant efforts of the Khorezmian rulers in organizing large-scale irrigation agriculture in the period beginning the Achaemenid Empire and after its end (early 4th century BCE).

The beginning of construction of the colossal fortress of Kalalygyr-kala I by the Iranian Achaemenid on the southern (left) bank of the Daudan riverbed, opposite the sources of the canal described above, took place between the 5th and 4th centuries BCE.[Note 144] According to S. P. Tolstov, the construction of the rest of this unfinished fortress was aimed to ensure the Achaemenid invaders access to the irrigation systems close to the Khorezm citadels, both on the left (Chermen-yab system) and the right banks of the South Daudan (the Kunya-Uaz canal). The attempt to erect the fortress could also be ascribed to the Achaemenids' system of events which are reported in the famous story about the Achaemenid king's irrigation policy (Tolstov 1955b:198; 1958:167).

In concluding the review of the Archaic irrigation systems on the left bank of the Khorezm, many similarities to the coeval irrigation of the right bank of the Amudarya should be noted. On the 'lands of ancient irrigation' of the left bank of the Khorezm, the progress in irrigation technology is also noticeable. Here main canals were identified as continuations of fading former riverbeds (around Kuyusaygyr), large (from 20 to 40 m) and short canals (around Kyuzeligyr-kala), connecting the east–west channels with head works excavated in the hard ground, and huge late Archaic systems (environs of Kunya-Uaz) tens of kilometers long, with angular 'sub-rectangular' layouts of lateral channels repeating the meanderings of delta channels. There are some clear differences: the branch-shaped systems of the Archaic Kelteminar type were not seen on the left bank, and on the right bank there were no wide north–south canals connecting the delta channels.

The most extensive construction of large irrigation systems took place in the Sarykamysh Delta territory and on the Khorezm right bank during the Kangju (4th century BCE–first centuries CE) and especially in the Kushan (until the 4th century BCE) periods. At that time began the process of reducing the number of individual small canals and combining them into more massive systems. The heads of canals were moved increasingly upward along the flow of large channels to the main riverbed of the Amudarya (Gulyamov 1957:67).

This process was clearly traced in the Chermen-yab basin, where the Archaic and early Kangju main canals, starting from the Northern Daudan, had a predominant north–south direction (Andrianov 1958b:327). On the other hand, late Kangju, Kushan and Kushan-Afrigid systems had their origin further upstream (see Figure 42.E) and had an east–west direction. Their sources extend to the contemporary cultivated zone.

In the basin of the Chermen-yab we identified canals of the early Kangju period, not united into a unique system and irrigating separate regions in the plateau between Kyuzeligyr-kala and Kalalygyr-kala I, as well as some areas located east along the bank of the Southern Daudan. There, the ancient irrigation merged, and partially was covered by Medieval irrigation networks (see Figure 42.I). A series of head works and an irrigation system, dated

to the early Kangju period, were discovered east of the Kalalygyr-kala I elevation in the low left bank of the Daudan. The period the canal functioned, may, coincide with the short-lived ancient settlement of Kalalygyr-kala I.

The layout of irrigation systems of this period was characterized by more advanced branch shapes which, however, maintained a certain angularity, typical of the Archaic irrigation. The irrigation was based only on freshets; wide and shallow canals depended only on channels water levels and floods. Much water was supplied to the fields while excess water was drained into lower areas and adjacent channels beds.

The head works were especially interesting. As mentioned above, the Archaic irrigation systems on the left bank of the Khorezm had complicated head works in the form of multiple riverbeds (*poisk* 130). Special dam-shaped constructions were found in the early Kangju canals east of Kalalygyr-kala I (*poisk* 224, 396, 398). The dams directed the channel water into the canal and protected its upper part from erosion (see Figure 42.H).

For example, the canal diverted from the riverbed 1 km east of Kalalygyr-kala I (*poisk* 396) had a general width of more than 20 m. Nowadays, it is marked only by vegetation and microrelief. Its central part is denser and darker. Near the bank slope of the Daudan, canal heads were designed with a dam-shaped construction in the form of a flat bank covered by a layer of clay *takyr*. This bank, 5 m wide, was built up on both sides of a funnel-shaped depression on the riverbed bank. It was used to direct running water into the canal and to protect its heads against erosion. The general bank width of the canal's head was 16 m. We traced it down to the southern end of Kalalygyr-kala I elevation (*poisk* 115,116), where we discovered rare finds of likely Kangju pottery on the banks of the canal. The general length of the canal was approximately 2 km.

Adjacent to them, another canal with a general width of 30 m was marked by a flat bank 1–1.2 m high above the surrounding *takyr*. In the middle and in its lower part, the canal was eroded and marked only by microrelief and vegetation. In some places bank remains 0.5-0.6 m high were preserved. Its head construction rose above a large wind eroded hollow in the shape of two horseshoe-shaped dams directing the river waters into the canal. Its length was 2.5 km. Branches were very rare.

An example of a multi headed canal source could be found in the Kangju canal head constructions on the right bank of the Daudan (*poisk* 209, 210). At least three heads, in the form of *takyr* clay strips (10-15 m wide) originated in the vast hollow of a dispersed riverbed, preserved here (see Figure 42.H).

The Daudan riverbed divides into the Northern and Southern Daudan 5 km east. There were very suitable conditions for the construction of main canal heads, in particular we discovered the head works of the Kangju-Kushan and Early Medieval systems, which irrigated the environs of Kunya-Uaz in Antiquity and the Middle Ages.

The remains of an Antique canal massive head work, situated on a steep bend on the banks of the Daudan (*poisk* 224, 215), were particularly interesting. Its head works were connected to outcroppings of hard rock. Here were the most optimal conditions for constructing the heads of flood canals. They conform to the conditions reported by Ya. G. Gulyamov, which are: 1) the stability and hardness of the ground; 2) constant and moderate water flow at this point; 3) correspondence of chosen point to some other old source (Gulyamov 1957:239). The head construction of canals irrigating the environs of Kunya-Uaz shows that, already in ancient times (in the first centuries BCE), the Khorezmian irrigators knew well the technique for hydraulic devices and for managing large main canal sources.

Thus, already at the beginning of agriculture, there appeared in Khorezm (see p. 148), skills in building head-water regulating facilities which were further developed in Antiquity, in the Archaic and, especially, in Kangju and Kushan periods. This is clearly illustrated by the above described structure, preserved by lucky coincidence on the banks of the dry riverbed of the Daudan. Most of the ancient head parts of canals

were either washed away by the stormy waters of the Amudarya, or rebuilt after the sources to the main riverbed moved, or were destroyed by farmers in the Middle Ages.[145]

Chermen-yab
Historical changes in the topography of irrigation systems in Antiquity, i.e. the reduction in number of single small canals and combining them into unique massive systems in the Kangju, and especially in the Kushan periods, were traced in the basin of the Chermen-yab. These are described as follows (see Andrianov 1958b:312-313: *Karta Drevnyaya irrigatsionnaya sistema Chermen-yaba* – Map of Chermen-yab ancient irrigation system).

Over different periods the canals of this system, followed branches of the dry riverbeds of the Daudan, the Tunydarya or the Kangadarya.[146] They pursued the northern edge of the Kyzylkum sands for their whole length of 150 km from the contemporary oasis (environs of Zmukshir or Zamakhshar) to Kangagyr, where are the ruins of the Kanga-kala fortress (mid-1st millennium BCE - 4th century CE).

West-northwest of the fortress, during the field work of 1954, directed by S. P. Tolstov, a small irrigation network, with the head in the main Kangadarya riverbed washing the Kangagyr promontory, was found. The extant period of this system coincided with the fortress. The irrigation network was characterized by a 'sub-rectangular' shape. The canals were 1.7 m wide, their direction west-northwest, and they were cut across by canals 2.4 m in size between banks (Tolstov and Kes 1960:191).

The archaeological-geomorphological investigation in the Kangadarya lower reaches showed that its channels formed a kind of small 'delta' and the whole plain was cut across by riverbeds and canals. The fortified settlements of that period coincided with the hills. These are the above mentioned Kanga-kala and Gyaur-kala I, built on slopes and elevations, and towering over the vast farming environs (Tolstov 1941, 1948a:95) (see Figure 54).

During our archaeological-topographical

research in 1952-1953, some Kangju and Kushan sites were discovered there, along with several small north-south early Kangju canals.[Note 147] These canals were connected and straightened by the later main canal branching from the channels of the Tunydarya, 15 km northeast of Gyaur-kala, and were built along the hills (see Andrianov 1958b:312-313: *Karta Drevnyaya irrigatsionnaya sistema Chermen-yaba* - Map of Chermen-yab ancient irrigation system)

The total system length of the lower Chermen-yab was approximately 40 km. In the upper part on the bank of the Tunydarya, the canal was marked by a massive bank 40 m in size (*poisk* 55). Small (1 m high) bank remains were also preserved. The width of the canal at this point was more than 100 m wide. The canal followed the riverbed bank in a southwest direction with only rare branches directed right. At 10 km from the head its overall width was more than 55 m and 25 m between banks, which were 2 m high. Then the canal encircled the elevation and changed its direction veering west and even west-northwest.

At this point (*poisk* 4, year 1952), a huge canal, 53 m wide and 32 m between banks, branched from the Chermen-yab. Its banks rose only 30-50 cm above the *takyr*. The canal began near the Chermen-yab, directed west, and drained into the main riverbed. The overall length was 2 km. The unusual relationship between length and width of this canal, as well as its topographic location, forced us to suppose that it was not an independent irrigation canal, but played the role of a regulating-diverting canal (*bedrau* or *mufriga*), to drain excess floodwater from the main bed of the Chermen-yab (see Gulyamov 1957:240). A. I. Glukhovskoy had already noted that "arbitrarily increasing or decreasing this consumption, a Khivan could use the amount of water required for irrigation" (Glukhovskoy 1893:18).

The draining canal was used to regulate floodwaters. It should be dated to the Kangju period. This is confirmed by archaeological finds, in particular by fragments of many Kangju *khum* dated to the 4th-3rd centuries BCE (Andrianov 1958b:319).

Here began the vast ancient 'oasis' of Gyaur-kala. Numerous traces of dwellings and farms, covered by Antique ceramics and remains of ceramic production, filled the entire space along the Chermen-yab riverbed and its numerous branches. In this location the Chermen-yab canal was 16-18 m wide between banks (Figure 46.25). The banks were no more than 2-2.5 m high. A lateral canal, making a large loop at the height of Gyaur-kala, branched off the Chermen-yab 3 km from the Gyaur-kala fortress. Then, after 1 km, a small straight canal, directly connected with the Gyaur-kala fortress began irrigating lands on the right.

A small canal of later origin, which cut across the main riverbed in several places, was built on the left along the main bed of the Chermen-yab. It was 10-12 m wide. The right bank of this canal was the bank of the main channel of the Chermen-yab (see Figure 43.24). Smaller distributors branched from this channel on all sides and in one (Point 17,3X), traces of ceramic production and remains of settlement were found. The archaeological materials collected were dated by M. G. Vorobeva to the 2nd-3rd centuries, and this enables us to date the construction of the narrow canal to the Kushan period.

Several canals branched right from the main bed of the Chermen-yab toward Gyaur-kala. The first canal was 12 m wide between banks, the second only 9 m and the third, by capacity, was not inferior to the main bed, and it was 20 m wide between banks. The system of canals branching from the main bed of the Chermen-yab north of Gyaur-kala I gradually ended in drainage canals and natural depressions. The last drainage was found 3 km from the fortress. The main bed continued further west toward the Tarymkaya piedmont. The main canal of the Chermen-yab near Tarymkaya was approximately 50 m wide and 25 m between banks (*poisk* 102). Below the promontory at the southwestern end of Tarymkaya, the canal broadened up to 70 m and 30 m between banks (*poisk* 105). Its banks were 2.5-3 m in preserved height. The canal flowed into the channel.

The irrigation network of the Gyaur-kala Oasis occupied more than 3,000 ha and was rebuilt several times. Our archaeological-topographical research revealed several stages: 1) the short main canals which can be linked to the Kangju period; 2) construction of the larger Kangju main canal with sources in the Tunydarya, and general expansion of the irrigated area; 3) the heyday period of the settlements of Gyaur-kala I and II, and the beginning of the gradual reduction of the irrigated area. At that time a narrow lateral canal (dated by archaeological finds to the Kushan period) was built, as well as the Chermen-yab main canal by merging together short independent systems. This over the vast territory from Gyaur-kala in the southwest to Kyuzeligyr and Kyunerli-kala in the east, within the limits of the contemporary cultivated zone for a total length of more than 150 km.

The Shakh-Senem sector of the ancient system, located next and upstream of the Chermen-yab, also preserved traces of several reconstructions. Its width was less than 30 m and 11 m between banks (Figure 43.23-24). Further upstream, the overall width of the canal narrowed to 25 m and 9 m between banks. The main canal preserved its width in the environs of the ruins of Shakh-Senem (see Andrianov 1958b:317-318, figs. 2, 7-11). In some places the banks were severely eroded and their height did not exceed 1 m. In contrast, in other places the banks were well preserved and rose 3-4 m high above ground. The canal bed was heavily covered with sediments and recorded much higher, 1.5-2 m than the surrounding *takyr*.

The fortress of Shakh-Senem was the center of this region, which was identified by S. P. Tolstov (1941:181, 1948a:107) as the Medieval city of Suburna, located, according to al-Asir, 20 *farsakh* (150 km) from Urgench (MITT 1939:404; Bartold 1965:171). He supposed that this name was of Turkic origin and could be deciphered as 'cape of water', corresponding to the topography of this narrow wedge of irrigated land surrounded by deserts (Tolstov 1958:219; 1962a:260). In 1952, a team of the Khorezm Expedition (directed by Yu. A. Rapoport) carried out archaeological

excavations in Shakh-Senem, ascertaining that the site arose in the late 1st millennium BCE and lasted until the end of the Kushan period. In the Afrigid period it fell into decay, but in the 10th-12th centuries it became again a city, that died after the Mongol invasion. There was an attempt to revive the city in the late 14th century (Rapoport 1958:419-420).

The ancient irrigated cultivated area of the Shakh-Senem Oasis started 4 km north-east of the fortress, where the main bed of the Chermen-yab is thoroughly covered by sand and only the highest banks are detected on the ground at a distance of 16-20 m from each other (see Andrianov 1958b:313-317). Here the main canal had a branch on the right, from which started a dense network of a secondary ditches (Figure 47).

The source of water in the beginning of farming culture was the lateral channel of the Daudan, in particular the Kangadarya riverbed. In vertical aerial photographs this riverbed could be traced quite clearly through the network of canals and small agro-irrigation layouts, in the shape of sinuous meanders marked by sand accumulation in the modern landscape (see Andrianov, Bazilevich and Rodin 1957:523, fig. 3). The canals branching was conditioned by the direction of the ancient hydrographic system: an arm of the channel flowed northwest, the other, turned southwest. The ancient canals followed their direction (see Figure 47).

The zone of irrigated lands broadened as it approached the ruins of Shakh-Senem, with ditches diverted from the right branch of the Chermen-yab canal. One kilometer from the fortress, almost at the point of contact with the main riverbed of the Chermen-yab, the right branch changed its direction deviating west. The triangle space between these two canals was covered with numerous ruins of ancient and Medieval dwellings.[Note 148]

The right branch of the Chermen-yab was less wide than the main riverbed: it was 19-20 m wide and 9 m between banks. The canal banks were not always well preserved, but often they exceed the banks of the main bed of the Chermen-yab.

The canal rounded the fortress from south. Here the canal was 19 m wide and 9 m between banks. West of Shakh-Senem (*poisk* 26) the canal was covered with sediments, but preserved its width of 9 m between banks and stood high above the surrounding terrain. At a distance of 1 km from the fortress, and branching from the canal was a largely eroded riverbed 3 m wide, dated by ceramics to the Kushan period. The surrounding area near the bed of this canal took the shape of a flattened elongated bank 0.7–1.25 m high above the ground. Three kilometers from Shakh-Senem, traces of the irrigating canal were lost and merged with banks of the old riverbed.

As mentioned above (see p. 81), soil studies were carried out in the environs of Shakh-Senem (see Andrianov, Bazilevich and Rodin 1957). Analysis of the ground test trench and archaeological survey enabled us to identify a series of agro-irrigation layers differing from each other. According to its morphological peculiarities, the lowest agro-irrigation layer was formed in the period of the primitive *kair* or estuary agriculture; the middle layer, the most massive, was linked to regular, irrigated farming during its functioning in the Kangju-Kushan period of the Chermen-yab. The upper layers were dated to the 10th–early 13th centuries. After that, the Shakh-Senem 'oasis' ceased to exist and the process of desertification began, turning the cultivated soil into *takyr* and sand (see Figure 41: Section 48). The soil morphology reflected the number of history stages and the landscape evolution going from active riverbed to marsh-*tugai kair*, to farming oasis and the final return to a desert landscape in the 'lands of ancient irrigation'.

North of the Shakh-Senem 'oasis' we discovered another massive main canal, perhaps functioning in the Kangju-Kushan period. It ran from northeast to southwest at the foot of a sandy elevation and could be traced for a length of 30 km. Its general width was 35–40 m. Its origins were located in the area of the Kuyusay plateau, where the ancient canal was deepened anew in the Middle Ages, so it had a typical double profile and high banks.

The general direction of the canal was parallel to the main bed of the Chermen-yab. It rounded the Shakh-Senem plateau from northwest and flowed southwest, reaching Gyaur-kala II after 10 km. Here it could be traced with great difficulty on the flat *takyr* surface. In the lower part, its overall width was 18–19 m and 9 m between banks. Moving upward, the canal widened. At *poisk* 14 (26 September 1952) its width was 24 m and 10 m between banks. The canal was largely eroded, its banks rose as low hillocks on one side or the other. The general direction was well outlined by vegetation. Upstream the width increased. At *poisk* 10 (26 September 1952) its overall width was 28 m and 15 m between banks. At the foot (bottom) of the Shakh-Senem plateau the canal was 32 m wide and 18 m between banks (*poisk* 4, 26 September 1952). In its middle part, north of the Shakh-Senem plateau, the canal widened to 37 m. Here traces of small canal networks were quite numerous. Individual canals could be traced for 1–1.5 km on both sides of the canal. Rare finds of Kangju pottery were found on the irrigation ditches (*poisk* 10-15, 26 September 1952; *poisk* 29, in 1953). Moving upstream, the relative level of the canal riverbed above ground increased. The irrigated territory were located mainly right of the riverbed, whose overall width was more than 40 m and 25 m between banks.

The main Kangju-Kushan bed of the Chermen-yab was traced east of the Kuyusay plateau, stretching in an east–west direction and almost merging with the high sands of the Kyzylkum. In the southern outskirts region of the Kuyusay plateau, in 1953, thanks to aerial photographs, we discovered a large fortification (Kuyusay-kala) which was dated by S. P. Toltsov from the 1st century BCE to the 2nd–3rd centuries CE (Tolstov 1958:70; Andrianov 1958b:313).

In its lower parts, the Kuyusay's section of the main canal was connected with the Shakh-Senem. The canal was almost fully covered with sediments and visible only as a low wall-shaped elevation (*poisk* 42). Its total width was approximately 30 m (see Andrianov 1958b:315:fig. 2). In the canal there were a large number of very small Antique ceramics of little importance. In several places it was crossed by narrow Medieval ditches, along which the ruins of Medieval bricks dwellings were found. Near them there was much Medieval pottery, fine glass and traces of ceramic production. Here the ancient Chermen-yab had a more definite shape. Its overall width was 32-33 m and 14-15 m between banks (*poisk* 8). In the area where roads crossed the canal (*poisk* 11, 4, 10, 1952), its overall width was 32 m and 12 m between banks (see Figure 43.24).

Upstream the riverbed of the Chermen-yab ran between huge sand dunes, at which bases there were rock outcrops. In the west rose the Kuyusay-kala plateau and in the east the high sand dune of the Karakum (*poisk* 2-4, 24 September 1952). Here the overall width of the Chermen-yab was more than 50 m and 18 m between banks (*poisk* 4, 24 September 1952). The canal was filled with sediments and rose 2-3 m above the surrounding *takyr*. In the upper part, between two elevations, the Chermen-yab had two beds, the largest exceeding 50 m in its overall width and 20 m between banks (see Figure 43.26), the smaller bed was 11 m wide between banks.

The investigation in the environs of Kyzylcha-kala revealed that the ancient badly damaged canal, situated at the foothill where the ruins rose, did not function in the Middle Ages (*poisk* 25, 6 October 1952). Its overall width was more than 37 m and 14 m between banks (see Andrianov 1958b:315:fig. 2). At 1.5 km south of Kyzylcha-kala, traces of an ancient settlement were found near the canal, and an almost intact vessel dating to the Kushan period was discovered (*poisk* 22, 6 June 1952). This finding, as well as the abundant surface ceramics collected were the evidence of intensive life in the environs of Kyzylcha-kala during the Kushan period.

As we noted above, in the Archaic and early Kangju periods, the main canal of the Chermen-yab had, apparently, its sources near Kyuzeligyr-kala. Throughout the Kangju period there was a reduction in the number of short independent canals, taking water directly from the Southern Daudan channels and the establishment of a longer main irrigation canal (see Andrianov 1958b:327; Gulyamov 1957:91).

In the basin of the Chermen-yab, numerous independent main canals, receiving water from lateral channels, were merged into one massive system, whose sources were later gradually moved further upstream. In particular from Kyuzeligyr-kala to the environs of Kalalygyr-kala I, to the bend of the Southern Daudan and perhaps even to the environs of Kyunerli-kala and the ruins of the Medieval city of Zamakhshar, within the limits of the contemporary cultivated zone (see Figure 42.E).

The process of reconstructing its irrigation systems can be traced not only in the middle and lower parts, but also in the Upper Chermen-yab, near Kyunerli-kala. This latter system was dated by S. P. Tolstov to the Kangju period (Tolstov 1948a:100-102). At the foot of Kyunerli-kala we recorded the bed of a highly destroyed canal 30 m wide, whose banks were barely preserved. Its central part, apparently the bed of a later canal, was 6 m wide (see Figure 43.9). Downstream, this canal was slightly better preserved in the form of a flat bank 50 m in size, covered with small Kangju pottery fragments (*poisk* 754, 761-765, year 1959). This canal clearly had its origin in a small meandering ancient channel with a general north-south direction (Zeykash?).

Through vertical aerial photographs of the heavy sand area, we were able to identify another, parallel, canal branching from the north-south riverbed (*poisk* 756-759, 761). About 4 km north of Kyunerli-kala, this canal was traced in the irrigation area of Khorezmshah and preserved in the shape of two banks 1.5 m high above the *takyr* and 16 m wide between banks. The general width was 45 m (see Figure 43.12). The canal was traced 5 km northwest of the fortress. The area of the Chermen-yab Medieval main systems was plowed. Perhaps this canal served for a certain period of time as a source of the Chermen-yab system.

The upper reaches of the massive Kangju-Kushan bed of the Chermen-yab were discovered and studied in the environs of the ruins of the Medieval city of Zamakhshar (Zmukshir), at which base there is a more ancient fortress (Voevodskiy 1938:235-245; Tolstov 1941). A series of badly damaged main canals of different periods were traced 1 km north of the fortress. In one of them (*poisk* 691) there were no Medieval ceramics but many fragments of Kangju-Kushan pottery. The general width was ascertained with great difficulty, since the banks were highly destroyed. The canal was no less than 30-35 m wide. The main bed of the ancient Chermen-yab was better preserved in some places in the west area. Its overall width was 40-60 m and 9 m between banks (*poisk* 732, 751). In this canal there were only Kangju and Kushan ceramics.

The study of individual sections of the Kangju-Kushan bed of the Chermen-yab around Kyunerli-kala and Zamakhshar, proved S. P. Tolstov's hypothesis that, during the Kushan period, the individual main canals were combined into one system.

Antique mesopotamian Systems between the Northern and Southern Daudan

On the right bank of the Southern Daudan Basin, during the Kangju and Kushan periods some reconstructions were carried out in the basin of a massive Archaic canal about 50 km long. These were evidenced by many individual system parts of that time, revealed east of Kunya-Uaz in the territory between the ruins of this fortress and the large dry riverbed of the Daudan. The whole territory was covered with moving sand, traces of a dense network of ancient and Medieval canals, small distributors and fields. The surface materials collected from these sites allow us to date traces of agricultural activities to life of the site even later.[Note 149] Excavations carried out by a team of the Khorezm Expedition (directed by E. E. Nerazik) revealed cultural layers from the Kangju, Kushan, Kushan-Afrigid periods, a break in the 6th-8th centuries and an upper layer of the 9th century (Tolstov 1948a:118; 1958:216; Nerazik 1958:369-393).

The study of ancient irrigation systems in the area between the two rivers, and between the channels of the Northern and Southern Daudan,

indicated that on this territory the most ancient canals (Archaic and Kangju), branching directly from the Northern Daudan, had mostly an approximate north-south direction, connecting the channels of the Northern Daudan with those of the Southern Daudan. On the contrary, the Kushan and Kushan-Afrigid main canals branched from the upper reaches of the Daudan, within the limits of the modern cultivated area, and were built in an east-west direction following the main slope and large riverbeds of the Sarykamysh Delta.

As seen above, nearly the same location of the early Kangju and the late Kushan main canals were recognized in the basin of the Chermen-yab.

Several main canals were investigated in the environs of Mangyr-kala located on the sandy hills of the Northern Daudan right bank. A fortress was dated by S. P. Tolstov approximately to the Kushan and Kushan-Afrigid periods (Tolstov 1953:22).[Editor 52] The north-south main canal (*poisk* 650, 654), beginning at the large bend of the Northern Daudan 10 km east of Mangyr-kala, was more than 40 m wide and 20 m between banks. Another canal, near which a large Archaic fortified settlement was revealed, started parallel to it (see p. 188). Traces of another, straight ancient canal were found on the opposite bank, very close to the Mangyr's elevation. It was preserved as a flat bank 50 m in size, covered with small pottery fragments dated to the Kangju and Kushan periods (*poisk* 344). Kushan finds prevailed, corresponding to the lifetime of the Mangyr-kala settlement, which was the center of a vast agricultural district. This canal was crossed by ditches of the Golden Horde period (*poisk* 341, 342, 345, 346, 349-351).

About 16 km southwest of Mangyr-kala there were ruins of the large Medieval settlement of Yarbekir-kala (see Figure 44), which, according to the stratigraphic excavations of 1958 (headed by N. N. Vakturskaya), was already built in the Kangju period (see Tolstov 1947b:92; 1953b:314; 1958:22; Vakturskaya 1963:41-45).

In its environment we identified many ancient canals, which mainly had an east-west direction (see Tolstov 1962a: *Karta Zemli drevnego orosheniya i nizovyakh Amudari i Syrdari* – Map of the 'Lands of ancient irrigation' in the Lower Amudarya and Syrdarya). They are an extension of systems, whose origins were located north of Kunya-Uaz (see Figure 45). The upper and middle parts of these systems were covered by later agro-irrigation layers, in particular they were cut across by the Uaz Medieval irrigation system.

A canal bed, 25 m wide and 15 m between banks (*poisk* 546) (see Figure 46.H), was revealed very close to Yarbekir-kala and northeast of the settlement. Approaching the fortress, the canal narrowed to 11 m between banks. Its left bank was preserved as a flat bank 10 m wide, and the lateral branch was marked by a dark line 2 m in size and an accumulation of Kangju ceramics (*poisk* 548), which allowed to date the system to this period. The canal was conveniently located close to Yarbekir-kala.

North of Yarbekir-kala and the system described above, there was a huge handicraft ceramic workshop and surface fragments show mainly Kushan ceramics (*poisk* 550, 894) and a number of settlements were dated to the same period (*poisk* 323, 558, etc.).

In the environs of Yarbekir-kala, a huge canal was dated to the Kangju-Kushan period, with a south-west direction connecting the two east–west riverbeds of the Daudan, the Northern Daudan and the channel of the Southern Daudan rounding Tuzgyr from the north (*poisk* 352, 362) (Figure 46.10). The overall width of the canal was more than 70 m and 16-20 m between banks. The bed was 1.5-2 m lower than the *takyr* and the banks were 5 m high. The canal head part in the Northern Daudan (*poisk* 334) was a wide decreasing funnel-shape, on whose slopes there were Kangju-Kushan and also Medieval glazed ceramics. In its lower parts the canal narrowed to 15-16 m between banks with clearly visible traces of a later deepening. The remains of Medieval rural settlements located along a small network of ditches, were clearly dated to the period of its secondary development (see p. 206).

In the Kangju-Kushan period the channels of the Southern Daudan, flowing from the north and south sides of the Tuzgyr plateau, still continued functioning, as in their lower reaches an extensive irrigated area and ancient settlements were discovered. During field work in 1964 on the southern cliff of the Tuzgyr plateau, the ruins of the large fortress of Tuzgyr-kala were found, and the poor surface materials were dated to the early 1st millennium BCE. Some kurgans were located next to it, along the whole southern and southeastern edge of the elevation on both sides of the fortress. The Kangju fortress of Akcha-Gelin-kala, dated by S. P. Tolstov to the 3rd century BCE, bordered them on the east (Tolstov 1958:26). The archaeological excavation of the kurgans, carried out in 1965 by a team of the Khorezm Expedition (directed by V.A. Lokhovits), revealed that the majority of the kurgans dated to the 1st–3rd centuries (Lokhovits 1968).

Traces of ancient fields and vineyards were found at the foot of the Tuzgyr elevation. An especially well preserved ancient vineyard (*poisk* 976) had dimensions of 65 × 50 m. The distance between ditches was 3 m and the canal size was 1.2 m. The contemporary desert vegetation covering the ridges, lively resembled greenish vineyards.

South of Tuzgyr, ancient canals, fields, vineyards and settlements were recorded over a large area (*poisk* 971-989). The largest one is Tuzgyr-kala 2, which had the structure of a fortified dwelling building. Seven kilometers south of the Tuzgyr-kala fortress, on the bank dumps of large canals, there were numerous fragments of mainly Kushan ceramic production (Figures 46.5-6). In the Kangju, and probably the Kushan period, the environs of Tuzgyr-kala 2 were likely well watered given that a small protective dike 1-2 m wide was found on the right bank of the channel in front of Tuzgyr 2. According to the remains from the Kushan settlements, the neglect of this area occurred in the late 1st millennium (Tuzgyr 2; *poisk* 974, 977, 989).

The remains of Kangju-Kushan settlements, fortifications and irrigation systems were also discovered in the area between the Daudan and the Daryalyk. On the western margin of this area there was the massive Kangju fortified settlement of Kandum-kala, surveyed by Ya. G. Gulyamov (1957:82-83). In 1958, the archaeological-topographical unit, with the participation of architect D. S. Vitukhin, mapped the settlement plan and found out in its environs remains of Antique settlements (*poisk* 659-661, 668). Here it was possible to trace remains of a massive east-west canal irrigating the fortress environs. Ground measurements showed that its general width was 40 m and 7.5-8 m between banks. In many places, the canal was intersected by Medieval and modern canals.

As we have seen, during the Kangju and Kushan periods some channels of the Daudan still continued functioning and the Daudan riverbed was even more full of water. This was shown by the widespread fortified settlements of that period, mostly on Dev-kesken and Butentau-kala 1 and 2 elevations. Irrigation traces of that time are very few and it was probably due to the very intensive development of these lands during the Middle Ages. Just west of the Kangju settlement of Kurgan-kala (*poisk* 500), in 1955 we discovered the remains of a largely destroyed canal, 40 m wide and 8 m between banks (Tolstov and Kes 1960:192).

* * *

Thus, on both banks of the Amudarya, during the Archaic, Kangju and Kushan periods the irrigation systems underwent substantial changes. They were primarily associated with the complexity of the irrigation general scheme, the appearance of new links and the improvement of various types of head works. This shows progress in hydraulic skills, and good knowledge of local conditions by irrigators and their efforts to create gravity canals for a long-term exploitation with less riverbed silting.

The most ancient small Archaic systems in the Chermen-yab basin were based on the fading southern channels of the Daudan. They still did not exceed a length of 15-20 km and a width of 20 to 40 m between banks. Their scheme was the typical: lateral river channel - head canal - main

canal - feeder - field. The numerous head works of short and wide canals of the Archaic period were similar to the early Kangju period structures. Like some Archaic canals, the early Kangju canals also had their sources in different channels of the Southern Daudan and they irrigated small plots. In contrast, on the north bank of the Southern Daudan, the massive Kunya-Uaz canal, 50 m long and dated to the late Archaic period, was a vast system with large distributors. They diverted from the main canal (in cases where they did not follow branches of the ancient riverbed) very often at right angles to the side opposite the Daudan riverbed. Its plan with its 'sub-rectangular' lateral distributors recalls some of the major systems of the Khorezm right bank.

During the Kangju, and especially the Kushan periods, active farming activities in the Chermen-yab basin led to the gradual fading of the lateral Southern Daudan channels and, consequently, to the need of shifting canal sources irrigating the environs of Gyaur-kala I and II. First, the shift took place upstream at the Shakh-Senem plateau, then at Kuyusaygyr, in the environs of Kyuzeligyr-kala and Kalalygyr-kala I and, finally, at Kyunerli-kala. Then the unification of local irrigation systems into one network took place.

The Shakh-Senem and Kuyusay plots had a total length of 50 km and were combined in the Kangju period. In the Kushan period, the Shakh-Senem and Gyaur-kala parts were connected by a narrow bed (10-12 m wide). In the Kushan period, the Chermen-yab cut in an east-west direction all the different earlier systems. S. P. Tolstov rightly noted the difficulty in solving the questions about the origins of the Chermen-yab in the Kushan period and wrote that "it was possible that it had the sources in the Amudarya like the right-bank canals" (Tolstov 1958:114). However, the fact that the sources of the large Kunya-Uaz canals were, for a long time (until the Khorezmshah period), located in the middle part of the Daudan, at the 'fork' of the Southern and Northern Daudan, 20 km above the Kangju ruins of Kyunerli-kala, suggests that the sources of the Chermen-yab in the Kushan period (whose bed was found north of the Zamakhshar ruins) were still located on the Daudan riverbed and not in the Amudarya.

During the Antiquity, in the struggle to prevent the silting of canals and to find more efficient and economic water supply to the fields, by the end of the Kangju and, especially, in the Kushan period, the irrigators replaced the 'sub-rectangular' irrigation system layout with a tree-shaped one, and increased the frequency of lateral distributors and branching. The canals cross-sections were also changed. Several well preserved main canals provided a basis to conclude that changes in the transverse sizes of ancient canals, moving away from their sources, were judged to be most efficient. Thus, moving from the origins of the Kangju canal, located north of Shakh-Senem plateau, over a length of 30 km, the width gradually decreased from 22 m to 10-9 m between banks in its terminal part. The huge Archaic canal in the environs of Kunya-Uaz started as 45 m wide between banks and only 12.5 m wide in its lower part near the site (*poisk* 572). We have a similar picture in the Chermen-yab, which was 20 m wide between banks at Kuyusaygyr and 9 m wide to the of Shakh-Senem, near the Kushan site (*poisk* 47).

The Kushan main canals were already narrower (8-15 m wide), but with more frequent distributing networks. By their size, the Antique main canals of the Khorezm left bank (as well as the main canals of the right bank) were similar to the canals of ancient Mesopotamia, where R. J. Forbes reported a width between banks up to 25 m (see Forbes 1955, Vol. II:18). It is highly possible that, with the appearance of more frequent branching and distributors, the co-efficient of land use within the irrigation basins significantly increased. On the 'lands of ancient irrigation' on the left bank of the Khorezm, fewer traces of vineyards and gourd plantations were found than on the right bank. This may be attributed to the peripheral, marginal location of preserved ancient plots in relation to the central, mainly developed areas of Khorezm, which included the whole Southern Akchadarya Delta.

Medieval Irrigation Works

The rise and expansion of irrigated agriculture and the urban growth in the Khorezm Oasis (and in other areas),[Note 150] occurred during the formation and flourishing of the Great Kushan state in Central Asia (in the first centuries CE), followed by a worsening of economic conditions caused by political decentralization, migrations and invasions of pastoralist tribes surrounding the agricultural oases. Also a worsening of socio-economic divisions and social riots (see Tolstov 1948a:50; 1962a:233-244; Nerazik 1966:121-125).[Note 151] S. P. Tolstov and E. E. Nerazik connected the death of some cities of the left bank of the Khorezm with the Sassanid-Chionite wars in the mid-5th century. They also noted 'barbarization' of the farming population of oases outlying districts on both the right and left banks of the Khorezm (Nerazik 1966:125). Probably all of this contributed to the desolation of the terminal parts of many canals and the failing exploitation of the single irrigation systems in the 4th-5th centuries (see p. 171).

According to S. P. Tolstov, the decline of irrigation systems on the left bank of the Khorezm in this period led to a significant increase in waters in the Sarykamysh riverbeds and a short-term flow of water into the Uzboy, when Igdy-kala was built (Tolstov 1955a:109-110; 1962a:233-235). Archaeological work at Kunya-Uaz (see Nerazik 1958), Turpak-kala, Mangyr-kala and some other places in the Sarykamysh Delta revealed cultural layers and finds dated to the Kushan-Afrigid period, which, together with surface fragments collected on some irrigation systems in these environs, enabled us to ascertain the functioning of these systems at that time.[Note 152] However, Afrigid settlements and canals of the 7th-8th centuries were not found on the left bank as they were on the right one. Here, as on the right bank of the Amudarya, the process of ending irrigation development according to old traditions, the gradual decay of ancient irrigation systems, and the emergence of new principles of irrigation techniques are clearly visible. This is due to some extent to the fundamental changes in social relations, in particular with the collapse of the "old patriarchal slave-owning traditions, which limited the development of the feudal economy" (Tolstov 1948b:232; see also1958:107ff.; Gulyamov 1957:114-123).

The political crisis during the Arab conquest of Khorezm in 712 CE, the struggle between the 'illustrious' brothers Chigan and Khurrazad, and the rivalry of the old economic and political center of Kyat, rising north of Urgench, the new center, contributed to the revival of irrigation in the fading parts of irrigation systems. At that time, Khorezm was politically not united. Even V. V. Bartold reported on the disintegration of the country into two states during the Arabs' rule (Bartold 1963, Tom I:202, 323ff.; 1965:163; Gulyamov 1957:123).

The Arabs found Central Asia divided into many small states independent from each other (Tolstov 1948b:205-207). The union of Khorezm and Kerder (possession in the Amudarya Delta) occurred, according to E. E. Nerazik, in the 9th century (Nerazik, 1966:128). The beginning of political stabilization and a noticeable increase of irrigated areas dated to the 9th-10th centuries, when, after a long struggle, in 995 CE the ruler of Gurganj conquered Kyat. This is confirmed by single finds of Afrigid-Samanid ceramics (9th-11th centuries) at the settlements near Zamakhshar, Kunya-Uaz and Yarbekir-kala (*poisk* 135-137, 147, 154, 160, 162, 167, 227, 260, 694, 696, 704, 713, 721, 772, 774, 876, 886).

The excavations of M. V. Voevodskiy in 1934 in the ceramic kilns at Zamakhshar, a city mentioned for the first time in the late 10th century by al-Makhdisi, provided material from the 11th century (Vakturskaya 1959:265). The finds of glazed ceramics of Zamakhshar type from the 9th-10th and 10th-11th centuries, were found in the middle area of the Chermen-yab (*poisk* 135, 137, 147, 154). They suggested that the canal was already being revived from Zamakhshar to Shakh-Senem, where Yu. A. Rapoport's dig also provided objects from the 10th-12th centuries (Rapoport 1958:412-415, fig. 13).

According to descriptions of Arab geographers

(Ibn-Rust, al-Istakhri, al-Makhdisi, etc.), which Ya. G. Gulyamov combined with archaeological data, up to the 9th century the Amudarya flowed through an existing riverbed and it watered the delta's (Daukara), eastern basin where the Amudarya waters merged with those of the Janydarya in the Syrdarya basin (Gulyamov 1957:125-197). The Khorezm farming regions were mainly located in this southern part, in the basin of the Daudan channels.

Southern Khorezm was irrigated by canals already constructed in the early Kangju period. The Khazarasp fortress, was located on one of them, its ancient walls were built no later than the early 4th century BCE (Vorobeva, Lapirov-Skoblo and Nerazik 1963:198). This was mentioned in 1219 CE by al-Tabari at the turn of the 9th and 10th centuries. In this fortress lived Yakut, and also Djuveyni (13th century) and Mir-Khvand (15th century) wrote about it (MITT 1939:179, 187, 189, 438, 443, 540). Another canal, the Kheykanik (Palvan-ata) reached Khiva (Gulyamov 1957:143). The third canal, Madra, matched with Gazavat and irrigated the environs of Zamakhshar.

On the right bank of the Daudan were located the canals Vadak and Buve (Vedak and Buvve) which had the same direction of the Shakhabad (Shavat), a canal of the 17th century (see Glukhovskoy 1893:11, 17). In the 9th century the Vadak reached Gurganj (Urgench). According to Ya. G. Gulyamov, the artificial canals Palvan, Gazavat and others first appeared in connection with floods of the Daudan and then by regulating these floods, they were built along its banks (Gulyamov 1957:91).

In the Middle Ages different and significant farming centers arose in Northern Khorezm: in the Mizdakhkan region (Mazlum khan-slu, Gyaur-kala), in the northwestern delta (Khakim-ata) and in the eastern delta (Khayvan-kala, Tok-kala). In the 9th century the Amudarya riverbed moved in a northwest direction, its water rushed into the western part of the delta, in the Aybugir lowlands. The main riverbed flowed between Mizdakhkan and Gurganj (Gulyamov 1957:133:fig. 10). The Kerder canal, corresponding to the modern

Kegeyli, was built in the bed of the old river bottom. As shown by the study of Ya. G. Gulyamov, the river was gradually moved in the direction of Gurganj. At the time of Ibn-Rust (913 CE), the river ran 6-7 km from the city between Urgench and Mizdakhkan and, in the period of al-Makhdisi (985 CE), its water already washed the urban walls (MITT 1939:150, 187-188). According to Yakut (early 13th century), the inhabitants of Gurganj built a dam and separated the river from the city (MITT 1939:420).

Medieval Chermen-yab

In the period of the Great Khorezmshah, when Khorezm became a major world kingdom,[Note 153] the huge Chermen-yab canal functioned in its territory (over almost the whole length from Zamakhshar to the ruins of Gyaur-kala) (see Figure 42.I). The very lively political and economic links of Khorezm with Khorasan and other western regions of Central Asia, led to the revival of the so-called Shakhristan road, which led from Khorezm southwest along the canal (Gulyamov 1957:145). As shown by the archaeological-topographical study of the settlement of Suburna (Shakh-Senem), a narrow Medieval canal was built along the Antique bed of the Chermen-yab (see Figure 42.I). Its general width was 15 m and 7 m between banks in the environs of Gyaur-kala I. This ditch had several small branches directed left, two of which crossed the wide (up to 250 m) riverbed of the Kangadarya. At 5.5 km from the Chermen-yab, the ditches approached the vast ruins of a caravanserai located at the foot of the Niyaz-Khan hill.

Traces of Medieval agriculture are especially abundant in the environs of the once prosperous city of Shakh-Senem (Suburna),[Note 154] which, according to Yakut, who visited it in 1220 CE, was at the end border of Khorezm in the direction of Shakhristan (MITT 1939:423).

The majestic ruins of the city of Shakh-Senem rose above the dead oasis, covered with destroyed farms and separate buildings, and cut across by a complex network of ancient irrigation canals (see Figure 47). Adjacent to the fortress on the south there was a highly effective and wide park complex (with a cross-shaped layout and divided into 16 squares by irrigation ditches), with tower-shaped corner pavilions and a central building (Orlov 1952b:154-159). According to the very weak 'striped' layout of separate squares, they appear to have been used for vineyards, fruit and garden products.

The main canal irrigating the environs of Shakh-Senem was highly eroded near the southwestern corner of the fortress for over 200 m. Apparently, the destruction of this irrigation canal should be attributed to the last period of existence of Shakh-Senem (Suburna, Suburli), the period of the Mongol invasion when, according to Djuveyni (1226-1283 CE), Suburna was flooded. About Suburna, Djuveyni wrote: "It was a city, which is now under water" (MITT 1939:445). According to S. P. Tolstov, the flood was caused by a breakthrough in the Amudarya along all the Sarykamysh riverbeds, due to the Mongols' destruction of the dams (Tolstov, 1962a:26; see also Bartold 1914:88; Gulyamov 1957:165-166).

The Medieval part of the Chermen-yab, at the northwestern end of the Shakh-Senem plateau, was more than 40 m wide and 17 m between banks (*poisk* 42). Its banks rose 4-4.5 m above the *takyr*, around which there were many Medieval and Kangju-Kushan ceramics, indicating the long existence of this farming region.

Between Shakh-Senem and Kuyusay, the ancient main canal of the Chermen-yab was rebuilt in the Middle Ages.

As already noted, in the Middle Ages the large Antique canal was deepened again in its upper part (over 7 km from the northwest end of the Kuyusay plateau). The width of the narrow Medieval canal here reached 12 m between banks (see Andrianov 1958b:315:fig. 2.28). The Medieval canal turned sharply from the Antique bed south and southeast and ran further into the Shakh-Senem part of the Chermen-yab canal. At the point where the canal turned south, small distributors branched off. On one of them the ruins of a Medieval farm were found with the preserved layout of a main living building with multiple rooms and an adjacent garden-park complex with ruins of a *kaptarkhana*.

The surface ceramic dated to the 12th-14th centuries (*poisk* 19, 24 September 1952). The survey in the Kuyusay-kala area indicated that the upper part of the Antique Chermen-yab, situated east of Kuyusay-kala, was abandoned in the Middle Ages. This was evidenced by the abundance of Antique pottery and few Medieval ceramics, as well as the relief of its bed, almost entirely covered by later sediments.

The ancient irrigation systems in the environs of Kuyusay-kala were rebuilt in the Middle Ages. Here an unusual canal was created, on whose sharp meanders there were large Medieval glass workshops (see Trudnovskaya 1958; Andrianov 1958b:312-313: *Karta Drevnyaya irrigatsionnaya sistema Chermen-yaba* - Map of Chermen-yab ancient irrigation system). These workshops were dated to 12th-early 13th centuries (Trudnovskaya 1958:428). In its lower part, the Kuyusay canal was 59 m wide and 16 m between banks.

North of Kuyusay our team researched the wide fan of Medieval canals starting from Medieval Chermen-yab, 6 km from Kyzylcha-kala, and merging into one large canal rounding the Kuyusay plateau from northwest. Their overall widths diverged from 20 to 40 m and 10 to 20 m between banks (*poisk* 7-20, 6 October 1952). In the surrounding area of these Chermen-yab canals, ceramics of the 12th-early 13th centuries were dispersed everywhere: dark-gray fragments of jars, *khum*, small fragments of glazed ceramics, glass pottery, etc.

The main bed of Medieval Chermen-yab was traced in its upper part 3 km north of Kyzylcha-kala. It was well recorded for a considerable distance between Kyzylcha-kala and Kuyusay-kala. In some places the canal profile could be clearly seen. Its overall width was 36-40 m and 18 m between banks (*poisk* 32, 7 October 1952). In the canal, and around it, Medieval ceramics of the 10-14th centuries predominated.[Note 155] In its upper part, the bed of the Mediaeval Chermen-yab was clearly seen in aerial photographs and on the ground up to the environs of Daudan-kala and Zamakhshar (Zmukshir).

Here the width of the main canal ranged from 25 to 35 m even up to 45 m, and 10-15 m between banks (Figure 48.1-10). The banks were preserved in some places for a height of 3 m. Northeast of Daudan-kala the main bed of the Chermen-yab was more than 43 m wide and 13 m between banks (*poisk* 153). One bank was well preserved and raised 3-4 m over the *takyr*. The whole area, from west of this site up to the modern cultivated zone, was covered with ceramics and ruins of individual farms and dwellings. The thickness of subsequent agro-irrigation (approximately 1 m) was evidence of an intensive and long-term cultivation. The agro-irrigation deposit partitions clearly corresponded to two main stages of agriculture: the Kangju-Kushan (formed by a layer 50 cm thick) and the Medieval (the upper layer).

The survey of a complex connection in the region of two large farms northeast of Daudan-kala revealed two reconstructions of the Chermen-yabn upper sections (*poisk* 162, 156). The older canal heads were traced in the area of Zamakhshar (see Figure 42.I) and they led to the Gazavat system (the Medieval Madra canal). The later and wider 40 m riverbed had its source in the Shavat basin (*poisk* 166). This bed of the Chermen-yab crossed the main riverbed of the Daudan, marking the obliteration of the channel water flow. At this time, Ya. G. Gulyamov wrote that the Chermen-yab, was supplied with water from the channel of the Daudan, branched from it 15-16 km southeast of the Ilyaly, and then, in all probability, also from the Gazavat canal. The team traced the later bed leading to the basin of the Shavat. It was very heavily damaged, and only a series of *takyr* terraces and strips were located on the right bank of the Daudan, covered with 12th-14th centuries ceramics.

In the environs of Zamakhshar the parallel bed of the Chermen-yab canal was preserved, even if highly damaged. Here, as on the Gavkhore Basin on the right bank of the Amudarya, smaller size canals were constructed in that period. North of the settlement the Chermen-yab was 8 m wide between banks (see Figure 48.9-10), and the parallel bed was 18 m wide and 8 m between banks. The banks were 2.5 m high above the highly damaged *takyr*. The whole territory was full of deflated hollows, intricately eroded remains and numerous sand dunes.

During the survey of 1959 around Zamakhshar, O. A. Vishnevskaya researched many Khorezmshah rural sites (see Vishnevskaya 1963). In this region, in association with the prevailing Khorezmshah settlements, there were also earlier Medieval ones.[Note 156]

As in the environs of Zamakhshar, in the middle and lower reaches of the Medieval Chermen-yab we investigated several agro-irrigation layouts, which were of two main types: watered quadrangular plots surrounded by low ridges, and gourd plantations and vineyards in different configurations with long *gryad*. Gourd plantations of a Khorezmshah farm (*poisk* 717) were 32 × 35 m in size. Each *gryad* (*karyk*) was 2.2 m wide and the space in-between was 1.2 m. In the environs of Kaz-kala, northeast of Daudan-kala, a vineyard with a *karyk* of 3.5 m was recorded (*poisk* 164). In the middle part of the Chermen-yab, near the ruins of a Khorezmshah farm, traces of a small vineyard 25 × 15 m in size, with *karyk* approximately 4 m wide, were found on a lateral irrigation ditch (*poisk* 128).

Concluding the description of Medieval Chermen-yab, it should be noted that the Afrigid-Samanid, and especially the Khorezmshah, periods were the most important in the history of this irrigation basin. The water-lifting devices were widely developed in this period and the area of crops increased by 30-35% in relation to the whole territory covered by irrigation facilities. We estimated this territory to be 25,000-30,000 ha (Andrianov 1958b:328) and 10,000-12,000 ha cultivated at the same time.

Based on the size data of the Medieval bed of the main canal, we can calculate the labor cost needed for its construction and the creation of small irrigation networks. The total length of the canal, from the environs of Zamakhshar to Niyaz-Khan, was 150 km. The area stretching for 100 km from Shakh-Senem to Zamakhshar was subjected to a radical reconstruction. Here, the average cross-section of a canal was 8-12 m². The total volume of excavated soil from the main canal was not less than 1,000,000 m³ and roughly the same amount of soil was removed for digging the small irrigation network.[Note 157] At a rate of 3 m³ per day, 12,000-14,000 diggers were needed to achieve this goal in 50 days. For canal cleaning, the labor of 5,000-6,000 men was necessary annually. It means that 20-30 working days were required for 1 ha, which corresponded to the labor costs in Khorezm in the early 20th century (see Tsinzerling 1927:208).

How many people could live in the basin of the Medieval Chermen-yab?

The Medieval rural settlements in this region were not studied sufficiently. Therefore, archaeological material alone may not provide us with data to determine the overall size of the farming population.[Note 158] In such cases, without reliable statistic data, the information on population density maps usually comes by extrapolating and using different indirect data.[Note 159] Such an attempt was made by us.

On the basis of archaeological-topographical large-scale maps, the whole area of the Khorezmshah Chermen-yab was divided into three sections: high density, assumed by analogy to the Kavat-Kala Oasis (12th-early 13th centuries), with 80-150 people per 1 km²; medium density, with analogy to the Southern Khorezm in the early 20th century, with 75 people per 1 km²; low density, as in the Turkmen part of the oasis in the early 20th century, with 25 people per 1 km². The maximum density was allocated to the environs of Zamakhshar, Daudan-kala and Shakh-Senem; the medium density to the territory between Kuyusaygyr and Daudan-kala; and the lowest density in the remaining areas. The whole Chermen-yab area of 300 km² and the total population was estimated by us at approximately 20,000-30,000 people. The average number for each hectare of cultivated and irrigated land accounted for 2 to 3 people.[Note 160] The male working population probably was 5,000-8,000 people, which alone was already sufficient for irrigation work, without the involvement of outside workers.

The intensive development of irrigated agriculture in the Khorezmshah period, so clearly traceable on both the right and left banks of the Khorezm, was interrupted by the Mongol invasion. On its devastating consequence on the agricultural oases of Central Asia a lot has been written by several researchers, in particular A. Yu. Yakubovskiy, S. P. Tolstov, B. N. Zakhoder, A. M. Belenitskiy and I. P. Petrushevskiy. A large section devoted to this issue was included in the book by I. P. Petrushevskiy *Zemledelie i agrarnye otnosheniya v Irane XIII-XIV vekov* (Agriculture and farming relations in Iran in the 13th-14th centuries) (Petrushevskiy 1960:36-46).

The author directs attention to the total massacre of the farming and craftsman population in many regions of Central Asia and Iran, or the taking into captivity of people of these areas.[Note 161] Khorezm did not escape this fate (Bartold 1963, Tom I:500-504; Tolstov 1948b:289-295; Gulyamov 1957:163-167; etc.). Here, as in other oases, the huge population loss, the destruction of head works, dikes[Note 162] and dams caused the decline of irrigated agriculture. Not only the right bank of the Khorezm became depopulated (see pp. 179-180), but also large tracts of the vast territory on the left bank of the Amudarya, where the densely populated oasis of Chermen-yab was no longer a land of ancient irrigation (Tolstov 1948a:51; Gulyamov 1957:163-167).[Editor 53] Other territories, in particular the environs of Urgench, where the Mongols destroyed the dam ('bridge') on the Daryalyk, were flooded (Bartold 1963, Tom I:503; 1965:50, 172, 548; Gulyamov 1957:165-166). The neglect of canals and destruction of dams during the Mongol invasion caused a temporary breakout of the Amudarya waters west, and the rising level in the Sarykamysh Lake (see Tolstov, Kes and Jdanko 1954, 1955).[Note 163]

With the formation of the Mongol Kingdom of the Golden Horde, most of Khorezm became part of it. The southern edge of Khorezm became part of the possession of the Chagatai Khanate (Bartold 1965:548). By the end of the 13th century,

and especially in the 14th century the desolate lands were widely developed anew. The cities and several handicraft centers revived, especially in the northwestern part of the oasis. Urgench, the former capital of Khorezm, again became a major trade and craft center, strengthening and developing its relations with the Volga region and other parts of the vast Mongol state, and the irrigated area was also expanded (see Tolstov 1948b:308; Vakturskaya 1959:300-301; 1963:53; Vishnevskaya 1963:63-72; Vaynberg 1960).

During the Golden Horde period, in the 13th-14th centuries, a number of new canals were built in Khorezm (Gulyamov 1957:168-176). Particularly significant were the systems, based on the Daryalyk riverbed, which, according to Ya. G. Gulyamov, received a large amount of Amudarya waters in the 14th century (Gulyamov 1957:177:fig. 12, 178). Most of these systems, in the area between the Daryalyk and the Daudan, had a southwest direction. Some of them were re-deepened in the 19th century and irrigated large territories in the Uaz basin, At-krylgan in the environs of Dev-kesken (Vazira) and along the Daryalyk banks (see p. 212).

A complex history characterized the eastern main canal, constructed in the Khorezmshah period, abandoned in the 15th century and newly restored in the 19th century, when it received the name of 'Shamurat' (see Figure 42.K). This system was characterized by a complex interweaving of ancient Medieval and Turkmen canals of the 19th century (see Glukhovskoy 1893:177). Its total length, from the sources in the Daryalyk (northeast of the ruins of the Kandum-kala) to the environs of Tuzgyr, was 75-80 km. In its lower part there was Yarbekir-kala, where the excavations of N. N. Vakturskaya revealed Antique, Khorezmshah and Golden Horde cultural layers (Vakturskaya 1963). Here, numerous ruins of Khorezmshah and Golden Horde periods were also discovered.[Note 164]

The Shamurat, in its lower part, was 30 m wide and 15 m between banks (*poisk* 300) (see Figure 48.11-12). Further upstream it was 32 m wide and 9 m between banks. North of Turpak-kala (*poisk* 281, 282) the complex crossing of the

Shamurat with the Antique systems was found. One of the Medieval branches of the Shamurat was 37 m wide and 4.5 m between the inner banks. Glazed ceramics of the Mongol period were found beside the canal. North of the abandoned Turkmen settlements of Uaz, behind high sandy ridges, Medieval canals dated to the Golden Horde period, according to farmsteads located on its small branches, were also found together with Kangju-Kushan canals.

The Medieval canal (*poisk* 630) had an overall width of 20 m and 6 m between banks. In its environs there are many ruins of individual buildings, in the form of substantial hills with weak layout traces and covered with fragments of baked bricks (the predominant size being 27 × 27 × 5 cm), unglazed and glazed pottery of the Golden Horde period (*poisk* 625, 628, 631). Near a farmstead (*poisk* 631) was the layout of a vineyard, whose *gryads* had the dimensions of 4, 4.2 and 4.5 m with 1 m between rows. Upstream, the total width of the Shamurat increased to 50 m, which is associated to the reconstruction dated to both the Middle Ages and the 19th century. The Turkmen bed was 7 m wide (see Figure 48.14-16). The canal cut across the large riverbed of the Daudan. It is difficult to ascertain the time in which the waterway shifted from the right bank of the riverbed to the left, because the Daudan channels were continuously supplied with water during the Middle Ages. The lack of settlements of the 15th-17th centuries in the Shamurat basin suggests that the neglect of this system occurred at the end of 14th century after Timur's devastating campaigns, who ordered to raze Urgench and to sow barley in its stead (Bartold 1965:174; Gulyamov 1957:173-175).

Indications of Medieval irrigation facilities can be traced further west along the Northern Daudan, in the environs of the large Medieval settlement of Yarbekir-kala, and also along the banks of the Daryalyk, irrigating Urgench, Shemakh and Vazir (Dev-kesken). The remains of intensive farming activity, was suggested by numerous canals, small-scale irrigation networks, fields, vineyards and gourd plantations with ruins of rural unfortified farms, densely covering the whole territory up to the outskirts of Sarykamysh.[Note 165]

In the Golden Horde period a series of ancient canals were restored in this region. Thus, a large Antique canal located northwest of Yarbekir-kala was deepened once again (the remains of Medieval settlements and farms in the environs of Yarbekir-kala, spread along a small network of ditches, clearly date the time of its secondary development; *poisk* 318, 326, etc.). In one section of the Antique canal (*poisk* 332) the overall width was more than 20 m (see Figure 46.10) and it was greatly deepened. A rich accumulation of the Golden Horde period ceramics was on the banks. Downstream, the canal gradually narrowed (at *poisk* 320 its width was 16 m between banks) (see Figure 46.11).

The peculiar character of the irrigation works in the lower reaches of the Northern Daudan lay in the area of the lakes Tyunyuklyu and Tarymkaya. They strongly resembled the late Karakalpak canals of the Amudarya Delta. For example, west of Lake Tarymkaya (*poisk* 382) the canal maintained a negative form. Its banks were no higher than the surrounding terrain. There was a meandering head work, opening in the bed as a sloping depression. There was a sand spit in front of its outfall. On the bank there was black swampy soil covered with an ashy-gray, thin *takyr*-shaped crust. The ground was soft and contained abundant salt efflorescences. From the canals branched small irrigation ditches, on which the fragments of Mongol gray-ware and glazed ceramics were found (*poisk* 382, 389-393).

The fields and the remains of settlements of this region bear traces of heavy floods, which covered the cultural horizons with sediments. It is necessary to note that most of the entire region of the Daudan, west of Yarbekir-kala, preserved traces of Medieval flooding, as indicated by the nature of the soil and vegetation and a widespread salty area (Letunov 1958:75, 139; Konobeeva 1965:127). *Keurek* and black *saxaul* vegetation prevail on these *takyr* of medium and heavily salty soil. Large areas are covered with plump salt marshes (Andrianov 1955:358-359).

The irrigation facilities in the environs of the small unfortified settlement of Shekhrlik of the 13th-14th centuries, disclosed by N. N. Vakturskaya in 1961, are of great interest to describe the irrigation in the peripheral areas of Khorezm in the 13th-16th centuries. She identified it as the settlement of Yany-Shekhr mentioned in the written sources (Vakturskaya 1963:53). During archaeological excavations carried out on site by N. N. Vakturskaya, approximately 500 coins mainly of the Golden Horde period (minted in Khorezm from 1301 CE to 1388 CE) were found. According to N. N. Vakturskaya, the base of Shekhrlik was a Khorezmshah period settlement. Its main culture was dated to the Golden Horde period, and its decay to the period of Timur's conquest (Vakturskaya 1963:53).

Numerous finds (ceramic, jewelry and coins) from rural settlement ruins in its territory confirmed irrigation systems functioning at that time (see Note 165). In the Khorezmshah period, gravity irrigation apparently prevailed, since the region was watered well enough by the Northern Daudan. By the end of the 13th century, the population suffered great difficulties because of water lack due to the cessation of water flow from the Daudan, so at that time there were widespread *chigir* installations, and these left large pits (up to 10 m in diameter) in almost all the canals from which very small canals streamed (Figure 49.5-9). The bank of pits and canals were covered with countless vessel fragments used with the *chigir*.

Among the Khorezmshah settlements, the better preserved was a large farm (*poisk* 903) located west of the Akerkek-yab canal. The canal was more than 40 m wide and 12 m between banks, which were 2.5 m high. The canal dated to the Kangju period but it was reconstructed many times in the Middle Ages. The environs of the farm were irrigated by a distributor reaching a total width of 13 m.

The Khorezmshah complex consisted of a well-planned settlement with: a fortified central farm with quite impressive walls and a moat (36 × 40 m), several large buildings with many rooms (8 to 12) on both sides of an alley with two parallel ditches, leading to a large garden (200 × 140 m) surrounded by a low adobe wall and ditches.

A number of vast garden-park compounds of the Golden Horde period, located very close to the production center of Shekhrlik, were studied. South of the settlement (*poisk* 922-924) a fence 160 × 160 m in size and 30 cm high was preserved (see Figure 10.C). Some of its corners were fortified by earthen pillars. Ceramics of the Golden Horde period were found inside the fence. To the west there was a large and deep canal (18 m wide between banks) connected with a riverbed, which bent sharply in this area. In some places the riverbed banks were strengthened by dikes and the bottom closed by dams in three places (see Figure 10.D). The channel was 30 m wide and 2 m deep.

On the opposite side of the riverbed there was a second layout (110 × 80 × 100 × 80 m), with two small ditches close by. In the upper part of the riverbed, a canal 10 m in size branched off. On its banks, *chigir* pits and remains of sites were dated to two periods: Khorezmshah and Golden Horde. The canal preserved reconstruction traces. In the Khorezmshah period the irrigation was by gravity. In the Golden Horde period there was lack of water and the canal was deepened again and water lifting facilities (*chigir*), whose wells survived, were dug on the banks. The *chigir* pits (6 × 6, 10 × 15 m etc.) and scattered *chigir* jars were found in many other places northwest and west of Tuzgyr. In the 14th-15th century, a number of riverbeds of the Daudan (in this part of the Sarykamysh Delta) received water from the Daryalyk Basin. Water flowed from irrigation systems located northeast. Since the riverbed flow was slow, dams with different water intake works were built in order to raise the water level, as a reminiscence of the Syrdarya Medieval facilities (Figures 49 and 62).

Thus, a large hydraulic engineering site, in the environs of a Golden Horde village (*poisk* 959 and 960), consisted of a retaining dam and several head structures in the form of canals, from which water was supplied into a small irrigation network surrounding the settlement (see Figure

49.5). This was made up of individual debris of small dwellings (with two or three chambers). The hills of ruins (about 10) were 1.5-2 m high and covered with pottery of the 13th-14th centuries. The canal irrigating the fields had a width of 5 m between banks. The canal had two heads: one was 150 m above the dam and the other close to it. Through a by-pass canal, water was fed into other canals in the lower part of the riverbed. Among agro-irrigation layouts, rather large sized fields prevailed. A measured field (*poisk* 964) was $37 \times 58 \times 32 \times 58$ m in size (see Figure 49.7). It was irrigated using a *chigir* installation, whose pits and fragments of *chigir* jars were preserved.

Between Shekhrlik and Ak-kala, the irrigation works of the later period were found on the dry channels of the Northern Daudan (*poisk* 817, 818, 820, 827, 859, 861, 869, 870, 935, 936, 938, 965). According to ceramics found in the rural settlements, they were dated by N. N. Vakturskaya to the 15th-16th centuries. At *poisk* 936, the settlements survived in the form of houses and a small (25×25) sub-square fortification with built earthen walls rising 2 m above the surroundings plain. The overall width of the canal was approximately 20 m and 7 m between banks. There were many *chigir* pits on its banks. Between the canal and the fortification there was a reservoir 60×70 m in size, supplied with water by a *chigir*.

In a number of this region's settlements (*poisk* 915, 859, 861, etc.), remains of yurts bases were revealed close to the ruins of mud brick dwellings covered with Medieval ceramics. Southeast of the Kazanly-Auliya graveyard, the base of a yurt 10×9 m in size was found (*poisk* 859). A canal with remains of a settled dwelling and ceramic kilns was located close by. N. N. Vakturskaya dated the pottery to the 16th century. This was probably a Turkmen settlement (see also Gulyamov 1957:183-191).

As mentioned above, most of the canals and riverbeds in the lower reaches of the Northern Daudan preserved traces of heavy Medieval flooding. In some places the banks were completely washed away, and on the bottom were typical loamy clay and sand sediments. The environs of the Gosha-khavuz site settlement are very characteristic in this respect (*poisk* 919). The settlement center was a large square enclosure oriented according to cardinal axes. A village with ruins, often dwellings, was nearby. The canal, along which many *chigir* pits of different sizes (mainly 6×10 and 10×15 m) were found, was encircled by a fence from northwest. On its slopes, and along ditches and into fields, there were visible traces of flooding. They were probably connected to the Amudarya's waters breakthrough along the riverbeds of the Sarykamysh Delta at the end of the 14th century, during the campaign of Timur against Khorezm (1372-1388 CE).

Timur destroyed Urgench and the surrounding settlements and devastated the entire area north and northwest of Khorezm. This campaign led to the destruction of dams, dikes on canals and riverbeds of the Amudarya. The river supplying water to the riverbeds of the Sarykamysh Delta again filled the Sarykamysh depression and the Uzboy for a very short period, and probably reached the Caspian Sea (Tolstov, Kes and Jdanko 1954; Tolstov 1962a:17-26, and references).

In the 15th-16th centuries, the Turkmen tribes settled on the shores of the Sarykamysh Lake, and built, on the gradually drying slopes, very unusual irrigation facilities in the form of walled-aqueducts, catchment basins and connecting canals up to 6-7 km long.

The Sarykamysh irrigation systems were studied in detail by the field team of the Khorezm Expedition, led by S. P. Tolstov and described in the aforementioned publications (Tolstov and Kes 1956; Tolstov, Kes and Jdanko 1954, 1955; Tolstov and Kes 1960:29-31; Tolstov 1958:116-142), thus there is no need to dwell on them. It should only be noted that the question about the method of water lifting from the reservoirs to the ducts of the walled-aqueducts is not yet clear. This uncertainty is caused by the lack of *chigir* jar fragments. S. P. Tolstov supposed that the system used here was a shaduf, i.e. a water lifting crane (Tolstov 1958:138-139). He mentioned an example similar to an estuary or irrigation type, near lakes of the Kazakhs in the early 19th century described by A. I. Levshin (Levshin 1832:199-206; Bregel 1961:61).[Note 166]

* * *

Completing the description of the most typical Medieval irrigation systems that have survived in the 'lands of ancient irrigation' of the Sarykamysh Delta, we should note the sharp contrast between Medieval irrigation and the irrigation systems of the previous, Antique period. These differences emerged in the decrease of main canals cross-sections (from 20-15 m to 12-5 m), in the significant layout change of irrigation systems, and in the appearance of complicated branching systems with dense small-scale irrigation networks which, in some places, with their small branching 'veins', deeply penetrated the whole area adjoining a main canal.[Note 167]

The spreading of complicated branching systems was connected with farmers' attempts to reduce canals silting, to reduce the amount of cleaning labor and, at the same time, to irrigate new plots within the limits of irrigation basins, and thus increase the coefficient of irrigated lands, which rose up to 20-35% in the Middle Ages. According to the data of the Khorezm Expedition, the wide appearance of the *chigir* irrigation in Khorezm can be dated to the 9th-11th centuries.[Note 168]

According to the prominent specialist in irrigation V. V. Tsinzerling, in the recent past the *chigir* irrigation in the floodplain and delta of the Amudarya was the most perfect method of irrigation from a technical point of view. The use of *chigir* irrigation reduced water waste up to 30-50% than the gravity system; significantly reduced waterlogging and salinization risks; reduced the amount of work for cleaning small irrigation networks; improved hydrology and salt conditions; reduced soil erosion; increased harvests. Finally, a more compact and rational organization of the lands became possible.

The significant use of water lifting devices was due, in particular, to the lowering water level in irrigation canals, as a result of their deepening and to the rising level of cultivated lands because of agro-irrigation sediments (see Andrianov 1951:327; Gulyamov 1957:259). Ya. G. Gulyamov, in the special section of his work on the history of irrigation in Khorezm, rightly linked the development of water lifting devices to the general social and economic transformation of society (Gulyamov 1957:246-259). The development of this technology is confirmed by the documentation outlined by R. J. Forbes (1955, Vol. II:30-41), who showed a clear scheme of the development of water lifting devices (see Table 5).[Note 169]

In the example of Medieval Chermen-yab it can clearly be seen how the spreading of *chigir* facilities and branched small irrigation networks made possible the reduction of the general area occupied by irrigation facilities and, at the same time, it increased the area of irrigated lands. The coefficient of irrigated area in that period increased at least twice in comparison to the Antiquity.

The development of irrigation in the Middle Ages (especially in the Khorezmshah period) went through the improvement of large flood systems by the following scheme: river - head works (*saka*) - main canal - distributors of the 1st and 2nd orders - feeder - *chigir* - fields. An example of such large canals can serve not only the above described Medieval canals such as the Chermen-yab and the Shamurat, but also the large irrigation systems of the Southern Khorezm in modern time such as the Palvan-Ata (Kheykanik), the Gazavat (Madra) and some others which already existed in Antiquity and the Middle Ages (Gulyamov 1957:129).

The large canal in Southern Khorezm was 100 to 150 km long and 30 m wide in the upper part and 2 m deep (MRSA 1926, Kn. 2, Ch.2:10-19; Gulyamov 1957:200). B. M. Georgievskiy, writing about the head and middle part width of main modern canals ("not less than 20-25 m and 60-70 m in some places"), noted that during the construction they had a much smaller width. As a result of erosion, many became wide resembling somewhat meandering river channels (Georgievskiy 1937:113).

It is very difficult to judge the true size of the cross-sections sources of large Medieval main canals. The investigation on the well preserved bed of the Medieval Chermen-yab showed that its section went from 15 m to 7 m in its middle and lower parts over a distance of 120 km from Zamakhshar to Shakh-Senem. The Shamurat was traced for a distance of 80 km and its middle part was 12 m wide whilst the lower part was only 9 m.

Together with the Khorezm traditional large irrigation systems in the 'lands of ancient irrigation' of the 14th-16th centuries, there were, particularly in the environs of Shekhrlik, local systems watered twice a year based on lower channels of the Daudan and the Daryalyk. A variety of retaining dam, water distributors and water-lifting facilities, indicated the introduction of different irrigation principles brought by herding and farming peoples from the Syrdarya Basin.

Irrigation Works of the Modern Period

For the Khorezm Oasis and the whole territories of the Sarykamysh Delta the 16th-18th centuries were a period of general decay, particularly evident in the tendency of feudal fragmentation and disintegration, which reinforced different civil wars and nomadic invasions into the settled farming regions (see IUZ 1955:429). The formation process of a strong feudal monarchy was interrupted by the events of the 13th-14th centuries and could not be restored under later conditions (Tolstov 1950:5-6). In this period, the demise of feudal Khorezm, divided between quarrelling sultans, emerged very clearly.

Soon after Shaybani Khan (born in 1415 CE, ruling during 1488-1510 CE) conquered Urgench in 1505 CE with the fall of the Timurids, Khorezm came under Persian control. The Persian governor ruled Khorezm for a very short time and was driven out by his sons, Berke Khan's sons - Sultan Ilbars (ruling in 1511-1518 CE) and Bilbars, who both had previously led a nomadic life in the steppes of the Aral Sea and the lower reaches of the Syrdarya. Since that time, the massive resettlement of nomadic

Uzbek tribes into Khorezm began and, like the next migrations of the Karakalpak tribes in the 18th-early 19th centuries, they started from east, from the Syrdarya Basin (Zadykhina 1952:321ff.; Andrianov 1958:39-47). From the west, the migration of Turkmen populations increased and they were forced to leave for the drying shore of the Sarykamysh Lake and the Lower Daryalyk (see Vaynberg 1960:115-117; Bregel 1961:21-22). A cruel struggle for power in the oasis, and for its land and water, began between Uzbek and Turkmen feudal lords (Gulyamov 1957:199; OITN 1954:161-260).

The stormy events of political life in that period, dissensions and wars of feudal lords, did not favor settled irrigation farming, and the outskirts of oases, where the semi-nomadic population frequently changed, particularly suffered from these events.

During that time, the farming region on the Daryalyk, with its center in Vazir, which repeatedly served as capital and residence of the Khorezm *khan* in the 15th-16th centuries, played an important role in the oasis (Tolstov 1948b:315). In the 16th century the Daryalyk water flow gradually ceased (see Bukinich 1926). This means that the irrigation system revival of south Khorezm in the 15th-16th centuries accelerated shifting east the area of 'intensive watering' by the Amudarya. The continued exploitation of the irrigation system of Medieval Kunyadarya's farming oasis led to a rise in ground level and a flow reduction in the Daryalyk. At the end of the 16th century, this process ended with the Amudarya 'turning' east, causing an overall change in the hydrography of the lower reaches of the Amudarya (Andrianov 1951, 1954, 1958a-b).[Editor 54]

The water level rise in the Aral riverbed created an unstable situation, in which the river supplied water to the riverbed of the delta (Kokozek) eastern channel and directed a huge amount of water towards the Daukar depression, where a vast flood-land was formed ('Taukara-Tengizi' in Khiva's chronicles). In the Daryalyk the water first ceased to reach the Sarykamysh Lake, whose level fell dramatically and quickly (leading to the formation of the Daryalyk canyon). Then there was a lack of water in the environs of Medieval urban centers of Northern Khorezm, such as Adaka, Vazira and Terseka (Shemakha-kala) and, finally, Urgench.

The sharp flow decline in the Daryalyk riverbed was noted in the 1550s by A. Jenkinson (Rubinshteyn 1938:177; Bartold 1902:102; Bukinich 1926). [Editor 55] In 1570 water did not reach Urgench at all (Bartold 1902a:107). [Editor 56] In reality, in the late 16th century the flow was resumed, but only for a very short time (Bartold, 1902b:correction 31). [Editor 57]

The drainage of a huge region, and the formation of ancient irrigated lands in a significant part of the territory of the Sarykamysh Delta, was a major disaster for the agricultural population of Khorezm.

The political union of Khorezm began in the 1640s century during the reign of Abu al-Ghazi Khan, who made a number of important acts in order to strengthen the central power (MITT 1938:324, 328; OITN 1954:130-210; Zadykhina 1952:335). [Editor 58]

In the 17th-early 18th centuries there was a new watering of the Daryalyk, but Uzbek feudal rulers, following a hostile policy toward the Turkmen, blocked the riverbed with dams (about which the Turkman Khodja Nepes informed Peter I; Glukhovskoy 1893:47; Popov, 1853:329; Vaynberg 1960:117).

In the late 18th-early 19th centuries the Uzbek-Aral dynasty of Kongrat came to power in Khorezm. The period of its rule was characterized by a centralized strengthening and some progressive changes in economy and culture (see Andrianov 1958a:77-78; Bregel 1961:193; etc.). [Note 170] The rise in economic life was apparent in particular with the cleaning and reconstruction works of the abandoned Medieval systems. In the north and northeastern lands of ancient irrigation in the Sarykamysh Delta, plots were apportioned to Uzbek feudal and Turkmen rulers (Gulyamov 1957:221-225; Bregel 1961:96-111). The head works of these systems were shifted from the Daryalyk to the main riverbed of the Amudarya.

In the first half of the 19th century, during cultivation of new lands in this area, the Turkmen used some Medieval canals (the Sipay-yab, the Shamurat, etc.), which were cleaned by the Karakalpaks and the Uzbeks as labor conscription by order of the central power (see Gulyamov 1957: 220-222; Andrianov 1958a:82-83; Vaynberg 1960:119-123; Bregel 1961:60-61, 111-112, 191-196). In order to lift water, the Turkmen constructed a series of dams on the Kunyadarya riverbed, including those of Tash-bent, Ushak-bent, Salak-bent and the massive dam in the environs of the Medieval settlement of Vazira (Egen-klych; see Kaulbars 1881:405-419). On the banks of the riverbed, the Turkmen-Yomuts erected the fortifications of Mashryk-Sengir, Bada-Sengir, Atalyk-Sengir, etc. on both sides of this dam (Vaynberg 1960:121-122; Bregel 1961:99). Popular legend dated the construction of these fortifications to the beginning of the 19th century, during the conflicts between the Yomuts and the Kazakhs. The ethnographer V. G. Moshkov linked the construction of Mashryk-Sengir to this period (see OITN 1954:274).

A central place among the Turkmen fortifications on the Daryalyk was occupied by Mashryk-Sengir (I, II and III), of which the most significant is Mashryk-Sengir I, located on the right bank of the Daryalyk, 5 km south of the Ustyurt plateau, between two parallel branches of the canal originating in the riverbed and flowing to the northeastern wall of the settlement (Andrianov and Vasileva 1957, 1958).

Mashryk-Sengir I and II are located directly on the riparian banks on the right and left sides of the river. They were erected close to the massive Egen-klych dam (Tolstov 1962a: *Karta Zemli drevnego orosheniya i nizovyakh Amudari i Syrdari* - Map of the 'Lands of ancient irrigation' in the Lower Amudarya and Syrdarya). The dam is preserved for a length of 250 m, where the river bends and the left bank of the dam has been washed away. The width of the breakthrough was approximately 40 m. The dam was 6.5 m high from the river bottom, 12-13 m wide in the middle and 37-40 m in the lower part. This dam was included in the complex defense system of the whole region by the Turkmen, and was similar to the Karakalpak dam on the Mayliuzyak, where the fortress and three significant fortifications (Khatyn I, II and III) were connected to the dam crossing the riverbed (see p. 243).

A number of canals irrigating the environs of Mashryk had their origin from the dam of Egen-klych. The main canal, supplying water to the fields of Mashryk I, was 25 m wide and 9 m between banks; the preserved mounds are 1.6 m high (*poisk* 455) (Figure 50). The canal was dug in a small width riverbed and the high level of standing water indicates a good watering of the Daryalyk in the early 19th century.

During this period several Medieval canals were restored above the dam, for example in the environs of Vazir where, together with the old swollen Medieval canal, deepened canals were also recognized. From these, other canals branched out to irrigate the Turkmen fields and gourd plantations, still preserving their ridges and high *gryad* (*karyk*). Especially numerous were gourd plantations. The high number of gourd plantations was one of the most characteristic features of the environs of Mashryk I. The preserved layouts (for example at *poisk* 459, 465, 467) differ in their very high *gryad* (*karyk*) 2.2-3.6 m wide, and deep ditches (*salma*). They form a large continuous terrain, seldom mixed with the common *atyz*, i.e. irrigated plots without *gryad* and an average size of 20 × 40 and 20 × 50 m (*poisk* 464).

In the area of Turkmen irrigation of the 19th century, we also found small subsquare and trapezoidal fields (3.6 × 4.5 × 3.7 × 4.6 m, etc.), bordered by flat walls and forming large arrays (see Figure 50). Large quadrangular fields (for example, 6 × 10 × 7 × 9 m), and fields with sides more than 15-20 m were also widespread. These fields, mainly for grain crops, were entirely irrigated by water inlets. We should also mention the huge fields surveyed by the archaeological unit in the environs of Er-burun, where a regular system of rectangular fields occupied a vast territory west of Buten-tau. The owners of these fields lived in semi-earthen houses which were usually located in a compound (three to four semi-earthen houses), closely connected to each other. The measured irrigated plots had the following sizes: 45 × 78 × 45 × 76 m and 40 × 56 × 40 × 47 m etc. Probably, these settlements and fields were dated to a later period (late 19th-early 20th centuries), at the time of the breakthrough on the Daryalyk, when some Turkmen population groups were engaged in agriculture (see Vaynberg 1960:123).

Describing the irrigation works of the modern period on the 'lands of ancient irrigation' on the left bank of the Khorezm, and in particular the Turkmen irrigation in the basin of the Daryalyk, it is necessary to note that the majority of canals were not constructed anew, but were old canals preserved from earlier periods (see Andrianov and Vasileva 1957:105; Andrianov 1958a:82-83; Bregel 1961:60-61). Naturally, the irrigation methods of the Turkmen farmers in these lands (especially along the Sipay-yab and the Shamurat) differed very little from the traditional methods of the Khorezm farmers. According to Ya. G. Gulyamov, vast areas belonged to the Khivan high dignitaries, as reported by the court chronicler Bayan (see Gulyamov 1957:221; Tolstov 1958:20).

Describing Turkmen irrigation in the basin of the Daryalyk, it is necessary to mention the peculiarities of hydraulic devices, ranging from grand water-lifting dams, head structures with semi-dams, head-constructions with two and three branches (topographically included in the general defense of the *sengir*), and ending with a water raising system thanks to an ingenious systems of storing water in reservoirs (see Figure 50).

This system consisted of two parallel, interconnected canals, on one of which three large reservoirs connected to each other were built (*poisk* 474). These reservoirs were linked with the second canal by a system of ditches with dams. This canal, and when necessary also a second one, was used to simultaneously fill the reservoirs (I and II). Water in the reservoirs was raised up to high levels and then let into the fields by gravity method. These original hydraulic works resemble the Medieval systems of the 15th-17th centuries on the lower reaches of the Daudan watered twice a year and also the Medieval irrigation of the 11th-13th centuries on the Inkardarya in the Lower Syrdarya (see p. 237).

The desolation of the vast Turkmen farming region on the Daryalyk, and the settlement of the Turkmen and Uzbeks in Khanabad was the result of feudal conflicts in the mid-19th century, the Yomuts revolt against the Khiva khan in 1855-1857 (see Gulyamov 1957:22-228; Andrianov 1958a:96-99; Bregel 1961:197-228). The Khiva rulers also barraged dams causing the migration of some Turkmen to other regions (see Kaulbars 1881:410).

According to B. I. Vaynberg, some groups of Turkmen used temporary water breakouts from the Daryalyk and farmed there till the beginning of the 20th century (Vaynberg 1960:123). N. A. Dimo in 1913 wrote that in the district of Uaz there was water until 10 years ago, i.e. 1903 (EOZU 1914:389).

The Turkmen irrigated farming was very labor-intensive. According to archive data, at the beginning of the 20th century, in the Middle Amudarya, the Turkmen spent on average 80-120 working days a year per farmer to clean their canals and main irrigation systems (TsGA TSSR, Fund 616, op. 1, d. 70, l. 34). The time spent in cleaning canals in the Khorezm oasis was limited,

as the main gravity-flow canal could only be cleaned when the water level was low. There were other extremely labor intensive earthen works such as opening and closing *saka*, to strengthen bank dikes, etc. According to the economist S. K. Kondrashev, the farming economy of an oasis could exist "only with extreme labor intensity" (Kondrashev 1916:24).

According to the irrigation specialists V. V. Tsinzerling and I. N. Shastal, in the Khorezm oasis at the beginning of the 20th century, the maximum area of ditches and fields, in the administrative system sphere, was no more than 30-45%, whereas less than 30% of the territory was occupied by drainage lakes and *solonchak*.

A very typical example of modern day Khorezm irrigation is the Palvan-Ata Canal. According to 1920s data, it was 102 km long, 31 m wide in the upper part and 2 m deep. The layout of distributors and feeders was branched; the number of distributors was 51 for a total length of 913 km. The Palvan-Ata drew water from the Amudarya with eight *saka*. Of the total administration area (181,500 ha), only 24.5% (45,600 ha) was actually irrigated. In 1924-1925, in the basin of the Palvan-Ata, 141,000 people lived and the density of population was 331 people per 1 km² of plowed fields, or 247 people per 1 km² of territory occupied by irrigation facilities (MRSA 1926, Kn. 2, Ch. 2:10-19).

Chapter 5
The Lower Syrdarya

Boris V. Andrianov

In the remarkable epic poem *Shahnameh* by Firdausi, telling the struggle of the rulers of Iran with Afrasiab, and the evil king of Turan, there is a story about the origin of the fantastic city of Kang (Siyavushgird, 'the city of Siyavush'), located beyond the Chin River ('daryay-e Chin') at a distance of a month's march, where "the limitless space: desert, wherever you throw the gaze", where "there is blue space of the sea on one side and high mountains on the other" (Firdausi 1960:199-201, 588: note 6381).[Note 171]

Many scholars identify the Kang of Firdausi's poem (the city which was erected by Siyavush in ancient days on a steep rock), with the Kangkha of the *Avesta*, mentioned in connection with one of the episodes of the struggle between the people of Airyana Vaējah ('the Iranians') and the people of Turan ('the Tura'), and also with Kangdez or Kangdiz of the Pahlavi texts (*Bundahishn, Datistan i Menoke khrat*, etc.), and finally with Kangju of Chinese sources (see Tolstov 1945b:218-284; 1948a:20-26; 1948b:145ff.; 1958:72-73; 1961:143-144; Staviskiy 1961b:113; Klyashtornyy 1964:167-179).

In the *Bundahishn*, Kangdez is mentioned in connection with the environs of Lake Vorukash (the Aral Sea?); in the *Datistan i Menoke khrat* it is located on the border with Airyana Vaējah (see Geiger 1882:52; Tolstov 1948a:20:note 3). J. Marquart convincingly proved that the Kangju were Iranians, and localized Kang in the Lower Syrdarya (Marquart 1903:60-71; see also Shcherbak 1959:370). V. V. Bartold, and also G. V. Ptitsyn, Ya. G. Gulyamov and S. G. Klyashtornyy, placed the main territory of the Kangkha of the *Avesta* (the Kangju in the Chinese sources) in the Middle Syrdarya.[Note 172]

In 1945, S. P. Tolstov assumed that the term 'Kangkha' had the same meaning as 'Khorezm', and later he linked the original territory of Kang - Kangju with the vast cultural area of the two interlocking deltas of the Amudarya and Syrdarya. According to his words, "Khorezm was only a part of this vast area, though a very developed part, which became the political leader of the entire Kangkha" (Tolstov 1961a:143-144).

S. P. Tolstov based his point of view on the information of the Chinese Zhang Qian, who travelled in the Central Asian countries in the 2nd century BCE and reported that the headquarter of the ruler of the Kangkha-Kangju state was placed 2,000 li (800-1,000 km) northwest of the capital (the city of Ershi in the Fergana valley) of the Davan state (Tolstov 1948a:21; see also Klyashtornyy 1964:173). If this information is accepted, then the center of Kangkha-Kangju could not have been located in the Middle Syrdarya, but placed either in Khorezm (as S. P. Tolstov originally supposed) or in the Lower Syrdarya (as J. Marquart thought). The latter point of view is supported by some archaeological evidence, in particular the similarity of cult burial complexes seen in the Lower Syrdarya (Tagisken structures of the 9th-5th centuries BCE) and worship buildings in Khorezm, especially at Koy-Krylgan-kala (see Rapoport 1962: 79).[Editor 59] The mud bricks mausoleums of the early Tagisken, were very grandiose for that time (see also Gryaznov 1966), suggesting that the sacred center of Kangkha was also placed near there. The zone of Kangkha apparently connected the lower reaches of the Syrdarya and the Amudarya and, in the 1st millennium BCE, the people of Khorezm who populated the Lower Amudarya, and the Saka the Silis ('Syr'), i.e. the Lower and Middle Syrdarya, they could call themselves Kangju (see names of the Syrdarya in Levshin 1832:215-264; Klyashtornyy 1953:189-190; 1964:73-76).[Editor 60]

The Kangju state and its people were known in the Chinese written sources starting from the 2nd century BCE (Bartold 1963, Tom II, Chast 1:175; Tolstov 1948b:144; McGovern 1939:40-41; Klyashtornyy 1964:173; Maksimova et al. 1968:8). In this period, according to Ya. G. Gulyamov, the Khorezmian kingdom disappeared from the historical arena and was included into Kanga (Gulyamov 1957:98). The archaeological excavations of Koy-Krylgan-kala (4th century BCE-4th century CE), and the religious center of ancient Khorezm, revealed traces of fire preceding the period of desolation (2nd-1st centuries BCE), which was linked by researchers to steppe tribes migrations.[Note 173]

Chinese authors referred to Kangju as a big state. In Central Asia, besides the main territory of Kangju (from Talas east, Chach south, to the Lower Syrdarya north), the countries of five 'small rulers' (Suse, Fumu, Yueni/Yuyni, Gi/Ji, Yuegyan/Yuejian) were subordinate to the Kangju ruler. Their geographical location is open to question, and only the identification of Yuegyan with Gurganj, i.e. Khorezm, can be accepted. The Yancai people (Sarmatian Alans tribes of the Caspian Sea), living northwest, were also submitted to the people of Kangju. The Kangju, like their neighbors (Yuezhi and the Wusun), led a nomadic life, although they also had cities (see Bartold 1963, Tom II, Chast 1:175-177; Tolstov 1948b:143-147; Bernshtam 1952:208-216; Akishev and Kushaev 1963:9-24; Klyashtornyy 1964:171-174; Maksimova et al. 1968:8-9, 243-248).

In the 3rd century, the Kangju state entered a period of decay; in the 4th century in the Aral Sea area (the country of Sude?) the Chionites controlled it, while concerning Kangju, in the mid-5th century, only a small domain was reported under the Hephthalites' rule (Klyashtornyy 1964:174; see also Nerazik 1966:125; 1968:202). In 563-567 CE the Hephtalite state was destroyed by the Turks, and by 571 CE the border of the Turkic Khaganate was established in the west along the Amudarya. From that period commenced the process of Turkization of the Central Asian peoples, which became very important for all later cultural history of Central Asia.

In the first centuries BCE and the early 1st millennium CE, the Aral Sea area and the vast territories of the Eurasian steppes became the arena of significant nomadic migrations. The chain reaction of 'the great migration of peoples' from east to west was caused by internal social processes, development of economic-cultural type of pastoralist nomads, and features of the political history of major states in Asia and Europe, which experienced a deep crisis in the mid-1st millennium CE.

The importance of the movement of steppe tribes into the settled farming areas of Central Asia was written by V. V. Bartold: "The migration of peoples from north to south through the cultivated areas of Turkestan was, as usual, only a consequence of the more frequent and significant migrations of nomads through steppes from east to west; not all the nomadic hordes appearing in the Central Asian steppes were moving south, and the political consequences of this last movement were relatively insignificant" (Bartold 1963, Tom II, Chast 1:179).

These conditions can explain the extreme stability over a long period of time (from the 1st century CE to the 8th-9th centuries) of the material culture of the three main areas: Lower and Middle Syrdarya - Djety-asar, Otrar-Karatau, and Kaunchi. In her detailed study of the ceramic complexes of the Syrdarya, L. M. Levina quite rightly explained the pottery similarities in the three mentioned areas by their semi-settled economy, which combined rain-fed or estuary (based on primitive irrigation) farming with settled herding on deltaic cane grasslands (Levina 1967:16-17; Maksimova et al. 1968:243-245).

In the 8th century the oases of the Syrdarya (at that time called 'the Kanga River') were part of the Pechenegs tribes union, meaning 'people (or men) of the Kanga', which are probably the Kengeres in the Orkhon inscriptions (8th century), later known as Kangary by Constantine VII Porphyrogenitus (10th century), and then the Khangakishi mentioned by al-Idrisi in the 12th ntury (Klyashtornyy 1964:178-179; see also lstov 1948a:23).

Starting from the Kimaki-Kipchaks' tribes conquering the Pechenegs-Oguz's lands on he Syrdarya in the early 2nd millennium, the name of Kangar (*kengeres*) turned into the name 'kangly' (Tolstov 1957b:101, 1961a:145). The most important Medieval center of this region was Farab-Otrar, which was identified by S. G. Klyashtornyy as 'Kangu Tarban' from ancient Turkic runic inscriptions, (Klyashtornyy 1964:155-161).

Returning to the issue of Kanga, it should be noted that, at the moment, the solution depends on the collection of archaeological materials characterizing the sites of nomads and the areas of irrigated farming of the Aral Sea region, starting from the 1st millennium BCE until the Mongol invasion, which interrupted the cultural-historical and ethnical traditions of Great Kangkha in the *Avesta*. The comparison of these areas from Khorezm to Turkestan, chronologically of different periods and geographically isolated from each other, on modern archaeological maps makes us doubt any direct geographical identification of Kangkha - Kanga - Kangju, i.e. the definite attribution of all these three historical-cultural terms to only one local area. The magnificent fortresses of Khorezm of the Kangju period, and the no less impressive fortified constructions of 'the Asars' of the 1st millennium CE, far superior to fortifications on the Middle Syrdarya, support such doubt.

Since the mid-1st millennium BCE, Kanga, the Siyavushgird of Firdausi, was the mighty Khorezm (with its massive fortified settlements on 'steep rocks'), together with settlements of herding and farming tribes of the Saka-Massaget on the Lower Syrdarya. The fortified settlements of herders and farmers on the Lower and Middle Syrdarya developed at the turn of our era and existed until the 8th-9th centuries, belonged to the Kangju.[Note 174] Thus, the historical-geographical concept of Kanga changed, in relation to the political and ethnical geography of the Aral Sea area. It was also connected to the advancement of irrigated farming into the zone of steppe

herding tribes, and the formation of large irrigation centers in the 1st millennium BCE. Initially these cradles appeared in the most ancient channels of the Lower Amudarya, then on the periphery of Khorezm and on the ancient channels of the Syrdarya (the Inkardarya) merging in the Amudarya Delta, and later, in the 1st millennium, on the younger channels and the main riverbed of the Lower and Middle Syrdarya and its tributaries (Djety-asar, Otrar-Karatau and Kauchin oases).

Natural Conditions

The 'lands of ancient irrigation' of the Syrdarya Basin, studied by the Khorezm Expedition, occupied a significant part of the left bank of the ancient delta in the shape of a huge triangle. It started in the place where the river emerged from the corridor between the Karatau ridges and the Kyzylkum plateau into the vast space of the Turan lowland. Almost equal to the other side, triangles were formed by the contemporary riverbed of the Syrdarya, the system of dry riverbeds of the Janydarya and the Inkardarya, connected with the ancient delta of the Akchadarya in the southwest. This vast territory occupied more than 400 km east-west and 200-250 km in the north-south direction. The deltaic plain gradually lowers from east to west; the absolute height is 140-151 m at the Chiili and Turtugay stations, 100 m at Djusali and 55 m at the seashore.

The borders of this region are the shores of the Aral Sea and its vast massif of sand ridges, crossed by valleys with north-south ancient riverbeds (Dayrabay, Ashinysay, etc.) in the west. The high desert sands of the Kyzylkume south, and the watered zone of the contemporary riverbed of the Syrdarya, full of lakes and swamps, in the north. Here is concentrated the contemporary farming population of the area. In the swampy depressions there are crops of rice interspersed with thickets of reeds, cattail and tamarisk. In fallow and abandoned areas there is *solonchak* and bush thickets, on the low watersheds there are other agricultural crops (millet, barley, etc.). On the river banks there are a few *tugai* forests of silverberries, poplars, willows and bushes (Borovskiy and Pogrebinskiy 1958, Tom I:31, 211-213).

The vast fertile alluvial plain of the Lower Syrdarya differs comparatively little in its farming economy capability from the Khorezm Oasis, placed in the lower reaches of the Amudarya. However, the Syrdarya is less water abundant than the Amudarya, which is two to three times more copious. The Syrdarya water is lighter, less silted and sandy,[Note 175] but the territory, on which the Syrdarya sediments lie, is 1.5 times greater than the general area of the Amudarya delta. These peculiarities in hydrography and geomorphology of the Lower Syrdarya, if compared to the Amudarya, reduced its quantity of alluvial sediments and agro-irrigation deposits (Fedorovich 1950:211; Borovskiy and Pogrebinskiy1958, Tom I:460.[Editor 61]

The same general patterns defining the relief formation of the whole landscape of the Amudarya Delta were also characteristic for the Lower Syrdarya. In its lower part, the Syrdarya flowed through the gently sloping elevation of its sediments, due to which even a small rise of river water level overflowed the banks and flooded a vast area. Natural river overflows took place both in summer flood time and in winter during the ice blockage. Due to mixed feeding the maximum water flow shifted from July to June (water flow in June was 1,300 m³/sec, in January it was 340 m³/sec and on average it was 600 m³/sec). At the beginning of December, and just below Kyzyl-Orda, the river froze and it did not thaw until early April.

According to B. A. Fedorovich and A. S. Kes, the earliest prehistoric riverbeds of the Syrdarya (pre-Syrdarya) flowed considerably south of the contemporary deltaic plain, in the territory of the Kyzylkum, where the former riverbeds of the Darya-sai, etc. were preserved (Fedorovich 1950:212-213; Kes 1958).[Editor 62] Perhaps, in that remote time, the pre-Syrdarya joined the pre-Amudarya (Tolstov and Kes 1954:141-145; Kes 1958). At a later period (at the beginning of the early Quaternary period, according to B. A. Fedorovich), the filling of the Kyzyl-Orda basin and the river outbreak into the Aral Sea (at 70 m above sea level) took place (Borovskiy and Pogrebinskiy 1958, Tom I:19). Then began the long-term process of formation of deltaic sediments, the shift of numerous riverbeds, and this region climate conditions were about the same as today, i.e. arid (Gerasimov 1937:37; Fedorovich 1950:206; Ilin 1946:225).[Editor 63]

According to geographers, before the development of irrigated agriculture, the entire delta was a huge area with numerous lakes and swamps, including meandering riverbeds full of water. Water slowly flowed down northwest on a very wide front, making its way through the sand ridges of the Eastern Aral massif then north. These features forming the ancient hydrography are expressed in the contemporary landscape of desert plains, cut across by meandering dry riverbeds with thickets of *saxaul*, a few remains of hilly sands, and with bare spaces of clay *takyr*, a mosaic interspersed with patches of grass. Judging from the remains of ancient irrigation works and sedentary agricultural settlements, fortresses and cities, this desert space, with a total area of 2.5 million ha, was exploited by farmers in the past, but now it is used only as pasture for grazing cattle. The area of the contemporary cultivated oasis along the Syrdarya is less than 100,000 ha.

The interpretation of aerial photographs and the complex field survey disclosed the ancient hydrographic network of the Lower Syrdarya, characterized by a complex interplay of riverbeds both east-west (mainly the eastern part of the delta) and north-south (prevailing in the southwest). Among the most significant and best marked in relief are the riverbeds of the Janydarya and the Kuvandarya, which are traced over almost their whole length, from the contemporary cultivated Syrdarya area to the shore of the Aral Sea. The Janydarya and the Kuvandarya run in parallel to two other riverbed systems of the Inkardarya and the pre-Kuvandarya (Northern Kuvandarya), preserved as separate parts not covered by later stratifications. Now, these rivers do not have direct links with the Syrdarya.

Irrigation Works of the First Quarter of the 1st Millennium BCE

Prehistoric culture sites in the lower reaches of the Syrdarya have not been studied enough, and ancient irrigation works of that period are not fully identified. Encampments of the late Kelteminar culture dated to late 3rd-early 2nd millennia BCE were discovered on the northern shore of the Aral Sea. They were in the environs of the Saksaulskaya railway station (Formozov 1945, 1949; Vinogradov 1959), and in the area of Jalpak (Vinogradov 1963), in the lower reaches of the Inkardarya, which S. P Tolstov considered the delta main water artery at that time (Tolstov, 1961a:116-117; 1961c:4). The Inkardarya riverbed could not possibly contain all the Syrdarya water and there is reason to suppose that there was another large channel, which would have a similar direction to the modern Syrdarya riverbed.

The Expedition discovered herder encampments of the Bronze Age in the sands surrounding the banks of the Inkardarya. Among them, a graveyard with 70 kurgans revealed through the aerial survey of 1959 on the plateau of Tagisken, was particularly interesting. Thanks to archaeological digs, part of the complex burial structures from the northern group graveyard was dated to the 9th-8th centuries BCE. The Saka kurgans of the southern group, like the kurgans of the neighboring hill of Uygarak, were dated to the 6th-5th centuries BCE (Tolstov 1962a:79-88; 1962b:127-138; Tolstov and Itina 1964; Gryaznov 1966).

The excavations, directed by S. P. Tolstov and M. A. Itina, provided materials which allowed to characterize the Tagisken culture (Tolstov 1962a-b; Tolstov, Jdanko and Itina 1963:36-47; Tolstov and Itina 1964; OIKK 1964, Tom I; Gryaznov 1966).[Editor 64] The rich collection of Tagisken vessels and other finds, such as large sized mud bricks, led us to conclude that the settling of these tribes involved breeding of cattle, goats and farming (Tolstov 1962a:137). The spread of farming with the Tagisken culture bearers can be discerned by a few sites scattered along the Inkardarya riverbeds east of the Tagisken graveyard, discovered by the Expedition in 1960-1961. Besides ceramic fragments similar to the Tagisken type, ceramic slags and calcareous millstones were found in this region. At one of the settlements (*poisk* 6, year 1961), remains of settled type dwellings and some weak traces of irrigation ditches, branching from the large riverbed nearby, were found.

Better preserved irrigation works were discovered in 1959 in the Middle Inkardarya, around the so-called 'slag kurgans' dated to the 6th- 4th centuries BCE by three-winged bronze arrowheads of the early Scythian period (Tolstov 1961a:138-142; Tolstov, Jdanko and Itina 1963:43). Due to the proximity of kurgans, some sites represented by clusters of rough handmade pottery were discovered.[Note 176] Traces of irrigation facilities related to these settlements were recorded on the bank of the Middle Inkardarya in an area extending for 12 km (*poisk* of the year 1959: 11, 14, 17, 18-21, 37).

During the survey of 1959 on the right bank of the riverbed we discovered for the first time a small irrigation system 600 m long, and the settlements topographically linked to this system which, according to finds, could be attributed to the 'slag kurgans' culture. The canal was preserved just as a dark band shape 3.5 m wide, which started in the ancient floodplain river terrace and was covered with sand and vegetation (Figure 51.A). In the lower part of the canal bank remains in the shape of hills 3-3.5 m wide were found. The general canal width was apparently no more than 10 m. Lateral branches were rare and generally had a primitive outlook with angular, and 'subrectangular' layouts.

Site 11 (*poisk* 11, year 1959) was located north of the canal and it was preserved as a large accumulation of ceramics which, according to L. M. Levina, were similar to the pottery found at another site, located east, right on the southern bank of the riverbed (*poisk* 18). This site was fortified with a low wall, now preserved only as a very small ridge (see Figure 51.A). In the environs of this site, on both banks of the Inkardarya faint traces of canals, small agro-irrigation layouts, and quadrangular field plots were found, indicating farming activities of the ancient inhabitants.

Another site, with similar findings, was discovered on the opposite and northern side of the Inkardarya (*poisk* 20), where traces of agro-irrigation layouts were faintly expressed in relief, but were sufficiently well indicated on the area large-scale aerial photographs.[Note 177] In this place several irrigation systems were discovered, two of them branched from a lateral channel of the Inkardarya and two others from the main riverbed (see Figure 51.A).

The most significant canal by size was situated 3 km northeast of the area described above (*poisk* 23, year 1959). The canal was preserved as a dark line, 9 m wide and covered with *biyurgun* (Figure 51.B). In its eroded banks, pottery of the 'slag kurgans' culture was found. The total width of the canal seems to have exceeded 20-25 m. It branched from the Inkardarya and was traced for a length of 1.5 km. The remains of a site, evidenced as an accumulation of ceramics, occupied an area of 30 × 20 m near the head works of the canal.

About 4 km east of the kurgans, the riverbed of the Inkardarya branched: the narrower channel (up to 40-50 m wide) was slightly sinuous and marked in relief by a shallow bed becoming flat in some places; the wider channel (100 m wide), on the left, abruptly meandered. In the narrow lateral channel water flow was probably less rapid and powerful, and the water inlet could have been regulated from the main riverbed. Therefore, the heads of canals were built on the lateral channel, where it merged with the main Inkardarya riverbed, as well as on the main riverbed, but lower than at their confluence.

At the confluence of the Inkardarya lateral channels, remains of a small island (see Figure 51.A) dividing the riverbed into two narrower parts (30 m wide) were visible. Possibly this island played an important role as a contingent dam used when necessary to regulate the water level. Reeds and earthen bundles were enough to narrow channel sizes in order to force the flowing water to rise to the needed level to irrigate the fields. Such structures are known in the contemporary period among the Karakalpaks as *kysme-bugut* or *kysme-*

saga, and are described by N. N. Belyavskiy.[Note 178]

Concluding the description of the remains of irrigation of the 6th-4th centuries BCE preserved in the lower reaches of the Syrdarya, it should be noted that this small farming oasis in the Middle Inkardarya existed only for a very short time. A regularly settled, irrigated agriculture, based on the use of main canals, had not yet become the main activity of inhabitants of the Lower Syrdarya at that time. Herding and the primitive *kair* farming still prevailed in their economy, which continued Bronze Age traditions (Tolstov 1959b:145).

Significant plots with remains of irrigation works of a later period (4th-2nd centuries BCE) were found in the Middle Janydarya, near the ruins of Chirik-Rabat (which, according to S. P. Tolstov, was the capital of the local Massagets or Saka tribes at this time), Babish-Mulla and Balandy, some fortified sites of the Saka. The richest farming traces and the largest number of rural sites dated to the 4th-2nd centuries BCE were found in the environs of Babish-Mulla (Tolstov, Vorobeva and Rapoport 1960:40-43; Tolstov 1961a:123-126; 1962a:156-158).[Note 179]

Irrigation in the Environs of Babish-Mulla

At that time, this entire area was an inner 'delta': a wet, swampy plain crossed by massive sand ridges and numerous meandering rivers, on which the inhabitants' economy was based (Figure 52). The deltaic areas were known to be characterized by the high variability of their hydrographic regime. The rate and flow capacity of the individual channels changed, and the location of irrigation riverbeds and flood changed correspondingly (see Andrianov 1958a:17-35). The agriculture of these places was developed under frequent and variable water level conditions in riverbeds and generally speaking in small channels and floods (see Tolstov, Jdanko and Itina 1963:74).

The remains of ancient irrigation works were recorded in the territory around Babish-Mulla, located north of a large bend in the main Janydarya riverbed resembling the shape of a Scythian bow. Its 'upper' part reached Akkyr, the eastern part started at the Tagisken elevation and the western part approached Chirik-Rabat. The distance from Tagisken and Chirik-Rabat was approximately 65 km in a straight line. The whole space in between was cut across by small meandering riverbeds with mainly north–south and southwest directions.

The archaeological-topographic research of the Middle Janydarya ancient irrigation works indicated that irrigation, in that period, was based on the wide use of fading riverbeds and former small channels beds of the inner 'delta' which formed a highly branched system. Irrigation was carried out according to the scheme: river – former riverbed – feeder – field. Small ditch networks had 'sub-rectangular' branching (see Figure 51.D-E). Many canals were without branches and characterized by steep sloping (to 1 cm per 1 m) from riverbed to field (see also Bowen 1958:45). Farming settlements were usually placed on fluvial ridges, next to riverbeds adapted for irrigation. In some cases the large main canals, 10-20 m wide, were at present marked on the surface by light *takyr* lines running on the highest points along the former rivers and deltaic riverbeds.

In the environs of Babish-Mulla the irrigation system consisted of three large riverbeds diverting from Akkyr in a fan shape, 20-40 km west and northwest (see Figure 52). The middle channel supplied canals irrigating fields around the settlement of Babish-Mulla. The northern channel supplied water to numerous north–south canals flowing along ancient riverbeds and sand ridges through inter-ridge depressions. The south channel, after joining the middle one, continued far west through a large canal. The southwestern group of sites was placed on this channel (see also Tolstov, Vorobeva and Rapoport 1960:40-41).

As shown by a trench, the '*takyr*' riverbeds were formed with a sand thickness, clay loam and clay, and these sediments were embedded in upward slopes along the banks. In the past, very still water slowly flowed in such riverbeds, forming tranquil back-waters.

Test trenches on one of the riverbeds in the environs of Babish-Mulla north of Akkyr (*poisk*

390) produced the following results: at the bottom of the riverbed, at a depth of 1.2 m, were found alluvial sand and two thin layers of layered loam above it, indicating the beginning of the fading channel process. Above that there was again alluvial sand, which, at the depth of 1 m, changed into a massive (70 cm) layer of dense dark-brown loam (layer 4). The loam border with a lower sand layer was very apparent, indicating a hydrographic regime change from fast flowing water to slow, semi-stagnant water. The thickness and uniformity of loamy clay sediments were evidence of a period of relative continuity, when water in the fading channel started to be regulated by dams and used for fields irrigation.

The upper light-colored layer of loamy clay sediments was connected with the fading riverbed, already abandoned by farmers when the flow was not sufficient. Precipitation brought fine earth-material from the banks which started accumulating in depressions. The contemporary loamy surface was covered with *takyr* crust devoid of vegetation. The riverbed was no more than 25 m at low water level and approximately 40 m at high, which corresponded to the average size among the main larger canals of Khorezm.

Most of the riverbeds of the Babish-Mulla Oasis were preserved in the shape of very flat banks with gentle slopes, with an average height 70-150 cm above the surroundings. In the highest parts, the profile levels revealed traces of canals in the forms of *takyr* bands. The northern riverbed, located 4 km northeast of the Babish-Mulla site (*poisk* 255), had an overall width of 80 to 100 m.

Among the three major canals of the Babish-Mulla Delta, the middle channel was the most important for the life of the whole surrounding rural area. It supplied water to canals on which all adjoining sites were based (see Figure 52). On this riverbed the peculiar system of several reservoirs was used, by which the water level was regulated and the stability of the irrigation source ensured over a long period. The deepened parts of the former riverbed were interconnected by small canals, and water accumulated in it during freshets. It reached a high level, and when the water level in the main bed fell, reservoirs were used as needed.

The most significant reservoir in the environs of Babish-Mulla started 600 m northwest of Akkyr in the shape of a very meandering deepened riverbed, 50-60 m wide and 2 m deep (Figure 53.A). The length of the deepened riverbed section adapted for the reservoir was 3 km. If the average width (40 m) and depth (2 m) are accepted as correct values, then its general size was approximately 200,000-250,000 m³. Short canals supplied fields with water and branched from both sides. The banks were dotted with fragments of pottery, stones from fireplaces, pieces of quartzite implements, millstones, etc. The reservoir ended where the riverbed was connected with a neighboring parallel riverbed by several canals. In one of these canals a large rural site (*poisk* 242) was found and it will be described in detail as follows (see also Tolstov, Vorobeva and Rapoport 1960:41) (see Figure 51.E).

The settlement was located 300 m west of the northern channel and 7 km east of Babish-Mulla. The canal was preserved as a *takyr* band 4 m wide. The dwellings appear on the *takyr* surface in the shape of contoured areas with vegetation, rich ceramic accumulations, quartzite implements, fragments of millstones, stones from fireplaces and bones (see Tolstov, Vorobeva and Rapoport 1960:43:fig. 31).

The settlement 'sub-rectangular' topography corresponds to some extent to the agro-irrigation layout. Lateral ditches branched off the canal on one side only, and thus resemble the Archaic systems of Khorezm. Of greatest interest were traces of a field, crossed by irrigation ditches in the middle, with a perimeter sides 45 × 46 × 52.5 × 43 m (see Figure 51.E). In its size and character of layout, this field was very similar to the one described above, i.e. a field layout near a rural settlement of the 4th-2nd centuries BCE in the environs of Djanbas-kala (*poisk* 631), which is similar to the field mentioned in *Videvdat* (see p. 164). It is difficult to say whether we were dealing with an accidental morphological coincidence or with the adoption of irrigation techniques from Khorezm.

Below this site and the reservoir, the middle riverbed became a very flat bank gradually rising above ground and reaching a height of 130 cm 6 km from Akkyr. The leveling survey revealed that the central *takyr* strip, 22 m wide, corresponded to the bottom of the canal. The banks had a total width of 80-100 m and were dotted with ceramics. The slope of a network of small ditches, branching from the main canal, had an average of 130 cm per 100 m. The middle stream preserved this character up to its branching into two riverbeds (2 km northeast of Babish-Mulla). The left riverbed, in the shape of a flat wall, continued south after some sharp bends and joined the southern riverbeds system of the 'delta'. It seems that the right riverbed ran out uniformly in the surroundings, but it was possible to trace a narrow ditch 1.5 m wide flowing into the next water intake basin. Like the first one, it was approximately 40 m wide, 1.5-2 m deep and 700 m long. Its capacity was about 45,000-55,000 m³. It was connected to the third canal reservoir, which was preserved as a *takyr* strip 6-7 m wide and 30 m long. The third reservoir had a size very similar to those described above. Its length was approximately 900 m long and ended 250 m north of the site (see Figure 53.A). The estimated capacity was 65,000-75,000 m³.

All three reservoirs were in a chain along the middle channel. They could retain a significant amount of stored water and provide a stable irrigation source for the settlements: the Babish-Mulla fortress and a number of unfortified settlements which stretched along the central canal, with its source in the last depression 1 km northeast of the site.

The question of this canal origins is rather complicated. The 1957 archaeological-topographical research revealed that the main canal of Babish-Mulla was connected with both the middle (from north) and southern riverbeds of the inner 'delta' starting at Akkyr. Initially, it was assumed that its main source was south, 6 km from Babish-Mulla (Tolstov, Vorobeva and Rapoport 1960:41).

However, the next analysis of archaeological-topographical materials clearly revealed a system of reservoirs on the middle riverbed, forcing us to consider that the northern end, originating in the reservoir basin, was the main source. This deduction does not contradict S. P. Tolstov's idea expressed in 1957 that, using a very slight slope, the irrigators of the Babish-Mulla 'oasis' would have been able to use the middle or southern channels as a water source.

The farming oasis, which was based on the main canal, extended for 4 km from northeast to southwest. During a detailed survey it was found that the complex of sites of the 4th-2nd centuries BCE bearing the name 'Babish-Mulla' included the ruins of the large settlement-fortress of Babish-Mulla I; the ruins of a large burial structure of Babish-Mulla II, located 150 m to its west; a number of unfortified settlements stretched like a chain along the canal (Tolstov 1947c:180; 1948b:57-58; 1949:254; 1952:30-31; 1961a:124-126, 1962a:154-170; Tolstov, Vorobeva and Rapoport 1960:40-59).

The canal, as already mentioned above, had its origin from the channel, 1 Km north-east of Babish-Mulla. It was preserved in the shape of a light *takyr* band 9-10 m wide (see Figure 53.A-B). Initially it ran along the fluvial levee, following the riverbed meanders, and then it turned directly southwest; forming, near the settlement of Babish-Mulla, some short canals branching from it on either sides and several irrigation ditches supplying the fortress with water. About 1 km from the last mentioned site, the canal changed its direction from southwest toward south and it was traced very clearly for 1 km from there. The bottom of the canal, preserved as a *takyr* band, became 10-11 m wider then it narrowed to 7 m. A darker band 3 m wide was well detected in the middle by *biyurgun* bushes. The central group of settlements was most significant in its large pottery accumulation, and visible layouts of separate dwellings (see Figure 52). There were more than a dozen dwellings or groups of dwellings in this hamlet.

The southern group of settlements on the Babish-Mulla canal started 1.2 km south of the

settlement. The ceramics scattered here belong to seven main clusters. There are characteristics sherds of different pottery vessels, where those manufactured with a potter's wheel prevail. Rough implements made of quartzite, Scythian bronze arrowheads, dated back mainly to the 4th-2nd centuries BCE were also found there (Tolstov, Vorobeva and Rapoport 1960:44).

There were many traces of small branches from the main canal near these settlements. The majority had a peculiar typical configuration (see Figure 53.A). In its lower part the canal branched out almost at a right angle into two canals, and each of them branched at right angles into a series of small irrigation ditches supplying water to fields. As already mentioned above (see p. 170), this irrigation system with a 'sub-rectangular' configuration is typical of the early stages of irrigation development. They were found in Khorezm in the Tazabagyab period (mid-2nd millennium BCE) and in the Archaic systems of the 7th-5th centuries BCE. The principle of irrigation system construction with right angle branches from the main riverbed was fairly widespread in different countries of the ancient Orient (see p. 160).

The head works of distributors, as a rule, had two or three heads. For example, a small system head structure northeast of Babish-Mulla, started in the riverbed with two sources (*poisk* 261). A 1 m wide band was clearly distinguishable on the ground by the soil dark color and *biyurgun* bushes. Ten meters from the riverbed they joined into one wider line, representing the remains of the bottom of the ditch, traced for a distance of 50 m. In the lower part of the canal no embankments were preserved, but in the depressions pottery fragments of the 4th-2nd centuries BCE were found.

Another, adjacent, significant farming oasis of the 4th-2nd centuries BCE, in the shape of a large number of single settlements and irrigation works was found by the field team during the 1957 trip from Chirik-Rabat to Babish-Mulla (Tolstov, Vorobeva and Rapoport 1960:41). This southwestern group of sites was located 11-20 km southwest of Babish-Mulla on a latitudinal part of the southern channel, revealed now by a flat bank 60-80 m wide, towering over the area for 90-100 cm in height. On the bank slopes and on lateral branches up to 20 separate settlements, with a preserved accumulation of ceramics were recorded. The distance between the settlements was 40-800 m and the total length of the oasis was 9 km. In some of the settlements, low mounds (30-50 cm high) were visible: they were the remains of dwellings with basements and room layouts. Near the settlements there were traces of small ditches branching from the diked riverbed. Canals were traced on the *takyr* light surface as a band of darker soil or accumulations of sand and regular rows of *saxsul* and *biyurgun*.

At one of the settlements, the eastern one (*poisk* 244), we recorded the upper head sections of four canals, the most extensive of which was preserved for 2 m and the distance between the four canals was between 6.6 m and 8 m. They branched from a diked riverbed. In the center there was a canal bed in the shape of a light *takyr* band, 5 m in size and dotted with a few pottery fragments. The settlement, which was preserved as a large accumulation of pottery fragments, was located east of the ditches.

Numerous fragments of wheel-made and hand-made pottery, fragments of quartzite tools, three-winged Scythian arrowheads (dated to the 4th-2nd centuries BCE), millstones, ceramic and iron slag and ingots were collected on the southwestern group sites.

As mentioned above (see p. 219), the northern channels of the Babish-Mulla Delta, supplied a series of canals with water, occupied the inter-ridge depressions of mainly north-south direction (see Figure 52). One of these canals, the most western, had a reservoir similar to Babish-Mulla (*poisk* 305-306). The canal length was approximately 8 km long. It had its origin in the northern channel of the delta and it joined the former riverbed, adapted as a reservoir 300 m long, by three canals (Figure 54). The canal sources with three canals, two of which were the widest (8 and 6 m), starting in the western edge

of the hollow (basin), and the third branched far upstream, directly from the channel, and supplying water to the reservoir.

The ancient irrigation ditches were preserved on the modern surface as *takyr* bands lighter than the surrounding areas; these lines converge together in the main canal bed 15 m from the hollow. The latter is 10 m wide and on the highest point of a flat bank, which is now 1 m over the surrounding area. This bank represents the remains of an ancient riverbed used for irrigation. In its middle, a *takyr* strip, 8-10 m wide in the upper part and 4-5 m in the lower part, is preserved. This marked the canal bed. In its lower part, small ditches branched out, forming several local systems, well visible on the surface and from aerial photographs. Adjacent to it, most of the pottery clusters were found (remains of non preserved sites). The largest site occupied a territory of 100×150 m in size (*poisk* 310).

Other major canals were discovered north and northeast of Babish-Mulla. One of them, starting 5 km north of the fortress, was approximately 8 km long (see Figure 52). The main watercourse that fed this canal was the northern channel, starting at the ancient Janydarya near the Akkyr elevation. The riverbed was preserved as a very flat bank, poorly marked by relief and vegetation. It was 175 m wide. The canal began with three ditches barely visible on the ground, in the form of light meandering of *takyr* (*poisk* 324). These bands merged into one 3.5-4 m wide 30 m from the riverbed. These are traces that the main canal ran along the riverbed.

About 5 km from the riverbed sources, the canal makes a very sharp bend and changes its general direction from east-west to north-south (*poisk* 319). The bottom of the canal was preserved as a light *takyr* band 3.5-4 m wide. Three kilometer north of the bend, the central *takyr* band broadened to 6 m.

The majority of irrigation ditches branched right from the main canal. They supplied water to fields, which were preserved as diked quadrangular plots revealed many times by vegetation. The central canal area of one such site was surveyed by the archaeological-topographical unit near a location, on a steep bend, south of the river levee (*poisk* 326). The site was preserved as a small cluster of ceramics visible on the *takyr* among hills covered with *biyurgun*. On the *takyr* below the settlement, traces of fields and a small irrigation network were found. The fields were small in size, of irregular, pseudo-trapezoidal shape with side perimeters 18 × 12 × 10 × 18 m in size. The ditches watering them were approximately 1 m wide.

Remains of agro-irrigation layouts were recorded in several places. At *poisk* 356, near the canal, traces of fields in the shape of rectangular layouts with sides measuring 11 × 6 m, 10 × 4 m, etc. were found. They began directly from the canal and were bordered on the other side by irrigation ditches running parallel to the canal (see Figure 52). The majority of sites and accumulations of rough ceramics of the 4th-2nd centuries BCE were found near the main canal. The largest settlement was found in the lower part of the canal (*poisk* 350). An accumulation of pottery, over an area of 300 × 50 m, stretched along the canal and sand dunes from north to south. The layout of dwellings could not be traced, but many fireplace stones were preserved.

The remains of ancient irrigation works of the 4th-2nd centuries BCE were also discovered far northeast of Babish-Mulla, in particular on the channel originating from the Janydarya, in the environs of the Medieval site of Djend. In the lower reaches of this channel, north of Akkyr, the riverbed was marked on the ground by a flat bank 100 m in width and 1.2 m high over the surrounding area. In the center of the bank there was a *takyr* band 30 m wide without vegetation (Figure 55.A). Traces of small irrigation networks were visible on both bank sides. The largest lateral branch was approximately 12 m wide (*poisk* 426).

The canal derived from the diked riverbed at a right angle. Small mounds 3-5 cm high above the banks and traces of the ditch bed, in the form of light *takyr* band 3 m wide, were preserved (Figure 55.D). There was an accumulation of sand, *saxaul* bushes and *biyurgun* on the banks, which were

dotted with ceramics of the 4th-2nd centuries BCE. As a rule, the canals branched off at right angles and had 'sub-rectangular' branches. For example, at *poisk* 402, 16 km northeast of Babish-Mulla, traces of a ditch 1.5 m in size, were clearly visible on the *takyr* as a light clay band without vegetation. Lateral branches diverted from the ditch (as seen in Figure 55.B) at right angles, mainly in one direction. Attention is drawn to the unclear traces of the canal broadening, indicating the possible presence of holes, whose purpose was not entirely clear. Perhaps the water was raised using primitive water lifting devices (leather baskets, shovels, etc.; see also Forbes 1955, Vol. II:30-31).

The archaeological-topographical research in the environs of Babish-Mulla revealed traces of 150 sites, among which about 100 could apparently have existed simultaneously. Assuming an average of 20-30 inhabitants/ village, the farming oasis could have been populated by 2,500-3,500 people (including the population of the Babish-Mulla settlement, which scarcely exceeded 500 people).

In this oasis there were no large watering systems and irrigation was based solely on the high-water of the Janydarya, accumulated in specially adapted reservoirs of former deltaic riverbeds and small and short ditches. Measurement of these systems, by the archaeological-topographical surveys, enables us to determine the general volume of excavated soil equal to 150,000-200,000 m³. In order to achieve this, the labor of 1,500-2,000 diggers for 50 days (at a rate of 3 m³ per day) was necessary. This would have exceeded the labor resources of the local population (800-1,125 working men). However, the local force would have been sufficient for the annual cleaning and irrigation maintenance. Such work could have been done by 500-800 diggers. For one hectare of irrigated land, on average 100-120 working days were spent.

In the environs of Babish-Mulla, traces of irrigation ditches, fields and settlements were recorded over an area of 10,000 ha. If 500 ha (5% of the entire territory) were cultivated simultaneously, the annual millet harvest (1,500 kg/ha) would have provided 750 tons of grain. It means that 5,000 people could have been fed from one harvest to another (at a daily rate of 400 gr of millet per person). This calculation enables us to conclude that the harvest was sufficient to feed the whole population of the oasis.

Irrigation in the Environs of Chirik-Rabat
South of Babish-Mulla and the Janydarya riverbed and up to the Kyzylkum sand massifs the plain is occupied by meandering latitudinal channels of the Middle Inkardarya. Based on vertical aerial photographs, particular *takyr* light spots alternate with dark-gray patches overgrown with camel thorns, and cut across by very few sand ridges. A series of less massive riverbeds, which in turn branched out into smaller arms, derived from the main course of the Janydarya. In the past, these arms watered numerous irrigation systems visible in aerial photographs and detectable by the vegetation marking them.

East of the Chirik-Rabat ruins, the ancient capital of the Massaget union or Saka tribes,[Note 180] irrigation systems, in close proximity to the settlement, were based on a lateral channel of the Janydarya adapted for irrigation. It started south of Akkyr and had a general southwest direction, parallel to the main riverbed; its overall width was approximately 60 m between banks (Figure 55.E); its slopes were dotted with ceramics and overgrown with *saxaul*. Apparently, in many places the lateral channels were strengthened by embankments.

The majority of canals branched to the right and were preserved as flat banks 30-40 m in size, with clearly marked ridges and a depression between them 12-15 m wide (Figure 55.G). Thus, 8 km northeast of Chirik-Rabat (*poisk* 521), the canal was preserved as a flat bank 15 m in size and dotted with pottery. Upstream, the canal banks were better preserved; they were 28-30 m wide and 10-12 m between banks.

A canal 30 m wide (*poisk* 524) and 15 m between banks was found 10 km northeast of Chirik-Rabat; individual ceramic clusters of the 4th-2nd centuries were found on the canal. Most

of the settlements were located along the canal in the form of significant ceramic accumulations, but building layouts were almost invisible (*poisk* 514, 517, 519-531, 533-536, year 1958). Settlements were usually characterized by wheel-made pottery. There were finds of bronze socketed arrowheads (dated to the 4th-2nd centuries BCE).

In the described area, small irrigation networks and agro-irrigation layouts almost did not survive. Among the large irrigation works, except for the canals, of great interest were the reservoirs, obtained from former riverbeds. One such former riverbed 1.5 km long had been surveyed east of Chirik-Rabat (*poisk* 517). As may be seen in its profile (Figure 55.F), the width between the fluvial banks was approximately 20 m. The banks of the former riverbed were strengthened with dikes, whose walls are sometimes preserved. Water was supplied to it by a canal, which started in the already described Janydarya lateral channel. The use of former riverbeds as reservoirs was well known in the environs of Babish-Mulla (see p. 219). On the banks of the former riverbed, accumulations of the already known types of pottery typical of the sites in the 4th-2nd centuries BCE were discovered.

East of Chirik-Rabat, a series of sedentary farming settlements of the 4th-2nd centuries BCE was found in 1959 (*poisk* 12a, 16, 25-36, year 1959). The most significant among them was the area of Balandy, where two burial structures (Balandy 2 and 3) and a small fortification (Balandy 1). Adjacent rural settlements were found and dated on the basis of the 3rd-2nd centuries BCE excavations (see Tolstov, Jdanko and Itina 1963:67; Tolstov 1962a:178-180).

The fortified settlement of Balandy 1 was the center of a small agricultural oasis with a total area of 150-200 ha (see Tolstov, Jdanko and Itina 1963:68:fig. 32). The settlement was located on the lower reaches of a branch of the Inkardarya Delta, preserved as a clay bank 50-60 cm high above the *takyr* (Figure 56).

The canal irrigating the environs of the Balandy 1 settlement began 1.5 km from it in the above mentioned east-west channel. Initially, it followed the curves of the riverbed over 1 km, straightening abrupt bends, then it branched out into several minor irrigation ditches, approaching the site from northeast. In its upper part the canal was expressed by a flat bank and a *takyr* strip more than 10 m wide. In many places remains of banks were preserved as small mounds 30-50 cm high, with a distance between banks no more than 3.5 m. In the surrounding settlements the canal was washed away and indicated only by alluvial sand and rare *saxaul* bushes. The territory of the site had a general area of approximately 8 ha and was bounded by a large sand dune in the southeast and a vast *takyr* northwest. It was crossed by a network of narrow ditches covered with clusters of ceramics, and in some places the rectangular layout of dwellings (6 × 10 m, 8 × 10, etc.) was visible. The number of individual dwellings surrounding the fortification of Balandy was more than two dozen. Assuming that a family with an average of 7-10 people lived in each dwelling, then the total number of village inhabitants (together with site inhabitants) could have been 200-300 people. Among them there would have been 70-100 working men. In order to maintain a canal 5 km long, with a cross-section of 4 m^2, it was necessary to remove 20,000 m^3 of soil, which would have been possible for 200 diggers working 50 days. Thus the local labor resources to construct the irrigation system were not sufficient. However the canal cleaning could have been carried out by local labor; it required the work of 60 men for 50 days. The whole territory occupied by canals and fields measured about 150-200 ha, 15-20 ha of which were cultivated. It was possible to harvest 22.5-30 tons of millet (with a yield of 1,500 kg/ ha) from this area. These grains could feed the entire population throughout the year.

* * *

Concluding the description of ancient irrigation works in the Lower Syrdarya, it should be noted that the main problem of stability of the source of irrigation was solved here quite differently than in the Lower Amudarya. On the Amudarya, the main channels of the Southern Khorezm were enclosed by protective embankments and the flood water regulated by a complex system of dikes and head works already in the second half of the 1st millennium BCE. In the Syrdarya Delta, inhabited by the 'barbarian' Saka-Massaget tribes, more primitive regulation methods for flood management were used, however based on a good knowledge of local environmental conditions. At that time, there was a rather simple scheme of irrigation: riverbed - former riverbed (reservoir) - feeder - field. Irrigation was of the estuary-lake type, since the water was drawn from the former riverbeds during flood periods.

The ancient inhabitants (semi-pastoralists and farmers) adapted the deltaic fading channels and former riverbeds (like the Tazabagyab population in the third quarter of the 2nd millennium BCE in the southern delta of the Akchadarya). They also connected neighboring (usually east-west) channels by canals. They alternately used one or the other canal to maintain the high water level. Most interestingly, they used a very clever system of deepening some sections of former riverbeds (connected with each other by narrow canals) as reservoir-basins, allowing them to maintain the water level required to irrigate fields. As well known, water regulation of canals using reservoirs was one of the most widespread methods of irrigated agriculture in ancient oriental civilizations (see pp. 106, 116). R. J. Forbes, describing the irrigation of ancient Mesopotamia, reported that in cases where the flood did not reach the required level, water was used from reservoirs, often arranged in natural depressions on the desert border (Forbes 1955, Vol. II:4). The inhabitants of ancient Sumer, engaged in farming along the banks of deltaic channels, constructed a fairly complex system of dikes, weirs and reservoirs, where the water level accumulated during floods was maintained throughout the season (Forbes 1955, Vol. II:17).

In his study of Sumerian-Babylonian mathematics, A. A. Vayman (1961b:240-241)[Editor 65] produced data on the size of some artificial reservoirs based on cuneiform texts of the 2nd millennium BCE. One of the reservoirs was described with a square configuration with a side length of 180 m (30 GAR, 1 GAR = 6 m). Its size was 32,400 m² and 2 to 1 m deep. The total volume was 30,000-35,000 m³. Another reservoir had a stepped shape and the total volume was 10,800 m³.

Within the limits of Central Asia, the ancient reservoirs were discovered in Southern Turkmenia. For example, in 1962, G. N. Lisitsyna studied a reservoir in the Tedjen Delta (not far from Mullali-Depe) dated to the second half of the 4th-first third of the 3rd millennium BCE. Its capacity was estimated as 2,625 m³ (Lisitsyna 1965:100-102). Similarly, the fading and former riverbeds were adapted by the inhabitants of the basin of the Middle Janydarya.

Irrigation Works in the Djety-asar District (First centuries BCE-9th century CE)

North of the dry riverbeds of the Janydarya and the Kuvandarya there was a vast lowland area with numerous ruins of large fortified settlements called by the local Kazakh population 'Djety-asar' meaning 'seven cities' or 'seven settlements', even if the ruins were more: several tens. The results of research and excavations (at Altyn-asar, Djety-asar 9, etc.) were given in detail in many publications of the Khorezm Expedition (see Tolstov 1947a; 1948b:125-140; 1949:246-254; 1950a:521-531; 1952a:16-19; 1954:258-262; 1958:235-252; 1962a:186-198; Senigova 1953; Levina 1966, 1967).

The excavations of the 'Big House' of Altyn-asar revealed that the whole cultural stratigraphy was divided into two main horizons: a lower one, or 'grinding stones horizon', and an upper one, or 'millstones horizon'. The lower horizon was dated from the first centuries CE to the 4th century CE; the upper horizon to the 4th-7th centuries, and perhaps, the beginning of the 8th century CE (Tolstov 1949:241; 1952a:18; Levina 1966:54, 69).

The surface ceramic collected in the other settlements of Djety-asar were studied by L. M. Levina and showed that the sites were of different periods, although belonging to the same archaeological culture. Baybolat-asar (Site 14), Djety-asar 1, Kara-asar, Dolomak-asar, probably, Tompak-asar and also Bidaik-asar and Karak-asar (Figure 57) were related to the earliest Asar period. Rabensay, Djety-asar 8 and Djety-asar 11 provided ceramic material analogous to the upper horizons of the 'Big House' and dated to the 6th-7th, and probably, 8th centuries. Some settlements (for example, Djety-asar 11) were populated at a later period, in the 9th-10th centuries (Levina 1966:78). Thus the Djety-asar agricultural oasis existed over a long time from the first century CE to the early Middle Ages. Its inhabitants could be considered the people of 'Kanga' (the Kengeres, the Kangars lived in the Lower Yaxarte, according to J. Marquart) as part of the united Pecheneg tribes.

The archaeological study of the Djety-asar settlements enabled S. P. Tolstov to identify the main features of the material culture and economy of the Djety-asar inhabitants. They continued the traditions of Bronze Age primitive complex economy and preserved the semi-settled lifestyle of the other semi-nomadic Saka-Massaget tribes (Tolstov 1948b:128; 1952a:19-21; 1962a:186-195).

This culture was characterized by large fortified communal houses built with large size mud bricks, whose walls included the bones of domestic cattle and horses. In these dwellings we found pits for economic purposes and filled with millet and barley, millstones, fishhooks and fish net sinkers (Tolstov 1952a:19-21; 1961a:127, 143). Herding played a major role in the economy. Cattle, horses and camels were 35.4% of the herd at Altyn-asar (Tolstov 1952a:19:note 1). Along with herding, fishing and hunting, the inhabitants of Djety-asar were engaged in irrigated agriculture, as evidenced by numerous finds of barley and millet, the abundance of grinding stones and millstones in excavated rooms, and also by remains of a variety of irrigation facilities.

Most of the riverbeds in the Djety-asar area were preserved as gentle banks 0.5-1.2 m high above the plain and reaching 30 to 100 m wide, with traces of canals in the central part as a *takyr* band 5-10 m wide. The lower part of this riverbed was surveyed in 1958 by the archaeological-topographical unit on the edge of the southwestern part of the Djety-asar Oasis, located 0.5 km east of Bidaik-asar I. A flat bank 60 m wide with typical lines of banks had a north–south direction and, in its central part, a clearly visible *takyr* strip 10 m wide was all that remained of the canal (see Figure 51.G). Darker, narrow bands 1 m wide were traces of small ditches branching from the *takyr* strip. In one of them was found a millstone fragment made in sandstone.

The irrigation in the environs of Bidaik-asar I-II was carried out in the same way as in the environs of Babish-Mulla on the Janydarya, with the use of natural, highly branched and meandering riverbeds and former riverbeds adapted for irrigation. The riverbed west of Tompak-asar (*poisk* 461), with an overall width of 60 m, had a central and more elevated part 30 m wide, with typical banks and the preserved canal was a narrow *takyr* strip. On the slopes of the fluvial banks there was much pottery similar to the ceramics of Bidaik-asar I and II. In many places on the riverbed, short (40-50 m) ditches, expressed by soil color and vegetation, departed from the riverbed 1-1.2 m high above the *takyr*. As a rule, ditches departed from the riverbed at a right angle. At *poisk* 462, such a ditch was traced for a distance of 50 m. It is expressed on the ground as a dark strip 1 m wide, along which bushes of *saxaul* and *biyurgun* grow. Another ditch was located 3-5 m from it.

East of Tompak-asar we surveyed a small embanked riverbed, 20-22 m wide, which was preserved in the shape of a flat bank overgrown with *biyurgun*, with a central narrow *takyr* strip dotted with rare fragments of Asar type pottery (*poisk* 464). In the middle of the band there was another dark line covered with bushes, the remains of a narrow ditch 1 m wide of the latest period of the system functioning. Near this ditch, an intact jar with dark-gray slip, and a gray ware beaker were found. These vessels were very roughly manufactured and badly burnt on

the bottom; according to L. M. Levina, they are similar to the upper horizon pottery of the 'Big House'. The riverbed started 2 km northeast of Tompak-asar and it fed a series of small irrigation systems. One of the ditches supplied water to the former riverbed in the depression surrounding Tompak-asar, and also served as a kind of natural water-barrier for defensive purposes.

The settlement of Tompak-asar had an oval-shaped 'platform' and a fortified central residential area with a rectangular configuration 10 m high above ground. In its northern part there was an inner courtyard enclosed by a high wall. The platform and the central part were made of rough bricks of different sizes. The mud bricks alternate with blocks of *pakhsa*. These finds date Tompak-asar, and the settlements in its environs, to the first centuries CE (Levina 1966:78). To the same period was also dated Djety-asar 1, investigated by the Khorezm Expedition in 1946 and located 10 km east of Tompak-asar (Tolstov 1948b:128-130).

In the environs of Djety-asar 1, remains of irrigation works, in particular a ditch branching from the riverbed at a right angle (*poisk* 467), were recorded. The ditch was preserved as a dark strip of vegetation 2 m wide, with rectangular branches in its lower part. Two significant ceramic clusters and small mounds indicated the remains of a dwelling (see Figure 51.G). The whole space around the ditch was dotted with small pottery fragments. Weak traces of quadrangular fields were preserved in the areas without vegetation. The ditch had its origin from the riverbed and, very interestingly, it served as a reservoir-basin. The north–south part of this riverbed preserved traces of an earthen embankment in the shape of mounds along the banks. The hollow was 22-25 m wide. By its size, the diked riverbed was similar to the Antique main canals of Khorezm.

Vertical aerial photography[Note 181] of single areas of Djety-asar showed that former riverbeds, and their parts adapted as reservoirs, were widespread over the whole territory of the Djety-asar culture (see Figure 57). As can be seen in the plan of the environs of Tompak-asar, several canals supplying water to former-riverbeds, now

Lower Syr Darya [handwritten annotation in left margin]

basins, derived from the main east–west channel. From them, branched small irrigation systems from several hundred meters to a kilometer long.

The plan of Altyn-asar environs revealed a very complex picture (Figure 58), with many reconstructions of irrigation canals, a gradual improvement of the system of reservoirs and an increase in their numbers. It should be reminded that the life of this settlement lasted almost a thousand years: from the first centuries BCE to the 7th–8th centuries CE. The huge trapezoidal fortification covering 16 ha was constructed at the confluence of two canals: the main, with an east–west direction, and another coming from the northeast (see Tolstov 1952a:16:fig. 4).[Note 182] The flood waters of these channels were regulated by dikes, but when the water was not sufficient, individual links of these channels and former riverbeds were turned into reservoirs, from which small canals derived.

The environs of another large fortified site of the Djety-asar culture, i.e. Rabensay (Djety-asar 4), was also an example of irrigation combining small main canals drawing water from a large channel of the Kuvandarya through a system of former riverbeds and reservoirs. The site, constructed on a sharp bend of the river and surrounded by depressions, could be clearly seen on aerial photographs (see Tolstov 1948b:132:fig. 33). The fortress controlled approaches to the head works of the irrigation canals near it. Their lower parts watered two former riverbeds connected by canals and used as reservoir-basins. The small agro-irrigation layouts are barely preserved.

Similar irrigation was found in the environs of Djety-asar 8 (see Figure 58), where, together with the main irrigation canals (C), the former riverbed (A), was supplied with water from the main riverbed. As can be seen in plan, from the former riverbed departed small lateral irrigation systems with sparse 'rectangular' branches. The collector served as a canal entering into a large depression (E).

Djety-Asar 8 is a large strongly developed fortification but without residential buildings. It was a cattle enclosure. The fortress controlled the head works of a small irrigation system derived from the riverbed at a right angle. This system rounded the site southwest and the northeast walls, and then flowed northwest, with short 'sub-rectangular' branches directed mainly to the right side of the riverbed. The main canal (C) was 1.5 km long. In its lower part there was a depression (E) with traces of flooding, where the water of irrigation canals was discharged. The irrigation system at Djety-asar 8 had Archaic quadrangular shapes of early stages of irrigation development, and resembled the Archaic Khorezm systems of the 6th–5th centuries BCE.

The archaeological-topographical study of irrigation facilities at Djety-asar proved that the primitive methods of flood water regulation, known in the irrigation of the environs of Babish-Mulla and Chirik-Rabat (4th–2nd centuries BCE), were further developed in this region in the 1st millennium CE. Features of this 'deltaic' irrigation were the use of diked riverbeds, canal connections of east–west riverbeds, and the use of deepened parts of fading channel systems and former riverbed as reservoir-basins.

The natural conditions of the Lower and Middle Syrdarya were characterized by continuous changes of single channel hydrographic regimes such as: the level of water rising during freshets in the warmest season associated with the increased snow and ice melting in the mountains; or when ice blocked channels in winter. During this period the water widely overflowed from the main riverbed into the nearest depressions or former riverbeds. In the 10th century al-Masudi reported that in Farab the river flooded an area of 30 *farsakh*, i.e. 200–300 km (Bartold 1963, Tom I:234). From the general western sloping of the Djety-asar plain, the majority of freshet floods flowed west, thus the ancient farmers had only to spend relatively little effort to direct the water by ditches into former beds and riverbeds which could become reservoir-basins. Interestingly, the method employed by the farmers using natural flooding in the Lower Syrdarya clearly existed till the 18th–19th centuries. In particular N. Dingelshtedt reported this fact. Describing traditional irrigation methods of the Kazakhs, he noted that high-water was gradually drained from the main river through a series of diked places and depressions for a distance of 60 *verst* from the main river (Dingelshtedt 1893:369).

The most widespread system of reservoir-basins, apparently took place in the late period of the Djety-asar Oasis, when the general volume of water in the main channels of pra-Kuvandarya (the Northern Kuvandarya) was greatly reduced.

Irrigation Works in the Environs of Barak-tam (4th-beginning of 5th centuries CE)

West of the ancient alluvial plain, where the ancient channels of the Syrdarya merged with the Akchadarya, another cradle of irrigated-farming culture of the steppe tribes of the Aral Sea area of the mid-1st millennium CE was discovered. It was located near the ruins of Barak-tam, and, according to S. P. Tolstov, in the headquarters of one of the chiefs (rulers) of the Chionite-Hephtalite tribes, semi-settled pastoralists (Tolstov 1959a:32). The castle of Barak-tam was studied in 1945, 1948 and excavated in 1956 by a team directed by E. E. Nerazik (Tolstov 1946b:85; 1958:127ff.; 1959a:31-33; Orlov 1952a:135-152; Nerazik and Lapirov-Skoblo 1959). This castle had a very unique architecture, built of mud bricks with sizes and proportions typical of Khorezm in late Antiquity. The complex settlement was dated to the late 4th-early 5th centuries CE (Tolstov 1959a:32; Nerazik and Lapirov-Skoblo 1959:83). The inhabitants of the ancient Barak-tam Oasis were engaged in cattle breeding, as evidenced by the original architecture which used bones of domestic animals as wedges in the construction of vaults. Along with herding, irrigated farming was also extensively practiced.

In 1956 the archaeological-topographical unit discovered and researched approximately fifteen farm-houses,[Note 183] preserved in the form of mounds with traces of layouts, and also remains of irrigation works on one of the old riverbeds of the Akchadarya (Andrianov 1960a:174, note 15). From this riverbed derived several canals, among which the most significant was the main canal irrigating the environs of the Barak-tam I castle. It began 11 km northeast of the site. The instrumental survey carried out at its source by the team revealed the remains of complex head works installed in the riverbed floodplain terrace, having at this point a width of 80 m. Apparently, there were three canal heads, two of them preserved only as double mounds raising 1.5-2 m above the bottom.

The banks of the first structure were 10 and 9 m and the distance between them was 9 m. The second remaining structure was most significant (22 m wide between banks). It was built for the lowest water level. The difference in the levels of water intake between the first and second head constructions was 50-60 cm. The third structure was the highest but almost entirely destroyed; the riverbed was preserved only as a *takyr* strip on the bank and a modern road built on it. The character of these head structures, built taking into account the water intake on three different levels, suggested a relatively highly developed irrigation technique.[Note 184]

The main canal was badly preserved and could only be traced as a *takyr* band by soil and vegetation. About 4 km from the riverbed, trace of the canal turns towards the Barak-tam castle, in whose environs were recorded traces of small irrigation systems, small canals, irrigated quadrangular plots, and remains of *gryad* gourd plantations 3-4.5 m wide (see Orlov 1952a:139:fig. 3).

The second significant main canal was found in the environs of Barak-tam, 1 km south of the above described head structure. Its source was hidden under a huge sand dune cutting across the riverbed. Further south, sand covered the canal for a few hundred meters, and only where it sharply changed its direction, the washed away banks were visible on the *takyr* as a double line of vegetation and small mounds. The distance between banks was probably 10-11 m.

Near the canal a significant accumulation of ceramics and remains of dwellings were found. A number of farm-houses, in the form of protruding mounds almost razed to the ground, were found very close to the main canal head structures, and in the lower part, near the Barak-tam I, II, and III castles. According to E. E. Nerazik, most of these dwellings were dated to the late 4th-5th centuries CE, (Nerazik and Lapirov-Skoblo 1959:81). The irrigation works of the Barak-tam Oasis should probably be dated to the same period.

The question of the main source of irrigation in this region is very interesting. The location of head works, and the direction of canals south and southwest of the riverbed toward the castles of Barak-tam, convincingly demonstrate that the main source was the Akchadarya riverbed. It was not watered by the Amudarya, but by the Syrdarya through the lower channels of the Janydarya. These riverbeds of 'stagnant water' (al-Fahmi) with the ruins of 300 cities and villages are mentioned by Biruni in the late 10th century (Biruni 1966:95).

The irrigation in the environs of Barak-tam existed for a short time. Like the irrigation of Djety-asar, it was rather estuary-lake type, since its source was the semi-stagnant water flowing into the dry riverbeds of the Akchadarya from the Janydarya in the period of high water. The irrigation was carried under the scheme: river - former riverbed - main canal - distributor - feeder - field.

Medieval Irrigation Works (9th-16th Centuries)

The ruins of Medieval sedentary farming settlements of the 11th-16th centuries, and the irrigation facilities related to them, are fairly widespread along the Syrdarya. Between the 1st and 2nd millennia this area became the scene of important historical events. In the 8th-9th centuries, in the lower and middle reaches of the river, the local Pechenegs tribes (the Kengeres) struggled with the Oghuz, Karluk and Kimak tribes which perhaps were pressing them, as al-Masudi wrote (see MITT 1939:166). [Editor 66] A part of the Pechenegs tribes migrated west (Marquart 1903:60-78) while others were assimilated with the Oghuz (Tolstov 1947b:84-90; Jdanko 1950:108-109). According to al-Idrisi,

"the cities of the Oghuz are numerous, they stretch one by one from north to east" (MITT 1939:220).

In the second half of the 10th century the barbaric state of the Oghuz tribes was established and broadened its boundaries. Southeast to Taraz and Shash; southwest to the delta of the Amudarya and the outskirts of the Khorezm Oasis; northwest to the Ural foothills. However, in the early 11th century, after defeat in the struggle with Khorezm, the state fell into deep political and social crisis. According to S. P. Toltsov, this was the result of conflicts between the large cattle-owning nobility and members of the communes, the *yatukes*, preserving the natural herding-farming economy (Tolstov 1947b:100-102; year 1947). The nomadic aristocracy, headed by the Seljuks and supported by a group of the Oghuz tribes, interested both in broadening their pastures and predatory raids. They migrated southwest toward Bukhara, Khorezm, Khorasan (where in 1034 CE they subdued the state of the Ghaznavids), and finally to Asia Minor (Yakubovskiy 1947; Tolstov 1950b; Roslyakov 1951).[Editor 67] The Oguz tribes remaining on the Syrdarya were headed by a political enemy of the Seljuks (the son of the Oghuz Ali Yabgu), Shah Malik the ruler of Djend, who even conquered Khorezm for a short time (1041-1043 CE) (Bartold 1963, Tom I:365).

In the 12th century, the political power of Khorezm was strengthened during the reign of Atsiz b. Muhammad (1127-1156 CE), and Djend became a possession of the Khorezm state. After a prolonged period of frequent wars and civil strife in the relations between the nomadic-pastoral 'barbaric' periphery (which was the territory of the Lower Syrdarya) and Khorezm (which in the 12th-early 13th centuries became the most extensive and mightiest state in the whole Near and Middle East) there was a period of peaceful trade and cultural relationships (Tolstov 1958b:274-289; Horst 1964:1-6).

This was the time of economic and cultural heyday of the Khorezm state, coinciding with the rapid broadening of the cultivated-irrigated territories along the Janydarya (al-Fahmi) in the 11th century. There, at the end of the 10th century, al-Biruni noted the ruins of 300 cities and villages (Biruni 1966:95; Toltsov 1962a:274; Tolstov, Jdanko and Itina 1963:82-83).[Note 185] The strip of cultural oases of the 12th-early 13th centuries, stretching from Khorezm to Farab on the Middle Syrdarya, restored for some time the disrupted (late 1st millennium) union of historical-cultural areas in the north territory between the great Central Asian rivers: Kangkha - Kanga - Kangju.

The archaeological-topographical work in the Janydarya basin identified a series of preserved Medieval pre-Mongol irrigation on the banks of the Lower and Middle Janydarya. The research showed that the most significant areas were located both in the Lower Janydarya and east of the well-known (in sources) Sag-Dere. It was situated 20 *farsakh* (120-140 km) away in the city of Djend (Jan-kala settlement) and further in the Upper Janydarya, up to the ruins of Asanas (or Ashnas or Eshnas), on the border of the modern cultural oasis of the Syrdarya (Tolstov 1948b:60-61).

Irrigation in the Environs of 'Swamp Settlements'
Medieval canals and ditches were also found on the shore of the Aral Sea in the environs of the so-called 'swamp settlements'. The largest of such sites was Kesken-Kuyuk-kala, whose rounded contours resembled those of Asar. This settlement was 500 × 700 m and stretched from east to west, its citadel with a rectangular layout differed for the irregular buildings made of mud bricks, similar in size to the Afrigid ones. The excavations of the Khorezm Expedition in 1963 (directed by B. I. Vaynberg) revealed that the citadel massive upper cultural layer dated to the 7th- 9th centuries.[Note 186] According to S. P. Tolstov, the history of swamp settlements ended in the 10th-11th centuries.

One of the swamp settlements was Djankent (Yangikent, Dekh-i-Nau, al-Karyat-al-Khadisa) which preserved its Early Medieval Turkic name and was the capital of the Oghuz in the 10th-11th centuries (Tolstov 1947b:56; Yakubovskiy 1947:49). The inhabitants of swamp settlements had a natural mixed economy, combining a sedentary life with semi-nomadic deltaic irrigated farming (millet), herding and fishing. Some literary volumes give us a representation of their pure nomadic lifestyle.[Note 187]

The archaeological-topographical work in the environs of swamp settlements revealed a highly developed irrigation network based on the deltaic channels of the Syrdarya, as well as farming in unfortified settlements. North of Kesken-Kuyuk-kala, we discovered traces of ditches in the form of dark bands of vegetation 1-2 m wide (*poisk* 5 and 6, year 1963), and ceramics found there were very similar to those from the settlement of the 7th- 9th centuries. Elsewhere (*poisk* 8, year 1963), canals were traced along riparian deltaic channels, preserved as a flat wall 20-25 m wide. Traces of two canals 4-5 m wide, indicated by a small depression, darker soil and vegetation color, were clearly visible. A few ceramic finds, similar to those of Kesken-Kuyuk-kala, indicated that the main part of the irrigation facilities existed simultaneously with the settlement, though some canals continued functioning in later periods. This confirmed the existence of the big settlement of Kesken-Kuyuk 2 also in later periods.

This settlement was located 2.2 km west of the Kesken-Kuyuk-kala site, on the bank of a small riverbed, expressed as a depression several meters wide. It stretched 300 m along a ditch, detached from the channel, which had here a north-south direction. The ditch was 5-6 m wide, and its typical profile had clear traces of later artificial deepening. Along with the layout of residential buildings, remains of ceramic production (ceramic, glass slag and kiln hillocks) were found at the settlement. Among the findings, the gray ware or the grey pottery prevailed, such as jars with rounded handles of the 12-13th centuries, dark gray plates and bowls with a polished surface. All these finds suggested that the irrigation network in this region existed until the 12-13th centuries. Later, apparently in the 18th century, it was restored for a short period of time by the Karakalpaks and then by the Kazakhs (Andrianov 1952a:570).

Irrigation on the Janydarya

Traces of a vast area of Medieval irrigation, in the southwest delta of the Syrdarya and in the Lower Janydarya, especially along the main riverbed of the Janydarya, were covered by later Karakalpak irrigation. At the southwestern edge of the Syrdarya Delta, on the Lower Janydarya, remains of a Medieval irrigation, widespread on vast areas, mainly along the Janydarya main riverbed, seem to have been covered by the later Karakalpak irrigation. Some Medieval settlements were also discovered far from the Janydarya, on its lateral channels: at Ak-Mambet, south of the *mazar* on the Zangar plateau; north of Aralbay-kala, and at Murzaly.[Note 188]

A group of *mazar* and the Ak-Mambet village cemetery are located on a Medieval settlement (*poisk* 67), with the remaining buildings preserved in the shape of mounds. In the environs there are traces of a branched irrigation network beginning in the Janydarya main riverbed. The majority of canals (five in number) were less than 1 km long and 2-3 m wide. On the fields, and especially in the ruins of buildings there were many ceramics dated by N. N. Vakturskaya to the 13th-14th centuries. South of the plateau and the *mazar* of Zangar, groups of Medieval sites dated to the 11th-14th centuries were discovered. In one of them (*poisk* 78) it was possible to identify traces of gourd plantations with dimension of 15 × 25 m, and *gryad* 3.2 m wide. The gourd plantations and fields around the settlements were irrigated by a small canal 5 m wide (2.5 m between banks).

Of great interest were the rural settlements of the Khorezmshah period in Murzaly (*poisk* 493-498), located on the lower part of a meandering riverbed 60 m wide. This riverbed was expressed on the ground by a flat bank, raised 50-60 cm high above the *takyr*, and used as the main canal. Its banks were strengthened by dikes and a canal, 5-7 m wide, was constructed on its bottom. Numerous small canals branched from the diked riverbed mainly on the right side. The settlement began in a place where a large sand dune was close to the canal on the north (Figure 59). Many fragments of glazed, unglazed, and gray pottery,

dated by N. I. Vakturskaya to the 12th-13th centuries, were found on the site ruins.

In 1946, east of Murzaly and close to the Karakalpak farm of Orunbay-kala (18th- early 19th centuries), a large Medieval fortification (100 × 300 m) was discovered. It had double walls, many oval-shaped towers, and a system of moats connected to the riverbed from the north. The settlement was surveyed under the direction of S. P. Toltsov and was called Beshtam-kala (Tolstov, Vorobeva and Rapoport 1960:16-18). The lower layers of this settlement were dated to the pre-Mongol period, though, judging by the findings of typical ceramics of the Golden Horde period, they existed also later. There were many late Karakalpak settlements around it.

The environs of Beshtam-kala were covered with a dense irrigation network, in which the survey revealed both earlier and later canals (Andrianov 1960a:187). The main Medieval canal began in the Janydarya riverbed, 100 m east of the settlement (*poisk* 167). It was approximately 12 m wide and 4 m between banks. The canal was diverted from the main riverbed at a height of 2.5 m above the bottom. It bore traces of its later re-deepening, probably Karakalpak, and it was locked by dams in several places. Another canal, drawing water 1 km east, was 11-12 m wide and irrigated a large territory south, where we found the remains of Medieval dwellings in the shape of flat light-gray mounds, placed 100-200 m from each other and dotted with Medieval ceramics. On the banks of the canal traces of swollen *chigir* pits were visible (*poisk* 179).

A significant area of Medieval irrigation was discovered in 1957 in the environs of the large domed *mazar* of Sarly-tam, dated to the 14th-15th centuries (Tolstov 1948b:56-57, fig. 16; 1962a:291-294; Tolstov, Vorobeva and Rapoport 1960:20-22). At that time, apparently, the Janydarya water level had already begun to decrease, and in the 15th-17th centuries the massive diversion dam and dike systems were built 11 km south of the *mazar*, in the tract of Besh-Chongul, where the main riverbed was 80-100 m wide (Figure 60.D). The dam was preserved only in the central part of the riverbed as two massive earthen constructions 80-100 m long and 20-30 m wide. The dam was separated from the bank by two large gullies (20 and 35 m). The eastern gully led to a large hollow 10 m deep, 40 m wide and 140 m long. Irrigation canals branched from the riverbed on both sides above the dam. The dam, apparently, was constructed to raise water in all the canals located in the upper part of the Janydarya, up to the environs of Sarly-tam.

On the bank of one of the irrigation facilities there were ceramics findings, dated by S. P. Tolstov to the 16th-17th centuries. Similar ceramics were found in the irrigation ditches running near the ruins. One of these ditches was 6 m wide and 2.5 m between banks. The largest canal surveyed in this area, 0.5 km east of the dam, was 25 m wide and 15 m between banks (*poisk* 193) (Figure 60.F). Along this canal were the ruins of dwellings and fragments of thick dark gray and black ware, resembling the findings in the environs of Sarly-tam which, perhaps, should be also dated to the 16th-17th centuries.

In its middle reaches, the Janydarya ramified into two channels between Chirik-Rabat and Irkibay. The southern one was blocked by a dam in the Middle Ages. Near this important irrigation center, the head works of several canals irrigating the large territory of Irkibay were constructed, and a significant iron-related production was located here (*poisk* 215) (see Figure 60.A). The river banks were strengthened with a dike, which was best preserved on the right bank in front of a Medieval workshop of metallurgists. The dam had a bent configuration and it was 15-20 m wide and 80 m long. On the left bank there was a complex four canal head structure; the fourth one, the closest to the dam, was blocked by it and the other three were connected by transverse canals. The largest canal was approximately 10 km long, a total width of 15-19 m wide and 9 m between banks (*poisk* 21) (see Figure 60.B). A series of small ditches branched off from this canal. There were many agro-irrigation layouts of different shapes. Of great interest were pit remains for the water-lifting facilities (*chigir*). They were rounded, oval-shaped

or pear-shaped basins, ranging from 7 to 10 m in diameter (see Figure 60.C). The water of these basin supplied fields of different sizes through small ditches. The smallest, a quadrangular *atyz* 4 × 4 m in size, covered a rather large area.

In 1958, in the lower reaches of a canal, 6 km southeast of the Irkibay well, the Expedition field team discovered a large site of the 12th-14th centuries (Tolstov, Jdanko and Itina 1963:83). Here single buildings and farms had stood along a small canal, preserved as small mounds 1-1.5 m high, showing few traces of room and fence layouts, and some ceramic accumulations.

Another significant irrigation center, consisting of water-retaining semi-dams, dams, and several head works, was discovered on the northern channel of the Janydarya, 15 km from Chirik-Rabat near the fortress of Irkibay-kala of the 12-14th centuries; it watered the Medieval irrigation systems on both banks of the river. If we consider its layout, very characteristic was a hydraulic water-retaining structure with a lateral dike on the left bank, a sluice, and a 60 m dike, similar to others in Medieval Orient. For example, in Mesopotamia in the huge Nahrwan canal of the Sassanid period, and in the environs of Samarra of the Abbasid Caliphate (750-1285 CE), according to Ahmed Sousa dams were built in stone and also with large (in the earlier stages) and small size bricks (see Willcocks 1903:10-13:8; Sousa 1948:plan 15) (Figure 61.A-B).

A settlement (*poisk* 509) with ceramics of the 13th-14th centuries was discovered in the environs of Chirik-Rabat, 1 km north of the funerary mausoleum of Chirik 2 of the 4-3rd centuries BCE. Like the settlement at Irkibay, it stretched along an irrigation ditch, but it was poorly preserved. At this settlement, approximately ten ceramic clusters of the 12th-13th centuries, and several hillocks, representing the remains of destroyed buildings, were found. The canal flowing there in the past was 4.5 wide and 2 m between banks. In many places there were irrigated, quadrangular plots enclosed by walls. One of the field was 30 × 30 × 33 × 21 m in size.

The system of ditches, irrigating the environs

of Chirik 2, was fed by the water of a main canal approximately 10 m wide and 6 m between banks. This canal was, apparently, deepened anew in the 18-19th centuries by the Karakalpaks. In the lower parts it is marked by a strip of vegetation 1 m wide. Over the whole territory covered by the irrigation system there were many finds of Medieval ceramics of the 12th - early 13th centuries. The settlement was located on the right bank of the Janydarya, while the ruins of Chirik-Rabat, ancient capital of the Saka tribes, were located on the left bank. At the south-west end of this site, in the 12th century arose a small settlement, known as Sag-Dere in Medieval written sources.[Note 189]

A number of small Medieval irrigation systems were found in 1959 on the Middle Inkardarya, in the section of Bayan, 35-36 km south-east of Chirik-Rabat (Tolstov 1962a:276; Tolstov, Jdanko and Itina 1963:35). Small canals (5-6 m wide between banks) derived from the riverbed of the Inkardarya, which in this area (already known for the irrigation structures of the 'slag kurgans' period) was 80-100 m wide between banks and 1.5-2 m deep in this place. Along the canal banks there were traces of deflated *chigir* pits, some canal branches and accumulations of ceramics of the 10th-12th centuries. A number of similar systems, far distant from each other, were discovered in the upper part of the Inkardarya. The irrigation here was of estuary type, and was based on the slow moving high-water flowing into the Middle Inkardarya from the Upper Inkardarya and the Janydarya.

Moving eastward from Chirik-Rabat, there were increasing ruins of Medieval settlements and more traces of Medieval irrigation. In the environs of the Uygarak mound (towards north and east), the survey teams of the Expedition discovered irrigation systems with a branched configuration, covering a very broad territory. They were predominantly directed east-west, and topographically associated with the east-west parts of former river beds, which, as a rule, crossed the rivers in a north-south direction. The majority of the old Inkardarya channels merged with canals into a single system, and some of them were diked and adapted as reservoir-basins.

A large handicraft settlement of the 12th-14th centuries was discovered in 1959 on one such riverbed, 6 km south of Uygarak (see Tolstov, Jdanko and Itina 1963:83-84, fig. 38). Archaeological excavations, directed by N. N. Vakturskaya, were carried out at the settlement named Uygarak-I. There were compact, but irregular, jumbled house remains, preserved only as basements, hidden under bulging earth mounds.

Close to the dam, the riverbed, 40 m wide between banks, approached the site from northeast. South of the dam the riverbed was filled and crossed by a latitudinal canal, along which were scattered the settlement basements and small walls of the large yurts belonging to the semi-settled farmers and herders of neighboring tribes. As shown by the excavation of N. N. Vakturskaya, Uygarak-I was an urban type settlement. There existed advanced handicraft productions: ironworks, pottery, as indicated by remains of pottery kilns, finds of iron ingots, and copper and pottery slags. Many copper coins (mainly dated to the 14th century), an iron sickle and a jar full of melon seeds were also found. The gourd plantations, fields and numerous small ditches, forming a dense complicated network, were well traced on both sides of the canal near the settlement.

The main canal was 12-15 m wide and 6 m between banks. On the banks there were many slightly covered pits, where the *chigir* was once constructed. Remains of farms of a semi-settled type were found near the fields. These consisted of mobile dwellings, which were preserved in the shape of circular basements and ground walls of yurts, traces of earthen houses and small household rooms. The walls of one of these recognized by us as walls of a yurt were 10-11 m in diameter and the basement was 21 m in diameter. The small earthen walls were preserved for a height of 60-70 cm and had a typical snail-shaped configuration.

Traces of Medieval irrigation, and ruins of

farms and fortifications stretched from Uygarak, along the south bank of the Janydarya, to the margin of the modern cultivated zone.[Note 190] The center of this wide region, populated in the 11th-14th centuries, was Djend (now the ruins of Djana-kala) the extreme northeastern outpost of the Khorezmshah state (Tolstov 1948b:60-61), which was captured by the Mongols in 1220 and turned by them into a Djuchi headquarter. In the 13th century Djend began to decline (Bartold 1965:230).

In the pre-Mongol period there was a significant farming population around Djend. The city neighborhood was irrigated by a large canal (which was 7-8 m wide between banks and 3-4 m deep) crossing the site from east to west (see Figure 51.3). Numerous hollows and traces of pits along the banks suggested that fields and orchards were irrigated using water-lifting devices (*chigir*). Finds of Medieval *chigir* pitchers indicated that *chigir* irrigation had already arrived in the period of this site and its environs. The canal was linked by lateral branches with a complex concentric system of moats, which accompanied the fortified structures of Djend. The main canal derived from the Janydarya, which here was wider than in other places, 11 m between banks and 5-6 m deep. In the center, there were clear traces of later re-deepening related to the Karakalpak period of life of the settlement. Below the settlement the canal was greatly washed away and in some places resembled a natural channel.

The main array of the Medieval irrigation land, as already written above, was located on the left bank of the Janydarya, which, around Djan-kala, turned sharply south and, after several large meanders, turned again in an east–west direction. At this place on the same bank, there was the Medieval fortification of Kum-kala, first surveyed by the Expedition in 1946, when its plan was drawn and fragments of surface material were collected.

The fortress had an irregular rounded layout (resembling Beshtam-kala) with well-preserved heavy walls forming a broken line and fortified by massive towers and a deep moat. Kum-kala controlled the irrigation systems coming from the

river at its walls. Adjacent to the fortification was a large Medieval *rustak* represented by rural farms, fences, and some buildings scattered among fields and gourd plantations for tens of kilometer around. The most significant of these settlements was Kum-kala 2 (Eastern Kum-kala). It was characterized by a combination of residential buildings (with rooms arranged in a row, *aywan*, and inner courtyards) with preserved round flat areas.

Southeast of the site there were three vast rectangular layouts. The largest had an almost regular rhomboid shape with sides 210 × 210 × 210 × 215 m. Two smaller-sized enclosures (200 × 100 m) adjoined each other and were 50 m south of the aforementioned settlement. According to the layout, these were great gardens.

Moving from Djend east and, especially, southeast (there were large settlements linked to the banks of the Inkardarya channel which flowed among huge sand massifs), the overall cultural landscape appearance of irrigated lands changed.

In this region, during the 1959-1962 Expedition field work, a very peculiar group of fortified settlements was discovered. The largest of them were Sarly-tam-kala and Zangar-kala in the west, Khodja-Kazgan I, II, III in the center, and the settlement of Asanas in the east (Tolstov 1962a:276-282; Tolstov, Jdanko and Itina 1963:79).

Many ruins of fortified rural sites, preserved as square layouts enclosed by a wall and moat, were discovered on the banks of channels of the Upper Inkardarya (Suvorov 1955). The settlements were close to the channel, on which abutted the extremities of outer fences with a round or oval shape. Around these settlements there were many small canals drawing water from the Inkardarya channel, as well as from reservoir-basins and dams. All these irrigation works were typical of regions with unstable deltaic agriculture and populated by semi-settled farming-pastoralist tribes. According to S. P. Tolstov and T. A. Jdanko, the sites belong to one of the Oghuz tribe groups who settled here in the 9th-11th centuries (Tolstov, Jdanko and Itina 1963:79; Toltsov 1962a:276).

The complex of settlements of Sarly-tam was 60 km east-southeast of Djend, on the middle of the three east–west channels of the Upper Inkardarya. It consisted of a large circular (250 m in diameter) fortress with adjacent settlement and *mazar* built on a high sand hill. West of the fortress there was, apparently, a large handicraft settlement, with remains of houses and different workshops, such as pottery kilns and kilns for firing bricks. Around this settlement there were many traces of fields in the form of quadrangular diked plots, and gourd plantations with typical *gryad* layouts enclosed by a wall (see Tolstov, Jdanko and Itina 1963:80:fig. 36; Tolstov 1962:278, fig. 181). In the environs of Sarly-tam-kala there were many small ditches 1–2 m wide drawing water from the Inkardarya riverbed.

At 250 m southwest of the fortress stood the Sarly-tam *mazar*, adorned with carved terracotta tiles typical of sites of the Karakhanid period Syrdarya (Tolstov 1962a:281). Among these, was the Aysha-Bibi mausoleum (11th– 12th centuries), faced with terracotta tiles and carved ornamentation (IKSSR 1957, Tom I:95). North of Sarly-tam-kala, on the opposite bank of the Inkardarya and 2.5 km from it, there was another large settlement called Zangar-kala (Tolstov, Jdanko and Itina 1963:82).

A series of retaining dams, semi-dams, catchment basins, and other small irrigation works were discovered south of Sarly-tam, on the southern channels of the Inkardarya deeply penetrating into the Kyzyl-Kum sands. The functional period of these facilities was identified by remains of dwellings and ceramic clusters.[Note 191] Fragments of surface ceramic material collected were poorer than in the sites of Sarly-tam-kala and Zangar-kala, but their character was completely identical to the above finds thus the settlements could be dated to the same period.

The largest and most complicated irrigation facility was discovered at the end of field work in 1962 in the tract of Bes-Molla, located 25 km southwest of Sarly-tam (*poisk* 13, 9 October). This irrigation site consisted of several significantly sized water-catchment basins along the riverbed,

separated from each other by sluices or dams. It was fed from northeast through a canal derived from the upper part of the riverbed, and following its bend. The straightened part of the riverbed was isolated by a diversion dam. The irrigation canals were placed south of the basins on both sides. To lift up basins water into ditches, *chigir*, were used. These were preserved on the banks as round pits and accumulations of *chigir* pitcher fragments. One of these pits on the west bank was now used as a water-catchment basin.

The irrigation facilities at Bes-Molla represented an improved variant of the system of reservoir-basins well known in Asar irrigation. The peculiarity of the system topography, in particular the presence of a diversion dam on the riverbed, showed an almost complete fading of the Southern Inkardarya channel below Bes-Molla in the 13th century. Water was lacking in this place and irrigation was possible only by water discharged from the middle and northern branches of the Inkardarya. Retaining dams, reservoir-basins and *chigir* were used to raise the water level. A series of Medieval retaining dams and small basins were found on the southern channels of the Inkardarya to the south and southeast of Sarly-tam-kala (*poisk* 3–7, 7 October 1962; *poisk* 8–10, 8 October 1962; *poisk* 13 and 14, 9 October 1962). For example, the tract in the Zeket district, the riverbed 35–40 m wide was closed by two dams, which were well preserved on the lateral sides adjacent to the banks (*poisk* 5, 7 October 1962) (Figure 62.A). The dam was approximately 2 m high and 3–4 m wide. Fragments of wheel-made Medieval vessels were found on it.

A more complex structure was found on the same riverbed 4 km east (*poisk* 6, 7 October1962). It consisted of a large water catchment basin cut across by a dam 50 m long (Figure 62.C). Above the dam, on the opposite banks, there were head structures of two canals beginning from *chigir* pits. Irrigation from the right bank system was possible only through a full riverbed flood. On the left bank the riverbed was re-deepened 1 m, and the irrigation from the left bank of the canal was carried out at a lower water level. Downstream along the river, a reserve basin 20-40 m wide was dug parallel to it, and water was stored here in case of a sharp reduction of its level. When the water level dropped, the dike and dam were closed and the water flowed into the *chigir* pits from the reserve basin.[Note 192]

Among a number of other irrigation structures on the Southern Inkardarya, discovered and studied in 1962, the system of two diversion dams in Ketty-Kazgan (*poisk* 10) should also be mentioned (Figure 62.D). One dam (like the dam at *poisk* 6) provided lifting the water level for two *chigir* pits placed on opposite banks. Below the dam there was a large basin with one more diversion dam and bypass ditches to drain excess water.

North and northwest of Sarly-tam-kala, on a northern channel of the Inkardarya, with an east–west orientation, the field team identified a series of fortified settlements, near which there were many small, meandering canals. There were no retaining dams at this place and the irrigation appeared to be gravity type.

A series of Medieval fortified settlements (of sedentary and semi-settled farmer-herders), and the related remains of works were widespread along the Inkardarya riverbed and further north-east of Sarly-tam-kala up to the ruins of Asanas (on the margin of the contemporary cultivated zone of the Syrdarya). The field research in this region in 1960-1962 indicated that the irrigation pattern east of Saykuduk (15 km northeast of Sarly-tam-kala) was somewhat different. Above this tract, the northern riverbed of the Inkardarya turned into the main canal of Asanas-Uzyak. It was more than 60 km long and it followed entirely the intricate bending of the Inkardarya channel, along which it was constructed. This canal flowed in the central part of the riverbed and then ran along riparian banks. Its width was three-four times narrower and the riverbed was 10-15 m between banks and its depth was 3-5 m. In many places the remains of *chigir* pits, semi-dams and other irrigation facilities were visible. At Saykuduk, in the lower part of Asanas-Uzyak, the canal was 11-15 m between the banks and 5-7 m deep.

Reservoir-basins located below the fortified site of Saykuduk-kala were very interesting (Figure 63). They followed one another at a distance of 100-200 m and they had roughly the same dimensions: 60 m long, 30 m wide and 7 m wide at the bottom, 7-8 m deep. It was easy to calculate their total volumes. The average capacity of each basin was 5,000 m³, thus 10 reservoirs collected up to 50,000 m³. The volume of earthworks to create multiple basins was more than 50,000 m³. At the average rate of 3 m³ per day, 15,000-17,000 people per day were required to build them. A team of 1,000 workers could construct 10 basins in 15-17 days.

The reservoir-basins at Saykuduk were complex hydraulic structures. They demanded their makers not only a huge labor efforts, but also competent hydraulic knowledge and skills. Apparently, these skills were not introduced into the territory from outside, but they were the result of the historical process of improving the local 'deltaic' forms of irrigation. The beginning of this went back to adaptations of former riverbeds and ancient diked riverbeds for irrigation purposes, and lasted till the irrigation systems of the Karakalpaks and the Turkmen in the 17th- 19th centuries (see p. 241).

The purpose of the basins in Saykuduk was, however, not yet fully understood. According to one version, these basins could have been used to maintain the water level in the Asanas-Uzyak, because the irrigation was practiced mainly with the help of water-lifting facilities (*chigir*), which means that the direction of water flow did not play any role. With water flow reduction in the Syrdarya, basin sluices were opened and the water level in the Asanas-Uzyak raised over its whole length of 60 km. According to another possible version, the construction of the reservoir-basins provided a stable water supply to the settlements of Zangar-kala and Sarly-tam-kala, located 15-20 km southwest of the Middle Inkardarya. There was no permanent water flow in this riverbed.

The construction of a system of basins at Saykuduk also suggested that the riverbed of the Northern Inkardarya ceased functioning below the basins, and presumably the inhabitants left the northwest group of settlements on the Inkardarya, while the settlements located east of the basins yet continued to exist.

It was possible to date the period of construction of the Asanas-Uzyak and its final stage of basins system on the Saykuduk according to the fortification of Saykuduk-kala. By its layout, this fortress resembled other sites on the Inkardarya, in particular Sarly-tam-kala and Zangar-kala. It had an irregular shape and fortified walls and moats. The ramparts and central part of the complex were largely eroded and only 0.5-0.7 m remained above ground. According to N. N. Vakturskaya, fragments of ceramic material could be dated to the 11th-12th centuries. The fortified settlements of Khodja-Kazgan 2 and Khodja-Kazgan 3 should also be dated to the same period. They were both situated on the Asanas-Uzyak, with a width of 10-11 m between banks.

During the 1961 field work, the fortress of Khodja-Kazgan 1 was researched; it arose, according to S. P. Tolstov, at the same time of Sarly-tam-kala, Khodja-Kazgan 2 and Khodja-Kazgan 3. In contrast, it lasted for a longer period and was not located on the Inkardarya, but on the Upper Janydarya (Tolstov 1962a:281).

In the upper reaches of the Asanas-Uzyak, there was the Medieval settlement of Asanas (Ashnas or Eshnas), well-known from the written sources describing the Mongol invasion. According to Djuveyni, the Mongol army commanded by Djuchi, on the way to Djend after conquering Sygnak, defeated Uzgend, Barchylygkent and Ashnas (Bartold 1963, Tom I:236). The latter had the most stubborn resistance. Near Asanas (Ashnas) the field work revealed several fortified Medieval settlements, including the large settlement-enclosure (?) of Asanas 2, on the south bank of the Asanas-Uzyak fortified by two rows of ramparts and moats. The environs of these sites were watered by an irrigation network based on the water of the large main riverbed-canal of the Asanas-Uzyak.

* * *

The study of Medieval irrigation works of the vast Syrdarya ancient delta, through field work and aerial photography, provided an opportunity to highlight further developments of irrigation in this peculiar area, where a primitive semi-settled economy prevailed, combining pastoralism (mainly cattle breeding), irrigated farming on the former riverbeds, and fishing. This type of economy, linked to permanent settlements near water, confirmed the stability (up to the 8th-9th centuries) and composition of the ethnic population in the lower and middle reaches of the Kangar-Syrdarya. Low mobility of the poor Pechenegs people, the Yatuks (the Balykdaky; see Tolstov 1947b:71-75, 100-101), largely remained in the former habitats after the 9th century cruel wars (against the Oghuz, the Kimaks and the Karluks). They then entered into the already Oghuz and Kipchak Turkic tribes (Kangly is the Turkic variation of Kangar), and later, in the 15-16th centuries, became part of the Uzbeks, Karakalpaks and Kazakhs (see Tolstov 1948a:23-24; 1947b:87-90; Jdanko 1950:111-112; Klyashtornyy 1964:178; Levina 1967:18; Akhmedov 1965:16-17, 77-80).

During the Middle Ages, in the Lower Syrdarya primitive methods of single irrigation based on freshet floods and estuaries were still significant. The irrigation was carried out under the scheme: river - former riverbed (reservoir-basin) - main canal - feeder (and with *chigir* from the 10th-11th centuries) - field. In small irrigation systems water was drawn from courses or former naturally or artificially flooded riverbeds. The largest main canal of this area was the Asanas-Uzyak, which was substantially an improved version of the Medieval use of a diked riverbed adapted for irrigation. However, in the beginning of the 2nd millennium, in the Lower Syrdarya, as well as in Khorezm (see p. 208), there were significant changes in the layout, and the 'sub-rectangular' systems were replaced by branching layout systems and the widespread of water lifting constructions (*chigir*). The particularly high level of development was characterized here by several hydraulic works, such as semi-dams, dams, basin-reservoirs and different small-scale, but effective water-regulating devices. This was facilitated by the peculiarity of deltaic conditions, such as the frequent change in the channels hydrographic regime (sometimes flooding, sometimes fading), floods filling the former riverbeds, etc. The irrigation in this region was heavily dependent on these hydrographic changes, since the main riverbed of the Lower Syrdarya was not under the control of people living there.

In the Middle Ages the process was different on the Middle Syrdarya. The 1963 aerial surveys, and the following 1966 field research on several sites and massive irrigation networks on the left bank of the river, indicated that the irrigation in the largest settlements of the region, such as Kyr-Uzgent (Uzgend), Ak-Kurgan, Mayram-tobe, was based mainly on gravity systems derived through head canals from the main course of the river. Here the Syrdarya was probably closed by a dam, and several head canals were discovered on the contemporary riverbed as the sources of these systems. The irrigation systems had a branched configuration and a considerable length (30-40 km); the main canals were branched into many distributors of the 1st and 2nd order, and irrigation ditches.

The irrigation systems of the Middle Syrdarya resembled Khorezm's Medieval systems. Particularly impressive were the canals (10-12 m wide between banks) in the environs of Mayram-tobe and Kyr-Uzgent. Like the flood gravity systems of Khorezm, the scheme of the Middle Syrdarya systems was complex: river - head constructions (*saka*) - main canal - distributors of 1st and 2nd order - feeder - field. As in Khorezm, the irrigation was gravity type. The farming oases of this region were severely affected during the Mongol invasion in the early 13th century, and again in the 15th century at the time of wars between nomadic Uzbeks against the Timurids for the basin of the Syrdarya and Khorezm (Akhmedov 1965:124, 146-148).

Irrigation Works of the Karakalpaks and the Kazakhs (17th-early 19th centuries)[Note 193]
The archaeological sites of the modern period on the Lower Syrdarya were farming settlements abandoned by the Karakalpaks and numerous irrigation works on the banks of dry riverbeds of the Janydarya and the Kuvandarya. The archaeological and historical-ethnographic research of the Khorezm Expedition showed that the Karakalpaks, in the 17th-18th centuries, adopted the Archaic traditions of primitive agriculture integrated with a herding-fishing economy from the ancient inhabitants of the Middle and Lower Syrdarya Basin (Tolstov 1947b:99; 1947d:72; Jdanko 1952:466).

In the 17-18th centuries most of the Karakalpaks inhabited the basin of the middle and lower reaches of the Syrdarya, in the area which, according to the Karakalpak popular tradition, was called Turkestan (Ivanov 1935:38ff.; Jdanko 1950:134ff.). The northernmost and largest Karakalpaks' group in the late 17th century was in a vassal relationship to the Kazakh Tauke khan (1680-1718; Andrianov 1958a:8; 1964:134-135). F. Skibin and M. Troshin, ambassadors of Peter I to Tauke, reported that agriculture prevailed in the Karakalpaks' economy. They sowed wheat, millet, barley and also gourds (melons and pumpkins) on the irrigated lands (MIKK 1935:151; Andrianov 1964b:135).

The Djungar invasion in 1723-1725 devastated the farming oases of the Middle Syrdarya (Tashkent, Turkestan). The main irrigation works of the Karakalpaks on the Middle Syrdarya were destroyed and left derelict after the invasion. The majority of the population fled the area with developed irrigation for new, uncultivated lands, where they had to arrange again the irrigation network (see Jdanko 1950:137; Tolstov 1959b:17).[Editor 68]

Political and economic crises, oppression, heavy requisitions for the Kazakh feudal nobility, and frequent robberies (1748, 1760, etc.) were the reasons of the Karakalpaks' gradual migration west, to the lower reaches of the Amudarya (see Andrianov 1958a) in the Aral kingdom. It is reported by P. I. Rychkov in 1762: "According to the latest news, many Karakalpaks due to the Kirgiz-Kaysak's oppression have joined the population of the Aral Sea and live together with them" (Rychkov 1887:16).

In the first half of the 18th century, the Janydarya was dry in its middle and lower reaches. According to A. I. Levshin, the caravan routes of the mid-18th century, did not contain any data about the river, in spite of the very detailed descriptions of all the tracts. The first mention of the Janydarya in the Russian written sources is dated to 1774. In his 'wanderings', F. S. Efremov (1811:60) wrote that the river was "like the Kazanka in size (approximately 200 *verst* and 40 *sajen* wide)" flowing into the Aral Sea. In 1794, T. S. Burnashev on his way to Bukhara crossed the 'new' river (Burnashev 1818:45). The Janydarya, dry in the late Middle Ages, was watered for almost its entire length in the late 18th century, and Karakalpak settlements and fields were sited on its banks (Andrianov 1952a:570-574; 1958a:52-61).

In some areas, the Karakalpak irrigation used Medieval canals, re-deepened and reconstructed. The new watering of the Janydarya was the result of the Karakalpaks' irrigation works (Kaulbars 1881:222). The Karakalpaks also restored the fortifications abandoned in the Middle Ages. Traces of the late Karakalpak reconstruction were detected by the Khorezm Expedition at the sites of Chirik-Rabat, Beshtam-kala, Djend, etc. The construction of canals in the lower reaches of the Janydarya, in the Kly tract, etc. should be dated to the same period.

Systematic and detailed study of the Karakalpak irrigation on the Janydarya began in 1956 in conjunction with the study of the ancient oasis of Barak-tam, the territories from Barak-tam to the Choban-kazgan and Kly tracts, where several Karakalpak fortifications (Aralbay-kala, etc.) were found. In 1957 the investigations were continued from the Kly tract upstream of the Janydarya, to Orunbay-kala, Sarly-tam *mazar* (on the Janydarya), Chirik-Rabat and further up to the environs of Babish-Mulla (Tolstov, Vorobeva and Rapoport 1960:22; Andrianov 1960a:174). In 1958-1959 the team studied the Karakalpak settlements and canals north of Babish-Mulla on the Middle Kuvandarya, and in 1961 the Expedition discovered a whole group of Karakalpak fortifications on the banks of the Mayliuzyak (Buzuk-kala, Khatyn-kala I, II, III; Tolstov, Jdanko and Itina 1963:34; Toltsov 1962a:314).[Note 194]

Karakalpaks' Irrigation on the Lower Janydarya
In 1956, during the investigation near the well of Chagyr, 35 km east of Barak-tam, a small Karakalpak irrigation facility (*poisk* 28) was discovered. It consisted of two *chigir* pits (7 m and 6 m in diameter), from which the water-lifting wheel supplied fields. Nearby, there were traces of several pits, probably the remains of mounded earthen houses. The basins were connected with a shallow branch of the Akchadarya channel by a ditch 30 m long. The canal was 8 m wide, 2 m between banks and 2 m deep.

A number of such structures and narrow ditches, drawing water directly from the main riverbed, were recorded further northeast, on the Akchadarya channels and on the main riverbed of the Janydarya. In many cases, below the ditch headworks, some diversion dams, preserved in the shape of a clay wall 2-3 m high, were built across the riverbed. For example, the diversion dam placed 2 km southwest of the Choban-kazgan tract (*poisk* 32) was 65 m long (see Figures 51.K) and it dammed up a shallow bed 60 m long of one of the Akchadarya's northeastern channels. The clay moat of this dam was 6 m wide, 2-2.5 m high, and it lay at the bottom of this dam. In front of the dam, to the left and to the right, there were some head works. The left canal, whose banks rose only 0.3-0.5 m above the fields, was 6 m wide, 2 m between banks and 1-1.5 m deep. The transverse profile carried out along the ditch and along the riverbed, showed that the water was lifted onto the fields by *chigir*.

In the Kly tract, a Karakalpak cultivated oasis stretched from north to south along the inner sand dunes for a length of 25-30 km and a width of 10 km (Andrianov 1960a:fig. 5). In the southern direction, toward the Jaldybay tract, the cultivated zone became gradually narrower and limited to a narrow strip along the main riverbed.

The central place in the Kly tract was occupied by the large Karakalpak fortification of Aralbay-kala surrounded by a dense irrigation network, *chigir* pits, and fields enclosed by walls, among which rose sites with traces of yurts. The rounded pits, supplied water to fields with water-lifting devices (*chigir*), various in shape and size, but their prevailing diameter was 5 to 6 m. The majority of pits stretched along the axis of the ditches and they were coupled, i.e. at the end part of a ditch, as a rule, there were two pits (Andrianov 1960a:fig. 8) (see Figure 51.L).

There was a large variety of different fields: diked irrigated plots with an irregular quadrangular shape (so-called *atyz*) encircled by walls 30, 40, 31 m, 38 or 42, 33, 41 m, and 26 m etc. in size (Andrianov 1960a:fig. 8) (see Figure 51). The fields were located in the lowest and flattest places. The cultivated crops were mainly cereals (millet, barley, wheat). A. I. Levshin, describing the agriculture of Kazakhstan on the Syrdarya in the first third of the 19th century, reported that the Kazakhs sowed millet, barley and wheat, and also noted that their irrigation works were, apparently, quite similar to those of the Karakalpaks between the 18th and 19th centuries. According to A. I. Levshin, "If the arable land was far from water, then they carried it from a river or lake to a ditch; at its end they dug a pit 2.3 or 3 *sajen* in diameter and arranged a mechanism in it" for lifting water onto the fields (Levshin 1832, Chast III:201).[Editor 69] A. I. Levshin also produced a detailed draft of a small irrigation system resembling the water systems in the environs of Aralbay-kala.

Among the various Karakalpak irrigation facilities in the environs of Aralbay-kala, the head constructions of canals were particularly interesting. The whole irrigation network southeast of Aralbay-kala was based on a canal derived from the main riverbed of the Janydarya 4 km from the fortress (*poisk* 57). This canal massive head structure was placed on a steep curve of the riverbed, which there was 60-70 m wide and 4 m deep (see Andrianov 1960a:fig. 10).

Another head structure surveyed by the field team, 6 km southwest of Aralbay-kala (*poisk* 89), consisted of three heads derived from the Janydarya channel at different levels (+1.37 m,

+1.46 m, +2.07 m; Andrianov 1960a:fig. 11). A square hollow up to 3.8 m deep and 25 × 40 m in size was found in the riverbed in front of them. Downstream of the riverbed there were remains of a diversion dam in the lower part of the bed. The canals branching off at this place were not wide (2.5-3.5 between banks). Their depth was 0.7-1.0 m. Large quadrangular diked plots with mostly rectangular fields were recorded in the lower parts of these canals. One of the fields measured 37, 65, 26, and 60 m (*poisk* 91).

In the Jaldybay tract, low (50-60 cm) fences in *pakhsa* enclosed fields with irregular rectangular and oval shapes, once were sown with crops (*poisk* 127). Similar agro-irrigation layouts were discovered in other places, for example the environs of Zangar *mazar*, near Orunbay-kala, etc.

About 5 km southwest of Zangar *mazar*, similar layouts were topographically connected to the irrigation derived from the main riverbed of the Janydarya, 60-70 m wide. On the left, eastern, high bank there was the *mazar*, a cemetery, and nearby remains of a settlement. On the bank of the riverbed there was a huge hollow-drainage basin, bound on the sides by two dikes. One dam was 25 m long and 4 m wide at the base. A wide and deep ancient riverbed was present in a depression, along its slopes a 200 m canal was built repeating with its meanderings the unevenness of the plateau eastern slope near the river. The ditch branched off in its lower reaches. On its right, some flat low plots were laid out as quadrangular or rhomboid *atyz* (irrigated plots) with 10-12 m sides. In some cases, the *atyz* had a layout with the enclosing banks up to 40-50 cm high. Field irrigation in the depression was gravity type. The fields on the left branch of the ditch were irrigated by using water-lifting facilities.

The next significant area of Karakalpak irrigation was placed upstream in the Janydarya, 15 km northeast of Kly, in the Sazdy-kuduk tract and around Orunbay-kala (see also Andrianov 1960a:185-187). Orunbay-kala was one of the main living centers for the Karakalpaks on the Janydarya at the turn of the 18th and 19th centuries and it occupied a leading position

over the above-described irrigation systems. This
could be linked with the feudal relationships that
existed in that region in those periods between the
mighty Karakalpak feudal lord, i.e. the Orunbay-
biy from the Manghit tribe (end 18th-beginning
19th centuries), and the common Karakalpak
farmers (see Jdanko 1952:322; Andrianov 1952a:
580-581).

In 1957 a number of late irrigation facilities,
far from each other, were discovered by the survey
team northeast of Orunbay-kala and Beshtam-
kala. Some irrigation facilities located 10 km east
of Beshtam-kala (*poisk* 200) belong to the later
stages of cultivation on the Janydarya. At this point,
the deepest part of the riverbeds were closed by
high dams made of massive *saxaul* trunks, buried
in the middle of the clay. A preserved dam
survived, and it was 3.5 m wide and 45 m long.
Near the dam, in the riverbed, there was a ditch
on whose right bank there were visible traces of
yurts encircled by walls. Apparently, the dam
raised water in the ditch during the last period of
the Janydarya, when its level was low.

Another Karakalpak irrigation device, placed
15 km east of Beshtam-kala, was also connected
to low water level in the main riverbed (see
Andrianov 1960a:fig. 13). The head structure
of the canal (*poisk* 204) was 16 m wide and 2.5
m deep. Traces of a narrower ditch, 3 m wide
between banks, were visible in the center of this
construction. Two heads, deriving from a small
basin (30 × 40 m), supplied the canal with water.
The northern structure was 7.5 m wide and 3 m
between banks. The southern one was closed by a
dam, whose remains were found not far from the
source. Generally, the construction was a classical
Karakalpak head structure (written about, already
in 1887, by the hydrographer Stetkevich:1889:
Ch. 4, *Gidrograficheskoe opisanie* - Hydrographic
description)[Editor 70], with two head canals and a
basin with a small island in the middle called
by the Karakalpaks as *kysme-bugut* (see p. 218).
Judging from the double profile of this structure,
it had two historical stages: the first, earlier, with
a high water level (perhaps in the Middle Ages);
the second, later, with a lower level (18th-19th

centuries). The third and final stage of agriculture
in the area was associated with the development
by the Kazakh people of the already dry riverbed
of the Janydarya at the end of the 19th century.
The dry riverbed was entirely occupied by fields.

Karakalpaks' Irrigation on the Kuvandarya

Already during the 1946 aerial survey it was
noted that, approaching the Kuvandarya, the
ancient irrigation in the environs of Babish-Mulla
was altered by the late Medieval Karakalpak
irrigation (Tolstov 1947c:180-181). The 1958
survey revealed the Saka sites distribution borders
and those of the Asar culture, separated by the
fairly wide (30-40 km) zone of late Karakalpak
irrigation along the Kuvandarya riverbed (see
Andrianov 1962). About 25-27 km northeast
of Babish-Mulla, traces were found of a very
late irrigation in the shape of narrow fairly well-
preserved canals, polygonal fields enclosed by low
walls, slightly visible *chigir* pits, as well as traces of
yurt ridges (*poisk* 405-410).

The ancient riverbeds changed their aspect
here: south, in the area of ancient irrigation, they
were mainly expressed by flat banks 60-70 m
high above the surrounding *takyr*; in the area
of the Karakalpak irrigation, the riverbed was
lower (1.5-2.0 m) than the *takyr* surface with
traces of recent Kuvandarya flooding on which
the Karakalpak irrigation was based. In contrast
to the ancient irrigation, which used fast-flowing,
northwest, waters of the Janydarya's interdeltaic
channels (cutting across the massive north-south
sand dunes of the Kyzylkum), the Karakalpak
canals operated in a completely opposite way.
They drew their water from estuaries and
freshets of the Kuvandarya, flooding the north-
south riverbeds from the north. The Karakalpak
irrigation in this area was thus estuary type.

The eastern border of the Kuvandarya late
settlements area was the massif lands along
the Mayliuzyak. A channel branched from the
Janydarya not far from Djan-kala, and connected
to the Kuvandarya with a system of small
meandering riverbeds. The Mayliuzyak had a
general northwest direction, and its 50 km long

banks were dotted with late fields and semi-settled type sites (with traces of yurts ridges), cut across by numerous canals. The general size of the Mayliuzyak massif was 40,000-50,000 ha.

The southern outskirts of this oasis was Djan-kala and its surroundings. The Medieval irrigation systems of Djan-kala (Djend) preserved traces of a later re-deepening, and the Karakalpak period of the settlement was very clearly expressed in the later wall and the central citadel with an octagonal tower. At this site there were many mounds of earthen houses, traces of yurts and fences for livestock.

In the center of the Mayliuzyak massif there was a vast complex of semi-settled sites and fortresses generally known as 'Khatyn-kala' (Figure 64). This complex was discovered during the aerial survey in 1960 and researched by the Expedition field unit in 1961 (Tolstov 1962a:311; Tolstov, Jdanko and Itina 1963:89-90). There were three fortresses in the Khatyn tract. The biggest was Khatyn-kala I, standing where the river was blocked by a huge dam (see Figure 64). A number of canals, irrigating the land both on the left and right banks, began in this place. According to S. P. Tolstov, at the base of Khatyn-kala I there was an older Early Medieval fortress existing, with possible interruption, until the 18th century, when the Kuvandarya Oasis was populated and exploited by the Karakalpaks (Tolstov 1962a:312; Tolstov, Jdanko and Itina 1963:90). After a brief abandonment in the 1830s, the fortress was renovated by the people of Khiva, trying to fortify their position on the Lower Syrdarya at that time. According to historical and topographic sources, it was then called Khodja-Niyaz-kala.

Particularly interesting was the dam at Khatyn-kala I mentioned above. By its size, plan and profile it resembled the Medieval dams on the Janydarya (see p. 211). The total width of the riverbed here was 200 m. At this point, it was considerably wider than in its upper and lower parts, where the width did not exceed 100 m. The outer wall of the fortress and moat were linked to the river in order to defend the approaches to the dam, near which there was a perimeter system of canals and a large area clearly used for docking boats. The massive dam structure was 130 m long and 12-15 m wide at the base. In its central part the dam was washed away, but another side 80 m long was linked to the left bank of the river. Four main canals, branched from the dam at different levels (the difference in level was up to 2 m). This fact should probably be considered as proof that the irrigation system should be dated to different periods. The widening of the riverbed in this place up to 200 m can be related to the long-term use of the head works and dams. It is possible that the massive irrigation knot was constructed at the same time as the fortress in the Early Middle Ages, and functioned intermittently until the 19th century.

The study of irrigation in the environs of Khatyn-kala I carried out by the Expedition field team (including geomorphologist A. S. Kes), revealed that, in addition to the above irrigation knot at Khatyn-kala, downstream of the Mayliuzyak there was another, double, dam, built at the intersection of the Mayliuzyak with an older riverbed (see Figure 51.J). Considering the location of canals and dams, the individual links of the ancient riverbed must have been adapted as reservoir-basins at the time of the fortress. After the construction of the diversion dam at Khatyn-kala, the main riverbed below the fortress was watered only by excess water. This water first flowed into the by-pass canal and then into the canal through the canal running along the course of the Mayliuzyak. The canal reached the dam and was directed into the second reservoir placed across the riverbed, where the large main canal directed north began. The bed of the Mayliuzyak was used for fields irrigation ditches derived from the riverbed along its banks. Parts of the riverbed were separated by dikes, and fields were also located behind them. Fragments of ancient ceramics, gray pottery of the 11th-12th centuries and later ceramics of the 18th-19th centuries were discovered in the fields and *takyr* near the canals. In the latest period, when water was lacking, it was stored in separate reservoirs and strictly distributed by canals.

The irrigation knot in Khatyn-kala I, consisting of a retaining diversion dam with by-pass canals, with a very complex system of connecting canals and reservoir-basins, witnessed a high level of development of irrigation technology among the Karakalpaks in the 18th-19th centuries.

Concluding the description of the characteristics of late irrigation in the basins of the Janydarya and the Kuvandarya, it should be added that the study of irrigation works on the abandoned Karakalpak and Kazakh settlements indicated that, in the 18th-19th centuries, the traditions of the deltaic 'semi-nomadic' irrigation, was largely preserved using the former flooded riverbeds and the estuary overflows with the scheme: riverbed - former riverbed - main canal - feeders - *chigir* - field. The peripheral areas, west (Lower Janydarya and Akchadarya), and east (Kuvandarya), where water flowing in the riverbeds was slow, were characterized by local cradles of irrigation development, using diversion dams on lateral channels and short deep ditches with basins for water-lifting facilities. The bottom of a flat riverbed was covered with fields in some places and, apparently, their irrigation was basically estuary type.

In the central, most densely populated areas, i.e. along the Middle Janydarya, near Djany-kala

and on the Mayliuzyak, there were large territories of intensive agriculture, where the irrigation systems, drawing water from the Janydarya and the Mayliuzyak, differed in their complex and branched layouts. Here, the canals and head structures were designed for freshets and fast-flowing water. To construct the main and distributing canals, a highly branched network of meandering lateral channels of the Janydarya were skillfully used. The canals were built with a remarkable knowledge of slopes and complex uneven bank reliefs: along the riparian banks, or along bank slopes, or even along channel beds. The agro-irrigation layouts differed widely. However, gardens and vineyards typical of the Khorezmian 'lands of ancient irrigation', were not discovered here, possibly because of the short-term presence of the Karakalpaks in this area.

The archaeological-topographical study of the Karakalpak irrigation in the Lower Syrdarya confirmed the conclusion, based on historical-ethnographic materials, about the rather primitive character of the farming-herding economy of the Karakalpaks, which was the ethnographic relict of ancient economic structures typical of the semi-settled Medieval tribes of the Aral Sea area (Tolstov 1947b:70-71, 99-100; Jdanko 1961, 1964).

Conclusion

Boris V. Andrianov

The great Russian orientalist and scholar of Medieval written sources on the history of Central Asian irrigation, V. V. Bartold, once rightly criticized the widespread view (N. I. Veselovskiy, V. V. Radlov, etc.) that considered futile any attempt to develop the dry blasted lands of Turkestan, as if any attempt were hopelessly condemned to an inevitable defeat (Bartold 1965:97-98; 307-310). Irrigation works during the Soviet period turned many desolate areas of ancient irrigation into flourishing oases, and gave water and tillable soil to the working people of Turkestan. The rapid development of science and technology allowed the exploitation of natural resources of this arid zone. The modern heirs of irrigators, mathematicians, and astronomers of the past, with their scientific success in the field of hydraulic engineering and amelioration, provided a radical reconstruction of Central Asia ancient irrigation systems, obviating the hard work of canal cleaning with the full mechanization and a more effective use of water, the main life source in this region.

The success of hydraulic engineering science in the last decades had raised a number of scientific problems, in particular the problem of a comprehensive landscape study of irrigation systems. Single features (characteristics of the river flow, water intake, profiles of canals, configuration of the distributing and irrigating networks, effective land use and methods of preventing silting, waterlogging and salting, irrigation equipment, etc.) must be considered in close connection to the geographic environment as part of the whole irrigation systems, which is the backbone of the cultural landscape of arid zones (Dunin-Barkovskiy 1960:24-25, 69-70).

Closely connected with the question of a geographical approach to irrigation is also the problem of the historical development of irrigation works, whose solution the role of material culture is without any doubt important (see Bartold 1965:308; Tolstov and Andrianov 1957:5; Dunin-Barkovskiy 1960:64-65).

Which main conclusions can be drawn from the detailed study of the ancient irrigation systems of the Aral Sea area, and a general review of the history of irrigated agriculture?

Research in the Aral Sea area convinced us of the rightness of A. I. Voeykov's words, chosen as the epigraph for this book. In fact, irrigation can turn a desert into a blooming oasis, but without water, oases turn again into deserts. The preserved traces of ancient oases are an excellent source for the study of material culture, economy and lifestyles of ancient people. It is impossible to understand these unique ancient cultural landscapes, in which centuries of human activities have left their indelible traces, without the close association between historical and natural sciences, and without the combined use of different methods.

The remains of canals and field layouts, together with the ruins of rural settlements of different periods, spread over many tens and even hundreds of kilometers in the desert, became only relatively recently the object of an extensive archaeological research, in association with the fundamental implementation of archaeology with different methods of natural and technical sciences, in particular the use of aerial methods. As we discussed in detail, the latter played an important role in our work with complex archaeological mapping and study of irrigation in the 'lands of ancient irrigation' of the Aral Sea area. As explained above, in the process of archaeological-topographical study, we applied a method based on field and desktop interpretation of aerial photographs and mapping irrigation systems.

During this work we mapped the ancient irrigation systems of the Aral Sea area and improved methods of relative and absolute dating of canals (Andrianov 1958a, 1965). This archaeological-topographical research, which combined archaeological field surveys with deciphering aerial photographs and subsequent mapping, revealed the continuous evolutionary development of irrigation systems in the Aral Sea area. This occurred over four and a half thousand years: from their origin in Prehistory, through Antiquity, during the Early and Late Middle Ages until the 19th century (see Tolstov and Andrianov 1957; Andrianov 1958a, 1959a-b, 1960b, 1961, 1962, 1963, 1964a, 1965) (see Figures 27, 34, 42, 51).[Editor 71]

The archaeological study of ancient irrigation works in the Lower Amudarya allowed to clearly identify the main development stages of Khorezm type river systems, to reveal changes in the character of water drawing, system configurations, canal profile sizes, the relationship between working and idle parts, etc. Such information was available for other areas of Central Asia, for example the most ancient canals of Central Asia in the Neolithic Geoksyur Oasis investigated by G. N. Lisitsyna.

In Khorezm, at the Tazabagyab and Suyargan settlements of the 15th-12th centuries BCE, both similarly simple and more complex systems were uncovered. The irrigation was carried out according to the scheme: riverbed - head canals 'regulated' riverbed - distributors - feeders - fields (Figure 65; Table 11).

In the process of developing natural resources in the deltaic region of the Lower Amudarya in the Bronze Age, techniques of wetland reclamations, flood controls, and deepening fading deltaic channels and former riverbeds were improved based on primitive forms of *kair* and estuary agriculture. The short ditches, located along the banks of small fields, were detached from them. At the stage of diked riverbeds, people invented head water regulating facilities, which allowed to control riverbeds water flow and to maintain a level of flood water in the canals.

In the Amirabad period (9th-8th centuries BCE), the regulated riverbeds and former riverbeds, cleaned of sediments, were turned into small artificial main canals and short feeders in the Early Bronze Age. The lateral small canals (only a few tens or hundreds meters long) were replaced by longer distributors up to 1 km long. Thus by supplying fields with water from the riverbed, new links appeared in the irrigation system.

Along with improvements of irrigation techniques in Prehistory, there was a gradual process of increasing irrigated territories. Thus, qualitative changes (improved methods of water supply to the fields) were accompanied by quantitative changes (increase in canal sizes and irrigated areas). However, during Prehistory, irrigation depended entirely on hydrographic changes. The large Akchadarya Delta riverbeds were not yet controlled by people or embanked by dikes. So small canal heads were quickly washed away and frequently people had to move from place to place, and start-up cultivated plots in areas which had not previously been part of the irrigated territory.

The technical level of Antique irrigation, i.e. the construction of massive long-term irrigation systems with permanent water intake into large channels, a variety of head structures, the invention of the waste canal and, most important, the scale of the works, was highly different from prehistoric irrigation. If the early Archaic canals in the environs of Bazar-kala were the continuation of small lateral channels of the Akchadarya, later the heads of canals were drawn from the largest branches of the Akchadarya riverbed. Farmers created not only artificial 'rivers', 20-40 m wide between banks and several tens of kilometers long, but also artificial 'deltas' with typical 'sub-rectangular' and bunched branches, repeating the natural channel courses (Archaic Kelteminar, Archaic system of Dingildje, etc.). The small and sparse irrigation networks of this period were mostly based on main canals, which followed the riverbed meanderings. The early Archaic canals already had a different section in the upper and lower course, which provided the necessary speed for gravity irrigation to carry water to the canals.

During the Kangju, and especially the Kushan periods, the wide and shallow Archaic canals were replaced by deeper and smaller cross-section canals, (right bank of the Khorezm, basin of the Chermen-yab) (see Gulyamov 1957:89-90; Tolstov and Andrianov 1957:7-8). At the same time they began the process of reducing single local systems to combine them into more massive systems by moving the canal heads upstream.

By the end of this period, many system sources were probably moved to the main riverbed and unused idle parts increased sharply. The canals were lengthened up to 100 km with cross-sections of 25-50 m². Irrigation was carried out under the scheme: main riverbed head - main idle canal - distributors - feeders - field. The technique of water supply to fields was improved, and different water distributing devices appeared. The progress, and the process of crops cultivation itself, sharply increased the range of crops, which was reflected in the variety of agro-irrigation layouts discovered by us.

At the time of the Kushan-Afrigid transition period (4th-6th centuries), there was evidence of decline of traditional forms of Khorezmian culture and economic life, and the abandonment of many branches and terminal parts of ancient irrigation systems. In some areas (like the Dingildje 'oasis', etc.) we may have recognized the beginning of a basic reconstruction of irrigation systems. This on the basis of new, already Medieval hydraulic solutions. According to evidence, a fact strictly linked to general tendencies toward a breakdown of the economic and social life in the oasis already showing characteristics of a new feudal age. The significant reconstruction of irrigation systems on the right bank of the Khorezm was dated to the late Afrigid time, i.e. the 7th-8th centuries. Since the end of the 8th century, and especially in the 11th-12th centuries, the most pronounced activity of the Khorezm scientists was to develop a unique school for specialists in irrigation, mathematicians

and astronomers, who developed practical field needs for the irrigation of Khorezm.

Medieval irrigation differed in its more economic use of land inside basins. The percent of irrigated areas increased up to 30-40%, thanks to the appearance of complex branched layouts, density of feeders and the widespread of *chigir* water-lifting devices (from the 9th to the 11th centuries). This explains the paradox that, although with the significant progress of irrigated agriculture in Medieval Khorezm, the total area covered with irrigation works sharply decreased, the total irrigated area increased because, as we said above, the rural population in several areas also increased (Gavkhore, Chermen-yab Basin in the 12th-early 13th centuries, etc.).

In the western areas of the Sarykamysh Delta, where the irrigated lands of the Golden Horde and Timurid periods (14th-17th centuries) were preserved, irrigation was developed, according to the evidence, in two main directions: 1) the extensive, highly branched systems of traditional Khorezm type (exemplified by the Medieval Shamurat) continued to improve; 2) and local systems appeared based on the lower channels of the Daudan, watered after a temporary interruption with water deriving from the Daryalyk through canals), with various small retaining dams, water distributors and water-lifting facilities. They resembled the irrigation of the semi-settled Medieval population on the Syrdarya. Irrigation was carried out under the scheme: river - flooded former riverbed - distributors- feeders - field.

On the shores of the drying Sarykamysh Lake in the 14th-16th centuries, the Turkmen constructed canals and earthen-aqueducts with water collecting basins, using water-lifting systems such as the shaduf. The irrigation scheme was carried out according to the typical estuary type: lake - water-lifting devices - feeders - field.

We researched the remains of irrigation works of the 19th century in the 'lands of ancient irrigation' of the Sarykamysh Delta and in the environs of abandoned Turkmen settlements. Here we found large retaining dams, different head structures, systems of water reservoirs, gourd plantations, vineyards and wide fields. Some of the Turkmen hydraulic works had broad parallels with those on the Syrdarya.

By the 18th-19th centuries the irrigation technology of Khorezm reached a high development. Typical features of the late Khorezmian irrigation were excellently explained in Ya. G. Gulyamov's monograph (see Gulyamov 1957: 237-267).

The irrigation techniques in the Lower Amudarya, from the mid-2nd millennium BCE to the 19th-20th centuries, evolved through a series of stages: 1) *kair*-estuary; 2) regulation and adaptation of fading deltaic former riverbeds (as well as the introduction of head works); 3) construction of huge main canals on large channels during the formation of the ancient Khorezmian state; 4) improvement of the Antique flood gravity systems, implementation of the schemes and changes in the layout of distributors and feeders with more frequent branches; 5) emergence of more effective systems and *chigir* irrigation in the Middle Ages; 6) late Middle Ages (further development of water-lifting devices and water regulation).

Somewhat different was the Lower Syrdarya. The irrigation appeared here only in the mid-1st millennium BCE. It developed primitive methods of river overflow regulation and the use of former deltaic diked riverbeds as reservoirs. In general the irrigation was of the estuary-lake type, since water was drawn from the former riverbeds flooded during freshets. Irrigation was carried out under a simple scheme: riverbed – reservoir – regulated riverbed – feeder – field.

In the 4th-2nd centuries BCE, in the Middle Janydarya, in the environs of Chirik-Rabat and Babish-Mulla, irrigation was based on the extensive use of diked riverbeds and former riverbeds of the inner 'delta'. The deepened sections of former riverbeds were turned into reservoir-basins (quite a widespread method of irrigation in the ancient Orient).

In the Djety-asar 'oasis' in the 1st millennium CE, the same principles of deltaic irrigation were employed, using banked riverbeds and reservoir-basins. The latter were evolved in the Middle Ages when, according to the irrigation facilities on the Upper Inkardarya, a variety of retaining dams and semi-dams, and systems of artificial basins, which periodically watered the former riverbeds, were specially developed. However, the irrigation was generally primitive with an estuary-lake character since water was not supplied by the main riverbed, but from the former riverbeds flooded during freshets. Even the largest waterway, such as the Asanas-Uzyak (60 km long) built on the banks of the Inkardarya, was just an improved Medieval variation of a regulated riverbed. However, at that time, water-lifting facilities (*chigir*) were widespread there. The introduction of the *chigir* irrigation marked a new stage in the development of irrigation techniques.

In modern times, the irrigation of this area was also based on flooded former riverbeds. Local cradles of irrigation systems prevailed here with retaining dams, basins, and small branched systems with several water-regulating devices.[Note 195]

Thus, the rate of historical irrigation development in the Lower Syrdarya was quite slow. The primitive use of deltaic riverbeds for irrigation, typical of prehistoric Khorezm (mid-2nd millennium-8th century BCE), continued here in a modified form, almost to the end of the 1st millennium CE.

According to our research, the main issue of irrigation, i.e. the stabilization and regulation of river overflows, was solved in the Lower Syrdarya differently than in Khorezm. And although the volume of water flow in the Syrdarya was two to three times less than the Amudarya, its main riverbed was not strengthened by dikes in its lower reaches neither in Antiquity nor the Middle Ages. There were no improved extensive irrigation systems with watering main canals many kilometers long, typical of Khorezm.

On the Middle Syrdarya in the Middle Ages, beginning with the Karakhanid period (10th-12th centuries), the irrigation on the left bank of the area was based on massive main canals dependent on the Syrdarya riverbed. Considering the head works, branch configurations and considerable length (30–40 km), these systems resembled the Medieval system of Khorezm. They irrigated the environs of the large Medieval settlements of Kyr-Uzgend, Mayram-tobe, etc.

Thus, the irrigation of the Lower Syrdarya differed from both the irrigation systems of Khorezm and the systems of the Middle Syrdarya. The reason was not only due to natural hydrographic settings.

The development of intensive and long-term irrigated agriculture on these vast regions was possible only under conditions of contending against natural flooding over huge areas, and with the stabilization of the main riverbeds by means of protective dikes and massive hydraulic works tens and hundreds of kilometers long. The achievement of such undertakings demanded the presence of a local long-settled farming population with agricultural traditions and, as K. H. Marx and F. Engels repeatedly pointed out, the existence of a long-term, strong and centralized state (see Marx and Engels 1957, Tom 9:132; 1961, Tom 20:152, 183-185; etc.).

The historical prerequisites necessary for creating such a large-scale and highly developed irrigation economy did not occur in ancient times (1st millennium BCE) in the Lower Syrdarya among the Saka tribes, whose economy was based mainly on herding and partially on primitive farming, although state unions took place there from time to time. The historical and political conditions, typical of the valley and delta terrain of the Syrdarya, bordering the steppe area and often the object of frequent invasions by nomadic herders ('Turs' later Huns, Turks, Kimaks, Kipchaks, etc.), played an important role. Of course, economic conditions also had a certain importance: the specialization of this region was herding, where the livestock grazed. In this territory the tradition of an integrated herding-farming and fishing economy persisted since the Bronze Age, it served for a long time as the basis of patriarchal-kinship ties (see Tolstov 1947a; 1947b:87-90, 100-102; Jdanko 1964:17-21.[Editor 72] Not by chance the indigenous inhabitants of this area, the Karakalpaks, retained the social-

kin traditions till the 20th century (see Jdanko 1950, 1952, 1958a, 1964). Thus, the Archaic irrigation in the lower reaches of the Syrdarya, its particular exploitation of local water-collecting basins, estuaries, semi-stagnant water and small irrigating systems requiring an insignificant labor cost, etc., could be largely explained by the socio-historical development of the semi-settled herder-farmers inhabiting the area.

* * *

The construction and annual cleaning of large artificial 'rivers' and 'deltas' demanded enormous labor costs.

It was not for nothing that, at the beginning of the 19th century, the Russian prisoners called the Khiva Oasis 'hard land'. Irrigation work occupied most of the time of the workers of this area. "In order to clean, deepen and broaden the Khan's or main canals, and also to dig new ones, the workers, one from each house or hearth paying taxes and having a plot of land, are gathered from the whole khanate and dispatched annually in the beginning of spring; workers are divided into several groups and shifted according to the volume of work. Each shift is obliged to work for 15 days... those, who are lazy or underhanded, are cruelly punished; there were some examples, when they were clubbed to death" (ZJ 1838, no. 6:338). This heavy work associated with the construction and annual cleaning of canals from silt and sand sediments, and the construction of diked banks (rash) was directed by the Khiva khanate's central authorities, and carried out as 12-day labor conscriptions (begar), but actually often reaching 50-60 days.[Note 196] The work also included the cleaning of distributing networks (kazu), the construction and repair of protective dikes (kazu) (Gulyamov 1957:235ff.). For digging new main canals and for the significant expansion of irrigated areas, additional workers from outside were required.[Note 197]

In the 19th century, when useful iron implements aided the cleaning of main canals, approximately 1 million working days were annually spent for cleaning and the total volume of work related to irrigation reached 7-8.5 million working days (see Smirnov 1933:9).

In 1906, in the Khiva Oasis, of the total costs of agricultural economy, 42.7% was devoted to irrigation (including cleaning of canals and construction of dikes), 22% to fertilizing, 13.4% to plowing, 4.5% to sowing and 16.8% to harvesting (see Shkapskiy 1900). Such a ratio of work input was typical of many other countries of classical irrigated agriculture. In some areas of China, for example, such work required 58.5% of labor for irrigation, 10.2% for field cultivation, 9.2% for harvesting and 21.1% for other purposes (see Buck 1930:306).

According to the 1920s irrigation specialists, the population of an oasis had to annually discard 1/3 of the volume of the entire irrigation network. In other words, every three years they had to rebuild the network anew! (Bukhanevich 1925:992; see also VIR 1926, no. 2:23; etc.).

If in the 19th-early 20th centuries in the Khiva Oasis ('hard land'), people cleaning the canals had to reconstruct, anew, the irrigation systems every three years, it is difficult to imagine the labor costs required in ancient times, with less sophisticated irrigation systems (with large-sized canals) and less effective work implements.

Undoubtedly, in ancient times, the work input for irrigation increased to such an extent that it was probably beyond living within the limits of irrigation basins.

In connection with the above, there was a need to augment the farming population. In the historical context of that period, this was achieved by harnessing captive neighboring peoples, whose labor was not directly related to the main objective of field-crop cultivation (which ordinarily was carried out by the local farming community). This simple division of labor conditioned the emergence and development of the slave-owning mode of production (characterized by a communal-collective character, simple cooperation, prevailing manual labor, preservation of communal labor in field cultivation, and periodic to more or less permanent use of war prisoners for irrigation work,

together with community labor).[Note 198] However, the question of the quantitative ratio of communal and slave labor in agriculture production of the ancient world in general, and Central Asia in particular, is far from being solved (see also Masson 1968: 100-101). For a comprehensive solution of such controversial questions, more reliable paleoeconomic data are required (see V. Masson 1968:96-99). Additionally, we need sufficient important material covering the major 'parameters' of ancient economy within large historical-cultural areas, quantitative and qualitative characteristics of economic development, research on labor input (in irrigated and rain-fed farming, and herding), and human resources of the single historical-cultural areas, economic units, etc. For those countries with irrigated agriculture, where the work input for farming today amounts from ¼ to ½ of all agricultural labor, the material concerning the increase of labor productivity in irrigation, in connection with the development of irrigation facilities from their origin in Prehistory to modern days, is especially important. From this point of view, the material on the history of irrigation techniques of the Aral Sea area allows to outline three main stages:[Note 199] 1) The Bronze and Early Iron Ages were characterized by 'single' estuary and primitive small irrigation systems on the lower deltaic channels and adapted former riverbeds; 2) in the period of the 'communal slave-owning' class in the Khorezm state there appeared more complex systems with a regular water supply and with a variety of hydraulic facilities and massive main canals; 3) Medieval irrigation (especially in the 9th-11th centuries), differed from ancient irrigation systems in smaller sized canals, branching layouts, more sophisticated division of water distributors and irrigation networks, and more effective water-regulating devices, with the widespread use of *chigir*, which sharply reduced the amount of required digging to clean canals.

The progress in production is primarily a process of growth in labor productivity and labor cost reduction needed to obtain a certain product. In irrigated agriculture, the reduction of labor costs occurred mainly with irrigation and was connected, in Medieval Khorezm, with the reduction of main canals cross-sections, the improvement of water distributing networks (which produced the reduction of canal silting), the spreading of water-lifting devices (*chigir*) and a significant decrease of labor costs for constructing and cleaning canals. The oasis population had increased in that period by 2-3 times.[Note 200]

All this provided an opportunity for the local rural communities to carry out the whole cycle of seasonal irrigation works. Shifts in production resulted from changes in socio-economic relations and the rise of feudalism in the Middle Ages. However, the persistence of community in this area, due to the necessity for collective irrigation work, delayed the process of formation of more developed feudal relationships (Tolstov 1932:41).

* * *

The development of irrigated agriculture in the Aral Sea area was not an isolated process, it was closely connected to the general path of irrigated agriculture history (the history of plant-growing, development of implements, and of irrigation skills) in the wide arid zone of the Old World countries. We attempted to reconstruct the history of the Aral Sea area, identifying what was general and what was specifically local in the irrigation of the ancient farming centers of the world. Our research, first of all, specified the position of irrigation in the general evolutionary scheme of economic development accepted by modern scientific literature (see Sauer 1952, 1956; Masson 1966a:158).

C. O. Sauer assumed that agriculture began in the tropical forest areas of Southeastern Asia countries as a result of further development of intensive and regular harvesting, appearing initially in a slash-and-burn form (Sauer 1952:21; 1956:56). In Soviet agricultural literature, there is a prevailing opinion, starting with V. R. Vilyams, that irrigated agriculture was a late phenomenon (Vilyams, Tom 6:347-354); very close to this point of view is the supposition advanced in a small popular book *Vozniknovenie i razvitie zemledeliya*

(The emergence and development of agriculture).[Editor 73] The authors wrote: "While the first farmers sowed wheat and barley on mountainous slopes and on high plateaus, their descendants moved down from mountains to dry valleys and met different conditions", and began to use stream overflows and estuaries (VRZ 1967:25-26). However, the modern archaeological and paleobotanic research rejected this scheme. H. Helbaek, K. V. Flannery, D. R. Harris and other scientists confirmed the theory of ethnographers (C. D. Forde, etc.) and archaeologists (V. G. Childe, etc.) that the most ancient agriculture arose in zones with insufficient rainfall (where only 10% of land was useful for farming based on seasonal precipitation). We have no good reasons to believe that, in the 8th-7th millennium BCE, these areas had different landscape conditions (see Butzer 1964; Helbaek 1964a-b; Flannery 1965, von Wissmann et al. 1956:281).[Editor 74] Already in this area, the first stages of plant cultivation could be connected with both rain-fed and irrigated agriculture, based initially on natural overflows ('swamp' or estuary farming) and then on artificially irrigated plots (see also Drower 1954).

The above suggests that primitive irrigation skills appeared in the *sai*-brook form, especially in the particularly favorable landscape conditions of mountains and valleys of Southwestern Asia, even in the 'pre-agricultural' stage (10th-8th millennium BCE) among Mesolithic 'harvesting people'. The cyclic moisture changes, apparently, played an important role in this process.

Agriculture arose from highly organized harvesting, under particular historical and eco-logical conditions, at the turn of the Mesolithic and Neolithic, when the process of hybridization of the natural polyploidy, together with the effects of artificial irrigation, promoted the formation of numerous cultivated cereal species (wheat, barley, etc.). The development and improvement of irrigation played an important role in the early stages of the history of agriculture. It is possible to recall the words of F. Engels, who wrote that the middle stage of barbarism "in the East begins with the domestication of animals, on the West with the cultivation of edible plants through irrigation and the use of adobe (sun-dried mud bricks) and stone for building" (K. Marx and F. Engels 1961, Tom 21:30).

New archaeological and paleoethnobotanic research has generally proven the polycentric concept of crop origins developed by N. I. Vavilov, although some of his conclusions were reviewed (Titov 1962:14; Sinskaya 1966:30; Andrianov 1968b:17).

The research of the most ancient centers revealed that intensive harvesting was gradually replaced by crops and the new developed economy (with extensive exploitation of the territory) was transferred to production, which in turn resulted in changes of economic-cultural type and a sharp increase in population density.[Note 201]

The spreading of skills in irrigated agriculture was a complex historical-cultural process. Material on the history of agriculture from around the world provide no reason to consider it a simple process of mechanical skills in farming and irrigation methods transferred from one area to another, as a process of direct migration and direct diffusion historically spread out from a single center. It was, instead, a gradual and extremely diversified process in different ecological niche conditions of natural vegetation and water resources (Andrianov 1968a:28; 1968b:25). Irrigation technique features and the type of irrigation works depended to a large extent on the water from rivers with permanent flow and floods, lakes, temporary brooks, and underground water (Table 12).

Irrigated agriculture was the most ancient type of agriculture in Central Asia. Archaeological research revealed a picture of gradual and historic movement, in the 6th–3rd millennia BCE, of early farming cultures of Southern Turkmenia (based on estuary and mountain-brooks farming and herding) to northeast, into the area of Neolithic hunters, fishers and gatherers in the steppe and mountains.

At the end of the 5th–beginning of the 4th millennium BCE, farming tribes entered the delta of the Tedjen River, where archaeologists discovered remains of irrigation canals dated to the second half of the 4th–beginning of the 3rd millennium BCE, and investigated the Geoksyur Eneolithic sites of farmers (with typical crops of barley, common and club wheat). In the 3rd, and especially in the 2nd millennia BCE, i.e. in the Bronze Age, the area of productive economy shifted to the Zeravshan Basin (early farming sites of the Makhandarya), to the Lower Amudarya, to the Fergana Valley and some other areas.

Further development of irrigated agriculture and irrigation skills was connected to local geographical conditions and water resources. Irrigated agriculture in Central Asia was developed over different landscapes. In the piedmont zone and on the banks of river systems, the beginning of irrigation was associated with estuaries: artificial bordering of embanked areas of flood water required for growing crops.

In the mountainous valley areas, improvement of *sai*-brook irrigation led to the creation of mountain-terraced farming (with canals, aqueducts, basins and various water-regulating devices). The process of irrigated agriculture in the valleys and deltas of large Central Asian rivers was no less complicated. From estuary (marsh) farming, through regulation of water sources by building embankments for overflows and construction of protective dikes, to advanced freshet systems, regulation of seasonal river floods and large-scale use of water basins.

In some mountain areas (Kopetdag, Nuratau, Karatau, etc.) farmers used underground water and created *karez* farming. Forms of rain-fed (dry) farming were developed on the slopes sufficiently moistened by atmospheric precipitation. On the outskirts of the agricultural oasis, primitive *kair* and estuary agriculture, and in deserts crop cultivation and gourd plantations based on springs and seasonal rain water, persisted until the beginning of the 20th century.

* * *

The look of the 'lands of ancient irrigation' with dead ruins of ancient cities and settlements, of dry canal beds and traces of fields, often raised the question of the causes of demise of the ancient civilizations. Even now, some researchers tend to explain the neglect of the ancient oases solely as the effect of natural factors (a drying process in Central Asia, catastrophic climate changes, desertification, etc.). The famous geographer E. Huntington, a member of the American archaeological expedition in Turkmenistan directed by R. Pumpelly, as noted above, attempted to prove that the formation of the 'lands of ancient irrigation' in Asia resulted from the drying climate (see Huntington 1905, 1906, 1907, 1908a-b-c, 1910).[Editor 75] Criticism to such hypotheses,[Note 202] explaining the desolation of 'lands of ancient irrigation' by different physical and geographical factors (climate change, desertification, etc.), were advanced over the last 50 years by the remarkable Russian geographers L. S. Berg and A. I. Voeykov.

L. S. Berg wrote: "E. Huntington seeks to prove the idea that Asia, and even the whole Earth were, and are now, in a state of permanent drying process". Further, L. S. Berg dismembered one after another the arguments of E. Huntington, and concluded that "none of these reasons can convince us. The Near East and Central Asia are overcrowded by ruins dated to completely different epochs, cultures and periods. The causes of collapse of cultural settlements are very different and, for anyone familiar with the historical geography of Central Asia, it should be clear that it is unlikely possible to explain that through a change of climate conditions and the 'drying up'. The main reason for the fall of sedentary settlements was, of course, wars. In the 13th century, Genghis Khan and his successors destroyed a number of cities in Turkestan and the Near East, destroyed vast irrigation systems and killed a lot of people. As a result of the destruction of irrigation canals the population, deprived of opportunities to sustain itself, partly died and partly escaped" (Berg 1947: 63). Hypotheses, similar to the catastrophic drying of Central Asian plains proposed by E. Huntington, were raised by some other geographers and geologists, such as G. E. Grum-Grjimaylo, I. V. Mushketov and P. A. Kropotkin.[Editor 76]

A. I. Voeykov supported L. S. Berg in his debate with E. Huntington, I. V. Mushketov and P. A. Kropotkin. In 1912, he wrote that "Berg's research promoted, more than any other, to disprove the legend about 'drying Central Asia'" (Voeykov 1912). A few years later, A. I. Voeykov criticized the popular-geographical view of P. A. Tutkovskiy on the geographical reasons for Medieval nomadic invasions into the cultural oases. According to his words, "This is very unstable and unproved, and shows the author's little acquaintance with history when dealing with the movements of people" (Voeykov 1915 [1947]). [Editor 77]

Archaeological research in Central Asia fully confirmed the rightness of the opinion of L. S. Berg and A. I. Voeykov, and proved the wrongness of the popular geographical hypothesis. It turns out that the decline of artificial irrigation and desolation of the flourishing oases of Khorezm, Lower Zeravshan, Surkhandarya and Fergana were due primarily to socio-economic factors, such as wars and feudal fragmentation, which contributed to the drying up of canals, damage to distributing irrigation facilities, collectors networks and protective dikes bordering the oasis from devastating floods (Tolstov 1948a:43-56; Gulyamov 1949:9).

The general pattern of social upheaval resulting from the abandonment of oases, migration of the farming population, and the restoration and development of the desert landscape, was traced and reported by archaeologists in other areas. For example, a detailed study of ancient settlements and conditions of irrigation in the Diyala Basin allowed R. McC. Adams to identify changes in population over 5,000 years. In a period of political growth and development of irrigated agriculture the number of inhabitants of settlements and cities increased rapidly. On the contrary, wars and invasions of warlike neighboring tribes (Gutians, Elamites, etc.) adversely affected both population growth and irrigation. Particularly damaging was the invasion in 1258 CE by Hulagu Khan's army, during which, protective dikes and head structures were destroyed. After that destruction, the cultural oases of the Diyala could not recover until the 19th century. Research in Khorezm, Mesopotamia, and other areas of the Old World fully confirmed the conclusions of K. H. Marx on the main reasons for the decline of ancient irrigation cultures of the East. He wrote that "a devastating war was a means to destroy a country for centuries, and to deprive it of all its civilization" (Marx and Engels 1957, Tom 9:132; Marx and Engels 1962, Tom 28:216-223).

The desolation of oases in certain socio-historical conditions of crisis also contributed to the shifting of rivers and their channels, and the process of secondary salting, which destroyed crops.

In contrast to bourgeois, metaphysical science, dialectical materialism considers the relationship and interaction between natural and social phenomena, not as an eternal and immutable connection between absolutely the same things, but as a historical context, developing both in time and space. The proper solution to the problem of interaction between nature and society and its reflection in the combined methods of natural and social sciences, are of great ideological and practical significance in the development of historical-geographical issues. Questions on economic history are necessary to explain the historical dynamic of 'lands of ancient irrigation'. These lands are great monuments to the labor of many generations of farmers. These 'lands of ancient irrigation' have not only a past but also a future, and could be returned to their previous cultural status (if the soil were to be provided with water, fill dry canals and lay out new main irrigation canals). Of course, modern irrigation specialists could be helped by the millennial experience of ancient irrigators.

The archaeological and geomorphological studies carried out by the Khorezm Expedition showed that, within the limits of the ancient deltas of the large Central Asian rivers (Amudarya and Syrdarya), there were vast areas that preserved traces of ancient irrigation: remains of abandoned canals could be traced above cultivated fields and numerous remains of sites of different periods, from Prehistory to the Late Middle Ages.

The magnitude of these 'lands of ancient irrigation' in the Aral Sea area was approximately 5 million ha, i.e. almost three times the area covered by modern irrigation networks in the same territory. The whole area of 'lands of ancient irrigation' in the USSR covered 8-10 million ha, which is equal to the whole area irrigated at the present time. Their systematic comprehensive archaeological-geographical study, as well as their mapping based on aerial methods, can provide a significant economic effect. The information on ancient irrigation works can be widely used to design modern irrigation, and the large-scale archaeological maps can help forecast secondary soil salinization and the dynamic of water resources.

It must be said that the study and mapping of ancient irrigation works was carried out by only a small number of archaeological expeditions in the 'lands of ancient irrigation' of Central Asia, Kazakhstan, Caspian Sea area and Caucasus. Aerial methods and advanced natural scientific studies are also very seldomly used (see Tolstov 1961a-b-c; 1962a:315-322; 1962b; Andrianov 1964; Andrianov and Kes 1967:38).[Editor 78] Even V. V. Bartold, in his article *Budushee Turkestana i sledy ego proshlogo* (see Note 2), wrote that many of the issues in new developments "can be solved only by detailed studies in situ, with the participation of archaeologists, being acquainted with Turkestan's general history and settlements of its past, as the participation of technician-irrigators" is fundamental (Bartold 1965:310).

The great plan of watering and irrigating desert lands and arid regions of our country, promoted by the Program of the Communist Party of the Soviet Union, which was further developed in many decisions of the Central Committee, in order to sustain agriculture and to widen irrigation construction, provided colossal works for irrigating new lands in the Central Asian Republics, Kazakhstan, Caucasus, and the European parts of the USSR. In light of these complex popular-economic tasks, the comprehensive work of archaeologists, ethnographers, geomorphologists, and soil specialists in the study of the 'lands of ancient irrigation', a huge reserve of land suitable for modern priority irrigation acquire an extremely important practical meaning.

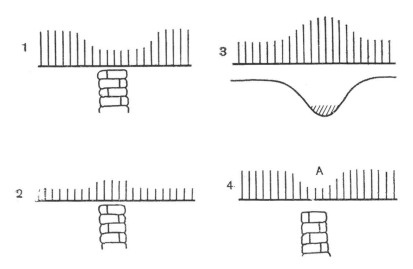

Figure 1. Ancient buried constructions disclosed by cultivated vegetation (according to Bradford 1957): 1) before harvest; 2) after harvest; 3) 30- 40 cm vegetation above a ditch; 4) light green color vegetation of (A), at maturity above a buried structure.

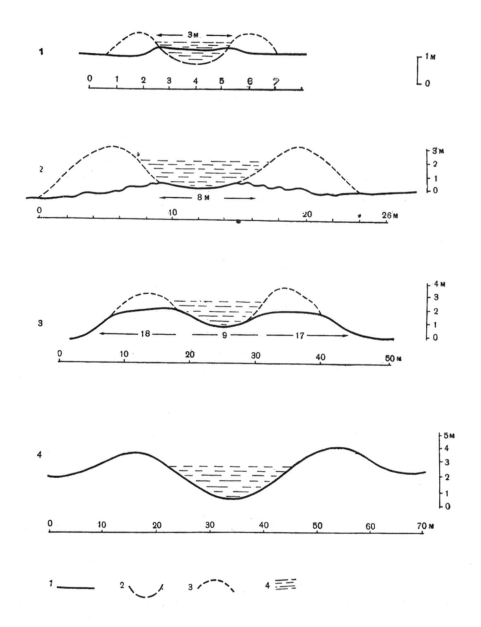

Figure 2. Profiles of ancient canals with different preservation conditions: 1) contemporary canal surface: at a Tazabagyab culture settlement (poisk 1609). Banks are completely destroyed; 2) profile of canal bed found in a trench: profile of Archaic canal near Bazarkala (poisk 423). Destroyed canal levees; 3) supposed canal levees: profile of ancient Kelteminar in its upper reaches (poisk 41). Canal levees preserved for a height of 2 m; 4) supposed water filled: profile of well-preserved canal in the environs of Yarbekir-kala (poisk 320).

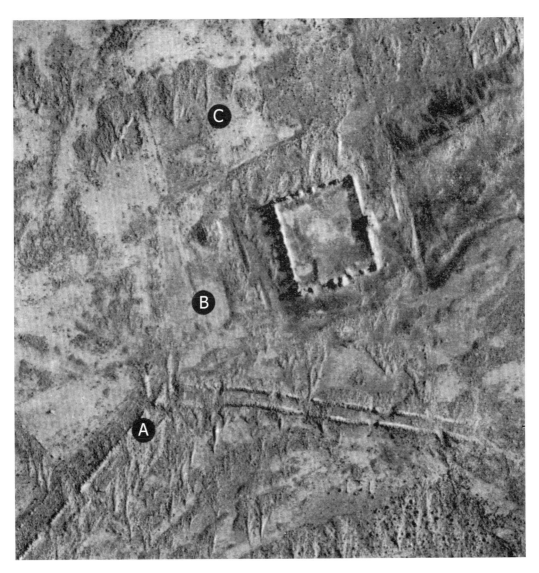

Figure 3. Environs of Big Kyrk-Kyz: A) the canal is detected by its levee shadows; B) small irrigation network; C) traces of vineyards (photo by N. I. Igonin).

Figure 4. Archaic canal (A) at Bazar-kala (photo by M. I. Burov) (see also the Profile 2 in Figure 2).

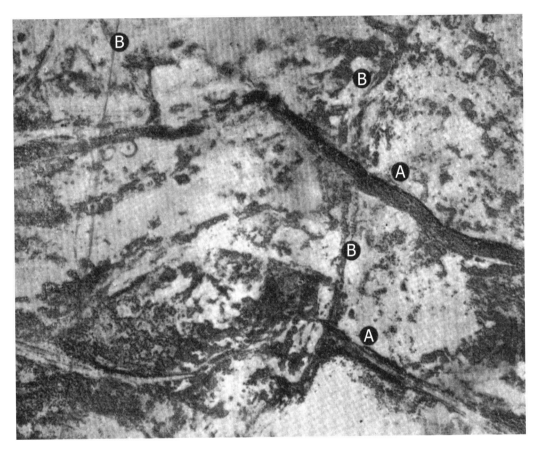

Figure 5. Prehistory (late Bronze Age). Predominance, in a landscape of typical takyr with biyurgun, with traces of wandering riverbeds in various areas, covered by sand. Rare small canals detached from natural channels; the systems do not exceed some kilometers in length; short feeders branch out at right angles; a small irrigation network is poorly preserved; the fields are along riverbeds and are highlighted by ground color and by vegetation. The contours of underground dwellings are visible thanks to aerial photos. Distinctive sectors: the area of the Tazabagyab settlement (third quarter of the 2nd millennium BCE) north of Djanbas-kala and the environs of Yakke 2 (9th–8th centuries BCE). On the aerial photo: Tazabagyab settlements north of Djanbas-kala; small dry riverbeds of the Akchadarya (A) highlighted by sand and vegetation accumulation; ancient ditches (B), highlighted by soil color and by vegetation (picture by N. I. Igonin).

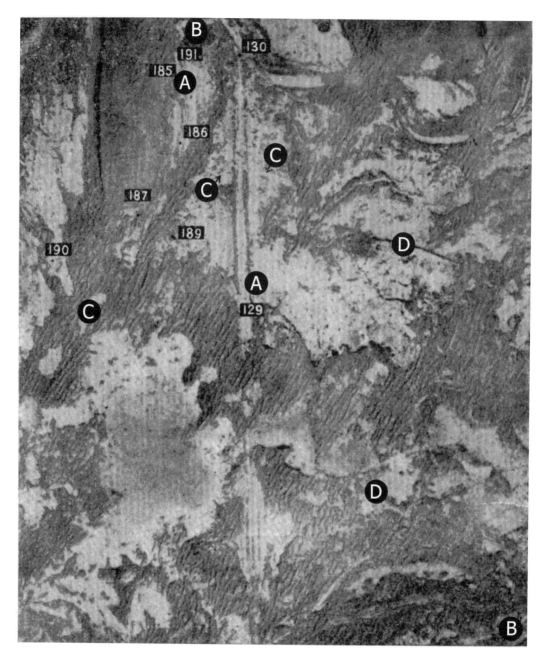

Figure 6. Archaic Khorezm (6th–5th centuries BCE). Predominance, in a landscape of typical takyr with biyurgun, of agro-irrigation layouts; massive canals (up to 50 m wide), stretched tens of kilometers, detached from lateral channels and repeating their outline, forming rare branches at obtuse and right angles, mainly in one direction. The irrigation systems are largely destroyed; the levees are not preserved. Rural settlements are rare and can be found only with field surveys. On the aerial photo: a north-south main canal (A) in the surroundings of Kyuzeligyr, detached from a riverbed (B); ditches (C) are highlighted by vegetation. More subsequent (Medieval) canals (D) cover the Archaic systems (picture by N. I. Igonin).

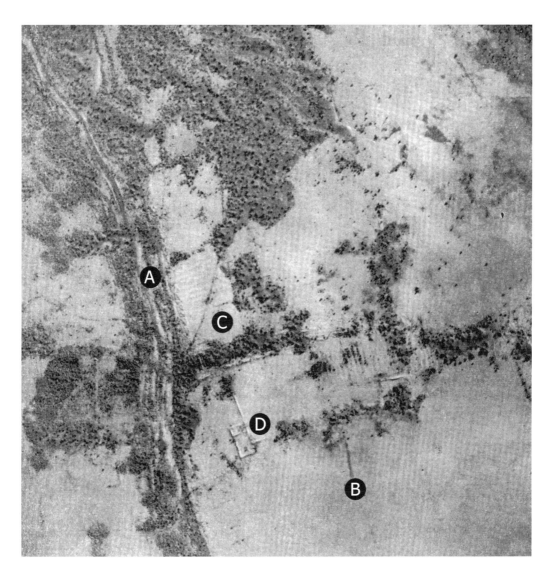

Figure 7. Kangju and Kushan Khorezm (4th century BCE–4th century CE). Predominance, in a landscape of typical takyr with biyurgun, of ruins of fortified cities and large unfortified rural settlements, surrounded by fields and vineyards; the combination of cities with garden-park complexes is characteristic. Main canals branch from riverbeds; they have a perfect tree-shaped layout with frequent canal networks. Vineyards are big, up to 100–200 m long and 60–80 m wide.[Editor 79] *Farms are located near the main canal. Distinctive sectors: the environs of Djanbas-kala. On the aerial photo: a main canal bed (A); a small irrigation network (B); traces of vineyards (C) and ruins of a Kushan farmstead (D) at poisk 611 (picture by N. I. Igonin; see also Figure 3).*

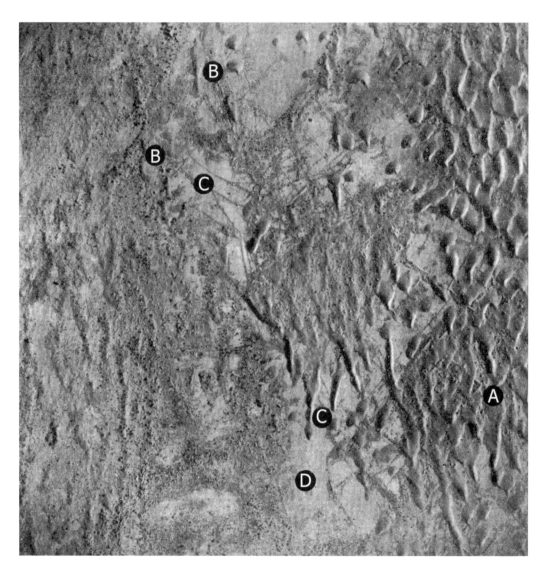

Figure 8. Afrigid period (5th-8th centuries CE). Predominance, in a landscape of typical takyr with biyurgun and saxaul, of fortified farm ruins located on lateral canal branches. The irrigation system differs from those of previous historical periods with a higher number of lateral branches. The canals are no more than 7-8 m wide between banks. The fields layout is very dissimilar. Distinctive sectors: the Berkut-kala oasis (7th-8th centuries CE), the environs of Adamli-kala and Karga-kala. On the aerial photo: the environs of the Afrigid farm of Dingildje (A; compare with plan in Figure 35); remains of ancient canals are seen through the sand (B); they are cut by Afrigid canals (C); traces of gourd plantations and vineyards are poorly visible (D).

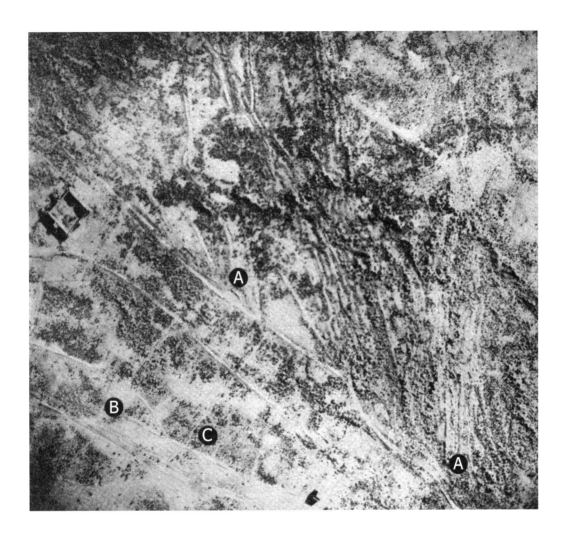

Figure 9. Khorezmshah period (12th-beginning 13th centuries CE). Predominance, in a landscape of typical takyr with biyurgun and saxaul, of numerous ruins of unfortified farmsteads and individual small fortifications among a dense network of canals and fields. The irrigation system is characterized by a complex branching structure of canals, along which the farmsteads are clustered. Distinctive sector: the Kavat-kala oasis. On the aerial photo (by N. I. Igonin): a series of main canals (A), a small canal network (B), fields (C); castles and farms completely preserved with walls that clearly stand out against the light takyr.

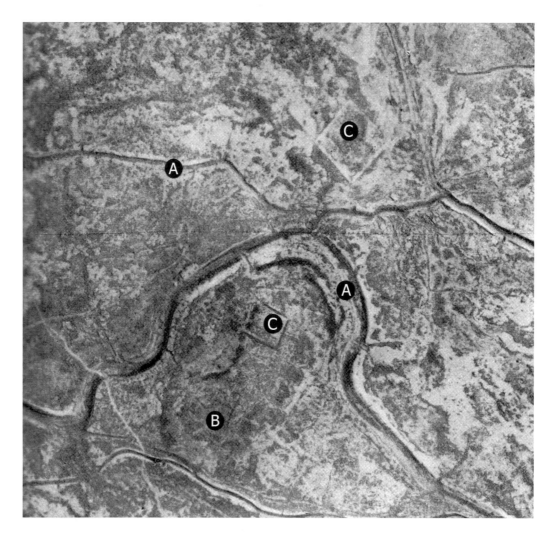

Figure 10. Golden Horde period (14th century CE). Predominance, in a landscape of takyr-shaped ground with keurek and saxaul vegetation, of agro-irrigation layouts, canals and individual ruins of unfortified rural settlements with sparse buildings. The irrigation system is characterized by complex, branched layouts. There are many large chigir pits and some large garden-park complexes enclosed by fences. Distinctive sector: the environs of Shekhrlik on the left bank of the Amudarya. On the aerial photo (by N. I. Igonin): riverbed and canal (A); a small ditch network (B); garden-park layouts (C).

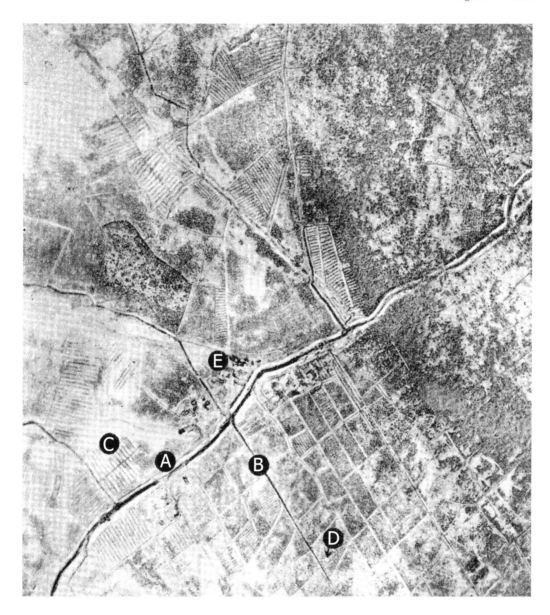

Figure 11. Abandoned settlements of the 18th-19th centuries. Predominance of agro-irrigation layouts in the landscape (on the takyr grounds with large solonchak, yantak and itsitek vegetation). On the aerial photo (by N. I. Igonin): the environs of Kattakar-Chardere, where, according to B. I. Vaynberg's data (1960:fig. 1), the Turkmen Sakar tribe lived along the Middle Sipay-yab. Here a Medieval main canal (A) and a regular small irrigation network (B) were rebuilt; many gourd plantations (C), traces of ridged yurts (D) and ruins of buildings (E).

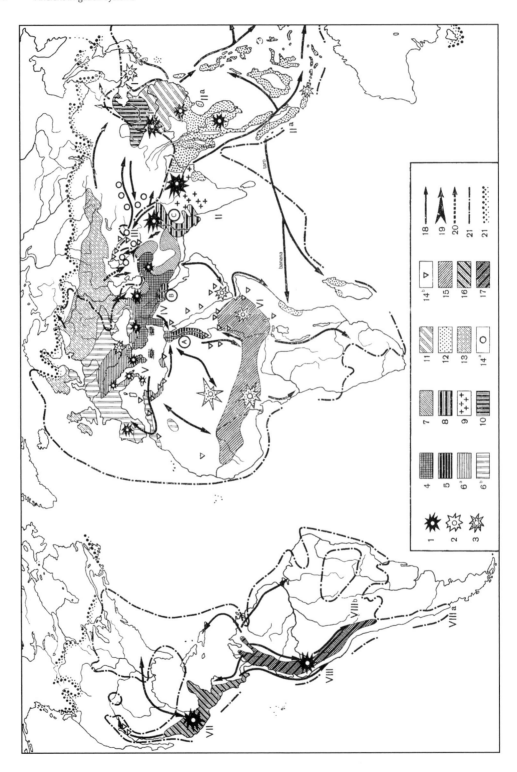

Figure 12.
World centers of plant cultivation and skills of irrigated agriculture (8th–2nd millennia BCE).
World centers (origin centers) of important cultivated plants according to N. I. Vavilov.

I - Chinese; II - Indian; IIa - Indo-Malayan; III - Central Asian; IV - Front Asian; V - Mediterranean; VI - Ethiopian (Abyssinian); VII - South Americana and Central Mexican: VIII - South American (Peruvian-Ecuadorian-Bolivian); VIIIa - Chilean; VIIIb - Brazilian- Paraguayan. 2) Centers of cultivated plants development according to N. I. Vavilov, which are not proven by archaeological study. 3) Independent centers of cultivated plants development: Equatorial Africa, the Black Sea area (?). 4) Areas of the Old World, where centers of stick-hoe agriculture appeared: irrigated and non-irrigated (with development of irrigation skills from mountainous sai and estuary irrigation to regulated river floods and self-flowing irrigation based on delta canals); main cultivated plants: wheat (einkorn (Triticum monococcum) and emmer (Triticum dicoccum)), barley, legumes, fruits; area of domestication of goat, sheep, horned cattle. 5) Area of hoe development (plow) irrigated agriculture in the 5th–4th millennia BCE; origin of urban civilizations (4th–3rd millennia BCE); types of irrigation: A - flood basin (Egypt); B - dammed delta canals (Mesopotamia); main cultivated plants: cereals (wheat and barley), legumes, fruits, oil-bearing and spinning plants (flax, etc.), vegetables; area of goat, sheep, horned-cattle, swine and mule (domesticated in Egypt) breeding. 6) Area of non-irrigated hoe (plow) agriculture in 5th millennium BCE (a) and 4th–3rd millennia BCE (b); main cultivated plants: wheat (einkorn (Triticum monococcum) and emmer (Triticum dicoccum)), two-rowed barley (Hordeum distichum), lentil, millet; area of sheep, swine, and horned-cattle breeding. 7) Area of stick-hoe irrigated agriculture (sai-brook) and non-irrigated agriculture in the 4th millennium BCE; main cultivated plants: wheat (soft (Triticum aestivum) and club (Triticum compactum)), barley, legumes, sesame, melons and gourds, fruits; area of presumed domestication of Bactrian camel and markhoor (Western Himalayas). 8) Area of hoe irrigated agriculture (based on floods) of the urban Indus civilization (2500-1700 BCE) (C); main cultivated plants: wheat (soft (Triticum aestivum) and club (Triticum compactum)), barley, legumes, sesame, cotton, melon; area of domestication of wild local oxen, buffalo and zebu-type horned cattle. 9) Area of irrigated hoe agriculture (with the use of wells and 'tanks') and non-irrigated (monsoon) slash agriculture of Central India and the Ganges Valley (?) in the 2nd–mid-1st millennium BCE; main cultivated plants: rice, wheat, legumes; area of horned cattle breeding. 10) Non-irrigated stick-hoe slash-and burn agriculture of ancient China (the 3rd millennium BCE?) with the main cultivated plants: millet (three types), kaoliang, legumes (soy bean, adzuki), bamboo, edible roots and tuber roots; zone of Asian pig breeding. 11) Non-irrigated and irrigated stick-hoe agriculture with predominant rice cultivation (3rd millennium?); area of cultivation of cereals, legumes, edible roots and tuber roots, vegetables, fruits; area of Asian pig domestication. 12) Area of tropical and sub-tropical stick-hoe agriculture (on monsoon precipitations) and partly irrigated agriculture; mountainous-terrace (2nd–1st millennia BCE); prevailing plants: edible roots and tuber roots (taro, yam), sugarcane, fruits (bananas, citrus), rice (upland and irrigated); area of swine and poultry breeding. 13) Area of non-irrigated slash-and-burn agriculture (in the West) and flood-land (in the East) hoe agriculture with prevailing millet and barley; area of sheep, cow and horse breeding and Bactrian camel breeding in the East (the 2nd millennium BCE). 14) Centers of irrigated agriculture in arid zones: a - the 2nd millennium BCE (including Khorezm); b - the 1st millennium BCE (including the Lower Syrdarya). 15) Area of non-irrigated stick-hoe tropical agriculture in Africa with cereals (fonio, sorghum, African millet) and edible roots (yam) in the 2nd–1st millennia BCE (?). 16) Area of irrigated (sai-brook) and non-irrigated slash-and-burn stick and hoe agriculture in Mesoamerica with cultivation of pea, gourd (pumpkins), pepper and maize (the 3rd–1st millennia BCE). 17) Area of irrigated (sai-brook) and non-irrigated (mountainous and tropical) stick-hoe agriculture in South America with legumes, potato, other tuber roots, cotton, maize (the 3rd–1st millennia BCE). 18) Further spreading of cultivated plants. 19) Advance of club wheat (Triticum compactum) in Khorezm. 20) Possible way of millet penetration into Khorezm. 21) Borders of maximal agriculture spread: a - by the 15th century; b - in the 20th century.

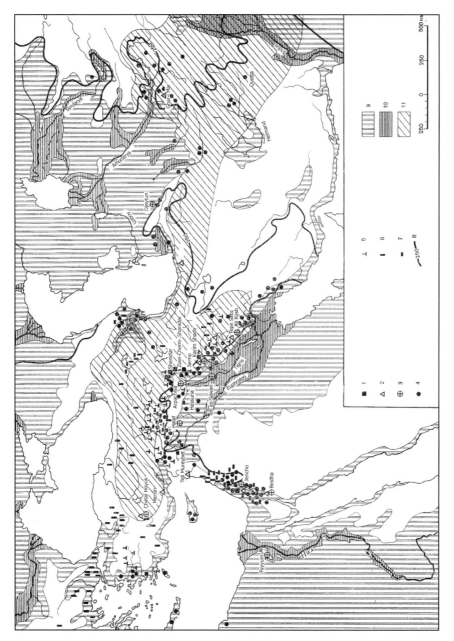

Figure 13. Southwestern Asia in the 11th–6th millennia BCE. 1) Pre-agricultural settlements of Mesolithic 'harvesting people' (11th–9th millennia BCE); 2) Sites with 8th–7th millennia BCE layers; 3. Findings of earliest cereals with weak traces of domestication in layers of the 7th–6th millennia BCE. Black points indicate recognized (according to Harlan and Zohary 1960) diffusion of: 4. Wild barley; 5. Wild one-grained wheat (einkorn – Triticum monococcum) in primary contexts; 6. Wild wheat (einkorn – Triticum monococcum) in secondary contexts; 7. Wild two-grained wheat (emmer – Triticum dicoccum); 8. Average annual amount of precipitation (250–300 mm); 9. Territories lower than 500 m asl; 10. Alluvial plains of rivers; 11. Reconstructed aerial of club wheat (Triticum compactum) origin according to N. I. Vavilov.

Figure 14. Primitive wooden and stone soil-digging tools. Old Egyptian hoes: 1,3) Middle Kingdom; 2, 4) New Kingdom; 5) 5th Dynasty (see Petrie 1920:pl. LXVIII; Wreszinski 1923-1936:taf. 97); 6) soil-digging stick of the Pima and Aymara American Indians; 7) small spades and knives of Andean Indians (see Casanova 1946:621:fig.50);[Editor 80] 8) small spades and knives of Andean Indians (see Bennett 1946:613, Pl. 131.b-d-e); 9) Australian woman root digger (according to G. Kunov); 10) Spanish soil-digger of the 19th century; 11) Chinese lei type forked soil-digger (Neolithic Lunshan culture); 12) socketed soil-digger with iron tip from Western Sudan. Kumans' tools (according to Nilles 1942-45): 13) soil-digger stick (2 m long); 14) hoe; 15) tip in hard wood; 16) adze; 17) axe; 18) wooden spade; 19) Maori soil digger.

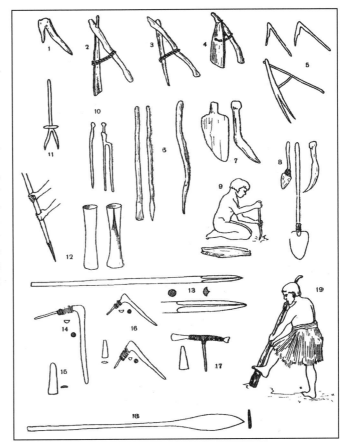

Figure 15. Paiute's cultivated lands (according to Steward 1933): 1) springs; 2) dam; cereals plot; 4) irrigation canals; 5) settlements; 6) irrigated plots; 7) pines.

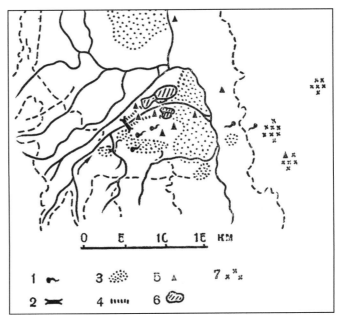

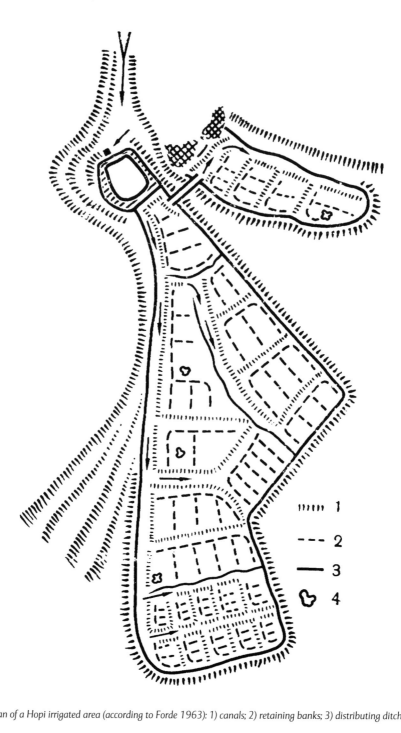

Figure 16. Plan of a Hopi irrigated area (according to Forde 1963): 1) canals; 2) retaining banks; 3) distributing ditches; 4) trees.

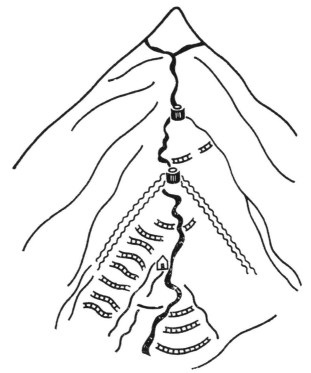

Figure 17. Scheme of Inca mountain-basin irrigation.

Figure 18. Peruvian Indians medieval irrigation fields.

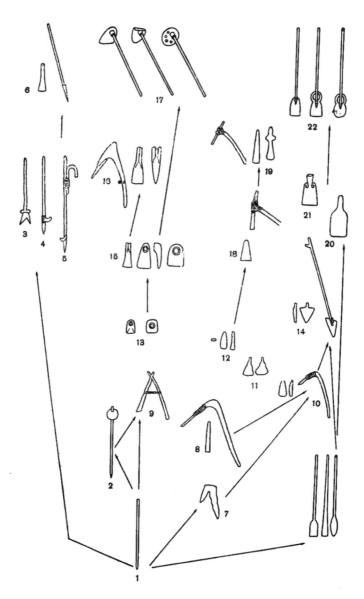

Figure 19. Scheme of development of soil-digging tools used in irrigated agriculture. Tools with a wooden or stone working part: 1) picket, shovel-digging stick; 2) digging stick with weight; 3) old Chinese digging stick; 4) Maori digging stick 5) Inca digging stick; 6) digging stick from Western Sudan; 7) wooden hoe; 8) wooden composite hoe; 9) old Egyptian hoe; 10) composite hoe with stone tip; 11) stone hoe from Hassuna; 12) hoes from Mersin (layer XXVII); 13) stone adze-hoe from Susa C and Tepe Gawra; 14) hoe-spade from Hama for irrigating furrows. Tools with a metal working part: 15) bronze hoe tips; 16) iron hoe tip; 17) Central Asian iron ketmen; 18) bronze hoe tip from Mersin XVII; 19) iron hoe tip; 20) bronze shovel from Chanhu-Daro (25th-17th centuries BCE); 21) 5th-3rd centuries BCE bronze shovel from China; 22) Central Asian iron shovels.

Figure 20. Contemporary Ethiopian digging sticks (according to N. I. Vavilov).

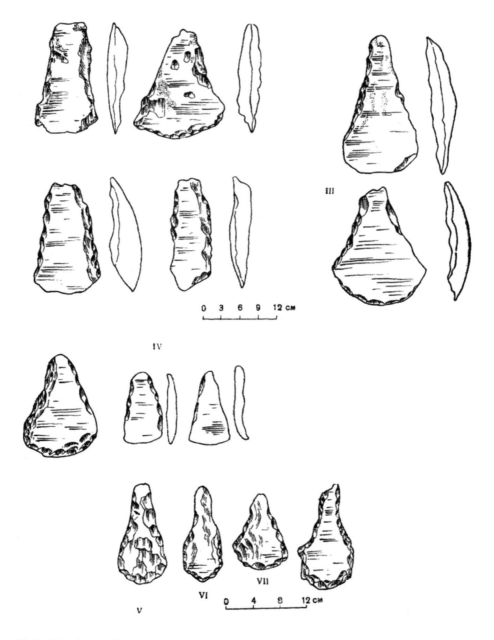

Figure 21. 6th-5th millennia BCE stone hoes: I) quartz hoes with traces of bitumen from Hassuna (layers Ia-II); II) sandstone hoe from Hassuna; III) obsidian hoe from Hassuna (according to Lloyd, Safar and Braidwood 1945); IV) hoes from Sialk I-II; V) Jemdet Nasr; VI) Ubaid; VII) Uruk (according to Christian 1940).

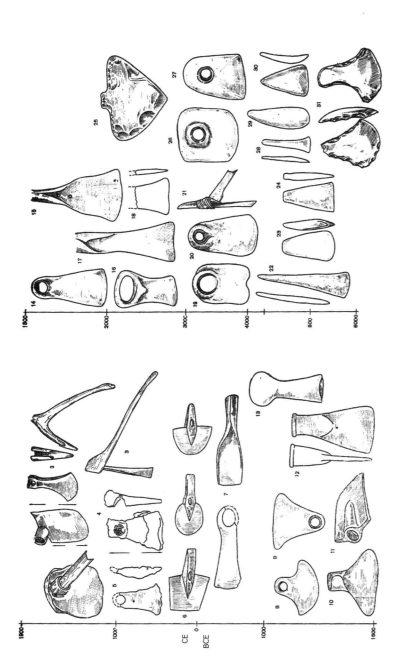

Figure 22. Development of working tools. Iron tools: 1) ketmen, Samarkand Province; 2) kaland Tajik hoe; 3) Tibetan hoe; 4) tools from Tok-kala, Khorezm (8th century CE); 5) hoe from Akbeshim; 6) Roman hoes; 7) Scythian hoes of the 6th century BCE (according to Shramko 1961). Bronze tools; 8) hoe from Tagiloni; 9) hoe from Colchis; 10) tip hoe from Bubastis (13th–12th centuries BCE); 11) celt spade; 12) celt from Kyrgyzstan (according to Kuzmina 1966); 13) celt from Troy (12th century BCE). Bronze tools:14) hoe from Maykop kurgan; 15) tip hoe from Byblos; 16) hoe from Mari (25th–22nd centuries BCE); 17) hoe from Ur (22nd–21st centuries BCE); 18) tools from Yalangachdepe (Southern Turkmenia); 19) hoe from Susa C; 20) hoe from Susa (27th–25th centuries BCE); 21) ancient Egyptian adze; 22) hoe from Mersin (layer XV); 23) hoe from Mersin (layer XVI); 24) hoe from Mersin (layer XVII). Stone tools: 25) hoe-spade from Hama for digging irrigation ditches (according to Steensberg 1964); 26) hoe from Susa C; 27) hoe from Tepe Gawra; 28) hoe from Mersin (layer XXIII); 29) hoe from Mersin (layer XXVII); 30) Neolithic tools from India (according to Sankalia 1964); 31) hoes from Hassuna (layer Ia).

Figure 23. Tops of digging sticks, maces and ritual scepters: A, B, C) petrogliphs (according to Stow 1930).

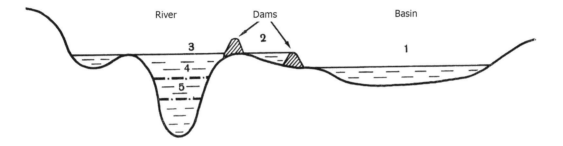

Figure 24. Simplest, basin irrigation system: 1) basin; 2) dikes; 3) highest level of river overflow; 4) medium level; 5) lowest level.

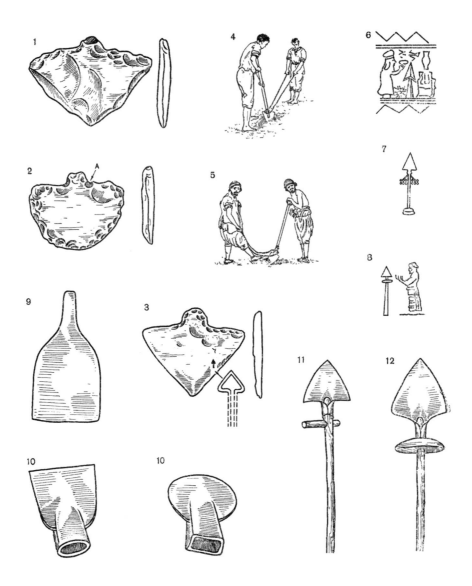

Figure 25. Hoes and spades used in irrigation work. Stone hoes from Hama (according to Steensberg 1964): 1) layer j6 (2300 BCE), 30 cm long, 20.2 cm high, 2-3 cm thick; 2) from Hama 3, 522 (2159 BCE), 23.5 cm long, 18.4 high, 2.8 cm thick; 3) from Hama 3, H45 (2150 BCE): sign of an arrow-shaped spade is visible on the blade; 4) A. Steensberg digging an irrigation furrow; 5) digging an irrigation furrow in Afghanistan (according to Vavilov and Bukinich 1929); representations of Babylonian spade-hoe on cylindrical seals: 6 (see Frankfort 1939:pl. XXXII.i), 7 (see Frankfort 1939:pl. XXXIII.d), 8 (see Frankfort 1939:pl. XXXVI.i); 9) bronze spade from Chanhu-Daro (25th-17th centuries BCE); 10) bronze Chinese spades from digs in Hunan province (5th-3rd centuries BCE); 11) modern iron spade from Afghanistan; 12) modern iron spade from Iraq (according to Hopfen 1960).

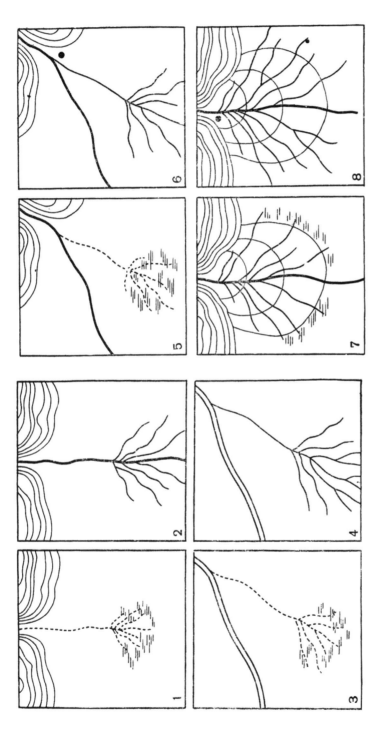

Figure 26. Scheme of irrigation development in the Fergana Valley (according to B. A. Latynin): 1) estuary flooding in the fading lower branches of a mountainous brook or sai on a plain; 2) gradual transformation of estuary flooding into earliest small artificial irrigation system by cleaning the riverbed and creating lateral canals; 3) estuary flooding in the fading lower deltaic branches of a river plain; 4) transformation of estuary flood into an artificial irrigation system by cleaning its riverbed and creating a network of lateral canals; 5) estuary flooding of freshet brooks from a mountain river entering the plain; 6) subsequent change of estuary flood by cleaning and building a head water-intake structure into the main canal with fan-shaped irrigation network (black dot in schemes 6 and 8 is a fortification for locking the irrigation system); 7) flood delta discharging channels of a mountain river, with a cone-shaped branching at the entry of the plain; 8) transformation of a delta, after cleaning of deltaic branches and the creation of lateral canals, in a large fan-shaped irrigation system typical of the Fergana Valley.

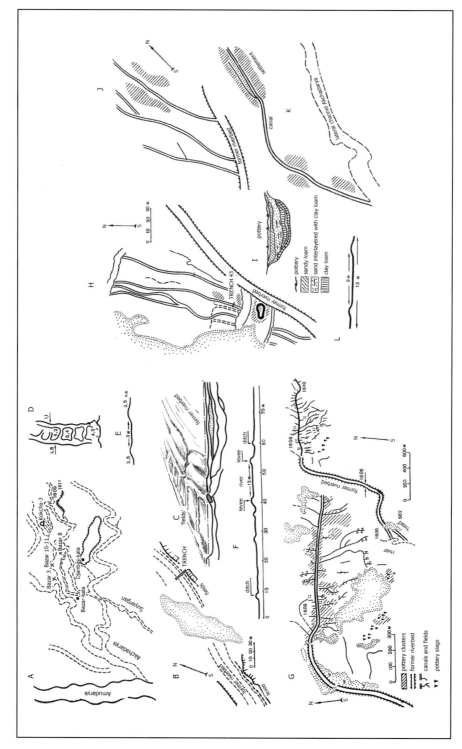

Figure 27. Historical development of irrigation systems in the Aral Sea area: Lower Amudarya. A) Southern delta of the Akchadarya (mid-2nd millennium–8th century BCE); B) 'fields' in the environs of Kokcha 3 graveyard; C) reconstruction of Kokcha system (Trench 35); D) 'fields' near the Kokcha 1 settlement; E) canal transverse profile at poisk 1609; F) transverse profile of former riverbed adapted for irrigation purpose near Kokcha 16 (poisk 1611); G) irrigation and settlements of the Tazabagyab culture north of Djanbas-kala (poisk 1605–1610); H) fields and canals of the Suyargan culture near the settlements of Bazar 3; I) trench section through a canal near Bazar 3; J) Amirabad period (9th–8th centuries BCE) irrigation near the site of Bazar 8; K) canal near the site of Bazar 10 and 11; L) profile of former riverbed.

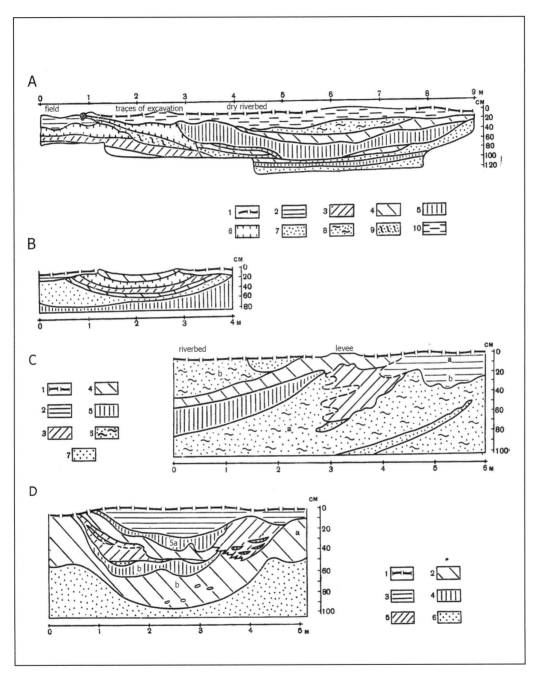

Figure 28. A) Kokcha 3. Trench 35; B) Canal trench near Bazar 2; C) Riverbed trench (poisk 1611); D) Canal trench (poisk 1609).

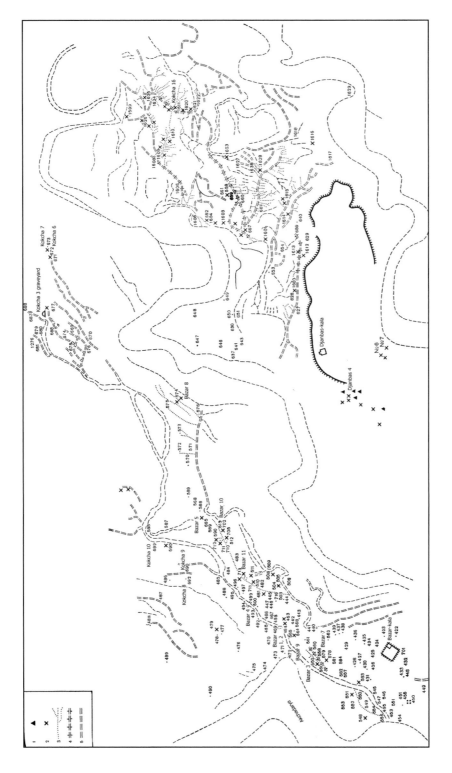

Figure 29. Prehistoric settlements in the environs of Djanbas-kala.

Figure 30. Settlement plan at poisk 611: pottery cluster and dwellings layout; 2) ceramic production area; 3) former riverbed adapted for irrigation purposes with bank remains in some places; 4) canals and fields (according to a photo by B. V. Andrianov and N. I. Igonin).

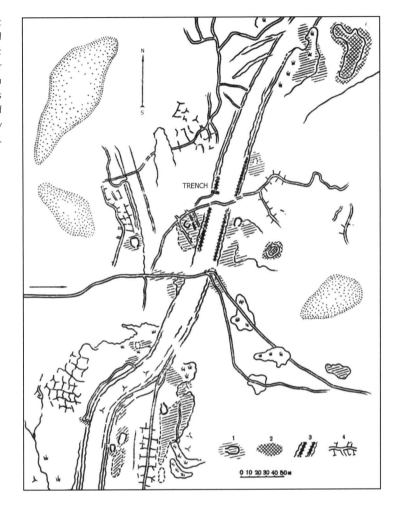

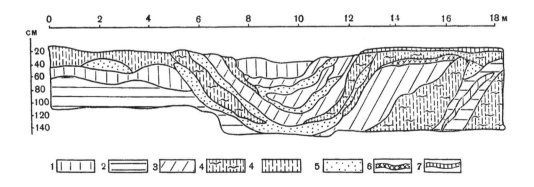

Figure 31. Trench 5 near the site of Yakke-Parsan 2.

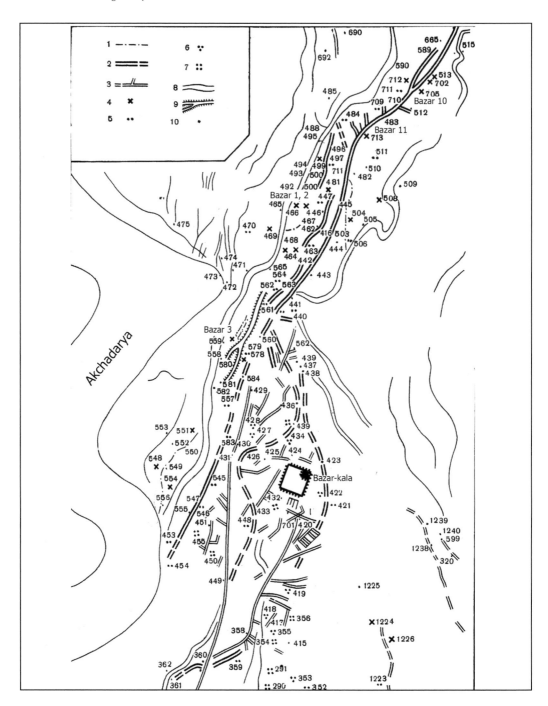

Figure 32. Canals in the environs of Bazar-kala: 1. Bronze and Early Iron Ages; 2. Archaic; 3. Kangju and Kushan. Settlements and large pottery dispersions: 4. Bronze and Early Iron Ages; 5. Archaic; 6. Kangju; 7. Kushan; 8. dry riverbeds; 9. former riverbeds adapted for irrigation purposes; 10. Poisk number.

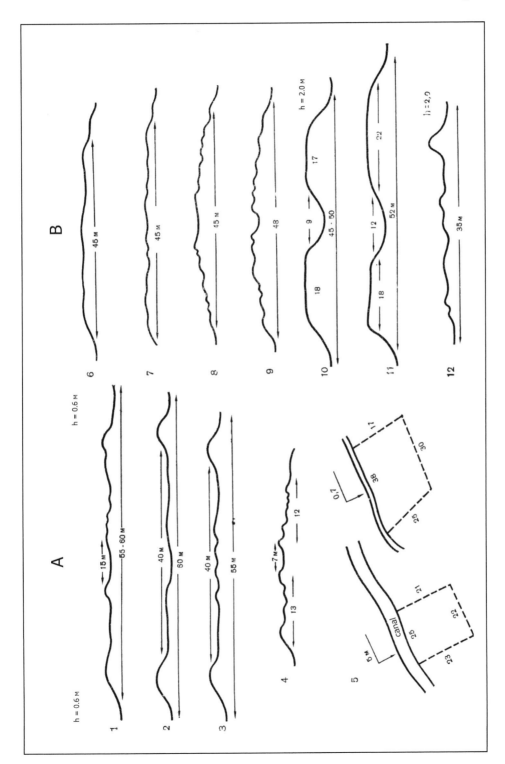

Figure 33. Canal profiles: A – 1) profile of Archaic canal near Bazar-kala (poisk 711); 2) poisk 443; 3) poisk 441; 4) poisk 453; 5) Archaic fields in Dingildje sector (at poisk 155 and 1205); B – 6) profile of Archaic canal of Kelteminar (poisk 367); 7) poisk 113; 8) poisk 38; 9) poisk 40; 10) poisk 41; 11) poisk 142; 12) profile of Archaic canal in the Dingildje sector.

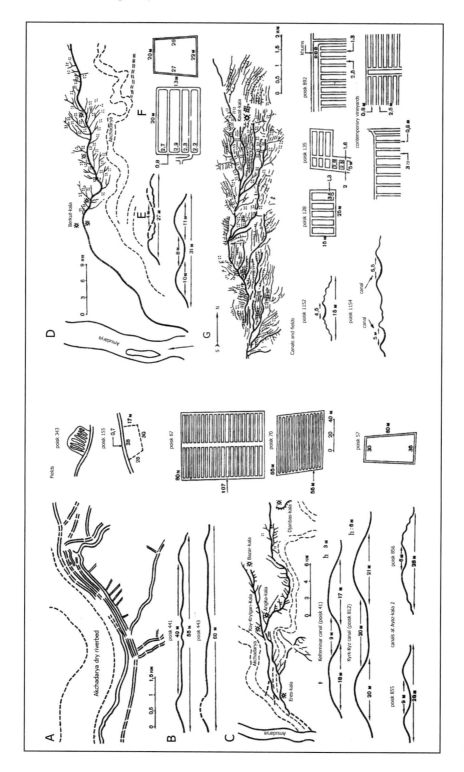

Figure 34. Historical development of irrigation systems in the Aral Sea area: Lower Amudarya and Southern Delta of the Akchadarya (7th century BCE–beginning of the 13th century CE). A) Archaic Kelteminar (6th–5th centuries BCE); B) Archaic period canal sections; C) Kangju-Kushan Kelteminar (4th century BCE–4th century CE); D) Afrigid Kyrk-Kyz (7th–8th centuries CE); E) Afrigid canals in the environs of Koy-Krylgan-kala and Big Kyrk-Kyz; F) Afrigid fields at Kum-kala; G) Khorezmshah Gavkhore (12th–beginning of the 13th centuries).

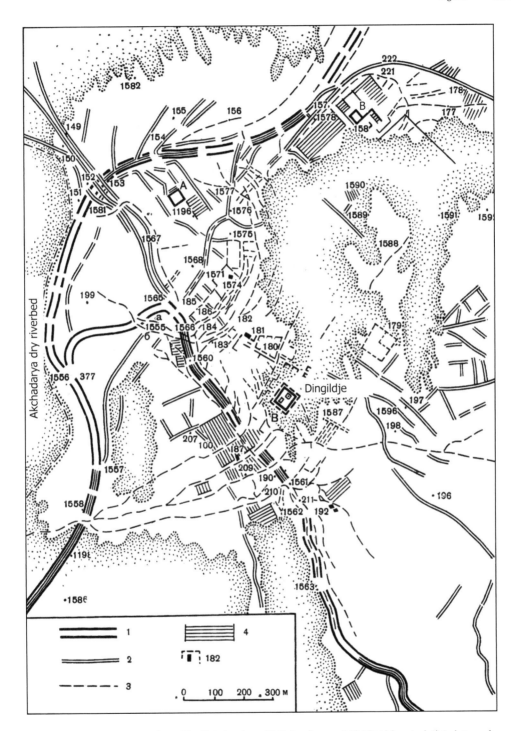

Figure 35. Canals in the environs of Dingildje. A) Archaic house; B) Kushan farmstead; C) Afrigid farmstead: 1) Archaic canals; 2) Kangju-Kushan canals; 3) Afrigid canals; 4) vineyards; 5) ruins of dwellings and archaeological poisk number.

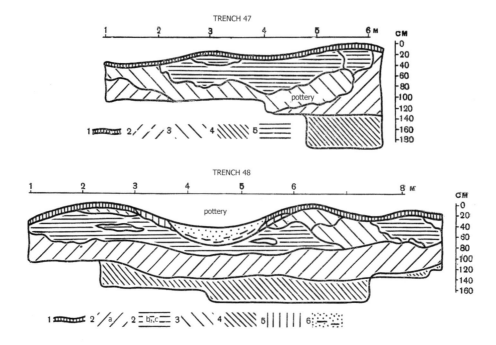

Figure 36. Sections of Trench 47 and Trench 48.

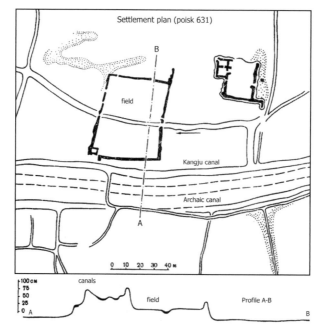

Figure 37. Plan of farm at poisk 631.

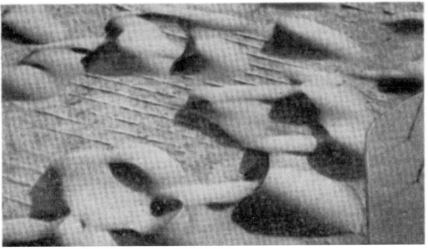

Figure 38. Vineyards in the environs of Koy-Krylgan-kala.

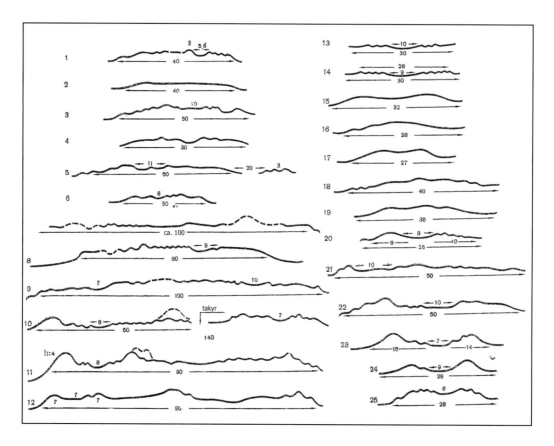

Figure 39. Profiles of Kyrk-Kyz canal: 1) poisk 870; 2) poisk 863; 3) poisk 769; 4) poisk 937; 5) poisk 934 and 935; 6) poisk 790; 7) poisk 797; 8) poisk 792; 9) poisk 838; 10) poisk 839; 11) poisk 844; 12) poisk 814. Profiles of Yakke-Parsan canal: 13) south of Narindjan; 14) north of Narindjan; 15) poisk 740; 16) poisk 753; 17) poisk 736; 18) poisk 757; 19) poisk 758; 20) poisk 1057; 21) poisk 1059; 22) poisk 1063; 23) poisk 1070; 24) poisk 855; 25) poisk 856 (environs of Ayaz-kala).

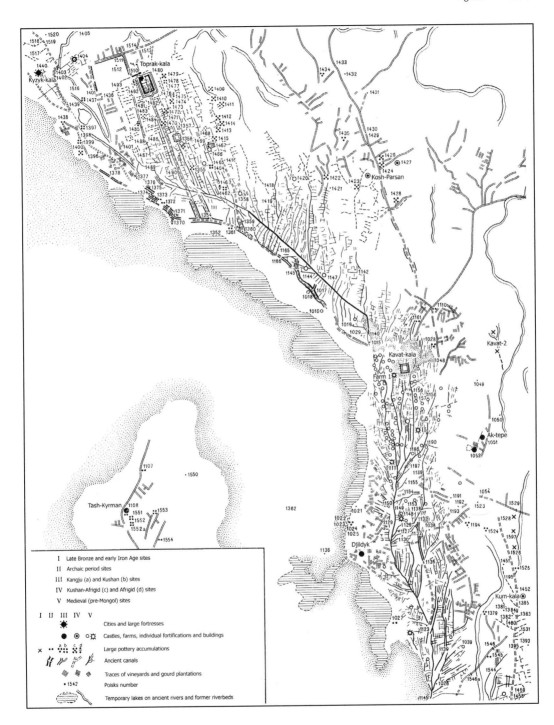

Figure 40. Ancient Gavkhore

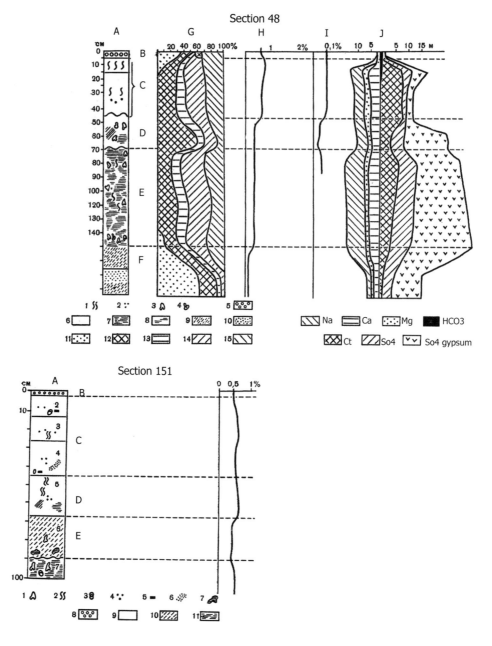

Figure 41. Section 48. A) morphological structure; B) takyr crust; C) agro-irrigated layer, uninterrupted cultivation; D) agro-irrigated layer (?), cultivation with interruptions; E) complex of alluvial clay of tugai buried soils with traces of waterlogging; F) complex of alluvial sandy loam of tugai buried soil; G) lithological column; H) humus distribution; I) phosphorus distribution; J) salt distribution: 1. salts; 2. charcoals; 3. rust spots; 4. freshwater mollusks shells; 5. of takyr origin; 6. non-stratified, heavy clay loam; 7. stratified, clay loam; 8. stratified, heavy clay loam; 9. stratified, loamy sand-sand; 10. stratified, sandy ground; 11. > 0.05 mm (sand); 12. 0.05-0.01 (large-sized powder); 13. 0.01-0.005 (medium-sized powder); 14. 0.005-0.001 (small-sized powder); 15. > 0.001 (silt).

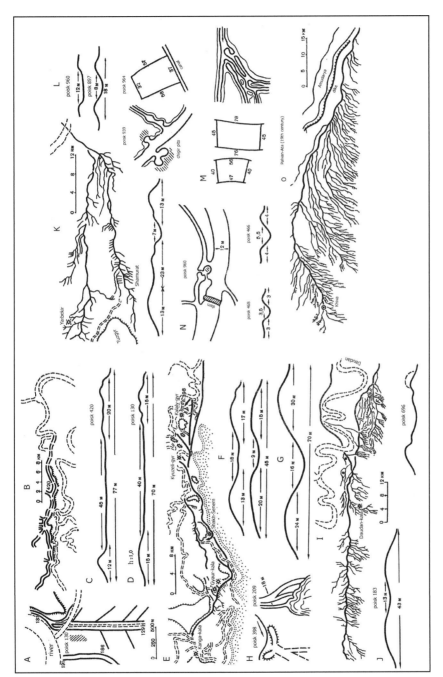

Figure 42. Historical irrigation development: Lower Amudarya: Sarykamysh Delta: A) Archaic canals (6th–5th centuries BCE) in the Chermen-yab system; B) Archaic canal on the southern Daudan; C) canal transverse profile (poisk 420); D) canal transverse profile in the Chermen-yab system (poisk 130); E) Kangju-Kushan Chermen-yab (4th century BCE–4th century CE); F) canal transverse profile at Gyaur-kala and at poisk 732; G) canal transverse profile at Mangyr-kala; H) head structures at poisk 398 and 209; I) Khorezmshah Chermen-yab (12th century–beginning of 13th century CE); J) canal transverse profiles (poisk 153 and 696); Medieval (post-Mongol) irrigation: K) Medieval Shamurat; L) canals and hydraulic structures in the environs of Shekhrlik; M) later systems (18th–19th centuries); N) canal profiles; O) irrigation system at Palvan-Ata (19th century).

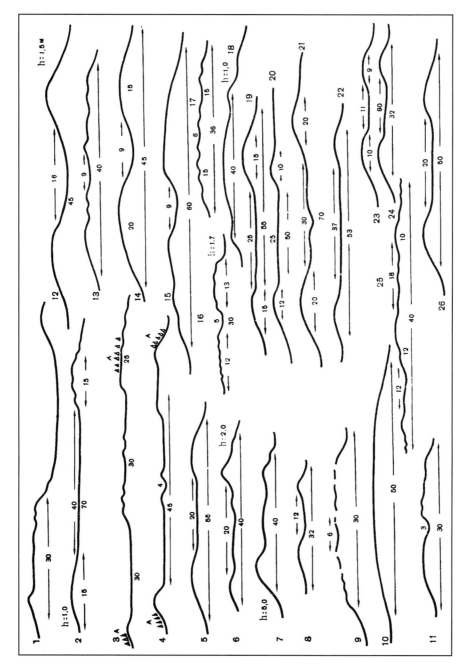

Figure 43. Canal profiles: 1) Archaic canal (poisk 35); 2) poisk 130; 3) poisk 259 (A - Archaic ceramics); 4) poisk 420 (A - Archaic ceramics); 5) poisk 580; 6) poisk 289; 7) poisk 579; 8) poisk 572; 9) Kangju-Kushan Chermen-yab at Kyunerli-kala (poisk 766); 10) poisk 754; 11, 12) canals north of Kyunerli-kala; 13, 14) poisk 732; 15) poisk 751; 16) poisk 175; 17) poisk 139; 18) poisk 55; 19) poisk 99; 20) poisk 102; 21) poisk 105; 22) drainage canal (poisk 4, year 1952; poisk 23 and 24) Kushan Chermen-yab (poisk 9 and 11, year 1952; poisk 25) Chermen-yab at Gyaur-kala; 26) Chermen-yab at Kuyu-Saygyr.

Figure 44. Yarbekir-kala. The ancient canal is visible in the upper part at left (picture by N. I. Igonin.).

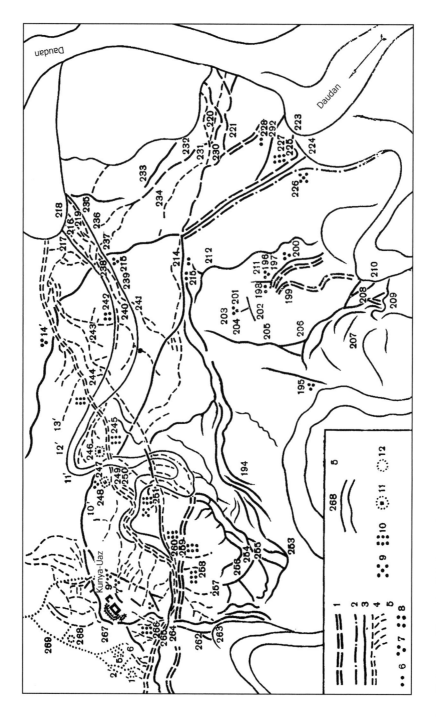

Figure 45. Environs of Kunya-Uaz: 1) Archaic canals; 2) Kangju and Kushan canals; 3) Medieval (pre-Mongol) canals; 4) Medieval (Mongol) canals; 5) numbers of poisk and dry riverbeds; 6) Archaic period sites; 7) Kangju sites; 8) Kushan sites; 9) Kushan-Afrigid sites; 10) Afrigid-Samanid sites; 11) Khorezmshah sites; 12) Golden Horde sites.

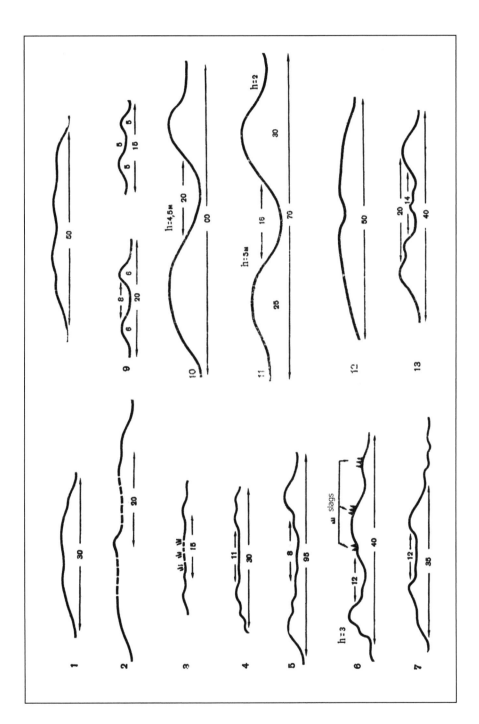

Figure 46. Canal profiles in the environs of Yarbekir-kala: 1) poisk 321; 2) poisk 324; 3) poisk 546; 4) poisk 547; ancient canals in the environs of Tuzgyr; 5) poisk 889; 6) poisk 981; 7) canal in the environs of Yarbekir-kala (poisk 984); 8) Kunya-Uaz oasis; 9) poisk 261 and 265; 10) canal in the environs of Mangyr-kala (poisk 332); 11) poisk 320; 12) poisk 343; 13) poisk 650.

Figure 47. Ancient Shakh-Senem 'oasis': 1) citadel; 2) garden-park layout; 3) ruins of Kangju-Kushan dwellings; 4) Medieval ruins; 5) Kangju and Kushan canals; 6) Medieval canals; 7) traces of ancient riverbeds; G = remains of glass production; P = remains of ceramic production.

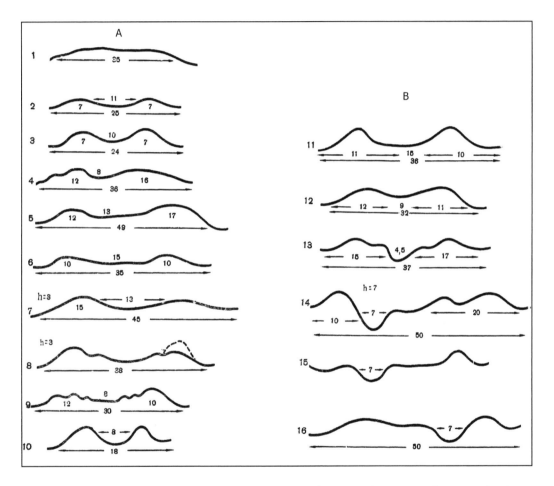

Figure 48. Medieval Chermen-yab. 1) poisk 136; 2) poisk 428; 3) poisk 429; 4) poisk 152; 5) poisk 253; 6, 7) poisk 162; 8) poisk 696; 9) poisk 690; 10) parallel bed of the Chermen-yab in the environs of Zamakhshar; 11) profiles of Shamurat (poisk 300); 12) poisk 290; 13) poisk 282; 14) poisk 647; 15) poisk 648; 16) poisk 621.

Figure 49. Medieval irrigation structures in the environs of Shekhrlik: 1) poisk 866; 2) poisk 882 and 883; 3) head structures (poisk 885); 4) poisk 891; 5) poisk 960; 6) dam and head of canal (poisk 866); 7) field (poisk 964); 8) poisk 939 (with visible chigir pits); 9) poisk 957.

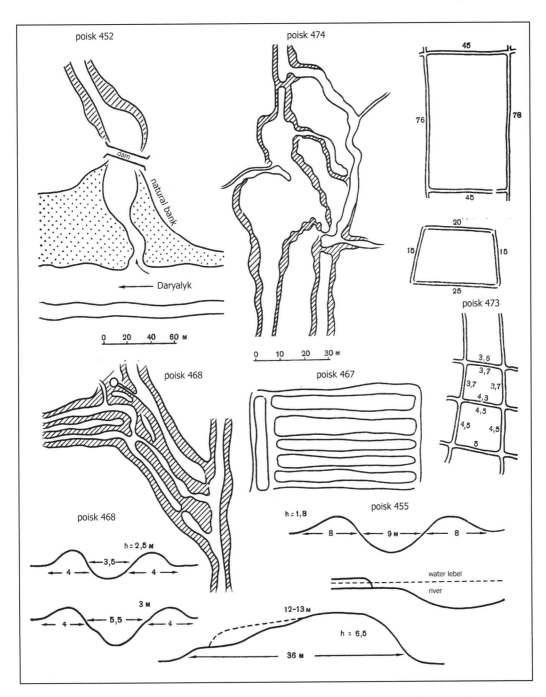

Figure 50. Turkmen irrigation of the 19th century in the environs of Mashryk-Sengir. Poisk 452: head structure of the main canal; Poisk 474: system of reservoirs to lift the water level; Poisk 473: agro-irrigation layouts in the Lower Daryalyk; Poisk 468: distributing structure of the main canal; Poisk 467: fields and gourd plantations in the environs of Mashryk-Sengir; Poisk 468: transverse profiles of the main canal; Poisk 455: main canal at Mashryk I; Poisk 481: Egen-Klych dam transverse profile..

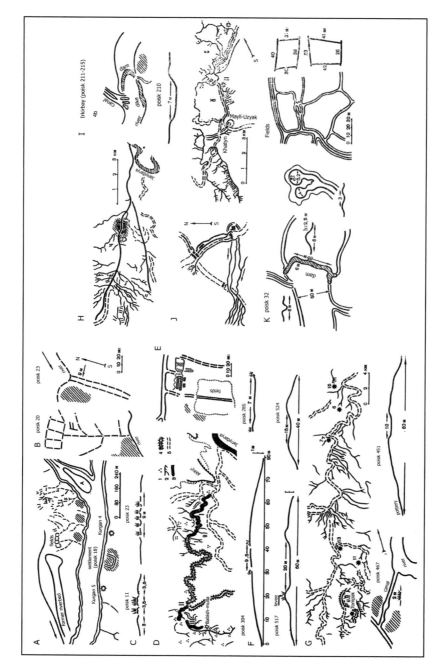

Figure 51. Historical irrigation development of the Aral Sea area: Lower Syrdarya. A) irrigation of the 6th–4th centuries BCE; canals and fields in the environs of 'slag-covered' kurgans; B) canals at poisk 20 and 23; C) canal profiles; D) irrigation of the 4th–2nd centuries BCE in the environs of Babish-Mulla: 1) remains of sites; 2) canals; 3) reservoirs; 4) embanked riverbed; 5) dry riverbeds highlighted by vegetation; E) field at a site in the environs of Babish-Mulla (poisk 242); F) profiles of embanked riverbeds (above), former riverbeds (below, left) and canals (below, right); G) Djety-asar irrigation (1st–8th centuries); H) Medieval systems of the 12th–14th centuries in environs of Djend; I) dam in the Irkibay sector; J) systems of the 18th–19th centuries in the environs of Khatyn-kala; K) dam, chigir pits, and fields in environs of Aralbay-kala.

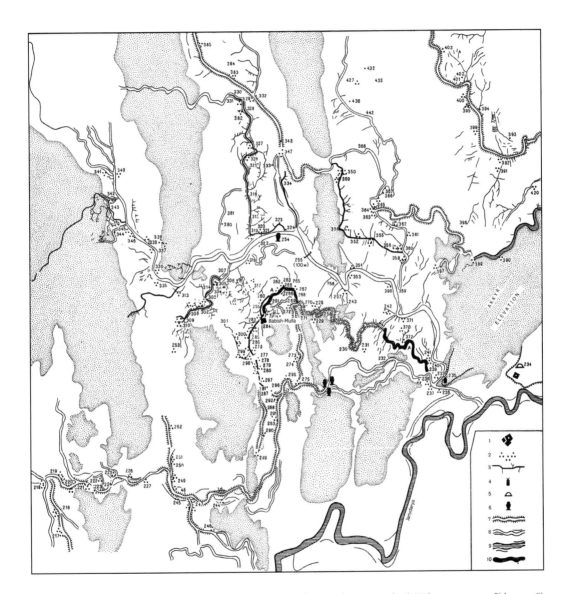

Figure 52. Environs of Babish-Mulla. 1) site; 2) remains of settlements; 3) traces of ancient canals; 4) 19th century mazar; 5) kurgan; 6) ceramic workshops. Ancient riverbeds: 7) raised above ground; 8) highlighted by vegetation; 9) preserving the negative forms of the relief; 10) reservoir-basins.

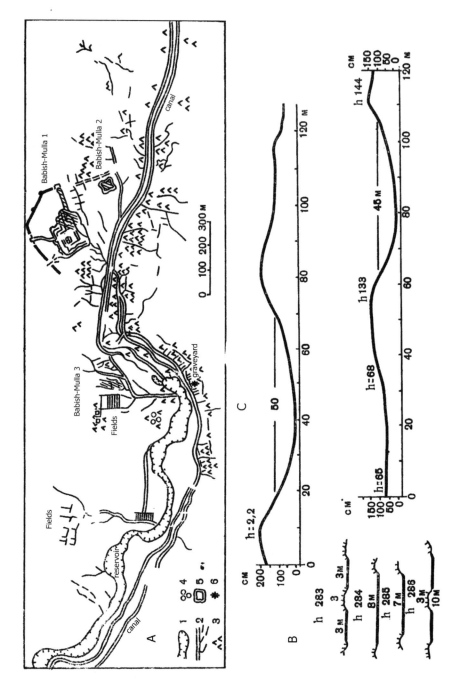

Figure 53. A – Irrigation structures in the environs of Babish-Mulla: 1) riverbed-reservoir; 2) canals and small irrigation network; 3) site remains in the shape of ceramic dispersions; 4) pottery kiln; 5) remains of fortifications and buildings; 6) burial ground; B – transverse profiles of Babish-Mulla canal (poisk 283,284, 285, 286); C–D – profiles of riverbed-reservoir (poisk 372 and 264).

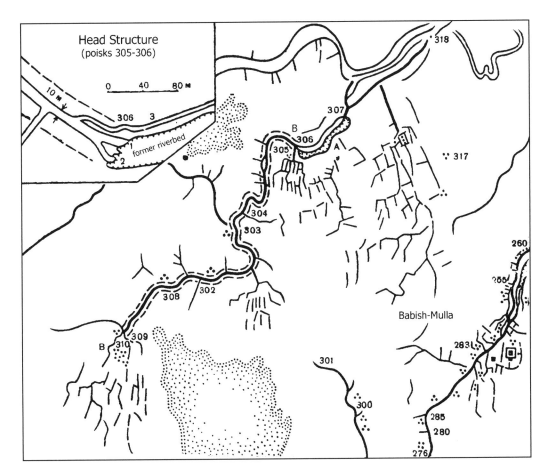

Figure 54. Irrigation system northwest of Babish-Mulla: A) riverbed-reservoir; B) head structures; C) remains of large site with pottery dispersion (poisk 310).

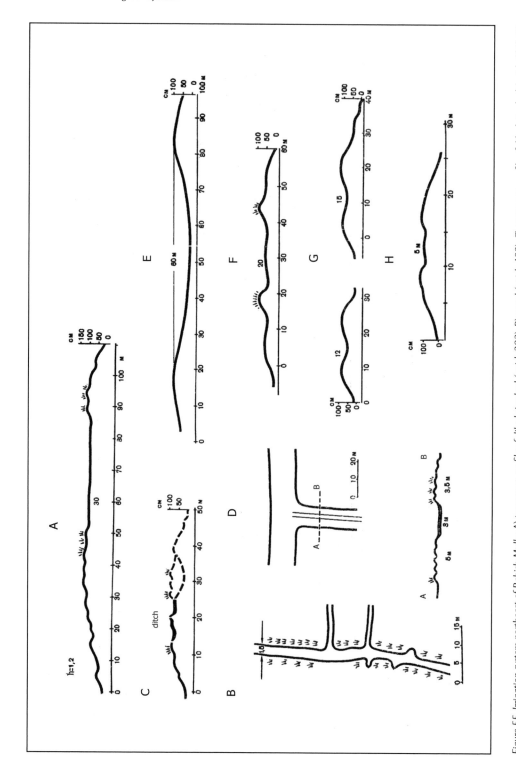

Figure 55. Irrigation system northeast of Babish-Mulla: A) transverse profile of diked riverbed (poisk 292); B) canal (poisk 402); C) transverse profile of diked riverbed (poisk 433); D) canal (poisk 426); E) poisk 523; F) poisk 517; G) poisk 522, 524; H) canal in the environs of Balandy (poisk 69).

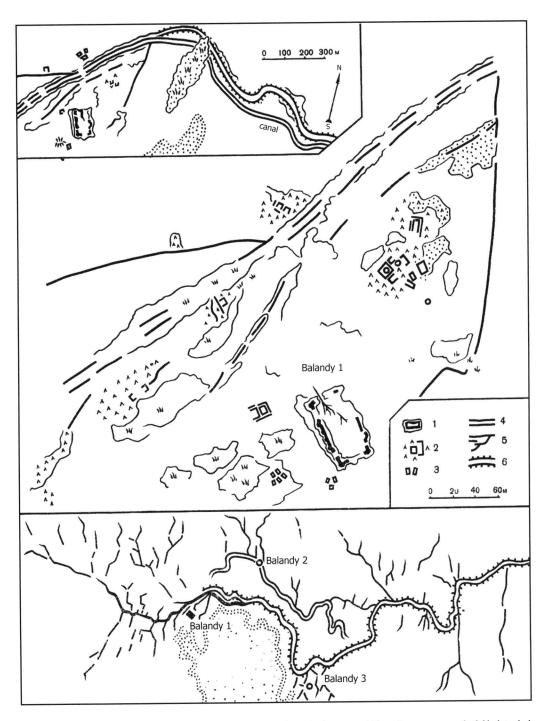

Figure 56. Plan of the Balandy sector: 1) fortified site; 2) dwellings plan; 3) burials; 4) main canal; 5) small irrigation network; 6) diked riverbed.

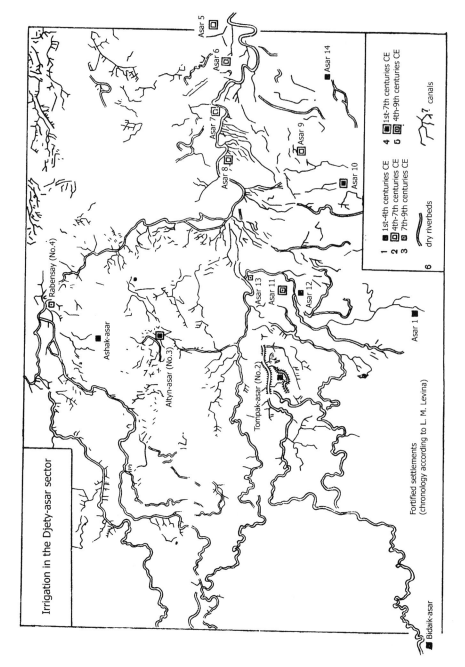

Figure 57. Irrigation in the Djety-asar sector. Fortified settlements (chronology according to L. M. Levina): 1) 1st–4th centuries CE; 2) 4th–7th centuries CE; 3) 7th–9th centuries CE; 4) 1st–7th centuries CE; 5) 4th–9th centuries CE; 6) dry riverbeds; 7) canals.

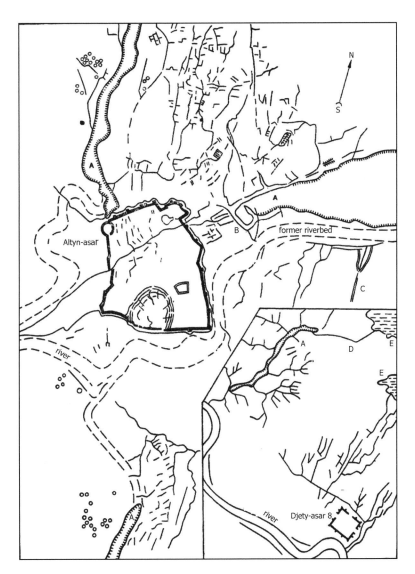

Figure 58. Environs of Altyn-asar and Djety-asar 8: A) riverbed-reservoir; B) retaining dam; C) main canal; D) connecting canal; E) drainage area.

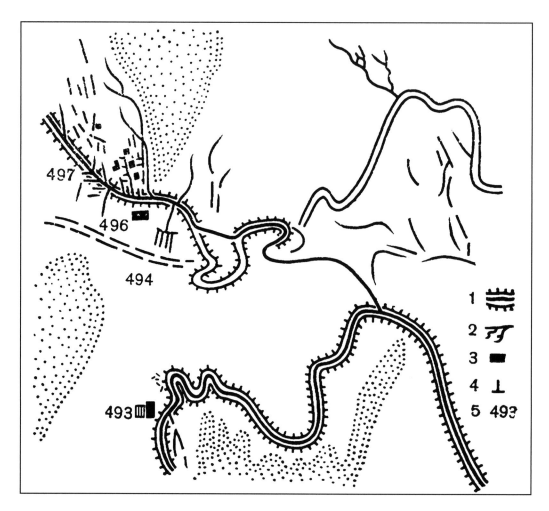

Figure 59. Schematic plan of the environs of a settlement of the 12th-beginning 13th centuries in the Murzaly sector: 1) diked riverbed; 2) small irrigation network; 3) remains of dwellings; 4) ruins of kaptarkhana; 5) 493, number of archaeological poisk .

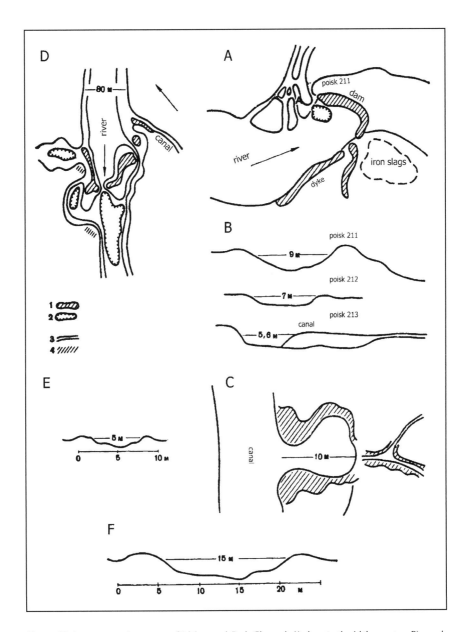

Figure 60. Irrigation in the sectors of Irkibay and Besh-Chongul. A) dam in the Irkibay sector; B) canals transverse profiles (poisk 211, 212, 213); C) chigir pit at poisk 213; D) Besh-Chongul dam: 1. remains of dam; 2. depressions; 3. canals; 4. Mediaeval pottery dispersion; E) transverse canal profile at poisk 194; F) transverse canal profile poisk 193.

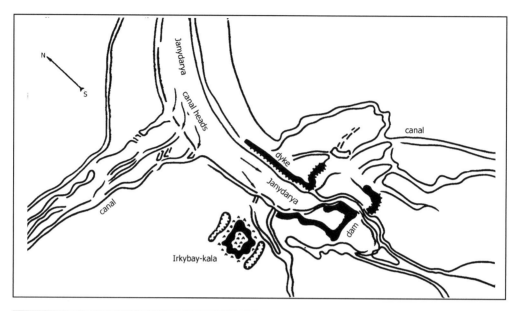

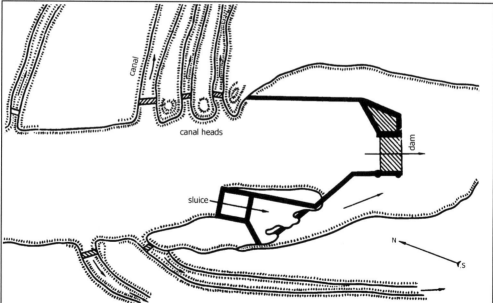

Figure 61. A) Plan of head structures and dams at Irkibay-kala; B) dam on the canal in the environs of Samarra (according to Sousa 1948).

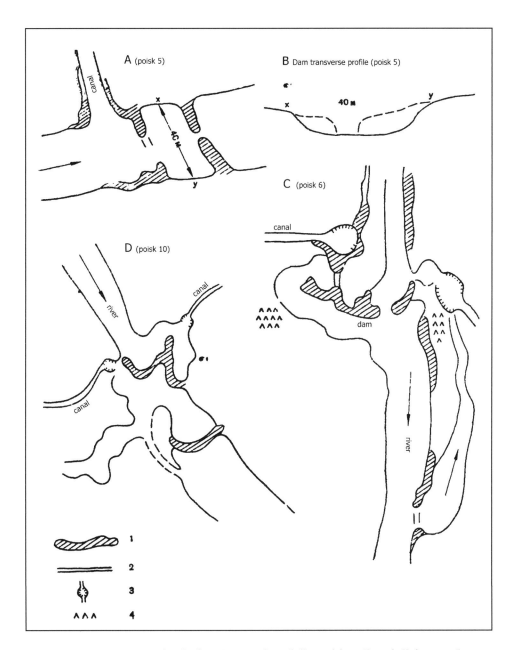

Figure 62. Irrigation structures on the Inkardarya. Retaining dams: 1) dikes and dams; 2) canals; 3) chigir pits; 4) pottery.

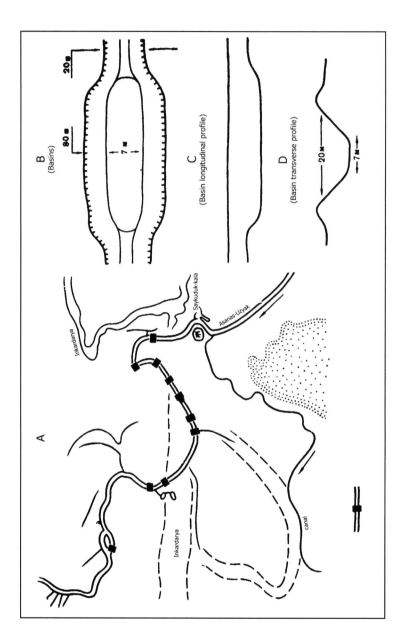

Figure 63. Irrigation structures in the Saykuduk sector. A) plan; B) basin sketch; C) longitudinal profile; D) transverse profile; E) Saykuduk-kala.

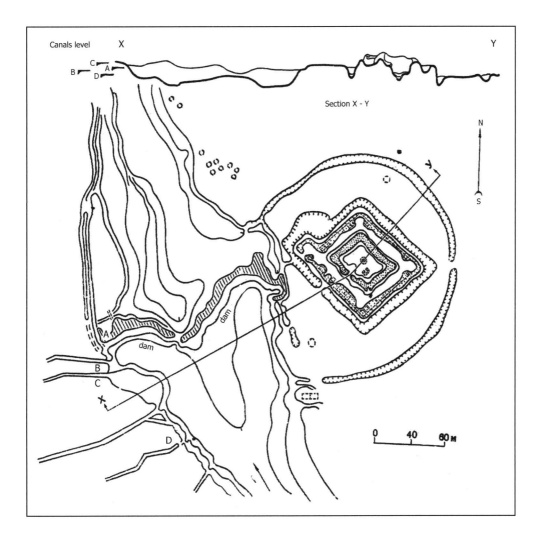

Figure 64. Plan of Khatyn-kala and dams.

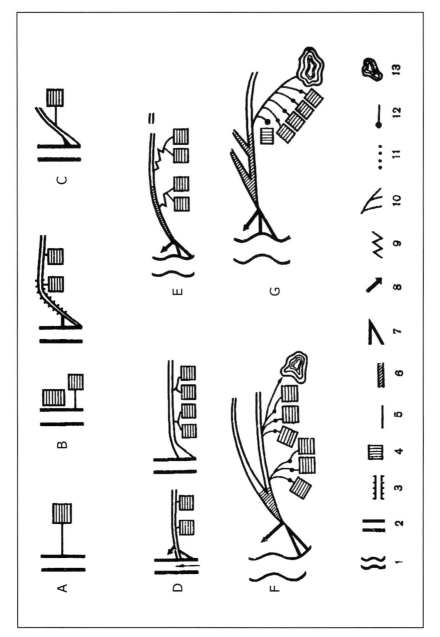

Figure 65. Development of irrigation techniques in deltaic areas (based on examples from Khorezm): A) Geoksyur (second half of the 4th–beginning of the 3rd millennia BCE); B) Kokcha 3 '1609' (15th–12th centuries BCE); C) Yakke 2 (9th–8th centuries BCE); D) Early Antique period; E) Late Antique period; F) Medieval (pre-Mongol); G – Late Medieval (according to Ya. G. Gulyamov). Explanation: 1. main riverbed; 2. lateral riverbed; 3. adapted riverbed; 4. field; 5. feeder; 6. main canal (the inactive part is shaded); 7. head; 8. drainage canal; 9. Ancient distributor; 10. Medieval branching distributor; 11. Second order distributor; 12. chigir; 13. overflow pond.

Table 1. Succession (change) of simultaneous vegetation

Irrigation cultures (according to S. P. Tolstov)	Structure of Soil Layer	Main Vegetal Association
Present-day	Irrigated meadows and meadow-*takyr*-shaped soils with patches of *solonchak*	Cultivation
Post-Medieval	*Takyr*-shaped soils with large remains of *solonchak*	*Yantak, itsitek, keurek*
Post-Medieval	*Takyr*-shaped sand soils with patches of residual humus *takyr*. Sand dunes and hills along the ancient channels and rivers	*Yantak, itsitek, keurek*
Late Medieval (after Mongols)	Complex of sand *takyr*-shaped soils and residual humus *takyr* in association with massive half-covered sand hills, sand dunes and with residual *solonchak*	*Keurek, saxaul*
Early Medieval (before Mongols)	Complex of typical *takyr* and *solonchak takyr*-shaped soils in association with half-covered sand hills, with desert sand soils and residual *solonchak*	*Biyurgun, saxaul*
Antiquity	Typical *takyr* with patches of *solonchak takyr*-shaped soils, half-covered sand hills, desert sand soils and residual *solonchak*	*Biyurgun*

Table 2. Classification of irrigated arid zones of the USSR

Geographical Zone	Sectors of River Basin			
	Riparian	Estuary	Peripheral	
Mountainous	River terraces / Upper river terraces	Alluvial fan	Slopes	
Piedmont plains	Lower river terraces / Upper river terraces	Alluvial fan a) mountain peaks b) wedge zone c) periphery	Wavy plains	
Desert low lands	Lower river terraces / Upper river terraces	Dry delta Seaside delta	Modern alluvial plains	

Table 3. Development of irrigated agriculture in Peru (according to R. S. MacNeish)

Years	Characteristic of Economy	Most Important Agricultural Cultures	Characteristic of Irrigation Techniques	Archaeological Analogies in the Old World
800 CE	Spanish conquest (1532 CE)	Maize, broad bean, cassava, peanut, potato, etc.	Development of mountain agriculture, based on large irrigation structures (canals, aqueducts, basins, etc.)	Uruk
	Incas			
	Moche Culture			
BCE / CE				
750 BCE	Final development phase of production economy	Maize, pumpkin, broad bean, pea, pepper, cotton	*Sai*-stream irrigation Transition from simple to complex irrigation	Ubaid, Jarmo, Jericho, Çatal Hüyük
			First attempts at regulating mountain stream floods	Jericho
			←	
				Çatal Hüyük
1200 BCE	Beginning of production economy			
Huaca Prieta Culture	Agriculture			
	Appearance of pottery			
2500 BCE		Pumpkins and broad bean	Regular harvest-gathering of natural watered plots	Mesolithic complexes of Karim-Shakir type
'Pre-cotton' Culture	Gatherers, fishers and hunters			
4000 BCE				

Table 4. Development of irrigated agriculture in Southwestern Iran

Humidity Change* and Climate**	Years BCE	Archaeological Sites	Hunting	Breeding	Gathering and Fishing	Characteristic Agriculture and Agricultural Cultures	Characteristic Irrigation Network
Atlantic, wet, warm	4000	Tepe Sabz — 4200 ±100; 4975 ±200	The meaning of hunting is preserved	Bovine and Capro-ovine, dog		Wheat: one-grained (einkorn), two-grained (emmer), hybrid species; Barley: naked-grained and six-rowed; Lentil, peas, flax and others	Formation of a continuous irrigation system
	5000						
	5200						Development of irrigated agriculture
	5700	Ali Kosh phase	Hunting of wild bull and onager	Development of herding (sheep and goat), prevailing on agriculture; Herding of goat and sheep (?)	Gathering of legumes	Some reduction in plant cultivation; Cultivation of two species of wheat and barley	Transition from simple to complex irrigation, sai-stream irrigation
Boreal, continental, dry	6000	Muhammad Jafar Phase — 6290 ±125; 6448 ±200			Regular gathering of Leguminosae and Graminaceae		
	6500	Bas Mordech phase — 6888 ±210	Hunting of gazelle, onager, wild bull and pig	Herding of goat (?)	Fishing	Cultivation of two-grained wheat (emmer) and two-rowed barley	First attempts at regulating seasonal floods
Pre-Boreal, cool, dry	7000	Pre-ceramic phase — 7950 ±200			Gathering of mollusks and freshwater turtles		
	8000	Ali Kosh					

* According to A. V. Shnitnikov.
** According to K. W. Butzer.

Table 5. Cultivated plants at Çatal Hüyük (according to H. Helbaeck 1964:121)

Cultivation	Area (A or E) and building-level						
	A II* 5797 ± 79	A III* 5807 ± 94	E III	E IV	E V** 5920 ± 94 5690 ± 91	A VI* 5781 ± 96 5850 ± 94	E VI
No. of samples (45)	8	4	1	4	1	6	21
Einkorn	XX	XX					XX
Emmer	XX	XX		XX		XX	XX
Bread wheat				X			X
Naked barley	XX		XX				XX
Pea	XX			XX	XX		XX
Vetch	XX	X					XX
Bitter vetch							XX
Crucifers						XX	XX
Taeniatherum		X					X
Eremopyrum		X					X
Scirpus		X					
Almond						X	
Acorn						XX	XX
Hackberry	X	X	XX	X	X	XX	XX
Pistachio							X

* Ehrich 1965:124.
** Mellart 1964:115.

Table 6. Development of irrigated agriculture in the Near East (in connection with the development of working tools)

Years	Some Archaeological Sites				Tools		Characteristics of Irrigation Techniques
3000 BCE	Mersin	XV	Susa	Sialk III		Bronze	Construction of protecting dams
			XV		Bronze hoes and shovels		
			XIV				
			XIII				
4000 BCE			XII	Hassuna			
			XI				Formation of continuous irrigation systems with heads and water-regulating structures
		XVI	X				
		XVII	IX				
			VIII				
			VII				
			VI				
			V	5301 ± 206			↑
5000 BCE			IV				
			III				
		XXIV	II				
		XXVII	I	C	Stone and wooden hoes		Transition from simple to complex irrigation
		5797 ± 79	I	B			
		XXXII	I	A Hassuna			
				Sialk I			↑
6000 BCE		6239 ± 257	Çatal Hüyük 6385 ± 101		Digging sticks with weights	Wood – stone	
		Jarmo 7371 ± 238	Jericho A 7114 ± 165				Mountain *sai*- stream irrigation. First attempts at regulating seasonal floods
7000 BCE			Jericho B 7257 ± 106				↑
8000 BCE		Natufian 8144 ± 277			Digging sticks Hoes Sickles		
9000 BCE		Zawi-Cheni Shanidar 9246 ± 309	8968 ± 309 Shanidar B		Stone pestles and mortars		Regular gathering of wild cereals
10,000 BCE			10,410 ± 412				

Table 7. Some data on the regime of 'historical' rivers

	Nile	Tigris	Euphrates	Amudarya
	Rain-fed	Mixed-feed	Mixed-feed	Snow and glacier feed
	The flood begins in June and gradually increases till September–mid October	The flood (resulting from snow melt) begins in early March with sometimes devastating results	The highest level is in winter (from rain) and in the spring (from snow melt), decreasing in summer	Two water rises: in spring from snow melt and in summer from glaciers melt
	The water rise is costant (approximately 7 m in the delta)	The water rise is up to 5 m	The water rise is up to 5 m	The water rise is about 2–3 m
	Flow rate at Assuan. Average: 2.600 m³/sec. Maximal: 1.500 m³/sec. Minimal: 1.400 m³/sec. Turbidity level: 1.800 g/m³	Annual flow rate at Baghdad. Average: 1.240 m³/sec. Maximal: 10.000 m³/sec. Minimal: 150 m³/sec. Turbidity level: 2.300 g/m³	Flow rate in May: 1.790 m³/sec in September: 260 m³/sec.	Maximum water flow rate in June–July. Turbidity level: 2.500–4.000 g/m³

Table 8. Development of irrigation in Mesopotamia

Years	Chronology*	Tools	Irrigation Techniques
2000 BCE	Ur III (2006–2113)		
	Akkadian period Sargon (2371–2316)		
	Early Dyanstic		Appearance of water-lifting devices (shaduf, etc.)
	III / II / I		
3000 BCE	Jemdet Nasr — Late	Bronze hoes and shovels	Irrigation by furrows
	Uruk — Middle / Early		Basin systems
	Eridu II, III, IV, V, VI, VII		Construction of dams and large irrigation systems
	Ubaid 4		Irrigation by flooding
4000 BCE	Ubaid 3 — Eridu VIII, IX, X, XI		Appearance of water-regulating and head structures
	Ubaid 2 — Eridu XII, XIII, XIV	Stone and wooden shovels	Small irrigation systems
5000 BCE	Eridu XV, XVI–XVIII		
	Hassuna — 5301 ± 206, Eridu XIX		
	V, IV, III, II, Ic, Ib, Ia		Discovery of gravity irrigation principle on deltaic channels
	5506 ± 73, Tell es-Sawwan (Layer 1)		Marshy (estuary) agriculture
6000 BCE			Embanked rivers

* According to Porada 1965; Lloyd and Safar 1947, 1948; Van der Meer 1955.

Table 9. Development of irrigation in China

Years	Chronology		Agriculture		Irrigation
1000 CE	Song (960–1234 CE)	Iron Age	Cotton	Construction of dams	Terraced agriculture
	Tang (618–907 CE)		Rice cultivation	Impotent canal	Irrigation canals
BCE / CE	Han (206–220 CE)		Irrigated agriculture		Drainage canals
500 BCE	Eastern Zhou (722–481 BCE)	Bronze Age			
1000 BCE	Western Zhou (1100–771 BCE)		Huang Ho: hemp, wheat, barley / Yangtze		
	Shang (Yin)		Non-irrigated agriculture		
	Protoshang (?)				
2000 BCE	Yangtze: Qinglian'gang, Lanzhou; Huang Ho: Yangshao, Longsheng	Neolithic	Upland rice / Millet / Non-irrigated agriculture / Origin of agriculture		

Table 10. Development of irrigation in Turkmenia

Chronology*			Tools	Irrigation		
Piedmont Plain	Southeastern Delta			Piedmont Plain	Southeastern Delta	
	Tedjen	Murghab			Tedjen	Murghab
CE — Namazga VI (IVa)		Yaz III	Iron tools	Small irrigation systems		Large piedmont dam
1000 BCE		Yaz II				Large irrigation systems (up to 50 km)
Namazga V	Altyn	Yaz I		Reservoirs		
		Takhirbay				
2000 BCE — Namazga IV (III)	Khapuz	Auchin		Mountain – stream irrigation		Irrigation on deltaic branches (?)
Kara Ia 2750 ± 220 (RUL2) 3000 / Namazga III	Geoksyur				Geoksyurian canals (up to 3 km), reservoirs	
3000 BCE — Namazga II (II)	Yalangach				Canals of the Yalangach period	
	Dashlydji		Wooden digging sticks with weighted hoes		Ilgynly canals (up to 1 km)	
4000 BCE — Namazga I (I)	Mondjukli			Retaining dams		
Anau Ia						
5000 BCE — Chopan III	Chagylly 5036 ± 110			Embankments		
				First attempts at flood regulation		
Chopan II				Estuary irrigation		
6000 BCE (?) — Chopan I / Djeitun				Origin of agriculture (?)		

(Piedmont Plain chronology also includes Anau III, II, I)

* According to B. A. Kuftin, A. A. Marushchenko, V. M. Masson; see also Masson 1966b:261.

Table 11. Some data on the development of irrigation techniques in the Lower Amudarya

Irrigation systems in different periods	Systems location in relation to rivers main courses	Characteristic of canal sources	Average canal dimensions	System layout characteristics	Correlation between exploited and unexploited	Reservoirs and regulating structures
Khiva oasis irrigation of the 19th–beginning of the 20th centuries, Palvan-Ata, Khanka-arna, etc.*	On main river	Multi-headed system (saka), retaining dams	Length: 25 to 60 km; width of main canals in the upper course: 10 to 30 m, depth: 2 m	Complex branch systems with first and second order irrigation distributors and feeders	Canal unexploited part is important	Large use of chighir, regulators on distributors and irrigators
Khorezmshah and Golden Horde systems of the 12th–14th centuries (Chermen-yab and Shamurat)	On main river	Multi-headed system (saka), retaining dams	Length: up to 100–150 km and more; width of main canals: 12–15 m in the upper course, up to 7–9 m in the lower course	Branch systems with irrigation distributors and feeders	Canal unexploited part is important	Large use of chighir, possibly with regulators on distributors and irrigators
Afrigid systems of the 7th–8th centuries (Kyrk-Kyz, environs of Adamli-kala)	On main river	Multi-headed system, retaining dams	Length: up to 60–100 km; width: 5–10 m	Development of branch systems	Canal unexploited part is important	Only gravity irrigation
Kushan systems of the 1st–4th centuries CE	On main or on large river	Multi-headed system, retaining dams	Length: up to 100 km; width: 20 m (upper course) up to 10 m (lower course)	Irrigation canals are detached at right and oblique angles	Canal unexploited part is important	Only gravity irrigation
Kangju systems of the 4th century BCE–beginning of CE (Chermen-yab)	On very large deltaic branch	Some head-structures, source locations are on mountain ridges	Length: 30–60 km; width: on average 20 m (upper course) up to 9 m (lower course)	Prevalence of rectangular offshoots	Canal unexploited part is important	Only gravity irrigation
Early Kangju and late Archaic systems (Kunyauaz canal)	On very large deltaic branch and its lateral offshoots	Some head-structures, source locations are on mountain ridges	Length: up to 50 km; width: up to 40–45 m (upper course) and 12 m (lower course)	Prevalence of rectangular offshoots	Canal unexploited part is important	Only gravity irrigation
Early Archaic systems (environs of Bazar-kala)	On lateral offshoots of the river	Only one head structure	Length: 11 km; width: 40 m on both the upper and lower course	Prevalence of rectangular offshoots	Canal unexploited part is important	Only gravity irrigation
Amirabad systems of the 9th–8th centuries BCE (at Yakke-Parsan 2, Bazar 8–10)	On fading lateral branch of the river lower course	Ancient riverbed adapted to main canal	Length: up to 3 km; width: 2–5 m	Irrigation distributors are detached at right and oblique angles		Only gravity irrigation
Tazabagyab systems of the third quarter of 2nd millennium BCE	On fading lateral branch of the river lower course	Adapted ancient riverbed	Length: 1–3 km; width: 2–3 m	Prevalence of rectangular offshoots	Canal unexploited part is important	Only gravity irrigation

* According to Ya.G Gulyamov and V.V. Tsinzerling.

Table 12. Scheme of irrigation development

Years	On mountainous rivers and brooks		In deltas and large rivers valleys				
	Southwest Asia	Peru	Southern Mesopotamia	Egypt	Lower Amudarya	Lower Syrdarya	Underground water
1000 CE	Development of terraced mountain irrigation	Development of terraced mountain irrigation. Complex systems (basins, aqueducts, canals)	Transfer to branching systems	Development of irrigation basins, development of water-regulating infrastructures	Chigir irrigation; Transition to 'branching systems'	Development of water-regulating devices	Well irrigation
BCE / CE	Large systems (canals, aqueducts, basins)	Sai-brook irrigation	Water-lifting devices	Water-lifting devices	Canal heads shift to main riverbed	Small canals on embanked riverbeds	Karez
1000 BCE	Development of water-regulating structures		Irrigation in furrows		Large systems	↑↑	Exploitation of springs
2000 BCE	Small systems		Construction of dams and large irrigation systems with head and water-regulating structures		Small canals on embanked riverbeds		Kair agriculture
3000 BCE	sai-brook irrigation	Water-management with dikes	Small systems		Fishers Hunters Gatherers		
4000 BCE	Estuary irrigation	Fishers, hunters and gatherers	Irrigation on embanked river beds		↑↑		
5000 BCE	Water-management with dikes		Discovery of the principle of gravity irrigation in deltaic channels	Estuary irrigation ?			?
6000 BCE			Estuary irrigation				
7000 BCE	Diked irrigation by 'Harvesting people'		↑↑				
8000 BCE			Fishers, hunters and gatherers				

Tools and agriculture (Southwest Asia): Iron tools · Bronze · Wood and stone; Hoe and Plow Agriculture · Stick-hoe

Sites (Southwest Asia): Settlements and states · Assyria and Urartu · Ugarit · Ali · Hassuna I-XV · Tepe Sabz · Kosh · Jarmo · Shanidar · Natuf

Tools and agriculture (Peru): Bronze tools · Wooden and stone tools; Stick-hoe Agriculture

Tools and agriculture (Southern Mesopotamia): Iron tools · Bronze tools · Wooden and stone; Hoe-plow agriculture · Stick-hoe agriculture

Sites (Southern Mesopotamia): States and settlements · Early Islamic · Sassanid · Achaemenid · Babylonia · Ur III · Eridu XIX-Eridu II · Uruk · Ubaid · Tell es-Sawwan

Tools and agriculture (Egypt): Bronze and iron tools; Hoe-plow agriculture

Key

↑↑ Plant cultivation adopted from adjacent areas

Pre-agricultural stage of irrigation practices

Origin of irrigated agriculture

Endnotes

Boris V. Andrianov

Note 1 For a detailed description of publications used see the bibliography at the end of the book.

Note 2 See the essays, *Budushee Turkestana i sledy ego proshlogo* (The future of Turkestan and traces of its past), *Dju-i Arziz. K voprosu ob istorii irrigatsii v Turkestane* (Dju-i Arziz. Questions on the history of irrigation in Turkestan), and others in V. V. Bartold (1965:274–276, 307–310).

Note 3 Ya. G. Gulyamov, active Member of the Uzbek Academy of Sciences of the Uzbek SSR, worked with the Khorezm Expedition as researcher in 1938–1939 and as deputy chief in 1940, 1945 and 1946.

Note 4 The *uy* mentioned by Ya. G. Gulyamov are not isolated sand hills, as he supposed, but ancient dispersed riverbeds (Tolstov and Kes 1960:39–45, 159–162).

Note 5 In her very interesting book about the irrigation of Southern Turkmenistan during the Eneolithic period, G. N. Lisitsyna was inaccurate in writing about Ya. G. Gulyamov's view on the origin of main canals in Khorezm. She connected Ya. G. Gulyamov's opinions with those of S. P. Tolstov on this issue. Ya. G. Gulyamov did not write about the "cleaning of natural riverbeds and their straightening", but rather on the agriculture which "developed here from the floods in the basin of lateral river channels" and the farmers "just retouched the natural thalweg", which was understood by the author as the lowering between sand hills, or 'long *uy*' (Lisitsyna 1965:9; Gulyamov 1957:60–67, 89).

Note 6 Very interesting thoughts about the importance of irrigation development in the change to the slave-based feudal formation were provided by I. S. Braginskiy. He appropriately divided the history of irrigation into two parts: the first corresponded to a period of 'large' irrigation with canals many kilometers long and, contemporaneously, the most developed slave-owning relations; the second corresponded to a period of 'complex irrigation' with development of water lifting devices (*chigir*) and more technically advanced methods of irrigation and husbandry, connected with the feudal period. Significant changes in productive forces, especially in irrigation, stimulated, according to him, the conflicts in the relations between productive guides, and slave-owners that led to their crises (Braginskiy 1956:164–167).

Note 7 Specific information on Central Asia's slave-owning system is given in Tolstov 1938a; Bernshtam 1947; Litvinskiy 1963; Rakhmanova 1964a–b; Masson 1968.[Editor 81] B. A. Litvinskiy, noted the uncertainty around the question of slave labor in ancient Central Asia; but, at the same time he supposed that this labor was widely used in agriculture, and in particular in irrigation works (Litvinskiy 1963:475–476).

Note 8 The question regarding the relationships between communal and slave labor in agricultural production in countries of the ancient world was again raised during recent discussions on the 'Asian Mode of Production', on a chronological seriation of history and historical development of Oriental countries (Yu. Semenov 1957, 1965; Vasilev 1965; Afanasev 1965; OOI 1966; Dyakonov 1966; Pechirka 1967; Masson 1967a; Nikiforov 1968; etc.).

Note 9 According to A. I. Tyumenev, the written sources from Sumer did not contain direct evidence of the involvement of war slaves in irrigation work (Tyumenev 1956:369). I. M. Dyakonov, summarizing the study of communities in the ancient Orient, rightly paid attention to the predominance of natural forms of social production in this period. According to him "slave labor in agricultural production cannot be considered as characteristic for the first class formation, which we call the slave mode of production" (Dyakonov 1963:16–17ff.). Abroad, the idea of the insignificance of slave labor in the agriculture of ancient Mesopotamia was developed, for example, by the famous archaeologist R. McC. Adams (1966:96–97, 102–104). The position of R. McC. Adams, mainly in relation to irrigation works, was recently criticized by K. A. Wittfogel (1967). As for Central Asia, the same point of view was shared by A. M. Belenitskiy in 1954 at the conference of historians in Tashkent (see Belenitskiy 1955:505–509) and in 1968 at the conference of archaeologists of Central Asia in Leningrad (see Belenitskiy 1968).

Note 10 The field diaries of the members of the expedition with the description of the research are stored in the archive of the Khorezm Expedition.

Note 11 This can be explained by the fact that, even in the most advanced area of topographic study of the earth's surface, during the interpretation of aerial photographs the questions on the methods of decoding cultural landscapes are simply excluded or limited to specific issues such as the study of modern human settlements, etc. (Bogomolov 1963:4; AEE 1967:50-56; Sokolov 1952; etc.).

Note 12 A similar conclusion (albeit in a conjectural and hypothetical way) was recently advanced by E. E. Nerazik (1966:15), who connected the farmer settlements of Khorezm during the Afrigid period with the peculiarities of Khorezm irrigation. As we shall see below, a clear connection between the topography of irrigation systems and the distribution of population can also be traced back to earlier periods.

Note 13 A very interesting method for determining the cultivated soil in the past was developed by the soil scientist N. G. Minashina (1960, 1962). She applied micromorphological analyses of soil structure in areas that have lost the outward signs of cultivation, with the aim of determining the extent and impact of irrigation on soil conditions. Valuable observations were made by N. G. Minashina on the historical dynamics of soil salinity, in connection with irrigation, followed by abandonment and new development of the 'lands of ancient irrigation' in the Murgab oasis.

Note 14 Soviet archaeologists have already made important advances on the matter of cataloging and recording data from archaeological sites within a coherent and rigorous system (Rybakov 1957). However, there are few special papers devoted to the methods of archaeological mapping in both the USSR and abroad (Dublitskiy 1927; Gulley 1960; Mongayt 1962), although archaeological mapping is quite widespread in recent years in archaeological research (Talitskaya 1952; AKK 1960; etc.). A. L. Mongayt rightly emphasizes that the mapping method in solving some complicated problems regarding the history of material culture have not yet been fully applied (Mongayt 1962:16).

Note 15 Most of the proceedings deal with the techniques of aerial photography, the equipment, the problems of processing and deciphering aerial photography for archaeological purposes. Noteworthy is the report by I. Scollar's (Federal Republic of Germany), which deals with the methods of decoding the archaeological sites in Central Europe disclosed by vegetation (see also Scollar 1965). The method of detection of ancient settlements, fortifications, roads, walls and canals in photographs of plowed and sown territories is of great importance for the archaeology of many European countries and in the European part of the USSR.

Note 16 Interestingly, the first experience in topographic detection of ancient irrigation in Mesopotamia took place in the mid-19th century. The well-known British irrigation specialist I. W. Willcocks collected these materials (maps, plans, and sections) in a book dealing with the question of ancient canals on the Tigris River (Willcocks 1903).

Note 17 In 1962-1963, during the study of the earliest irrigation work in the territory of the USSR dated to the Eneolithic period in the Geoksyur oasis, G. N. Lisitsyna focused her attention on the excavations of ancient canals and riverbeds. In this case, she successfully used the data provided by the main stages of flooding in the oasis. For example, the finds of ceramics and a fragment of a clay female figurine in a canal confirmed the artificial nature of the facilities whose discovery was made possible by using aerial methods (Lisitsyna 1965:figs. 4-11, figs. 25-36, 1966:99-100).

Note 18 Personal communication.

Note 19 According to al-Makhdisi, the legendary king of the Orient ruled a vast territory 'climatically' oriental , including Khorasan (Khorasan part) and Maverranahr (Khaytal or Hephtalite part; MITT 1939:185:note 6).

Note 20 According to Yakut, "in the language of the Khorezm population, *kas* means 'wall' (*khait*) in the steppe surrounded by nothing" (in MITT 1939:430, S. M. Bogdanovoy-Berezovskoy translated the word *khait* as 'farm'). According to V. V. Bartold, *kyat* is a space in the steppe encircled by a wall or a rampart (Bartold 1963, Tom I:199). S. P. Tolstov (1948a:11) translated this word as 'wall'. Ya. G. Gulyamov connected *kyat* with the 'dwelling fence' of an ancient Iranian settlement (Gulyamov 1957:45).

Note 21 E. Sachau, W. Geiger and V. Tomashek interpreted the toponym 'Khorazm' as 'land providing food and water', 'fertile land'. P. I. Lerkh, N. I. Veselovskiy and H. Kiepert supposed that Khorezm meant 'lowland'. A different interpretation was given by S. P. Tolstov, who, according to M. N. Bogolyubov, combined P. S. Savelev's and C. Bartholomae's translations as 'Land of (the people) of the sun', 'Land (country) of Hwarri or Harri people', 'Land of the Hurrians' (Tolstov 1948a:223; 1948b:80-87). M. N. Bogolyubov, based on the ancient Persian and Avestan terminology, suggests different interpretations, as 'country, with good fortifications for cattle', 'country, where the settlements have good walls', or 'country with good enclosures' (Bogolyubov 1962).

Note 22 See review of the opinions and the literature in the works by M. H. de Goeje, W. Geiger, G. Le Strange, A. Herrmann, W. W. Tarn, K. A. Inostrantsev, V. V. Bartold, S. P. Tolstov, Ya. G. Gulyamov, A. M. Konshin, etc.

Note 23 The most complete review of studies of the origin of agriculture is contained in C. D. Darlington (1963).

Note 24 The further development of this materialistic conception of Prehistory led to the refinement of some parts of

L. H. Morgan's scheme: F. Engels and Soviet scholars developed both social and economic-cultural aspects of this subject. See the works by S. P. Tolstov, M. O. Kosven, A. I. Pershits, B. B. Piotrovskiy, M. G. Levin, N. N. Cheboksarov, V. M. Masson, etc. (Andrianov 1968a:22:note 1).

Note 25 The question of the origin of cultivated wheat is still unclear. N. I. Vavilov (1967, Tom I:103–115) outlined two main centers: Southwest Asia, where the common wheat and club wheat (hexaploid) prevailed, and the Mediterranean area with the durum wheat (tetraploid). The data on the spreading of wild species led J. R. Harlan and D. Zohary to localize the process of domestication of the einkorn wheat in the piedmonts of the Zagros and Taurus mountains, and emmer wheat in the basin of the Jordan River (Harlan and Zohary 1966:1079). Emmer, a transition species close to the wild one, was found by H. Helbaek in Jarmo (8th–7th millennia BCE). J. Mellaart, during the excavations of East Çatal-Hüyük on the Konya plateau in Western Anatolia, discovered einkorn and emmer (Helbaek 1964a).[Editor 82] Common wheat was found in Southern Turkmenia in sites of the 6th–4th millennia BCE, particularly in the piedmont area sites of the Djeytun culture (Chagylly-Depe), in layers of Anau I B (Namazga I), in Tedjen, in the upper layers of Mullali-Depe (Yakubtsiner 1956:108; Khlopin 1964:93). In sites of the Namazga IV–V period (late 3rd–early 2nd millennia BCE) at Ak-Depe (Yakubtsiner 1956:109) and Namazga-Depe (Litvinskiy 1952), common wheat and club wheat were discovered. Club wheat, a typical mountain wheat, is represented in the irrigated plain of the Khorezmian oasis by a special subtype. According to P. M. Jukovskiy, this is the second form (Jukovskiy 1950:87). W. M. Bauden, and later H. K. Kihara, K. Yamashita and M. Tanaka, connected the origin of common wheat and club wheat with the process of natural hybridization of tetraploid samples with wild species of goatgrass (Bowden 1959; Kihara, Yamashita and Tanaka 1965:101).

Note 26 Wild two-row barley was found also in Southern Turkmenia by V. V. Nikitin and later by F. Kh. Bakhteev (1959). In the Neolithic site of Djeytun barley seeds were discovered (Yakubtsiner 1956:108). During the excavations at Mullali-Depe, 9,100 barley seeds and only 250 wheat seeds were found (Lisitsyna 1965:135). The finds by H. Helbaek in Jarmo of two-row barley with weak traces of domestication witness the fact that in Mesopotamia, between the 8th and the 7th millennia BCE, this process did not lead to visible changes, but by the 6th–5th millennia BCE two-row barley was replaced by six-row barley (Hordeum vulgare; Helbaek 1960a:116).

Note 27 In the Balkan Peninsula, in the pre-ceramic Neolithic layers of Argissa (6th millennium BCE), grains of three species of cultivated wheat (einkorn, emmer and spelt), two species of barley (two-row and six-row) were discovered as well as osteological remains of bovines and capri-ovines (Titov 1962, 1965a).

Note 28 The research of T. Harrison in Borneo revealed Neolithic layers of 4000 to 250 BCE with material culture remains which indicated a possible skill in plant cultivation (Harrisson 1964). Very interesting data about ancient agriculture were obtained recently from the mountainous regions of New Guinea, where, at the bottom of a dry marsh, stone polished axes, wooden digging sticks, shovels, the remains of drainage ditches and fields were discovered. The settlements are dated to: 2050±140 BCE to 350±120 BCE (Lampert 1967).[Editor 83]

Note 29 Citing materials of the Khorezm Expedition (Tolstov and Kes 1956; Andrianov 1951, 1955, 1959a, etc.), those facts proving the author's scheme have been quoted, while other facts and material contradicting it have been ignored.[Editor 84]

Note 30 Changes in moisture, in our view, existed and exerted a certain influence on agricultural activities over the centuries. They can be identified only with a very detailed archaeological and paleogeographical study of a territory, including calculations of land areas and determining the capacity of both water channels and irrigation system of the Aral Sea functioning simultaneously (Andrianov 1951, 1958a–b, 1965; Zadneprovskiy and Kislyakova 1965; etc.).[Editor 85]

Note 31 Digging sticks remained in use by North American Indians up to the 19th century. Describing methods of *sai-brook* of the Pima Indians, the American ethnographers noted that the main implement was a stick with a sharpened end (Castetter and Bell 1942:134; Bennett 1946:613, Pl. 131.b-d-e; Casanova 1946:621:fig.50; etc.).[Editor 86]

Note 32 We cannot agree with the conclusion of A. Ya. Bryusov on the limited spread of this earliest method of cultivation (Bryusov 1957:182). For example, A. P. Okladnikov (1962:418–431) rightly observed how the stick-hoe was widely spread in the earliest form of agriculture (Forde 1963:378–393; Darby 1956:198–199; Childe 1956:56; Masson 1966a:157–158; Nilles 1942–45:205–212; Damm 1954).

Note 33 According to C. A. Reed, the goat, the oldest domesticated animal in the Near East after the dog, is the first domestic animal in an economy where gathering was combined with the beginning of agriculture (Reed 1960:119;

Zeuner 1955). M. L. Ryder (1958:781) identified the process of gradual transformation of the wild short-haired goat (10th-9th millennia BCE) into the domestic goat (8th-7th millennia BCE) and, finally, into the domestic long-haired goat (4th-3rd millennia BCE).

Note 34 The earliest specimen of adzes were found at Tell Arpachiya (7th-5th millennia BCE) and described by H. H. Coghlan (1943:38). Robert J. Braidwood noted that some celt-shaped tools were used as hoes (Braidwood 1958). V. Christian, who published a series of celt-shaped stone tools from Uruk, Ubaid and Jemdet Nasr, considered them for working wood (Christian 1940:taf. 44.2-3, taf. 53.11-13, taf. 134.2). But, according to A. Steensberg, these tools were more convenient for uprooting roots, digging pits, leveling fields and furrowing for seed planting (Steensberg 1964:131).

Note 35 In Palestine (Jericho) the first metal artifacts were found in layers dated to the 6th millennium BCE. In Çatal Hüyük, Western Anatolia, in layers of the early 6th millennium BCE. J. Mellaart discovered copper and lead beads (Mellaart 1964:113-114). He also supposed that in the heyday of this settlement (layers II-IV) simple awls and borers were used. In Iran, metal was found at Sialk in layer I (early 5th millennium BCE). In Central Asia, the Anau center of metalworking is known. According to E. N. Chernykh, here, already in the period of Anau 1a (5th millennium BCE) and Namazga I (between the 5th and the 4th millennia BCE) personal copper ornaments and implements were widespread (Chernykh 1962).

Note 36 A chronological change of implements was fully revealed in Mersin, where, in the lower layers XXXIII-XXVII, stone adzes were found (Mersin XXVII: 5797+79 BCE) and in layers XVII-XV (end of the 5th-early 4th millennia BCE) bronze tools in the form of plates of different shapes were found (Garstang 1953:2; 30-31, figs. 13, 41, 95b, 80b, 129).

Note 37 The term 'urban revolution', as a synonym of the early stage of class society formation, is widespread among American and European archaeologists (see Braidwood and Willey 1962; Adams 1966; etc.). R. McC. Adams, for example, compared sites and cultures of the 'urban revolution' in early Mesopotamia and Mesoamerica (see Adams 1966:25-36, fig. 1). However, V. M. Masson (1966a:163) rightly observed that in this sense the use of this term does not appear appropriate.

Note 38 K. A. Wittfogel sharply criticized the position of R. McC. Adams, which considered the structure of early Mesopotamian society as archaic, and underestimated the significance of the centralized state power in organizing a large irrigation economy. K. A. Wittfogel rightly noted that R. McC. Adams' research gave little credit to the organizational and physical labor expended in the creation and maintenance of irrigation systems in Mesopotamia (Adams 1965:40-41; 1966:66-77; Wittfogel 1967:90-92).

Note 39 In Egypt's Neolithic sites, stone tops (pommels) are known with various shapes: elongated as a kind of pick, hammer-shaped, round, hexagonal, etc. (Brunton and Caton-Thompson 1928:LIII.1-15; Reisner 1910: 118). The tops of digging sticks of contemporary Ethiopia provide a very interesting ethnographic parallels. So, in a photo of N. I. Vavilov, a field deep tillage can be seen being done with digging sticks weighted with massive stone balls or discs (Vavilov 1962:161-162; Grum-Grjimaylo 1962:85, 91; see Figure 20).[Editor 87]

Note 40 About the Scorpion's mace, see the work of N. M. Postovskaya (1947:233-249; 1952:49-67), who carefully studied the image on the mace and advanced the important conclusion that the image can be viewed as a sign of the union of Upper and Lower Egypt (Postovskaya 1952:61).

Note 41 A. J. H. Goodwin, who studied the distribution of African digging stick tops, divided them into two groups: the large tops more than 15 cm in diameter (similar to the Bantu tribe's digging stick) were used as a weight, while the small ones as mace tops (Goodwin 1947:210). Neolithic 'maces' from Egypt are interpreted by some researchers as digging stick tops (see Drower 1954:50).[Editor 88] Hence, it is possible to assume that, before the invention of the hoe, the agricultural implements of the Nile Valley were digging sticks with a weight.

Note 42 The Scorpion holds in his hands a wooden hoe, consisting of two pieces of wood, forming an acute angle and joined in the middle by a cord, thus the angle between the handle and the working part can easily be changed. This is the most ancient image of the Egyptian hoe (mr). Its peculiar shape is well explained by Yu. F. Novikov, who proved the feasibility of the hoe structure as a universal feature used as a percussion farming implement and a tool to collect soil, build banks and dig irrigation furrows (Novikov 1964; Saveleva 1962:58-61; Reder 1962:165-170). D. G. Reder criticized B. Brentjes's supposition that the Egyptians borrowed the hoe from the Near East. The supposition is based on the comparison of mr (Egypt) and marr (Mesopotamia).

Note 43 In the Old Kingdom there was a title, which W. Helck translated as 'chief of the mouth' (Saveleva 1962:48).

Note 44 As supporter of the organizational theory of state formation, in his study on 'hydraulic' civilizations, K. A. Wittfogel highlighted the little difference between the labor of a member of a kin community and the forced conscriptions of class society (see Wittfogel 1957:24–25).

Note 45 Reporting on the construction of the famous Nile–Red Sea canal (begun during the reign of Senusret III and Ramses II), Herodotus remarks that in the reign of Necho II (7th century BCE) 120,000 Egyptians died working on that canal (Herodotus II, 158).[Editor 89]

Note 46 R. Taubenschlag cited data from a 83–84 BCE papyrus, where private dams existed and the schedule of workers were already implemented according to holdings size (Taubenschlag 1955:618, 42).

Note 47 A summary of archaeological finds and traces of floods in the layers of Kish, Ur and Shuruppak is given in the last paper by M. E. L. Mallowan (Mallowan 1964:62–63; Woolley 1938, 1954).[Editor 90]

Note 48 Such a combination of small riverbeds and narrow ditches has been identified also by G. N. Lisitsyna in the Eneolithic Geoksyur oasis (Lisitsyna 1965:41–74, 107–135).

Note 49 R. McC. Adams on the contrary, supposed that the emergence of statehood in Mesopotamia was not connected with the beginning of wide irrigation works (Adams 1966:68).

Note 50 A. Steensberg's thought that the wooden plough originated from such hoes-shovels (traction-spades), which were pulled by rope and which had an important function during digging irrigation ditches, is not a new one. Already in the late 19th century E. B. Tylor, and in 1931 P. Leser (Leser 1931), argued that such implement preceded the plow (see also Curwen and Hatt 1953). P. Leser identified a wide region where, in our time, such implements have survived: Korea, Japan, Central Himalaya, Baluchistan, Armenia, Arabia and many areas of Africa. Hoes-shovels are still widely used in irrigation works in the mountainous regions of Central Asia. N. I. Vavilov and D. D. Bukinich report the use of similar implements for irrigating furrows and for cleaning and digging canals in Afghanistan (Vavilov and Bukinich 1929:27:fig. 17, 168–171, fig. 125).

Note 51 In the Neo-Babylonian period this symbol of power was enhanced by a crossbar above the triangular blade. A similar implement was observed on a Syrian cylindrical seal of the 9th–8th centuries BCE (Frankfort 1939:Pls. XXIX.j-m; XLIII.b). Compare it with the modern shovels from Iraq (Hopfen 1960:fig. 21b, fig. 24a). On the seal of the first Babylonian dynasty, the triangular blade of a shovel with handle (?) is next to the image of Hadad, the God of Fertility, and Shamash, the God of the Sun. A. Steensberg supposed that the prototype of the 'spade-shaped' rod was not a simple shovel, but a special stone hoe-shovel used in irrigation work. Several types of shovels were published in 1957 by B. Brentjes. The figure from Susa of a man holding a shovel in his hand is well known (3rd millennium BCE). Shovels of this type, made of one piece of wood, were found in early agricultural sites of Southern Schleswig. They were also dated to the 3rd millennium BCE (Steensberg 1964:116–119)

Note 52 See more in J. Laessøe (1953:5–26), I. M. Dyakonov (1959:87:notes 115–116) and A. Salonen (1965:Pl. XCII). A Syrian relief of the time of Ashurnasirpal II (883–859 BCE) depicts a pulley with a bucket. Perhaps a water-lifting mechanism such as the *cêrd* existed already in Babylon. In Iraq the shaduf is called *dālia*, which goes back to the Akkadian 'lift' *dalū* (*dala* in Arabic). In Mediterranean countries the Arabic term for the water-lifting wheel *nā'ūra* changed to *noria* (Laessøe 1953:7).

Note 53 See the bibliography and a general overview in V. M. Masson (1966b:3–10) and T. I. Zadneprovskaya (1966). An attempt to summarize the archaeological materials on the history of irrigation in the Aral Sea area and the entire Western Turkestan was made recently by Robert A. Lewis (1966). He widely used the publications of the Soviet archaeological expeditions in Central Asia. The author rightly identified the main geographical centers and chronological stages of development of irrigation skills, from the primitive original forms in the piedmont to the extensive irrigation systems in the major river valleys (Lewis 1966:490–491).

Note 54 The material culture of hunters, fishers and gatherers of the Mesolithic and Neolithic periods, was found in the caves excavations on the Krasnovodsk plateau (Djebel, Kaylyu, Dam-Dam-Cheshme), in mountainous Tadjikistan (Chiluchor-Chasma, Oshkona, Tutakaul, etc.), in Kelteminar sites in the lower reaches of the Amudarya (Djanbas 4 and Kavat 7; Okladnikov 1953, 1956a-b, 1966), in Southern Kyzylkum (Darbaza-kyr I and II, Lyavlyakan 26) (Ranov 1960, 1961), and Osh-khona dated by 14C (RUL-280) to 7580±130 BCE; Tolstov 1941:156; 1948a:59–66; 1948b:65–74; 1957a:37–40; 1962a:27–41; Itina 1958; Vinogradov 1957a-b, 1958, 1960, 1963, 1968; Gulyamov, Islamov and Askarov 1966; see also the papers in Sections I and II in PASA 1968; etc.).

Note 55 In the southern outskirts of the Central Asian mountains of Afghanistan and Baluchistan, primitive forms of farming were discovered in cultural layers of the 5th millennium BCE (see Dupree 1964; Harris 1967:94).

Note 56 Before the discovery of the Djeytun sites, the most ancient farming culture in Central Asia was that of the painted ceramics of Anau, which, according to the stratigraphy of R. Pumpelly's expedition (Pumpelly 1908), was divided into four stages (Anau I-IV). The work by A. A. Marushchenko, B. A. Kuftin and V. M. Masson revised this chronology and identified local historical and archaeological periods (see Marushchenko 1939, 1956; Kuftin 1954, 1956; V. M. Masson 1956c, 1962b; Khlopin 1963; Sarianidi 1965; Lisitsyna et al. 1965; Berdyev 1966). The investigations of O. Berdyev revealed several development stages of the Djeytun culture before Anau Ia. The earliest is characterized by materials of Djeytun; next to it by the layers of Chopan-depe I; the middle by the upper layers of Chopan-depe 2, Bami 1, etc.; a later one in the upper layers of Bami 2 and Chagylly-Depe 2 (Berdyev 1966:24). The last stage was dated by14C analysis to 5036±110 BCE.

Note 57 Aerial photographs were taken in 1961 by a scientific member of the Khorezm Archaeological- Ethnographic Expedition, the engineer-geodesist N. I. Igonin.

Note 58 Later sites of the Bronze Age (Khapuz-Depe) were founded by migrants, coming from the Geoksyur oasis in the upper reaches of the Tedjen River, where lateral riverbeds, 60 m wide, still functioned and provided for farming maintenance (Sarianidi 1965:49).

Note 59 V. S. Sorokin, on the contrary, tends to see the Andronovo's tribes agriculture as a minor economic sector (Sorokin 1962:59).

Note 60 The publication of E. E. Kuzmina (1964a:151), on the comparative stratigraphy of anthropic layers of Southern Turkmenia sites, containing finds of rough ceramics of steppe type, clearly shows that this pottery already appeared there between the 3rd and 2nd millennia BCE (Anau III d-e, Namazga V).

Note 61 As seen in the sketch of irrigation development in different countries of the world, the pathway of origin of irrigation techniques was characteristic of Mesopotamia and China (see pp. 118-126).

Note 62 Compare with the description of irrigation work of the Arizona Indians (see Figure 16).

Note 63 Placing large fortifications near canal heads, to control irrigation networks, was noted in many other regions of Central Asia (see Bukinich 1945:194; Gaydukevich 1947:108-109; Tolstov 1948b:122; Latynin 1962:22).

Note 64 Modern field botanic research on the borders of the USSR, in the mountains of Pakistan, Afghanistan and Iran, has revealed an extensive area of distribution of wild wheat and goatgrass, the latter a natural hybrid which also produced hexaploid common wheat (Bowden 1959; Kihara, Yamashita and Tanaka 1965:2, 101) (see also p. 99).

Note 65 S. V. Kiselev supposed that the appearance of Tazabagyab tribes in Khorezm dates back to the second quarter of the 2nd millennium BCE, because the formation of Srubnaya and Andronovo cultures ended by the 17th century BCE (Kiselev 1965:57; Terenojkin 1965:64-65; Itina 1967:70-71).

Note 66 V. M. Masson described in details four independent cradles, or centers, of early farming cultures: 1) Jordan and Palestine (Jericho, etc.); 2) Syro-Cilicia and Southwest Turkey (Amuk-Mersin, Çatal Hüyük, etc.); 3) North Iraq (Jarmo, Hassuna); 4) Central Iran (Sialk) (Masson 1964:39-81; see also Andrianov and Kes 1967:27-28; 1968a:25).

Note 67 On Siyavush or Siavakhsh (Syavarsan in *Avesta*) see Ptitsyn 1947:309-310; Tolstov, 1948a:68, 202-205, 223; Dyakonov 1961:34-44; Klyashtornyy 1964:165-169; Tolstov and Vaynberg 1967:244.

Note 68 At their time, K. F. Geldner, F. Justi, and J. Darmesteter supposed that the birthplace of *Avesta* (Airyana Vaējah) was the country of Arran (between the Araks and Kura rivers); W. Geiger the Upper Zeravshan; J. Marquart, V. V. Bartold, F. C. Andreas, E. Benveniste, A. E. Christensen and S. P. Tolstov, the Khorezm oasis; M. M. Dyakonov Bactria; K. V. Trever, the Sogd; W. B. Henning, the area subjected to the ancient pre-Achaemenid Khorezm ('Great Khorezm'), in the valleys of the Tedjen and the Murgab; B. A. Livshits connected the Airyana Vaējah with the 'Great Khorezm' (see Marquart 1901:118, 155; Inostrantsev 1911:299-300, 307-316; Bartold 1965:544; Benveniste 1934:265-274; Dyakonov 1961:58-65, 361:notes 95-96; Struve 1948:5-34; Tolstov 1948a:20, 286-287, 341; 1948b:103ff.; Henning 1951:43; Livshits 1963:151-153; etc.).

Note 69 In the *Bundahishn* is preserved a list of ancient Iranian rivers, many of which, however, have not yet received a clear geographical identification (Müller 1880, Bundahishn:ch. XX:7; Ptitsyn 1911:315-316).[Editor 91] It is possible that the name 'Daraga' survived in modified form in a city name near the crossing of the Amudarya; it was called Dargan, Darugan, Darugan-Ata, and it was the most southern point on the territory of Khorezm in the 10th-11th centuries.

Note 70 Irrigated agriculture was known to the compilers of *Avesta*, but all mention of irrigation canals are limited to *Videvdat*, and therefore should probably be attributed to a later date (Dyakonov 1961:362:note 99).

Note 71 See, for example, the water flow in canals, the regime of the Djeykhun, and the measuring tools used in the construction of canals (Biruni 1957:286-290; Tolstov 1957b:10:note 17; Tolstov and Vaynberg 1967:252).

Note 72 Above the Tyuya-Muyun gorge, the river flows into the Duldulatlagan gorge, also known by the local population as Danysher, i.e. Dakhan-i Sher (in Iranian, 'Lion's mouth'), translated in Arabic as Fam al-Asad (see Gulyamov 1957:19; Biruni 1966:284:note 172). Al-Biruni also quoted the Khorezmian name of the gorges: Sikr ali-Shaytan, which means Shaytan's dam (Gulyamov 1957:19)

Note 73 Around the 8th century, a group of Pechenegs wandered into the Northern Karakum and in the Sarykamysh depression (see Yakubovskiy 1947:50-51; Tolstov 1950b; Klyashtornyy 1964:176-179). The riverbed of Mazdubast is the system of dry riverbeds of the Sarykamysh Delta (Daudan, Kangadarya, Daryalyk) and, likely, Uzboy (Gulyamov 1957:20-27; Tolstov and Kes 1960:9-17; Tolstov, 1962a:25).

Note 74 As we shall see below, the natural history of the three Amudarya's deltas (first, the dissection of the Akchadarya and later also the channels of the Sarykamysh), is reflected in the prevailing historical period physiographic processes: deflation processes that prevailed in the 'lands of ancient irrigation' on the right bank of the Amudarya, and the processes of accumulation on the left bank. This in turn affected the preservation of site materials (settlements and channels) in the two main areas of Khorezm: the right bank and the left bank of Amudarya.

Note 75 See the history of the study of the Akchadarya deltas and their geographical description in Tolstov and Kes 1960:35-66.

Note 76 This circumstance, apparently, also compelled Ya. G. Gulyamov (without good topographic materials and aerial photographs) to consider the Suyargan as the most ancient source of irrigation for the right bank of the Khorezm. He considered as Suyargan all the dry riverbeds of the southern delta (Gulyamov 1957:47, 62, etc.).

Note 77 In the Uzbek language *kul* means 'salt marshes', 'lake', as well as 'ash', 'ashes'; *kulcha* means 'flat cake'. In the Iranian languages *darya*, *daryache* (Persian) or *daryacha* (Afghan) means 'lake' or 'sea'; hence Amudarya or 'Darya-i Amus' (from the city of Amul or Amu; see Bartold 1965:319).

Note 78 The system of complex seasonal movement was typical, for examples, for the Nuer pastoralists, who harvested two crops a year (Evans-Pritchard 1940:76-81).

Note 79 For the Bronze Age sites see Tolstov 1939:174-176; 1948a:66-68; 1948b:76-78; 1958:90-91; Itina 1959a, 1960, 1961, 1962, 1967, 1968; Tolstov and Itina 1960b; Tolstov and Kes 1960:82-135.

Note 80 Settlements and encampments of the Bronze Age with remains of irrigation works in the environs of Bazar- kala are: Kokcha 1-10; Bazar 1-11; *poisk* 446-447, 464, 469, 477, 479, 481, 499, 504-511, 515, 548-549, 551-552, 560, 565, 567, 570-576, 583, 665, 671-675, 678-683, 688, 690-693, 702, 705, 713 (Figure 29).

Note 81 What was cultivated in these 'fields' is still unknown. The content analysis of burial vessels did not provide any results (Itina 1961:54).

Note 82 Geomorphologic research on the Tedjen River near the Bronze Age settlement of Khapuz-depe (3rd millennium BCE) also revealed the remains of an ancient artificial dike on the river bank (Masson 1967b:334).

Note 83 A clear example of the use of old riverbeds for irrigation in the 19th century is the formation of the Kuvanysh-Djarma system (Andrianov 1958a:88-90).

Note 84 Settlements and encampments of the Bronze Age, northeast of the ruins of Djanbas-kala, with remains of irrigation works are Kokcha 15 and 16; *poisk* of the years 1954 and 1964 were: 641, 647-653, 655, 656, 662, 663, 1599-1601, 1604-1607, 1609-1611, 1613-1616, 1620-1626, 1628, 1632, 1634.

Note 85 N. I. Igonin provided a topographic plan of the settlements of Kokcha 15 and 16 and of their fields, parts of which had diked areas measuring 16 × 10 m, 10 × 10 m, 12 × 10 m, 7 × 7 m, etc. (Itina 1967:74, 77:fig. 5;1968:77:fig.1, 79:fig.2, 81).

Note 86 The trench was described with the assistance of soil scientist F. I. Kozlovskiy, a member of the Aral Sea Expedition of the Institute of Geography.

Note 87 The settlements and encampments of the late Bronze and early Iron Ages between Little Kavat-kala and Yakke-Parsan are: Yakke 1, 2; Kavat 1, 2; *poisk*: 1363-1367; 1390-1392, 1449-1454, 1498-1508 (year 1957).

Note 88 By their size, the Amirabad semi-earthen houses are very close to the Tazabagyab ones (see Tolstov 1962a:49:fig. 19, 50, 69, 72:fig. 31; Itina 1959a:56; 1963:112).[Editor 92] This suggests that approximately the same number of people could live in underground dwellings, as well as in settlements (see p. 149). According to E. E. Kuzmina, in the late Alakul period (13th-12th centuries BCE), from 10 to 30 people could live in Andronovo semi-earthen houses 11 × 7 m in size (Kuzmina 1964b:106). K. V. Salnikov wrote that few tens of people lived in large Andronovo semi-earthen house (over 100 m²) (Salnikov 1965:28). M. I. Itina wrote that about 30-40 people lived in Amirabad dwellings in Yakke-Parsan 2 (Itina 1963:129).

Note 89 G. N. Lisitsyna provided similar calculations for canal digging labor costs in Geoksyur based on a standard of excavation (3 m³ per person per day) known from Sumerian documents (Lisitsyna 1965:128- 129), although this standard was good for workers with good metal shovels (see Smirnov 1933:10-12). We arbitrarily reduced it (due to imperfections of working implements) by one third.

Note 90 In this circumstance, perhaps, in terms of irrigation history in Central Asia and Near East, the coincidence of the term can be explained, both a natural channel and an artificial main canal: *djuy, djuybar, nahr* (pl. *ankhar*) meaning 'river' or 'channel' both natural and also artificial (see Bartold 1965:117). The genetic relationship between main canals and riverbeds is also confirmed by the Khorezmian term *arna*, which means 'canal', 'canal formed by river', as opposed to an excavated canal, *yap* or *yab* (Radlov 1893:303; 1905:259; Bartold 1965:118; Gulyamov 1957:243). S. P. Tolstov noted the proximity of the Khorezmian *arna* with the proto-kassites 'source', 'spring' (Tolstov 1948b:80-81).

Note 91 See also the hypothesis of D. D. Bukinich, who connected the beginning of agriculture in Central Asia with crops of forage grasses (Bukinich 1924). It is impossible not to recall that the Lower Amudarya is one of the regions of domestication of lucerne as the traditional food culture of Khorezm, which gives, with watering, 4–6 crops per year.

Note 92 According to V. M. Masson, during the Yaz 1 period, the Guniyab canal was 36 km long and the Gati-Akar canal 55 km long. They were 5-8 m wide and 2-3 m deep (Masson 1959:87-91).

Note 93 V. V. Bartold, and after him S. P. Tolstov, connected Herodotus' description on the Ak River with the Lower Amudarya (Tolstov 1948a:43).

Note 94 Finds of Archaic ceramics clusters in the environs of Bazar-kala: *poisk* 352, 359, 421, 422, 440, 448, 453,454, 470, 472, 484, 497, 500, 545, 546, 561, 562, 564, 583, 587, 589, 701, 709-711, 1223, 1225.

Note 95 Sites remains in the basin of the ancient Kelteminar with predominance of Archaic pottery: *poisk* 1, 2, 4, 6-7, 10, 12, 31, 39, 46-47, 50, 53-59, 72-73, 76, 80, 82, 94, 96-98, 100, 109-110, 114-115, 123-125, 165,203, 205, 232, 238-239, 276, 302, 304, 306, 334, 337, 345-347, 349, 365-367, 370-371, 373-374, 416, 501, 1243, 1272-1273.

Note 96 In the Dingildje 'oasis' the prevailing Archaic pottery came from Archaic Dingildje *poisk*: 138, 141, 150-151, 153-157, 160, 176-178, 191, 194-196, 199, 206-207, 223-226, 403, 1557-1558, 1570, 1579, 1584 (1198), 1586, 1597

Note 97 Grape was known in the Lower Amudarya already in the 5th century BCE. Finds of grape seeds were made in the Archaic house of Dingildje and during the excavations at Koy-Krylgan-kala and Toprak-kala. The ancient authors report that at the time of the campaign of Alexander of Macedon, the Greeks were surprised by the abundance of wine in Central Asia (Arrian IV, 21, 10; Curtius Rufus VII, 4, 2).

Note 98 M. G. Vorobeva held, however, a different point of view. She believed that vineyards were later and dated to the Kushan period.

Note 99 Remains of wheat and barley were found in Archaic Dingildje.

Note 100 Long-term excavations at Koy-Krylgan-kala contributed to clarify the chronology of Kangju and Kushan ceramic complexes of Khorezm. Kangju ware was recorded in the site lower architectural horizon, dated to the 4th-3rd centuries BCE. The middle and upper architectural horizons represented the development of a Kushan culture from the 1st century CE (and, possibly, the end of the second half of the 1st century BCE to the late 3rd-4th centuries CE). A special type of light slipped ceramics, connected with the appearance of new people in the oasis, was also found (Tolstov and Vaynberg 1967:19-20, 102, 310).

Note 101 On the problem of Kangju see p. 213.

Note 102 Remains of sites in the basin of ancient Kelteminar with a predominance of pottery are: 1) Kangju, *poisk* 3, 22, 52, 64-65, 239, 268, 335, 340, 523-525, 416; 2) Kushan, *poisk* 11, 21, 26, 36, 65, 67, 78, 81, 85-86, 88, 92, 95, 105-106, 116, 162-163, 171-172, 175 (202), 204, 236, 249, 252, 258, 261, 277, 284, 298, 307, 309, 311-312, 331, 335-336, 338-339, 533 (about the *poisk* 36, 252; 335, 339, 525 see Vorobeva 1961:165-166).

Note 103 Ya. G. Gulyamov also dated the construction of this canal to the 4th-3rd centuries BCE (Gulyamov 1957:81).

Note 104 Sites around Djanbas-kala with a predominance of pottery are: 1) Archaic, *poisk* 16, 319, 594, 1229, 1238; 2) Kangju, *poisk* 597-598, 609-611, 631, 641, 643, 645, 1222, 1227, 1230-1232 (for the *poisk* 643 see also Vorobeva 1961:167); 3) Kushan, *poisk* 320-321, 534, 600-601, 611, 616, 619-620, 1240; 4) finds of Medieval ceramics of the 9th-10th centuries in *poisk* 323-324, 544, 597).

Note 105 Approximately the same number (4,000-5,000 people) characterized the population of Artemita (Karastel) neighbors in Mesopotamia, where, along the 25 km long main canal, R. McC. Adams discovered the ruins of five Seleucid-Parthian villages and one urban center (see Adams 1965:61-62, fig. 4).

Note 106 Considering the average size of a cross-section of the Djanbas-kala canal as 30 m², we can easily calculate the general volume of excavated soil in 700,000-800,000 m³. To carry out this works, the labor of 5,000 diggers was required for 6 weeks with a rate of 3 m³ per person per day. For the canal seasonal cleaning at least 2,000-3,000 diggers were probably employed every year.

Note 107 Remains of millet, wheat, peach, grape and oleaster were found in the middle and upper Kushan layers of Koy-Krylgan-kala (Tolstov and Vaynberg 1967:51, 54, 56, 64, 69, 74, etc.).

Note 108 Irrigation ditches.

Note 109 Remains of sites with predominance of pottery in the Dingildje 'oasis': 1) Kangju, *poisk* 148, 1593-1596 (?); 2) Kushan, *poisk* 152, 158-159, 193, 197, 212, 215, 220, 1570-1571, 1574-1575, 1578, 1590-1591.

Note 110 Remains of sites with predominance of pottery along the Kyrk-Kyz canal (from Guldursun to Big-Kyrk-Kyz): 1) Archaic, *poisk* 724-726, 731, 767, 775, 867, 876; 882, 889, 892, 914, 939, 940, 944-947, 951; 2) Kangju, *poisk* 727, 768, 773, 796, 824, 827-828, 841, 846, 852, 854, 872, 876, 884, 887b, 915, 930, 955, 965, 970-972, 974, 975, 978, 980-984, 989, 991, 996, 1320, 1322-1323 (see also Tolstov 1948a:134; Nerazik 1966:9); 3) Kushan, *poisk* 734, 769-773, 778-779, 783, 790-791, 793-795, 801, 804, 813, 820, 829-830, 861, 868, 914, 934, 952, 954, 961-962, 964a, 971, 983-985, 1321, 1329 (see also Nerazik 1966:10; Vorobeva 1961:150-151); *poisk* 790-791.

Note 111 Remains of sites with predominance of pottery along the Kyrk-Kyz canal (between the Big-Kyrk-Kyz and Kurgashin-kala): 1) Kangju, *poisk* 1284, 1294 (?), 1296, 1302-1303, 1305-1306, 1310, 1312, 1314, 1316, 1330 (?), 1334-1338; 2) Kushan, poisk 1283, 1287, 1304, 1319, 1329, 1332, 1340, 1345. Note 112 Remains of sites with predominance of pottery: 1) Archaic, *poisk* 749, 754, 757-758, 762, 857, 1058, 1074, 1084, 1091, 1093, 1097, 1368, 1447; 2) Kangju, *poisk* 756, 1082, 1084; 3) Kushan, *poisk* 751-754, 768, 1060, 1065, 1082, 1086; 4) Kushan-Afrigid, *poisk* 759, 918, 1058, 1062, 1064-1065, 1067, 1071, 1083, 1444; 5) Afrigid, *poisk* 736, 738, 739, 741, 743, 745, 750, 760, 764, 766, 858, 902, 906, 917, 919, 1061, 1071, 1088, 1092.

Note 113 Remains of sites with predominance of pottery: 1) Archaic, *poisk* 1011, 1021-1022, 1027, 1032-1033, 1076-1077, 1110, 1113, 1161, 1362, 1393, 1525, 1545; 2) Kangju, *poisk* 1024, 1028, 1031, 1099, 1169, 1361, 1524; 3) Kushan, *poisk* 1052, 1114, 1192, 1194, 1379).

Note 114 Ceramic clusters and sites in the environs of Toprak-kala with predominance of pottery: 1) Archaic, *poisk* 1370-1374, 1443, 1512; 2) Kangju, *poisk* 1438; 3) Kushan, *poisk* 1397, 1399, 1437, 1462, 1464, 1478; 4) Kushan-Afrigid, *poisk* 1355-1356, 1400, 1404 (?), 1409(?), 1410-1415, 1422, 1471-1472, 1474, 1479, 1483; 5) Afrigid, *poisk* 1370, 1406, 1463, 1467, 1475, 1485, 1493).

Note 115 According to a preliminary definition by B. I. Vaynberg, the coins found are dated as follows: *poisk* 1355, two coins of the beginning of the 4th century; *poisk* 1356, one coin of the 4th century; *poisk* 1370, one coin of the 7th century; *poisk* 1399, one coin of the end of the 3rd century, one coin of the beginning of the 4th century, one coin of the 5th century, one coin of the 8th century; *poisk* 1410, two coins of the beginning of the 4th century, one coin of the 8th century; *poisk* 1411, one coin of the end of the 3rd century, two coins of the beginning of the 4th century, one coin of the 4th-5th centuries; *poisk* 1414, 10 coins of the end of the 3rd century, one coin of the 4th century, one coin of the 5th century; *poisk* 1462, one coin of the end of the 3rd century; *poisk* 1463, two coins of the 7th century (?); *poisk* 1464, one coin of the end of the 3rd century, one coin of the 8th century; *poisk* 1467, one coin of the 7th-8th century; *poisk* 1471, one coin of the 4th-5th centuries; *poisk* 1472, one coin of the end of the 3rd century, two coins of the beginning of the 4th century, one coin of the 4-5th century, one coin between the 4th and the 5th century, one coin of the 5th century, one coin of the 8th century; *poisk* 1476, one coin of the beginning of the 4th century, one coin of the 4th century, one coin of the 8th century; *poisk* 1478, eight coins of the end of the

3rd century, one coin of the early 4th century; *poisk* 1479, one coin of the 4th–5th centuries; *poisk* 1482, one coin of the 4th–5th centuries; *poisk* 1483, one coin of the 5th century.

Note 116 Ceramic clusters and sites with predominance of pottery: 1) Archaic, *poisk* 1121; 2) Kushan, *poisk* 1434–1435; 3) Kushan-Afrigid, *poisk* 1116, 1118, 1423–424, 1426, 1428).

Note 117 Ceramic clusters and sites with predominance of pottery: 1) Archaic, *poisk* 1107, 1108 (Tash-Kyrman), 1554, 1555; 2) Kangju, *poisk* 1108; 3) Kushan, *poisk* 1551–1553.

Note 118 Traces of ancient irrigation systems were not preserved in the contemporary cultivated zone adjacent to the Amudarya. Therefore it is quite difficult to answer the question of when the major irrigation systems headworks were moved on the main riverbed. According to S. P. Tolstov, it occurred in the Kangju period on the right bank of the Amudarya and in the Kushan period on the left bank (see Tolstov and Andrianov 1957:9–10).

Note 119 The term *saka* in modern Khorezm refers to water-intake place.

Note 120 Ceramic clusters and sites in the Kelteminar basin with pottery predominance: 1) Kushan-Afrigid, *poisk* 86, 92, 166, 236, 243, 245, 246; 2) Afrigid, *poisk* 30, 77, 79, 86, 93, 108, 168, 201, 208, 212, 236, 239, 242, 244, 300).

Note 121 Ceramic clusters and sites with pottery predominance: 1) Kushan-Afrigid, *poisk* 139, 182, 202, 298, 1556, 1572; 2) Afrigid, *poisk* 140, 179–184, 186–190, 192, 198, 208–210, 1559, 1572, 1573, 1587, 1589).

Note 122 Ceramic clusters and sites with pottery predominance : 1) Kushan-Afrigid, *poisk* 730, 772, 780, 788, 829, 845, 849, 953, 993; 2) Afrigid, *poisk* 722, 726, 776, 777, 781–783, 785–787, 802, 805, 807, 811, 820, 823, 824, 826, 830, 836–838, 841–843, 845, 847, 848, 851, 865, 866, 873, 877, 878, 881, 887a, 891, 911, 913, 935, 952, 964b, 968, 979, 995.

Note 123 On dating of the Afrigid farmsteads see Nerazik 1966:35–43.

Note 124 The people of Khorezm played a prominent role in the anti-Arab movement of the 'people in white', i.e. Abu Muslim and his followers: Ishaq al-Turk, al-Muqanna, Rāfi (Bartold 1963, Tom I:252–258; Gulyamov 1957:123).

Note 125 Ceramic clusters and sites of the Afrigid-Samanid period in the environs of Narindjan: *poisk* 746, 748, 899, 921.

Note 126 *Poisk* 1012, 1051, 1102, 1149, 1152, 1163, 1457, 1539, 1540.

Note 127 The term *rustak* (or *rotastak*, see p. 149) occurs in Iranian sources of the 7th–9th centuries and describes a group of villages (*volost*, according to V. V. Bartold) united by a single source of irrigation; the term *rustak* was mentioned by Medieval geographers describing the canals in the environs of Nishapur, Merv and Balkh (see MITT 1939:171, 173, 216ff.; Bartold 1963, Tom I:119:note 5).

Note 128 Ya. G. Gulyamov expressed doubts about the true width of the canal: 5 cubits, i.e. 2.5–3 m between banks (Gulyamov 1957:126). Perhaps, this figure possibly referred to the lower reaches, where indeed real canal dimensions were found.

Note 129 Ya. G. Gulyamov disagreed with Istakhri, who interpreted the name of 'Gaukhore' as 'cows' forage' (MITT 1939:179; Bartold 1963, Tom I:199; 1965:164–165). Perhaps Istakhri was right because it was an ancient name. In the *Avesta* language GAV (GAV-) meant 'cow' (Sokolov 1964).[Editor 93] The farmers in the *Avesta* were designated as 'supplying cows with forage' (see p. 154). It was also possible to call the canal 'cows' forage'. Beykhaki stated that in 1034 CE Harun al-Rashid gave the Seljuks 'pastures and the best places in Mash Rabat, Shurakhan and Gaukhore' (MITT 1939:305).

Note 130 Ceramic clusters and sites of the 12th–early 13th centuries with predominance of pottery: *poisk* 1010, 1012, 1016, 1018, 1039, 1045, 1047, 1079, 1105, 1106, 1124, 1141, 1142, 1144, 1146, 1147, 1149, 1152, 1153, 1160, 1162, 1165, 1186, 1384–1385, 1455, 1532, 1538.

Note 131 There is no demographic data describing family size and number of inhabitants in rural Medieval farms. As a very remote analogy, data from old Tashkent can be used for comparison, where in the 1920s researchers found that 75% of all houses had one room: with an average of 3.6–5 people living there (VIR 1925, no. 6:91–97). V.A. Shishkin suggested that the population density in the city environs of the Varakhsha site (6th–7th centuries) was 200 per 1 km² (Shishkin 1963:34). In the 1930s, in the most cultivated areas of the Khorezm oasis, the population density reached 80–100 or even 200 people per 1 km² (MRSA 1926, Kn. 2, Ch. 2:39). At present, some oases of Uzbekistan have a population density of up to 270 people per 1 km² (*Uzbekistan. Ekonomiko-geograficheskaya kharakteristika*, Tashkent 1950:67).[Editor 94]

Note 132 Also W. Willcocks, who revealed a decrease from 50 to 5–10 m in Mesopotamia, remarked on the huge cross section difference between ancient and later canals (Willcocks 1903:13).

Note 133 According to V. V. Tsinzerling, the coefficient of irrigated area in the Khiva oasis at the beginning of the 20th century was approximately 30%, which provided the necessary drainage for irrigated fields (Tsinzerling 1927:248–263).

Note 134 In Northern India *be-gar* means forced unpaid labor, an act of work conscription, etc. (see Platts 1884, published in USSR in 1959; see also the Greek-English Lexicon by Liddell et al. 1846).[Editor 95] In there, information on social labor conscriptions in Ptolemaic Egypt is provided (Taubenschlag 1955:618) (see also pp. 117–118).

Note 135 The development of irrigation and the expansion of irrigated areas in the 9th–11th centuries was observed in a number of oases of Central Asia: Dahistan, Ahal, Merv and Bukhara and also in Bactria-Tokharistan in the Upper Amudarya (see Bartold 1965:124, 128, 138, 187; Mandelshtam 1964:28; Gulyamov 1957:125–158).

Note 136 A Bronze Age site (*poisk* 437) with Tazabagyab (?) ceramics was discovered in 1954 on the bank of the Daudan, south of the ruins of Kalalygyr II.

Note 137 The image of the bull, which personifies the power of fertility, was linked with the cult of water not only in Central Asia, but also in India, Near East and the Mediterranean area. The most ancient evidence of this cult can probably be seen in a wall painting, dated to the 6th millennium BCE at Eastern Çatal Hüyük, depicting a bull on water (see Mellaart 1965b).

Note 138 S. P. Tolstov connected some religious sites on the left bank of the Khorezm, as Djumart-Kassab on the Mazlumkhanslu and the cemetery at Djumurtau, with the traditions of the first *Avesta* man ('man-bull' according to K. V. Trever): Gayomart or Gayumars living in Airyana Vaējah on the banks of the Dāitya River (Tolstov 1948b:87).

Note 139 *Poisk* where 'barbaric' rough ceramics were found: 13 (at a depth of 40 cm in a test trench), 53, 55, 82, 124, 126, 175.

Note 140 Remains of settlements and large pottery clusters with predominance of Archaic ceramic: *poisk* 1–2, 11 (24 September 1952); *poisk* 6–7 (25 September 1952); *poisk* 12–13, 17, 31–32, 35, 38, 51, 81, 84, 88–89, 112, 126a, 129–131, 173, 179, 185–187, 190, 424, 431, 435, 752, 753, 775 (1953 year).

Note 141 Ceramic clusters and site remains with predominance of pottery: 1) Antique (?), *poisk* 289, 410–411; 2) Archaic, *poisk* 198, 228, 258, 259, 270–271, 281, 412, 413, 417; 3) Archaic and early Kangju, *poisk* 572, 576, 579; 4) Kangju, *poisk* 288, 577; 5) Kushan and Kushan-Afrigid, *poisk* 414; 6) early Medieval, *poisk* 415–417a, 420.

Note 142 Comparing late Archaic ceramics from a house in Dingildje with early Kangju ceramics found in basement walls at Kalalygyr-kala I, M. G. Vorobeva divided them by half century periods, between the mid-5th and the end of the 5th or the beginning of the 4th centuries BCE (Vorobeva 1959a:78, 1959b:78).

Note 143 The 1964 aerial photographic coverage by N. I. Igonin and the aerial observations of the same year revealed places where the canal and dam coincided. It raises the question of the dam dating to the Archaic period, based on field work of 1953 (Tolstov and Kes 1960:185).

Note 144 Yu. A. Rapoport and M. S. Lapirov-Skoblo, in their article on the excavations at Kalalygyr-kala I, wrote that the basement walls and the palatial building were erected between the 5th and 4th centuries BCE (the radiocarbon analysis dates to 380±120 years BCE). After a very short abandonment the palace was inhabited at the very beginning of the Kangju period. Its use ended after a fire in the Kangju period. Later, in the 2nd–4th centuries CE, it was used just as an ossuary burial (Rapoport and Lapirov-Skoblo 1963:141–150).

Note 145 Remains of head works of different types of the 1st millennium BCE discovered in Khorezm are essential for the history of irrigation technology. Such material is absent in R. J. Forbes. Water regulating devices of the Sassanid period on the Nahrwan canals are known from measurements and publications by R. McC. Adams (1965:80, Figures 17–18). Plans of Medieval canal head parts in the environs of Samarra were published by Ahmed Sousa (1948:scheme 2a-b, 5, 25, etc.). Rashid al-Din wrote letters on the construction of heads and main canals on the Tigris (1247–1318) accompanied by a schematic plan (Petrushevskiy 1960:120–121, figs. 1–2). A section in Ya. G. Gulyamov's book includes a description of head works on freshet canals of Khorezm in modern time (Gulyamov 1957:239–242).

Note 146 The Khorezm Expedition in this region and their detailed archaeological-topographic and geomorphologic work (Andrianov 1958b) dispelled the geographers' doubts about the Chermen-yab canal, which they regarded as a natural riverbed (Murzaev 1940:109-110, etc.). Debating with S. P. Tolstov over the large sizes of ancient canals, E. M. Murzaev stated that, at that time, the people of Khorezm used former riverbeds as canals. As mentioned above, the historical stages of irrigation on adapted and embanked riverbeds in Khorezm was dated to Prehistory.

Note 147 Remains of sites with predominance of pottery in the environs of Gyaur-kala I and II: 1) Kangju, *poisk* 8 (3 October 1952); *poisk* 55(?), 108-109(?) [09 August 1953 (?)]; 2) Kushan, *poisk* 7, 10, 12-14, 17, 20 (3 October 1952); *poisk* 100-101 (7 August 1953) (see also Vorobeva 1961:163; Andrianov 1958b:319-320). Note 148 Ceramic clusters and site remains with predominance of pottery in the environs of Shakh-Senem: 1) Kangju, *poisk* 29 (21 September 1952), 8 (26 September 1952); 2) Kushan, *poisk* 6-7, 27-28, 33-35, 38-46 (21 September 1952); 56; I, 3 October 1952 (also *poisk* 47, 1953); *poisk* 14-15, 23 (?), 25, 42-42a, 44-46 (1953; see Vorobeva 1961:63, *poisk* 46).

Note 149 Ceramic clusters and site remains of sites with predominance of pottery in the environs of Kunya-Uaz: 1) Kangju, *poisk* 7, 9, 14 (8 October 1952); *poisk* 195-196, 201 (1953); 2) Kushan, *poisk* 4, 8 (8 October 1953), *poisk* 196, 200, 242, 247, 254 (?), 256 (?) (1953); 3) Kushan-Afrigid, *poisk* 226, 251; 4) Afrigid-Samanid, *poisk* 227, 260; 5) Khorezmshah, *poisk* 10, 12 (8 October 1952); *poisk* 224-246, 251a, 259, 261, 266 (1953); 6) Golden Horde period, *poisk* 1, 5, 11 (8 October 1952), *poisk* 268, 273 (1953).

Note 150 Archaeological work in the ancient Bukhara oasis, in Termez, in Samarkand, in the valleys of the Kashkadarya and Fergana provided much evidence of the significant expansion of irrigated areas, and the increase of cultural and economic standard of life in farming populations during the Kushan period (ITN 1963:369; Staviskiy 1961a; Shishkin 1963:229; etc.).

Note 151 The crisis in these oases in the 4th-5th centuries was noted by archaeologists in many places of Central Asia (see Masson 1945:5-6; Dyakonov 1953:292; Albaum 1955:70; Shishkin 1963:230; Masson 1968:100; etc.).[Editor 96] Even in the 'heart' of the Sassanid state in Mesopotamia (the Diyala Basin), archaeological work revealed, for this period, certain features of an irrigation crisis and the reduction in rural population (see Adams 1965:73-75).

Note 152 Sherds of Kushan-Afrigid ceramics were found at the following *poisk* 226, 251, 285 (?), 286, 414, 558, 578, 640, 670, 957).

Note 153 Among the recent work devoted to the Khorezmshah state, the book by H. Horst (1964) should be mentioned.

Note 154 Sites with predominance of Khorezmshah pottery: *poisk* 3-4, 13-14, 16-17, 24-25, 32, 37, 47-50 (21 September 1952), *poisk* 51-55, 57-61 (22 September 1952), *poisk* 19, 42a, 43 (year 1953).

Note 155 See *poisk* 23-24, 27-28, 31-32, 39, 42, 45-46 (7 October 1952) and *poisk* 128, 132-133, 135-138, 141-142, 144, 147, 150-151, 153-154, 160-162, 165, 168, 170, 407-408 (1953).

Note 156 Sites with predominance of pottery: 1) Afrigid, *poisk* 693, 699, 703; 2) Afrigid-Samanid, *poisk* 694, 969, 704, 713, 721, 772; 3) Khorezmshah, *poisk* 689-690, 700, 706, 707-709, 714, 716, 717, 722-725, 727-729, 734, 738, 746-747, 755, 767 (see also Vishnevskaya 1963:54-58).

Note 157 At the beginning of the 20th century, cleaning the small irrigation network ('inner *kazu*') took 1.5 more days than the main canals cleaning (see Smirnov 1933). In the 12th century, an irrigation network was rarer and we conditionally accepted a rate of 1:1.

Note 158 Participating in the 1959 work on the left bank of the archaeological-topographic unit, O. A. Vishnevskaya carried out a survey of the environs of Zamakhshar and Daudan-kala, but, as evident from the publication, the materials were not characterized for their quantity (see Vishnevskaya 1963).

Note 159 Such methods are widely used by geographers to determine the size of inhabitants and different populations (see Bruk 1958). Historians and archaeologists, reconstructing the number of inhabitants of ancient settlements, often apply analogy (in particular when comparing with contemporary data). It was by means of extrapolation (applying to the ancient city the contemporary urban population average in the Upper Diyala region), that R. McC. Adams provided the numbers of the inhabitants for all the ancient cities in the Diyala basin (Adams 1965:21-29, 112-116).

Note 160 V. V. Tsinzerling reported that in the Amudarya area there were 2.4 people for each *dessiatine* (1.0925 ha), in Egypt 3.8 and in China 13 (Tsinzerling 1927:213).

Note 161 R. McC. Adams noted that during this period the entire population of the Diyala Basin in Mesopotamia showed a sharp decline (Adams 1965:25, 106-110, 115).

Note 162 The laws of Manu, for example, say: "The king, who wants to conquer his enemies, must first destroy the dams on their territories" (see ICBIP 1965:34).

Note 163 Correctly noting that, as a result of the Mongol invasion, the Amudarya watered the Sarykamysh depression, Ya. G. Gulyamov, following V. V. Bartold, connected the formation of the new Daryalyk riverbed with this breakout (Bartold 1965:172; Gulyamov 1957:166). This is not supported by any archaeological or geomorphological research in this region (Tolstov and Kes 1960:192–196; Andrianov 1959b:182).

Note 164 Very characteristic of the Khorezmshah period was, for example, a huge public building near Yarbekir-kala (a caravanserai with 750 rooms) which was not completed because of the Mongol invasion (Vishnevskaya 1963:68–72). Remains of sites with predominance of Khorezmshah pottery were found at *poisk* 287, 298, 303, 311, 564, 633, 895; *poisk* with predominance of Golden Horde pottery: 4, 5, 275, 277, 282, 288, 293–294, 305–306, 539, 553–555, 563, 567, 571, 582–583, 594, 597, 598–600, 605, 610, 612, 621, 625, 627– 628, 630–632, 635, 636, 644.

Note 165 Sites with predominance of pottery: 1) Khorezmshah, *poisk* 346, 792, 796–799, 803–805, 808–815, 833, 838–840, 847, 860, 898, 900, 903, 924–925, 944; 2) Golden Horde, *poisk* 318, 326, 330–334, 341–342, 345, 349, 359(?), 368, 371–372, 375–376, 779, 781–784, 786–788, 793, 795, 800–802, 821–822, 824, 826, 829–830, 835, 837, 842–845, 848–849, 851–855, 863, 867–868, 872, 877–879, 883, 885, 887–889, 891, 893, 894, 906–907, 909–911, 913, 915–922, 931–934, 942, 947, 951–953, 957, 959, 961–964, 967).

Note 166 Describing the irrigation around the Turkmen lake in the 19th century, Yu. E. Bregel (1961:61:note 56) rightly noted that this issue was still little studied. However, one can hardly agree with the author who, following P. F. Preobrajenskiy, dated the beginning of lake farming by the Turkmen-Yomut to the mid-19th century. The Sarykamysh systems, and also Ya. G. Gulyamov's data (1957:60–61) on the wheat 'lake' crops under the Ustyurt ('Kuygun') in the 16th century, are proofs of the antiquity of such irrigation methods by the Turkmen. Also the 19th century Turkmen's tradition of irrigation, originating from the Syrdarya in the historical period, contrast with Yu. E. Bregel's deductions (see pp. 230–231).

Note 167 These dimensions were typical of Medieval main canals also in other Eastern countries, in particular in India, where, under the Mongols, they built the Kalsi canal, 177 km long and 9 m wide (see ICBIP 1965:42).

Note 168 The investigation of the Khorezm Expedition revealed vast territories with traces of *chigir* installations, which indicated a massive introduction of this system (Andrianov 1958b:325; Tolstov 1958:109).[Editor 97] N. N. Vakturskaya rightly linked the expansion of the massive ceramic production in the 9th–11th centuries with the spread of *chigir* irrigation and the increased need for *chigir* jars (Vakturskaya 1959:26).

Note 169 The principle of rotation in the construction of water-lifting devices was known in the ancient Orient at the beginning of the 1st millennium (Forbes 1955, Vol. II:39). The most extensively widespread water-lifting devices, such as the *chigir*, originated in Central Asia and in the Near East only at the end of the 1st millennium (in the feudal period). The transition from the shaduf (Egypt, Mesopotamia), *rati*, *denoli* (India), *nova* (Central Asia), *sepma* (Khorezm), to more complicated devices such as the *cêrd* (Mesopotamia and Egypt), *khurusu* (India) and finally the 'Persian wheel' (Iran), *charkha*, *sakiya* (Mesopotamia) and *chigir* (Central Asia) was an important stage in the development of irrigation techniques (Gulyamov 1957:251–253).

Note 170 It is noteworthy that, in the 19th century, an increment of irrigated lands occurred in some other areas, such as in the Diyala Basin in Mesopotamia (see Adams 1965:116).

Note 171 In his detailed study of the *Shahnameh* historical geography, G. V. Ptitsyn concluded that the Kanga country was situated in the Middle Syrdarya (see Ptitsyn 1947; Klyashtornyy 1964:165–166). It is possible to agree with Ptitsyn's deduction about the *Shahnameh*'s geographical reality. But, how to explain the absence of Khorezm from the list including Chach, Sogdiana, Samarkand, Bukhara, Sepidjab, and, finally, Kang, i.e. Afrasiab's main territory (Firdausi 1960:153)? Apparently, in this period, which saw the emergence of the mighty Kangju state, the Khorezm oasis was part of Kang (compare with Tolstov 1948b:145ff.).

Note 172 See also V. V. Bartold (1963, Tom II, Chast 1:175–176), G. V. Ptitsyn (1947), Ya. G. Gulyamov (1957:97–98), S. G. Klyashtornyy (1964:167–179). The last attempt to geographically identify the Kanga of Tarban (Turaband, Turar, Otrar) and the Kanga of Firdausi with the Kangkha in the *Avesta* cannot be considered successful. Archaeological investigation have not confirmed the early dating advanced by E. I. Ageeva for the Otrar oasis sites, based in turn on the inaccurate dates of the Kaunchi and Djety-asar cultures (see Levina 1967:15).

Note 173 After the neglect of Koy-Krylgan-kala, there appeared a peculiar hand-made, light-slipped ware with typical pottery features of steppe tribes. Such ceramics were found in other sites on the right bank of the Khorezm together with traditional wheel-made Khorezmian ceramics of the Kushan period; the light-slipped ware was not found on the left bank of the Khorezm (Tolstov and Vaynberg 1967:130-131, 310-311).

Note 174 The toponymic evidence for the historical shift from Kangkha to Kanga to Kangju could be identified by S. P. Tolstov west of Khorezm with Kangadarya and Kangagyr (with the fortress dated from the mid-1st millennium BCE to the 4th century CE) and east of the Early Medieval city of Otrar-Kangu Tarban (see Tolstov 1958:72-73; Klyashtornyy 1964:155-160).

Note 175 According to G. V. Lopatin, the average content of suspended material in the Syrdarya is 1 kg/m³ while in the Amudarya it is 2.7 kg/m³ (Lopatin 1952:162).

Note 176 Judging by the results of the desk study of archaeological finds from the Inkardarya unit in 1959, headed by L. M. Levina, different sites and funerary structures ('slag kurgans') of the same culture were widespread on a huge territory: from Sengir-Tam (the largest among the 'slag kurgans') in the west to the fortress of Kum-kala on the Janydarya in the east. The distance between these two sites is about 80 km.

Note 177 During the 1966 exploration on the Syrdarya further archaeological surface material was collected from Sites 14, 20 and 21. L. M. Levina, analyzing them, confirmed the sites chronology near the 'slag kurgans' to the 6th-4th centuries BCE.

Note 178 Semi-dams, or *kysme-saga*, were used to regulate natural channels in the delta of the Amudarya. N. N. Belyavskiy described the Karakalpaks' construction of a semi-dam at the mouth of the Shortanbay. Above the channel outfall, thick ropes made of cane were fixed to stacks, on which cane and brushwood were positioned and additionally covered with earth, finally all the structure was in the water. By mean of this immersion, the semi-dams increased their top. When it reached the high water level, the rope ends were tightly fixed. If the channel was narrow, then a closed dam was built and covered with river sediments (Belyavskiy 1887:126).

Note 179 *Poisk* 217-223, 225-228, 231-238, 240-242, 244-245, 247-255, 257, 259-264, 267, 269, 272-276, 278, 280, 290, 292, 295, 298-300, 303, 305, 308, 310-311, 313-314, 317-330, 333-337, 339-366, 368-373, 383, 385, 389, 391-393, 395, 397, 399, 400, 403, 419-421, 423-426, 428, 434, 441.

Note 180 The name derives from the post-Mongol Medieval period: from the mongol word *Tsirik* ('army') - Chirik (Shirik in Kazak) (Tolstov 1961a:121). The fortified citadel characterized by a complex of kurgans appeared no later than the second half of the 5th century and existed till the mid-2nd century BCE (Tolstov 1961a:122; 1962a:139).

Note 181 The vertical aerial photography of different sectors of the Djety-asar district was carried out in 1960-1961 by engineer-geodesist N. I. Igonin.

Note 182 The use of natural obstacles in deltaic regions, like ancient riverbeds, large river meanders, lake-shaped depressions and other water bodies, useful for defense, was widely spread in the ancient Orient. This is supported in topographic details of large fortified settlements of the Late Bronze Age, studied by W. J. van Liere in Syria, with vertical aerial photography. These sites, according to published maps, for their environs layout and topography closely resemble the Djety-asar district citadels. They are located on lateral deltaic branches, near large meanders in areas of branch confluences. They do not have true outer walls with round contours but have instead round platforms with massive central residential compounds (see Van Liere 1963).

Note 183 *Poisk* 3, 4, 4a-6, 8, 11, 13, 15, 20-25.

Note 184 It is possible that for the construction of this irrigation system, methods of the Khorezmian irrigators were used. In that period, cultural links between the inhabitants of Barak-tam and the Khorezm were consistent (see Tolstov 1959a:32).

Note 185 Idrisi wrote about the confluence of the Janydarya with the eastern branches of the Amudarya 10 miles from the Aral Sea (see MITT 1939:220).

Note 186 In the pottery of Kesken-Kuyuk-kala three jointed and coeval complexes of the 7th-9th centuries CE can be discerned. A proper Djety-asar complex (third phase), a Semireche one assimilated by the Djety-asar bearers, and partially an Afrigid complex (Levina 1967:9).

Note 187 Thus, in the manuscript found by A. G. Tumanskiy (*Hudud al-Alam*) it says: "the Oghuz do not have any towns, but the people, living in yurts, are rather numerous" (see MITT 1939:209-217).

Note 188 Sites on the Lower Janydarya with predominance of pottery: 12th-14th century CE, *poisk* 40, 48, 50, 52, 67, 70, 78-81, 102, 119-121 (Beshtam-kala), 124-126, 134, 136, 164, 166-168, 175-177, 181, 212-215, 486, 491-497, 500-501, 507; 15th-18th century CE, *poisk* 182-184, 187-189, 193-197.

Note 189 S. P. Tolstov, rather convincingly demonstrated that the Sag-Dare site, mentioned in a collection of documents of the 12th century (*insha*), located on a river at 20 *farsakh* (120 km) from Djend, fits with the the settlement ruins of the 12th century within the citadel of Chirik Rabat (see Tolstov 1948b:61).

Note 190 *Poisk* 12, 15-17, 26, 34 (1960); *poisk* 3, 6, 13, 16, 20, 21, 23, 31, 33, 52 (1961); *poisk* 2-5 (1962), *poisk* 3-7 (7 October 1962); *poisk* 8-10 (8 October 1962); *poisk* 13-14 (9 October 1962).

Note 191 *Poisk* 8, 12, 15-17, 26, 34 (November 1960); *poisk* 3-4, 13, 16, 20-21, 23, 31, 33 (1961); *poisk* 2-5 (1962).

Note 192 We noted such an original system with connected basins on the abandoned 19th century Turkmen settlements (see Figure 50).

Note 193 See also Andrianov 1960, 1964.

Note 194 *Poisk* of 1956: 28-37, 39, 41-47, 49, 51, 54, 55, 57-66, 71-74, 76-77, 82-99, 105-106, 108, 111-112, 114; *poisk* of 1957: 116, 118, 122-123, 131, 133, 138-152, 154-158, 163, 169, 170-173, 179, 180, 183, 200-208, 210-211: *poisk* of 1958: 374-379, 405-410, 444-447, 470, 472, 475, 504-506, 509-512, 515-516; *poisk* of 1961: Djany-kala, *poisk* 51, 54-69, Buzuk-kala, Khatyn I, II, III.

Note 195 Some similarity to Syrdarya irrigation can be found in India, where silted channels - *dunds* (dry channels within the limits of a river valley flooded during freshets) - and overflows - *seylseb* - are used (Ostrovskiy 1907:15-41; ICBIP 1965:23-42). On these *dunds* there were large canals, up to 80-180 km long and 10 m wide.

Note 196 The cleaning of networks and the reconstruction of distributing water structures took place only when water in the river was low, i.e. between November and the beginning of the following March (optimal for this operation was February - March) (MRSA 1926, Kn. 2, Ch. 2:15). I.M. Dyakonov suggests that in Babylonia in the 2nd millennium BCE, labor conscription for each community, or even for each adult men, was two or more months (see Dyakonov 1968:144-146).

Note 197 It is possible that especially in this consisted the economic condition (premise) of the conquering Khiva rulers at the beginning of the 19th century against the Karakalpaks. The latter were displaced from their own lands on the Janydarya to the lower Amudarya. In the 19th century, they had heavy duties to set up new agricultural lands and for cleaning the most important main canals of the central and Western Khanates (see Gulyamov 1949; 1957:212-235; Andrianov 1958a; Kamalov 1958:144-146).

Note 198 For example, at the time of the Greek-Macedonian expansion in the East, according to N. Pigulevskaya "the conquerors had large masses of prisoners, turned into slaves. They were used, always in a compulsory way, also for building and irrigation works" (Pigulevskaya 1956: 22). [Editor 98] According to a personal communication of V. A. Livshits, this possibility seems to be reported by written Khorezmian sources: the documents from the archive of Toprak Kala (mid-2nd-beginning of the 3rd centuries CE), in which slaves are mentioned (Livshits 1962:35-57; Perikhanyan 1952:25; ITN 1963:473-474, 568:notes 53-54).

Note 199 As the study of ancient sites and irrigation in Mesopotamia demonstrated, it is possible to recognize three main stages in the development of irrigation techniques. Researching the Diyala Basin, R. McC. Adams associated the first stage with the exploitation of riverbeds and small canals; the second with the state organization of large irrigation works (from the Neo-Babylonian to the Sassanid period); the third with the flourishing of irrigated agriculture during the Arab Caliphate (Adams 1965:112-166).

Note 200 The medieval 'demographical jump', i.e. the sudden growth of the rural population in Khorezm and in other areas of irrigated agriculture, is confirmed both by archaeological data (see Tolstov 1948a:155; 1965:25; etc.)[Editor 99] and by general estimates of demographers (see Ohlin 1965:2).

Note 201 According to estimated data, before the transition to agriculture, in the Nile Valley the tribes of Neolithic hunters, gatherers and fishers amounted to no more than ten thousand people. Their number increased in the next two millennia, with the development of irrigated agriculture, and there were 3-6 million people in the Old Kingdom period. The population density reached 200-400 people/1 km². The total world population grew some ten times. G. Ohlin estimated that the world population in the 7th-5th millennia BCE was about 5-10 million people, while at the turn of our era approximately 250-350 million (see Ohlin 1965:2, 9).

Note 202 See also criticism of the hypothesis of E. Huntington in the works of K. K. Markov (1951a), G. N. Lisitsyna (1965:12-21), R. LeB. Bowen (1958:83), G. Caton-Thompson and E. Gardner (1939), etc.

Editor's Notes

Simone Mantellini

Editor 1 Reference quoted without the year.

Editor 2 The author did not quote specific publications for 'Tolstov 1941' and 'Tolstov 1945' but for both 'a' seems to be most appropriate.

Editor 3 The author did not quote specific publications for 'Tolstov 1962' but both 'a' and 'b' refer to the subject.

Editor 4 The quotation 'Crawford 1929a' refers to the article "Desert markings near Ur" published in *Antiquity* (1929, N. 11, Vol. 3, p. 342) under the section "Notes and News". The article was not attributed to a specific author, so perhaps B. V. Andrianov considered that O. G. S. Crawford, editor in chief of the Journal, had authored the paper. Andrianov also quoted 'Crawford 1924', which was not found in the catalogues, thus not included in the final references.

Editor 5 The author also quoted 'Johnson 1930' which does not seem to exist.

Editor 6 The author did not quote specific publications for 'Tolstov 1946' but all three seem to refer to the subject.

Editor 7 The author did not quote specific publications for 'Titov 1965' but both 'a' and 'b' refer to the subject.

Editor 8 See Editor7.

Editor 9 The author did not quote specific publications for 'Markov 1951' but 'a' seems to be most appropriate.

Editor 10 The author quoted 'Kosov' without the year and without including any quotation by him in the final references. Both topic and pages fit very well with the 1965 publication.

Editor 11 The author quoted 'Solecki 1955' without including this work in the final references. Two articles were found which refer to the subject of interest.

Editor 12 The author wrongly quoted 'Garrod 1958' instead of 'Garrod 1957'.

Editor 13 See Editor 12

Editor 14 The author wrongly quoted 'SPF:21' instead of 'Kempton 1938'.

Editor 15 The author wrongly quoted 'Amiran 1965' instead of 'Amiran 1966'.

Editor 16 See Editor15

Editor 17 The author wrongly quoted 'Helbaek 1960c' (which does not exist) instead of 'Helbaek 1959b'

Editor 18 The author did not quote specific publications for 'Helbaek 1964' but 'a' seems to be most appropriate.

Editor 19 Reference quoted without the year.

Editor 20 The author wrongly quoted 'Baumgartel 1952' instead of 'Baumgartel 1955'.

Editor 21 The author wrongly quoted 'Caton-Thompson and Gardner 1937' instead of 'Caton-Thompson and Gardner 1934'.

Editor 22 The author wrongly quoted 'Diodorus I, 19, 51' instead of 'Diodorus I, 19, 5'.

Editor 23 The author wrongly quoted 'Boak 1925' instead of 'Boak 1926a'.

Editor 24 The author wrongly quoted 'Adams 1955', which does not exist, instead of 'Adams 1965' (*Land Behind Baghdad*), where Part 9 is devoted to the 'Configurations of Change in Irrigation and Settlement').

Editor 25 The author wrongly quoted 'Asil 1955' instead of 'Goetze 1955'.

Editor 26 The author wrongly quoted 'Willcocks 1911' instead of 'Willcocks 1917'.

Editor 27 The author wrongly wrote *GAN-GA* instead of *GAN-ID*.

Editor 28 The author wrongly quoted 'Forbes 1953' instead of 'Forbes 1955'.

Editor 29 See Editor 28

Editor 30 The author wrongly quoted 'DAT 1964:520' instead of 'DAT 1964:521'.

Editor 31 The author wrongly quoted 'Giles 1927' instead of 'Giles 1924'.

Editor 32 The author did not quote specific publications for 'Vayman 1961' but both 'a' and 'b' refer to the subject.

Editor 33 The author did not quote specific publications for 'Masson 1957' but 'b' seems to be most appropriate; the same for 'Askarov 1962', where both 'a' and 'b' refer to the subject.

Editor 34 The author did not quote specific publications for 'V. Masson 1956' but both 'a' and 'b' refer to the subject.

Editor 35 The author wrongly quoted 'Itina 1965' instead of 'Itina 1962'.

Editor 36 In relation to 'Sachau 1873' the author quoted 'Teil I:9-19', which could not be found. Therefore, the quotation reported here concerns either a work published in the same year, and referring to the same subject, or it is wrong.

Editor 37 The author quoted 'Shults 1948' without including this work in the final references.

Editor 38 In the original book the author quoted the publication *Karakalpakiya*. Here the same work is quoted with reference to the main editor, i.e. A. E. Fersman.

Editor 39 The author did not quote specific publications for 'Jdanko 1958'. The pages do not concern 'Jdanko 1958b' thus 'Jdanko 1958a' seemed to be most appropriate.

Editor 40 The author did not quote specific publications for the years 1952 and 1958 but for both 'a' seems to be most appropriate.

Editor 41 The author did not quote specific publications for 'Itina 1959'. The argument (Tazabagyab culture) does not fit 'Itina 1959c' while it probably refers to both 'Itina 1959a-b'.

Editor 42 Here the translation is faithful to the text, citing 'Trench 45', although in Figure 27.H it is reported as 'Trench 43'.

Editor 43 The author wrongly quoted 'Itina 1965' instead of 'Itina 1963'.

Editor 44 The author did not quote specific publications for 'Tolstov 1948' but both 'a' and 'b' refer to the subject.

Editor 45 The author did not quote specific publications for 'Tolstov 1962' but 'a' seems to be most appropriate.

Editor 46 The author did not quote specific publications for 'Tolstov 1953' and 'Tolstov 1962' but 'b' seems to be most appropriate for the former and 'a' for the latter.

Editor 47 Here the author quoted only the page, as opposed to all the other quotations from Müller, where he also specified the relevant parts and chapters.

Editor 48 The author quoted 'Saushkin 1949' without including this work in the final references.

Editor 49 See Editor 38.

Editor 50 The author quoted 'Litvinskiy 1968' without including this work in the final references. A work by Litvinskiy dated to that year, and referring to the subject, was found and is now included in the references.

Editor 51 The author did not quote specific publications for 'Tolstov 1947' and 'Tolstov 1955' but 'a' for the former and 'c' for the latter seem to be most appropriate.

Editor 52 The author did not quote specific publications for 'Tolstov 1953'. The pages mentioned do not refer to the two works in the final references under Tolstov for that year. Thus, either the citation is wrong or a third publication exists for that year but was omitted in the final references.

Editor 53 The author did not quote specific publications for 'Tolstov 1948' but 'a' seems to be most appropriate.

Editor 54 The author did not quote specific publications for 'Andrianov 1958' but both 'a' and 'b' seem to be most appropriate.

Editor 55 In this paragraph three of Bartold's quotations were reported in the original book. However, they do not appear in the final references. For that year, two monographs by Bartold were found and added in the final references as '1902a' and '1902b'. However, the corresponding pages mentioned by the author could not be verified.

Editor 56 See Editor 55.

Editor 57 See Editor 55.

Editor 58 The author wrongly quoted 'Zadykhina 1958' instead of 'Zadykhina 1952'.

Editor 59 The author quoted 'Rapoport 1962' without including this work in the final references. A work by Rapoport dated that year was found and is now included in the references.

Editor 60 The author wrongly quoted 'Levshin 1862' instead of 'Levshin 1832'.

Editor 61 The author did not quote specific publications for 'Fedorovich 1950' of the two reported in the final references.

Editor 62 See Editor 61.

Editor 63 See Editor 61.

Editor 64 The author wrongly quoted 'Tolstov and Itina 1965' instead of 'Tolstov and Itina 1964'.

Editor 65 The author did not quote specific publications for 'Vayman 1962'. The subject is treated in two of the three publications of that year but, according to the number of the pages, 'b' seems to be appropriate.

Editor 66 Here the original book contains also a second quotation, reported as '1964:177-178' but without the corresponding author. It might refer either to a second edition of the 1st Tom of MITT, first published in 1939 and then possibly in 1964 (a later edition of the 2nd Tom of MITT is dated 1968), or to an unknown work.

Editor 67 The author did not quote specific publications for 'Tolstov 1950' but 'b' seems to be most appropriate.

Editor 68 The author did not quote specific publications for 'Tolstov 1959': it is not 'a' while 'b' seems to be most appropriate.

Editor 69 The author wrongly quoted 'Levshin 1932' instead of 'Levshin 1832'.

Editor 70 The name of this Stetekvich is not reported and impossible to find.

Editor 71 The author did not mention specific publications for 'Andrianov 1959' and 'Andrianov 1960'. The former can refer both to 'a' and 'b', while for the latter '1960a' seems to be most appropriate.

Editor 72 The author did not quote specific publications for 'Tolstov 1947' but 'a' seems to be most appropriate.

Editor 73 Quotation reported as in the original. The book was edited by V. D. Blavatskiy and A. V. Nikitin (see VRZ 1967 in the Abbreviations).

Editor 74 The author did not quote specific publications for 'Helbaek 1964' but both 'a' and 'b' refer to the subject.

Editor 75 The author did not quote specific publications for 'Huntington 1908' but all three 'a-b-c' refer to the subject.

Editor 76 Although the author mentioned several times I. V. Mushketov and P. A. Kropotkin, there are no bibliographic references concerning their work.

Editor 77 The quotation 'Voeykov 1947' is not included in the final references. Apparently no works by Voeykov exist for that year, thus it possibly refers to a later edition of 'Voeykov 1915'.

Editor 78 Here the Author also quoted himself as '1969'. However, the only publication for that year is this book itself. According to the subject this quotation possibly refers to 'Andrianov 1965'.

Editor 79 Here the author wrongly reported '100 × 200 m' instead of '100-200 m'.

Editor 80 The author wrongly quoted 'HS, Vol. I:44a' instead of 'Casanova 1946:fig. 50', which is in Vol. 2 and not in Vol. 1 (see also Editor 14).

Editor 81 The author did not quote specific publications for 'Rakhmanova 1964' but both 'a' and 'b' refer to the subject .

Editor 82 The author did not quote specific publications for 'Helbaek 1964' but 'a' seems to be most appropriate.

Editor 83 The article published by Fowler and Evans in the same issue of *Antiquity* was wrongly quoted instead of that of Lampert.

Editor 84 The author did not quote specific publications for 'Andrianov 1959' but 'a' seems to be most appropriate.

Editor 85 The author did not quote specific publications for 'Andrianov 1958' but both 'a' and 'b' refer to the subject.

Editor 86 The author wrongly quoted 'HAI 1910:figs. 44, 131' instead of 'Bennett 1946' and 'Casanova 1946' (*Handbook of American Indians*, edited by F. W. Hodge). For more detail see the Plate 14.7-8.

Editor 87 The author quoted 'Vavilov 1962' without including this work in the final references. He probably refers to Vaivlov's famous work *Pyat kontinentov* (Five continents), published in 1962.

Editor 88 The author quoted 'Drower 1956:50'. This publication does not seem to exist, whereas that topic is treated by Drower in his article of 1954 (included in the final references), where, however, there is no 'page 50'. As partial confirmation, 'Drower 1954' is also quoted by the author in his Conclusion.

Editor 89 The author wrongly quoted 'Herodotus III, 158' instead of 'Herodotus II, 158'.

Editor 90 The author wrongly quoted 'Woolley 1955' instead of 'Woolley 1954'.

Editor 91 The author quoted 'Ptitsyn 1911' without including this work in the final references.

Editor 92 The author did not quote specific publications for 'Itina 1959' but 'a' seems to be most appropriate.

Editor 93 The author wrongly quoted 'Sokolov 1964' instead of 'Sokolov 1961'.

Editor 94 Quotation reported here, and not in the final references, as in the original book.

Editor 95 The author quoted 'Liddell et al. 1843' but, according to the Library of Congress, the first edition is dated to 1846.

Editor 96 The author quoted 'Albaum 1955' without including this work in the final references. A work by Albaum dated to that year was found and is now included in the references.

Editor 97 The author wrongly quoted 'Tolstov and Andrianov 1958a' instead of 'Andrianov 1958a'.

Editor 98 The quotation 'Pigulevskaya 1956' is not included in the final references. No article for that year and with that page number was found for this author.

Editor 99 The quotation 'Tolstov 1965' is incorrect since there are no publications by S. P. Tolstov dated to that year.

References

AN (Akademiya Nauk) = Academy of Sciences
Ch. (Chast) = Part
Izd. (Izdanie) = Publishing House
Kn. (Kniga) = Book
No. (Nomer) = Number
Otdel = sector
Ser. (Serya) = Series
Tom = Volume
Vyp. (Vypusk) = Issue

Proceedings of the founders of Marxism-Leninism

Engels, F.
1961a Iz podgotovitelnykh rabot k "Anti-Dyuringu". *Sochineniya*, Tom 20, K. Marx i F. Engels. Izd. Moskva. (2-oy izdanie).

1961b Dialektika prirody. *Sochineniya*, Tom 20, K. Marx i F. Engels. Moskva. (2-oy izdanie).

1961c Proiskhojdenie semi, chastnoy sobstvennosti i gosudarstva. *Sochineniya*, Tom 21, K. Marx i F. Engels. Moskva. (2-oy izdanie).

1962 Engels – Marksu: 6 iyunya 1853 g. *Sochineniya*, Tom 28, K. Marx i F. Engels. Moskva. (2-oy izdanie).

Lenin, V. I.
n.d. O gosudarstve. *Polnoe sobranie sochineniy*, Tom 39, V. I. Lenin. Mosvka.

n.d. Tovarishcham kommunistam Azerbaydjana, Gruzii, Armenii, Dagestana, Gorskoy respubliki. *Polnoe sobranie sochineniy*, Tom 43, V. I. Lenin. Mosvka.

Marx, K. H.
1957 Britanskoe vladychestvo v Indii. *Sochineniya*, Tom 9, K. Marx i F. Engels. Moskva. (2-oy izdanie).

References in Russian

Abaev, V. I.
1956 Skifskiy byt i reforma Zoroastra. *Arkheologicheskie Otkrytiya*, Tom XXIV.

Adykov, K. A., and V. M. Masson
1962 Drevnosti Tedjen-Murgabskogo mejdurechya. *Izvestiya AN Turkmenskoy SSR*, No. 2.

Afanasev, O. A.
1965 Obsujdenie v institute istorii AN SSSR problemy "aziatskogo sposoba proizvodstva". *Sovetskaya Etnografiya*, No. 6.

Ageeva, E. I. and G. P. Patsevich
1958 Iz istorii osedlykh poseleniy i gorodov Yujnogo Kazakhstana. *Trudy Instituta Istorii, Arkheologii i Etnografii AN Kazakhskoy SSR*, Tom 5.

Akhmedov, B. A.
1965 *Gosudarstvo kochevykh uzbekov*. Moskva.

Akishev, K. A. and G. A. Kushaev
1963 *Drevnyaya kultura sakov i usuney doliny reki Ili*. Alma-Ata.

Akulov, V. V.
1957   *Nekotorye dannye o delte Amu-Dari. Izvestiya Vsesoyuznogo Geograficheskogo Obshchestva*, Uzbeksk. filial, Tom III. Tashkent.

Albaum, L. I.
1955   Nekotorye dannye po izucheniyu Angorskoy arkheologicheskikh pamyatnikov (1948-1949). *Trudy Instituta Istorii Akademii Nauk Uzbekistana*, Vyp. 7. Tashkent.

Aliman, A.
1960   *Doistoricheskaya Afrika (Pervobytnaya arkheologiya i chetvertichnaya geologiya)*. Moskva.

Ali-Zade, A. A.
1956   *Sotsialno-ekonomicheskaya i politicheskaya istoriya Azerbaydjana XIII-XIV vv.* Baku.

Altunin, S. G.
1951   Regulirovanie rusla i nanosov Amu-Dari u Takhiya-Tashskoy plotiny. *Pravda Vostoka*, 16.V.

Altunin, S. G., and A. Buzukov
1950   Zashchitnye damby Kara-Kalpakii. *Materialy po proizvoditelnym silam Uzbekistana*, Vyp. I.

Anuchin, D. N.
1885   *K voprosu o sostavlenii legend k arkheologicheskoy karte Rossii (po doistoricheskoy arkheologii)*. Moskva.

Artamonov, M. I.
1947   Obshchestvennyy stroy skifov. *Vestnik Leningradskogo Gosudarstvennogo Universiteta*, 9.

Arutyunyan, N. V.
1964   *Zemledelie i skotovodstvo Urartu*. Erevan.

Askarov, A.
1962a *Pamyatniki andronovskoy kultury v nizovyakh Zeravshana. Istoriya materialnoy kultury Uzbekistana*, Vyp. 3. Tashkent.

1962b *Kultura Zaman-baba v nizovyakh Zeravshana. Obshchestvennye nauki v Uzbekistane*, 11.

1963   Poselenie kultury Zaman-baba. *Kratkie Soobshcheniya o Dokladakh i Polevykh Issledovaniyakh Instituta Arkheologii AN SSSR*, Vyp. 93. Moskva.

Avdiev, V. I.
1934   Selskaya obshchina i iskusstvennoe oroshenie v drevnem Egipte. *Istorik marksist*, No. 6.

1938   *Voennaya istoriya Drevnego Egipta*, Tom. I. Moskva.

1953   *Istoriya Drevnego Vostoka*. Izd. 2. Moskva.

1959   *Voennaya istoriya Drevnego Egipta*, Tom II. Moskva.

Bakhteev, F. Kh.
1959   Otkrytie 'Hordeum Iagunculiforme Bacht' na territorii Turkmenskoy SSR. *Doklady AN SSSR*, Tom 129, No. 1.

1960   *Ocherki po istorii i geografii vajneyshikh kulturnykh rasteniy*. Moskva.

1962   K arkheologicheskim raskopkam v Kyzyl-Kumakh. *Obshchestvennye nauki v Uzbekistane*, No. 11.

1966   Dalneyshee osushchestvlenie nauchnykh idey N. I. Vavilova v izuchenii zernovykh zlakov. *Voprosy geografii kulturnykh rasteniy i N. I. Vavilov*. Moskva - Leningrad.

Bartold, V. V.
1902a K voprosu o vpadenii Amu-dari v Kaspyskoe more. *Zapiski Vostochnogo odeleniya Imperatorskogo Russkogo Arkheologicheskogo obshchestva. Imperatorskoe Rsskoe Arhheologicheskoe obshchestvo, Vostochnoe odelenie*, V. R. Rozen. Sankt Peterburg

1902b Sredeniya ob Aralskom more i nizovyakh Amu-dari s drevneishikh vremen do XVII veka. *Izv. Turkestan. otr. Rus. Geog. Obshch. Nauchnye rezultaty Aralskoy ekspeditsii*. Tashkent.

1914   *K istoriy orosheniya Turkestana*. Moskva.

1963-1966 *Sochineniya*, Tom I-VI. Moskva - Leningrad.

Bashilov, V. A.

1967 Drevnie tsivilizatsii Tsentralnykh And. Moskva. (avtoref. kand. diss.)

Bazilevich, N. I., and L. E. Rodin

1955 K voprosu o genezise i evolyutsii *takyr*ov i printsipakh ikh melioratsii na primere Kizyl-Arvatskoy podgornoy ravniny. *Trudy Aralo-Kaspiyskoy ekspeditsii Soveta po izucheniyu proizvoditelnykh sil AN SSSR,* I.

1956 O roli rastitelnosti v formirovanii i evolyutsii *takyr*ov Meshkhed-Messerianskoy allyuvialno-deltovoy ravniny. *Takyry Zapadnoy Turkmenii i puti ikh selskokhozyaystvennogo osvoeniya.*

Belenitskiy, A. M.

1950 Istoriko-geograficheskiy ocherk Khuttalya s drevneyshikh vremen do X v. n. e. *Materialy i Issledovaniya po Arkheologii SSSR,* No. 15.

1955 O periodizatsii istorii Sredney Azii. *Materlialy nauchnoy sessii, posvyashchennoy istorii Sredney Azii i Kazakhstana v dooktyabrskiy period.* Tashkent

1959 Drevniy Pendjikent. *Sovetskaya Arkheologiya,* No. 1.

1968 O rabovladelcheskoi formatsii v istorii Sredney Azii. Problemy arkheologii Sredney Azii. *Tezisy dokladov i soobshcheniy k soveshchaniyu po arkheologii Sredney Azii (1-7 aprelya 1968).* Leningrad.

Belov, N. P.

1940 Pochvenno-geomorfologicheskiy ocherk Kunyadarinskoy ravniny. *Trudy Instituta Geografii AN SSSR,* Vyp. 35.

Belyavskiy, N. N.

1887 Dopolnitelnye svedeniya puti v Srednyuyu Aziyu ot zaliva Tsesarevich, po Ust-Urtu i Amu-Dare, 1885. *Sbornik geograficheskikh, topograficheskikh i statisticheskikh materialov po Azii,* Vyp. XXV. Sankt-Peterburg.

Berdyev, O. K.

1966 Chagylly-Depe: novyy pamyatnik neoliticheskoy djeytunskoy kultury. *Materialnaya kultura narodov Sredney Azii i Kazakhstana.* Moskva.

Berg, L. S.

1911 Ob izmenenii klimata v istoricheskuyu epokhu. *Zemlevedenie,* 1911, No. 3.

1947 Ob izmenenii klimata v istoricheskuyu epokhu. *Klimat i jizn.* Moskva.

1947-1952 *Geograficheskie zony Sovetskogo Soyuza,* Tom I-II. Moskva.

1953 O probleme usykhaniya stepey i pustyn. *Izvestiya AN SSSR, Ser. Geogr.,* No. 5.

Berlev, O. D.

1965 *"Raby tsarya" v Egipte epokhi Srednego tsarstva.* Leningrad. (avtoref. kand. diss.)

Bernshtam, A. N.

1947 Sredne-aziatskaya drevnost i ee izuchenie za 30 let. *Vestnik Drevney Istorii,* No. 3.

1949 Trudy Semirecheskoy arkheologicheskoy ekspeditsii. Chuyskaya dolina. *Materialy i Issledovaniya po Arkheologii SSSR,* No. 14.

1952 Istoriko-arkheologicheskie ocherki Tsentralnogo Tyan-Shanya i Pamiro-Alaya. *Materialy i Issledovaniya po Arkheologii SSSR,* No. 26.

Bibikov, S. A.

1965 Khozyaystvenno-ekonomicheskiy kompleks razvitogo Tripolya. *Sovetskaya Arkheologiya,* No. 1.

Biruni, A.

1957 *Izbrannye Proizvedeniya,* Tom I. Tashkent.

1963 *Izbrannye Proizvedeniya,* Tom II. Tashkent.

1966 *Izbrannye Proizvedeniya,* Tom III. Tashkent.

Blavatskiy, V. D.

1953   *Zemledelie v antichnykh gosudarstvakh Severnogo Prichernomorya*. Moskva.

Bogdanov, D. V.

1951   Kulturnye landshafty dolin severo-zapadnogo Pamira i vozmojnosti ikh preobrazovaniya. *Voprosy Geografii. Sbornik*, Sb. 24.

Bogdanovich, N. V.

1955   Nekotorye osobennosti pochvoobrazovaniya v delte Amu-Dari. *Trudy Instituta Pochvovedeniya AN Uzbekskoy SSR*, Vyp. I. Tashkent.

Bogolyubov, M. N.

1962   Drevnepersidskie etimologii. *Drevniy mir*, N. V. Pigulevskaya, D. P. Kallistov, I. S. Katsnelson, M. A. Korostovtsev. Moskva.

Bogomolov, L. A.

1963   *Topograficheskoe deshifrirovanie prirodnykh landshaftov na aerosnimkakh*. Moskva.

Boriskovskiy, P. I.

1961   K voprosu o drevneyshikh zemlekopnykh orudiyakh. *Issledovaniya po arkheologii SSSR*. Leningrad.

Borovskiy, V. M.

1955   Issledovanie vajnogo rayona orosheniya. K selskokhozyaystvennomu osvoeniyu nizovev Syr-Dari. *Selskoe khozyaystvo Kazakhstana*, No. 5.

Borovskiy, V. M., and M. A. Pogrebinskiy

1958   *Drevnyaya delta Syr-Dari i Severnye Kyzyl-Kumy*, Tom I. Alma-Ata.

Borovskiy, V. M., E. V. Ablakov, K. Ya. Kojevnikov, and K. D. Muravlyanskiy

1959   *Drevnyaya delta Syr-Dari i Severnye Kyzyl-Kumy*, Tom II. Alma-Ata.

Braginskiy, I. S.

1956   *Iz istorii tadjikskoy narodnoy poezii*. Moskva.

Bregel, Yu. E.

1961   *Khorezmskie turkmeny v XIX veke*. Moskva.

Bruk, S. I.

1958   Opyt sostavleniya etnicheskikh kart po materialam razlichnogo tipa. *Kratkie Soobshcheniya Instituta Etnografii AN SSSR*, Vyp. 28.

Bryusov, A. Ya.

1957   Retsenziya na sb. "Ethnographisch-archäologische Forschungen", 2. *Vestnik Drevney Istorii*, No. 2.

Bukhanevich, A. I.

1925   Kho eziskaya osblast. *Khlopkovoe Delo*, No. 11–12.

Bukinich, D. D.

1924   Istoriya pervobytnogo oroshaemogo zemledeliya v Zakaspiyskoy oblasti v svyazi s voprosom o proiskhojdenii zemledeliya i skotovodstva. *Khlopkovoe delo*, No. 3–4.

1926   Starye rusla Oksa i Amu-Darinskaya problema (V svyazi s voprosom obvodneniya Kelifskogo Uzboya i Daryalyka). *Biblioteka Khlopkovogo dela*, Kn. 4.

1940   Kratkie predvaritelnye soobrajeniya o vodosnabjenii i irrigatsii Starogo Termeza i ego rayona. *Trudy Uzbeskoy FAN SSSR*, Ser. I, Istoriya, Arkheologiya, Vyp. 2.

1945   Kanaly drevnego Termeza. *Trudy AN Uzbekskoy SSR*, Tom II, Ser. I. Tashkent.

Burnashev

1818   Puteshestvie ot sibirskoy linii do g. Bukhary v 1794 g. i obratno v 1795 g. *Sibirskiy vestnik*, 45.

Capot-Rey, R. [translated from original]

1958   *Frantsuzskaya Sakhara*. Moskva.

Chernykh, E. N.

1962   Nekotorye rezultaty izucheniya metalla anauskoy kultury. *Kratkie Soobshcheniya o Dokladakh i Polevykh Issledovaniyakh Instituta Arkheologii AN SSSR*, Vyp. 91.

1965   Spektralnyy analiz i izuchenie-drevneyshey metallurgii Vostochnoy Evropy. *Arkheologiya i estestvennye nauki*. Moskva.

Childe, V. G. [translated from original]

1949   *Progress i arkheologiya*. Moskva.

1956   *Drevneyshiy Vostok v svete novykh raskopok*. Moskva.

Chu Shao-Tan

1953   *Geografiya Novogo Kitaya*. Moskva.

Clark, J. G. D. [translated from the original]

1953   Doistoricheskaya Evropa. *Ekonomicheskiy ocherk*. Moskva.

Ding Ying

1958   Proiskhojdenie kulturnykh vidov risa i ikh differentsiatsiya v Kitae. *Agrobiologiya*, No. 1. Moskva.

Dingelshtedt, N.

1893   Opyt izucheniya irrigatsii Turkestanskogo kraya. *Syr-Darinskaya oblast*, Chast I-II. Sankt-Peterburg.

Djordjio, V. V.

1945   *Kolebanie rejima rek Uzbekistana*. Tashkent (avtoref. kand. diss.)

Djumaev, O. M.

1946   Osobennosti kaakovogo zemledeliya v Turkmenskoy SSR. *Izvestiya Turkmenskogo filiala AN SSSR*, No. 3-4.

1951   *K istorii oroshaemogo zemledeliya v Turkmenistane*. Ashkhabad.

Doskach, A. G.

1940   Geomorfologicheskiy ocherk Kunya-Darinskoy drevnealljuvialnoy ravniny (Sarykamyshskoy delty r. Amu-Dari). *Trudy Instituta Geografii AN SSSR*, Vyp. 35.

1948   Osnovnye etapy razvitiya idey o relefe peschanykh pustyn. *Trudy Instituta Geografii AN SSSR*, Vyp. 39.

Dublitskiy, B.

1927   *Programma po obsledovaniyu i naneseniyu na kartu pamyatnikov materialnoy kultury Djetysu*. Alma-Ata.

Dunin-Barkovskiy, L. V.

1956   Problemy irrigatsii i melioratsii oroshaemykh zemel v stranakh Blijnego i Srednego Vostoka. *Gidrotekhnika i melioratsiya*, No. 7.

1957   Sovmestnoe ispolzovanie vodnykh resursov basseynov rek Zeravshan i Kashkadarya i podacha vody iz drugikh istochnikov. *Selskoe khozyaystvo basseyna Zeravshana*. Tashkent.

1960   *Fiziko-geograficheskie osnovy proektirovaniya orositelnykh sistem (rayonirovanie i vodnyy balans oroshaemoy territorii)*. Moskva.

Durdyev, D.

1959   Gorodishche Starogo Kishmana. *Trudy Instituta istorii, arkheologii i etnografii AN Turkmenskoy SSR*, Tom V.

Dyakonov, I. M.

1947   Novye dannye o shumerskoy kulture. *Vestnik Drevney Istorii*, No. 2.

1952   Zakony Vavilonii, Assirii i Khettskogo tsarstva. *Vestnik Drevney Istorii*, No. 3.

1956   *Istoriya Midii ot drevneyshikh vremen do kontsa IV veka do n. e.* Moskva - Leningrad.

1959   *Obshchestvennyy i gosudarstvennyy stroy Drevnego Dvurechya, Shumer*. Moskva.

1963   Obshchina na Drevnem Vostoke v rabotakh sovetskikh issledovateley. *Vestnik Drevney Istorii*, No. 1.

1966   Osnovnye cherty ekonomiki v monarkhiyakh drevney zapadnoy Azii. *Narody Azii i Afriki*, No. 1.

1968   Problemy ekonomiki. O strukture obshchestva Blijnego Vostoka do serediny II tys. do n. e. *Vestnik Drevney Istorii*, No. 3-4.

Dyakonov, M. M.

1949   Retsenziya: "Opyt istoriko-arkheologicheskogo issledovaniya" na kn. S. P. Tolstov "Drevniy Khorezm". *Voprosy Istorii*, No. 2.

1950   Raboty Kafirniganskogo otryada. *Materialy i Issledovaniya po Arkheologii SSSR*, No. 15.

1953   Arkheologicheskie raboty v nijnem techenii r. Kafirnigana (Kobadian). *Materialy i Issledovaniya po Arkheologii SSSR*, No. 37.

1954   Slojenie klassovogo obshchestva v Severnoy Baktrii. *Sovetskaya Arkheologiya*, Tom XIX.

1961   *Ocherk istorii drevnego Irana.* Moskva.

Dyakonov, M. M., and V. A. Livshits

1960   Dokumenty iz Nisy I v. do n. e. *Predvaritelnye itogi raboty.* Moskva.

Eberzin, A. G.

1952   Paleontologicheskie issledovaniya v rayone trassy Glavnogo Turkmenskogo kanala. *Vestnik AN SSSR*, No. 10.

Efimenko, P. P.

1953   *Pervobytnoe obshchestvo.* Izd. 3. Kiev.

Efremov, F.

1811   *Stranstvovanie v kirgizskoy stepi, Bukharii, Khive, Persii, Tibete i Indii i vozvrashchenie ego ottuda cherez Angliyu v Rossiyu.* Izd. 3. Kazan.

Egorov, V. V.

1954   *Zasolennye pochvy i ikh osvoenie.* Moskva.

Ershov, N. N.

1960   *Selskoe khozyaystvo tadjikov Leninabadskogo rayona Tadjikskoy SSR pered Oktyabrskoy revolyutsiey (istoriko-etnograficheskiy ocherk).* Stalinabad.

Ershov, S. A.

1956   Chopan-Depe (Otchet o stratigraficheskom shurfe). *Trudy Instituta istorii, arkheologii i etnografii AN Turkmenskoy SSR*, Tom 2.

Eygenson, M. S.

1963   *Solntse, pogoda i klimat.* Moskva.

Eytkin, M. D.

1963   *Fizika i arkheologiya.* Moskva.

Fan Wen-Lan

1958   *Drevnyaya istoriya Kitaya.* Moskva.

Favorin, N. N., N. N. Ostrovnaya, and V. A. Timoshkina

1956   Rejim i balans gruntovykh vod oroshaemykh massivov nizovev Amu-Dari. *Trudy Aralo-Kaspiyskoy kompl. eksp.*, Vyp. VII.

Fedchina, V. N.

1967   *Kak sozdavalas karta Sredney Azii.* Moskva.

Fedin, N. F.

1951   Ob absolyutnom vozraste terras nizoviy reki Syr-Dari. *Izvestiya AN Kazakskoy SSR*, No. 114, Ser. Geol., Vyp. 14. Alma-Ata.

Fedorov, B. V.

1953   *Agro-meliorativnoe rayonirovanie zony orosheniya Sredney Azii.* Tashkent.

Fedorova, R. V.

1960   O zemledelcheskikh kulturakh v Drevnem Khorezme po dannym pyltsevogo analiza. *Materialy Khorezmskoy Ekspeditsii. Vyp. 4, Polevye issledovaniya Khorezmskoy ekspeditsii 1957 g,* S. P. Tolstov i. M. A. Itina. Moskva.

Fedorovich, B. A.

1946   Voprosy paleogeografii Sredney Azii. *Trudy Instituta Geografii AN SSSR,* Vyp. 37.

1950a   Drevnie reki v pustynyakh Turana. *Materialy po chetvertichnomu periodu SSSR,* Vyp. 3. Moskva - Leningrad.

1950b   Ob osnovnykh protsessakh relefoobrazovaniya Turana. *Problemy fizicheskoy geografii,* 15.

1957   Severnye (Zaunguzskie) Karakumy. Unguz i Sarykamyshskaya vpadina. *Geologiya SSSR,* Tom 22, Turkmenskaya SSR, Chast 1. Moskva.

Fedoseev, P. N.

1967   Idei Lenina i metodologiya sovremennoy nauki. *Problemy mira i sotsializma* 4(104).

Fersman, A. E.

1934   *Karakalpakiya: Trudy Pervoy konferentsii po izucheniyu proizvoditelnykh sil Karakalpakskoy ASSR,* Tom I-II. Leningrad.

1939   *Uspekhi i problemy aerosemki v dele izucheniya proizvoditelnykh sil SSSR. 20 let sovetskoy geodezii i kartografii.* Moskva - Leningrad.

Firdousi

1957   *Shahnameh,* Tom I. Moskva.

1960   *Shahnameh,* Tom II. Moskva.

Formozov, A. A.

1945   Ob otkrytii kelteminarskoy kultury v Kazakhstane. *Kazakhskiy Filial AN SSSR,* No. 2.

1949   Kelteminarskaya kultura v Zapadnom Kazakhstane. *Kratkie Soobshcheniya o Dokladakh i Polevykh Issledovaniyakh Instituta Istorii Materialnoy Kultury AN SSSR,* Vyp. 25.

1951   K voprosu o proiskhojdenii andronovskoy kultury. *Kratkie Soobshcheniya o Dokladakh i Polevykh Issledovaniyakh Instituta Istorii Materialnoy Kultury AN SSSR,* XXXIX.

Freyman, A. A.

1927   Sredne-persidskiy yazyk i ego mesto sredi iranskikh yazykov. *Vostokovedcheskie Zapiski Leningradskogo Instituta Jivykh Vostochnykh Yazykov,* Tom I.

Frezer, D.

1928   *Zolotaya vetv,* Vyp. III. Moskva.

Gafurov, B. G.

1952   *Istoriya tadjikskogo naroda v kratkom izlojenii,* Tom I. Izd. 2. Moskva.

Ganyalin, A. F.

1956   Tekkem-tepe. *Trudy Instituta Istorii, Arkheologii i Etnografii AN Turkmenskoy SSR,* II.

Gaveman, A. V.

1937   *Aerosemka i issledovanie prirodnykh resursov.* Moskva - Leningrad.

1939   K voprosu o teorii deshifrirovaniya aerofotosnimkov. *Izvestiya Vsesoyuznogo geograficheskogo obshchestva,* Tom 71, Vyp. 3.

Gaydukevich, V. F.

1947   Raboty farkhadskoy arkheologicheskoy ekspeditsii v Uzbekistane v 1943-1944. *Kratkie Soobshcheniya o Dokladakh i Polevykh Issledovaniyakh Instituta Istorii Materialnoy Kultury AN SSSR,* XIV.

1948 K istorii drevnego zemledeliya v Sredney Azii. *Vestnik Drevney Istorii,* No. 3.

Gegeshidze, M. K.

1965 Irrigatsiya v Gruzii. *Sovetskaya Etnografiya*, No. 5.

Gelman, X. V.

1891 Nablyudeniya nad dvijeniem letuchikh peskov v Khivinskom khanstve. *Izvestiya Russkogo Geograficheskogo Obshchestva*, Tom 27, Vyp. 5.

Geltser, G. P.

1923 *Ketmen, omach i plug.* Tashkent.

Georgievskiy, B. M.

1937 Yujnyy Khorezm. *Geologicheskie i gidrogeologicheskie issledovaniya. 1925-1935 gg.*, Chast I. Tashkent.

Gerasimov, I. P.

1931 O takyrakh i protsessakh takyroobrazovaniya. *Pochvovedenie*, No. 4.

1937 Osnovnye cherty razvitiya sovremennoy poverkhnosti Turana. *Trudy Instituta Geografii AN SSSR*, Vyp. 25.

1954 Cherty skhodstva i razlichiya pustyn. *Priroda*, No. 2.

1961 Pogrebennye pochvy i ikh paleogeograficheskoe znachenie. *Materialy po izucheniyu chetvertichnogo perioda*, Tom I. Moskva.

1968 Nauchnye problemy preobrazovaniya prirody Sredney Azii dlya razvitiya oroshaemogo zemledeliya i pastbishchnogo jivotnovodstva. *Problemy preobrazovaniya prirody Sredney Azii.*

Gerasimov, I. P., E. A. Ivanova, and D. I. Tarasov

1935 Pochvenno-meliorativnyy ocherk delty i doliny r. Amu-Dari (V predelakh Kara-Kalpakskoy ASSR). *SOPS, Ser. Kara-Kalpakskaya, Vyp. 6, Trudy Kara-Kalpakskoy Kompleksnoy Ekspeditsii.* Moskva - Leningrad.

Glazovskaya, M. A.

1956 Pogrebennye pochvy, metody ikh izucheniya i ikh paleogeograficheskoe znachenie. *Voprosy Geografii: Sbornik statey dlya XVIII Mejdunarodnogo geograficheskogo kongressa.* Moskva.

Glebov, P. D.

1938 *Kurs irrigatsii.* Leningrad - Moskva.

Glukhovskoy, A. I.

1893 Propusk vod r. Amu-Dari po staromu ee ruslu v Kaspiyskoe more i obrazovanie nepreryvnogo vodnogo Amudarinskogo puti ot granits Afganistana po Amu-Dare, Kaspiyu, *Volge i Mariinskoy sisteme do Peterburga i Baltiyskogo moray.* Sankt-Peterburg.

Goldman, L. M.

1958 Tsvetnaya aerosemka razlichnykh landshaftov. *Voprosy Geografii.* Sbornik, Sb. 42.

1960 Primenenie tsvetnoy aerosemki dlya izucheniya mestnosti. *Trudy TsNIIGA i K*, Vyp. 137. Moskva.

Gospodinov, G. V.

1957 *Osnovy deshifrirovaniya aerofotosnimkov.* Moskva.

1961 *Deshifrirovanie aerosnimkov.* Moskva.

Grigorev, G. V.

1932 K voprosu o tsentrakh proiskhojdeniya kulturnykh rasteniy. *Izvestiya Gosudarstvennoy Akademii Istorii Materialnoy Kultury*, Tom XIII, Vyp. 9.

1940a *Kaunchi-tepe (raskopki 1935 g.).* Tashkent.

1940b Gorodishche Tali-barzu. *Trudy Otdela Istorii Kultury i Iskusstva Vostoka Gosudarstvennogo Ermitaja*, Tom II. Leningrad.

Grum-Grjimaylo, A. G., and M. E. Grum-Grjimaylo

1962 *V poiskakh rastitelnykh resursov mira.* Moskva - Leningrad.

Grum-Grjimaylo, G. E.

1896   *Opisanie puteshestviya v Zapadnyy Kitay,* Tom I. Sankt-Peterbug.

Gryaznov, M. P.

1966   Vostochnoe Priarale. *Srednyaya Aziya v Epokhu Kamnya i Bronzy,* V. M. Masson. Moskva - Leningrad.

Gulati, N. D.

1957   *Oroshenie v raznykh stranakh mira.* Moskva.

Gulyaev, V. I.

1966a *Drevnie tsivilizatsii Mezoameriki (proiskhojdenie i razvitie vysokikh kultur Meksiki, Gvatemaly i Gondurasa po arkheologicheskim dannym).* Moskva. (avtoref. kand. diss.)

1966b Novye dannye o proiskhojdenii zemledelcheskikh kultur Mezoameriki. *Sovetskaya Etnografiya,* No. 1.

1966c Proiskhojdenie tsivilizatsii mayya. *Sovetskaya Arkheologiya,* No. 3.

Gulyamov, Ya. G.

1945   Iz istorii orosheniya v Kara-Kalpakii. *Byulleten AN Uzbekskoy SSR,* No. 9-10.

1948   K vozniknoveniyu irrigatsii v Khorezme v svete dannykh arkheologii. *Sbornik materialov nauchnoy sessii AN Uzbekskoy SSR (9.VI-14.VI. 1947 g.).* Tashkent.

1949   *Istoriya orosheniya Khorezma s drevneyshikh vremen do nashikh dney.* Tashkent (avtoref. d-rskoy diss.)

1950   Biruni ob istoricheskoy geografii nizovev Amu-Dari. *Biruni - velikiy uchenyy srednevekovya.* Tashkent.

1955   K izucheniyu drevnikh vodnykh soorujeniy v Uzbekistane. *Izvestiya AN Uzbekskoy SSR,* No. 2.

1956   Arkheologicheskie raboty k zapadu ot Bukharskogo oazisa. *Trudy Instituta Istorii i Arkheologii AN Uzbekskoy SSR,* Vyp. 8.

1957   *Istoriya orosheniya khorezma s drevneyshikh vremen do nashikh dney.* Tashkent.

1965   Issledovanie istoricheskoy gidrografii nizovev Kashka-Dari i Zeravshana. *Istoriya Materialnoy Kultury Uzbekistana,* Vyp. 6. Tashkent.

1968   Kushanskoe tsarstvo i drevnyaya irrigatsiya Sredney Azii. *Obshchestvennye nauki v Uzbekistane,* No. 8.

Gulyamov, Ya. G., U. Islamov, and A. Askarov

1966   *Pervobytnaya kultura i vozniknovenie oroshaemogo zemledeliya v nizovyakh Zeravshana.* Tashkent.

Gumilev, L. N.

1967   *Drevnie tyurki.* Moskva.

Gushchin, G. G.

1938   *Ris.* Moskva.

Hurst, H. E. [translated from original]

1954   *Nil: Obshchee opisanie reki i ispolzovanie ee vod.* Moskva.

Harrison, G. A., J. S. Weiner, J. M. Tanner and N. A. Barnicot [translated from original]

1968   *Biologiya Cheloveka.* Moskva.

Igonin, N. I.

1965   Primenenie aerofotosemki pri izuchenii arkheologicheskikh pamyatnikov. *Arkheologiya i estestvennye nauki.* Moskva.

Ilin, M. M.

1946   Nekotorye itogi izucheniya flory: pustyn Sredney Azii. *Materialy po istorii flory i rastitelnosti SSSR,* Vyp. II. Moskva - Leningrad.

Inostrantsev, K.

1911   O domusulmanskoy kulture Khivinskogo oazisa. *Jurnal Ministerstva narodnogo prosveshcheniya,* Chast XXI. Sankt-Peterburg.

Isachenko, A. G.

1953   *Osnovnye voprosy fizicheskoy geografii.* Leningrad.

1960   Ponyatie "tip mestnosti" v fizicheskoy geografii. *Vestnik Leningradskogo gosudarstvennogo universiteta*, No. 12.

1965   *Osnovy landshaftovedeniya i fiziko-geograficheskoe rayonirovanie.* Moskva.

Itina, M. A.

1959a   Novye stoyanki tazabagyabskoy kultury. *Materialy Khorezmskoy Ekspeditsii. Vyp. 1, Polevye issledovaniya Khorezmskoy ekspeditsii v 1954-1956 gg.*, S. P. Tolstov i M. G. Vorobeva. Moskva.

1959b   Pervobytnaya keramika Khorezma. *Trudy Khorezmskoy Arkheologo-Etnograficheskoy Ekspeditsii. Tom IV, Keramika Khorezma*, S. P. Tolstov i M. G. Vorobeva. Moskva.

1959c   Pamyatniki pervobytnoy kultury Verkhnego Uzboya. *Trudy Khorezmskoy Arkheologo-Etnograficheskoy Ekspeditsii. Tom IV, Keramika Khorezma*, S. P. Tolstov i M. G. Vorobeva. Moskva.

1960   Raskopki stoyanok tazabagyabskoy kultury v 1957 g. *Materialy Khorezmskoy Ekspeditsii. Vyp. 4, Polevye issledovaniya Khorezmskoy ekspeditsii 1957 g*, S. P. Tolstov i M. A. Itina. Moskva.

1961   Raskopki mogilnika tazabagyabskoy kultury Kokcha 3. *Materialy Khorezmskoy Ekspeditsii. Vyp. 5, Mogilnik bronzovogo veka Kokcha 3*, S. P. Tolstov. Moskva.

1962   Stepnye plemena sredneaziatskogo mejdurechya vo vtoroy polovine II - nachale I tysyacheletiya do n. e. *Sovetskaya Etnografiya*, No. 3.

1963   Poselenie Yakke-Parsan 2 (raskopki 1958-1959 gg.). *Materialy Khorezmskoy Ekspeditsii. Vyp. 6, Polevye Issledovaniya Khoresmskoy Ekspeditsii v 1958-1961 gg*, S. P. Tolstov i A. V. Vinogradov. Moskva.

1967   O meste tazabagyabskoy kultury sredi kultur stepnoy bronzy. *Sovetskaya Etnografiya*, No. 2.

1968   Drevnekhorezmiyskie zemledeltsy. *Istoriya, arkheologiya i etnografiya Sredney Azii i Kazakhstana*, A. V. Vinogradov, M. G. Vorobeva, T. A. Jdanko, M. A. Itina, L. M. Levina, Yu. A. Rapoport. Moskva.

Ivanov, P. P.

1935   Ocherk istorii karakalpakov. Materialy po istorii karakalpakov. *Trudy Instituta vostokovedeniya*, Tom VII. Moskva - Leningrad.

1954   Khozyaystvo djuybarskikh sheykhov. *K istorii feodalnogo zemlevladeniya v Sredney Azii v XVI-XVII vv.* Moskva - Leningrad.

Jdanko, T. A.

1950   Ocherki istoricheskoy etnografii karakalpakov. *Rodo-plemennaya struktura i rasselenie v XIX - nachale XX veka.* Moskva - Leningrad.

1952   Karakalpaki Khorezmskogo oazisa. *Trudy Khorezmskoy arkheologo-etnograficheskoy ekspeditsii. Tom I, Arkheologicheskie i Etnograficheskie Raboty Khorezmskoy Ekspeditsii 1945-1948*, S. P. Tolstov i T. A. Jdanko. Moskva.

1955   Istoriko-etnograficheskiy atlas Sredney Azii. *Sovetskaya Etnografiya*, No. 3.

1958a   Patriarkhalno-feodalnye otnosheniya u poluosedlogo naseleniya Sredney Azii. *Materialy Pervoy vserossiyskoy nauchnoy konferentsii vostokovedov v g. Tashkente.* Tashkent.

1958b   Byt kolkhoznikov - pereselentsev na vnov osvoennykh zemlyakh drevnego orosheniya Kara-Kalpakii. *Trudy Khorezmskoy arkheologo-etnograficheskoy ekspeditsii. Tom II, Arkheologicheskie i Etnograficheskie raboty Khorezmskoy Ekspeditsii 1949-1953*, S. P. Tolstov i T. A. Jdanko. Moskva.

1961   Problema poluosedlogo naseleniya v istorii Sredney Azii i Kazakhstana. *Sovetskaya Etnografiya*, No. 2.

1964   Kara-kalpaki (Osnovnye problemy etnicheskoy istorii i etnografii). *Doklad po opublikovannym rabotam, predstavlennym na soiskanie uchenoy stepeni doktora istoricheskikh nauk.* Moskva.

Jukovskiy, P. M.

1950   *Kulturnye rasteniya i ikh sorodichi (sistematika, geografiya, ekologiya, ispolzovanie, proiskhojdenie).* Moskva.

Jukovskiy, V. A.
   1894   *Drevnosti Zakaspiyskogo kraya. Razvaliny starogo Merva.* Sankt-Peterburg.

Kabo, R. M.
   1947   Priroda i chelovek v ikh vzaimootnosheniyakh kak predmet sotsialno-kulturnoy geografii. *Voprosy Geografii.* Sbornik, Sb. 5.

Kamalov, S.
   1958   Narodno-osvoboditelnaya borba karakalpakov protiv khivinskikh khanov v XIX v. *Trudy Khorezmskoy arkheologo-etnograficheskoy ekspeditsii. Tom III, Materialy i issledovaniya po etnografii karakalpakov,* T. A. Jdanko. Moskva.

Karazin, N. N.
   1875   V nizove Amu-Dari. Putevye ocherki. *Vestnik Evropy,* mart, No. 2.

Karpov, G. I.
   1945   Turkmeny-oguzy. *Izvestiya Turkmenskogo filiala AN SSSR,* No. 1.

Kastalskiy, B. N., and E. M. Timofeev
   1934   Gidrostroitelstvo i arkheologiya v Sredney Azii. *Vestnik Irrigatsii,* Sb. 1. Samarkand.

Kaulbars, A. V.
   1881   Nizovya Amu-Dari, opisannye po sobstvennym issledovaniyam v 1873 g. *Zapiski Russkogo Geograficheskogo Obshchestva,* Tom IX. Sankt-Peterburg.

   1887   Drevneyshie rusla Amu-Dari. *Zapiski Russkogo Geograficheskogo Obshchestva,* Tom XVII, No. 4. Sankt-Peterburg.

Kel, N. G.
   1937   *Fotografiya i fotogrammetriya.* Moskva – Leningrad.

Kes, A. S.
   1939   Ruslo Uzboy i ego genezis. *Trudy Instituta geografii AN SSSR,* Vyp. 30.

   1954   Razvitie relefa Sarykamyshskoy vpadiny. *Trudy Instituta geografii AN SSSR,* Vyp. 62.

   1957   K voprosu o drevnechetvertichnoy istorii sistemy Amu-Darya – Sarykamysh – Uzboy. *Trudy Komissii po Izucheniyu Chetvertichnogo Perioda,* Vyp. 13.

   1958   Prirodnye faktory, obuslovlivayushchie rasselenie drevnego cheloveka v pustynyakh Sredney Azii. *Kratkie Soobshcheniya Instituta Etnografii AN SSSR,* Vyp. XXX.

   1959   Znachenie aerofotosemki dlya vosstanovleniya paleogeografii nizoviy Amu-Dari. *Trudy Laboratorii aerometodov AN SSSR,* Tom 8.

Ketskhoveli, N. N.
   1964   K istorii proiskhojdeniya kulturnykh rasteniy v Gruzii. *VII Mejdunarodnyy kongress antropologicheskikh i etnograficheskikh nauk.* Moskva.

Khanykov, Ya. V.
   1901   Poyasnitelnaya zapiska k karte Aralskogo morya i Khivinskogo khanstva s ikh okrestnostyami. *Zapiski Russkogo geograficheskogo obshchestva,* Kn. V. Sankt-Peterburg.

Khasanov, A. S.
   1959   Gidrogeologicheskie usloviya pravoberejnoy chasti nizovev Amu-Dari. *Materialy po proizvoditelnym silam Uzbekistana,* Vyp. 10. Tashkent.

Khlopin, I. N.
   1963   Eneolit yujnykh oblastey Sredney Azii: Pamyatniki rannego eneolita yujnoy Turkmenii. *Arkheologiya CCCP: Svod Arkheologicheskikh Istochnikov, BZ-8, Chast I.* Moskva-Leningrad.

   1964   *Geoksyurskaya gruppa poseleniy epokhi eneolita.* Moskva – Leningrad.

Khudayberdyev, R.
   1962   Ostatki rasteniy Chustskogo poseleniya epokhi bronzy. *Doklady AN Uzbekskoy SSR,* No. 10.

Kink, X. A.
1964   Egipet do faraonov. *Po pamyatnikam materialnoy kultury*. Moskva.

Kiselev, S. V.
1951   *Drevnyaya istoriya Yujnoy Sibiri*. Moskva.

1965   Bronzovyy vek SSSR. *Novoe v sovetskoy arkheologii*. Moskva.

Klingen, I. N.
1898   Sredi patriarkhov zemledeliya. Chast I, *Egipet*. Sankt-Peterburg.

1960   Sredi patriarkhov zemledeliya narodov Blijnego i Dalnego Vostoka. *Egipet, Indiya, Tseylon, Kitay*. Izd. 2. Moskva.

Klyashtornyy, S. G.
1953   Yaksart – Syr-Darya. *Sovetskaya Etnografiya*, No. 3.

1964   *Drevnetyurkskie runicheskie pamyatniki kak istochniki po istorii Sredney Azii*. Moskva.

Kolchin, B. A.
1965   Arkheologiya i estestvennye nauki. *Arkheologiya i estestvennye nauki*. Moskva.

Kolesnik, S. V.
1952   Uchenie o landshaftakh v svyazi s preobrazovaniem prirody v SSSR. *Izvestiya Vsesoyuznogo Geograficheskogo Obshchestva*, Tom 84.

Kondrashev, S. K.
1916   *Oroshaemoe khozyaystvo i vodopolzovanie Khivinskogo oazisa*. Moskva.

Konobeeva, M. G.
1965   Evolyutsiya pochv drevnikh oazisov pustynnoy zony. *Geografiya i klassifikatsiya pochv Sredney Azii*. Moskva.

Konshin A. M.
1897   Razyasnenie voprosa o drevnem techenii Amu-Dary, po sovremennym geologicheskim i fiziko-geograficheskim dannym. *Zapiskii Russkogo geograficheskogo obshchestva po obshchey geografii*, Tom 33, No. 1. Sankt-Peterburg

Konshin, M. D.
1954   *Aerofototopografiya*. Moskva.

Kosambi, D. D.
1968   Kultura i tsivilizatsiya drevney Indii. *Istoricheskiy ocherk*. Moskva.

Kostyakov, A. N.
1951   *Osnovy melioratsii*. Moskva.

Kosven, M. O.
1960   K voprosu o voennoy demokratii. *Trudy Instituta Etnografii AN SSSR*, Tom IV.

Kovalevskiy, G. V.
1931   Kulturno-istoricheskaya i biologicheskaya rol gornykh rayonov. *Priroda*, No. 2.

Kovda, V. A.
1947   *Proiskhojdenie i rejim zasolennykh pochv*, Tom II. Moskva – Leningrad.

Kovda, V. A., V. M. Egorov, A. G. Morozov, and Yu. P. Lebedev
1954   Zakonomernosti protsessov solenakopleniya v pustynyakh Aralo-Kaspiyskoy nizmennosti. *Trudy Pochvennogo In-ta im.* V. V. Dokuchaeva, Tom X/IV.

Kramer, S. N.
1965   *Istoriya nachinaetsya v Shumere*. Moskva.

Krivtsova-Grakova, O. A.
1948   Alekseevskoe poselenie i mogilnik. *Trudy Gosudarstveennyy Istoricheskiy Muzey*, XVII.

Kruglov, A. P., and G. V. Podgaetskiy

1935   Rodovoe obshchestvo stepey Vostochnoy Evropy. *Izvestiya Gosudarstvennoy Akademii istorii materialnoy kultury*, Vyp. 119. Moskva - Leningrad.

Kryuger, O. O.

1935   Selskokhozyaystvennoe proizvodstvo v ellinisticheskom Egipte. *Izvestiya Gosudarstvennoy Akademii Istorii Materialnoy Kultury*, Vyp. 108.

Kryukov, M. V.

1960   Inskaya tsivilizatsiya i basseyn reki Khuankhe (k voprosu o spetsifike obshchestva In v Kitae). *Vestnik istorii mirovoy kultury*, No. 4.

1964   U istokov drevnikh kultur Vostochnoy Azii. *Narody Azii i Afriki*, No. 6. Moskva.

Kryvelev, I. A.

1965   *Raskopki v "Bibleyskikh" stranakh*. Moskva.

Kuftin, B. A.

1954   Raboty YuTAKE v 1952 g. po izucheniyu "kultury Anau". *Izvestiya AN Turkmenskoy SSR*, No. 1.

1956   Polevoy otchet o rabote XIV otryada YuTAKE po izucheniyu kultury pervobytno-obshchinnykh osedlozemledelcheskikh poseleniy epokhi medi i bronzy v 1952 g. *Trudy Yujno-Turkmenistanskoy Ekspeditsii AN SSSR. Tom VII, Pamyatniki kultury kamennogo i bronzovogo veka yujnogo turkmenistana*, M. E. Masson. Ashkhabad.

Kushner (Knyshev), P. I.

1951   *Etnicheskie territorii i etnicheskie granitsy*. Moskva.

Kuzmina, E. E.

1958   Mogilnik Zaman-Baba. *Sovetskaya Arkheologiya*, No. 2.

1964a   O yujnykh predelakh rasprostraneniya stepnykh kultur epokhi bronzy v Sredney Azii. *Pamyatniki kamennogo i bronzovogo vekov*. Moskva.

1964b   Andronovskoe poselenie i mogilnik Shandasha. *Kratkie Soobshcheniya o Dokladakh i Polevykh Issledovaniyakh Instituta Arkheologii AN SSSR*, Vyp. 98.

1965   Nekotorye obshchie zakonomernosti ranney metallurgii medi v Sredney Azii i na Kavkaze. *Materialy sessii, posvyashchennoy itogam arkheologicheskikh i etnograficheskikh issledovaniy 1964 goda v SSSR (tezisy dokladov)*.

1966   *Metallicheskie izdeliya eneolita i bronzovogo veka v Sredney Azii*. Moskva.

Kyzlasov, L. R.

1959   Arkheologicheskie issledovaniya na gorodishche Ak-Beshim v 1953-1954 gg. *Trudy Kirgizskoy arkheologo-etnograficheskoy ekspeditsii*, Tom II. Moskva.

Latynin, B. A.

1931   Informatsionnoe soobshchenie o rabotakh Ferganskoy ekspeditsii 1930 g. *Soobshchenie Gosudarstvennoy Akademii istorii materialnoy kultury*, No. 1-2. Leningrad.

1935a   *K voprosu ob istorii irrigatsii. Sb. AN SSSR - Akademiku N. Ya*. Marru. Moskva - Leningrad.

1935b   Raboty v rayone proektiruemoy elektrostantsii na r. Naryne v Fergane (otchet o rabotakh). Izvestiya Gosudarstvennoy Akademii istorii materialnoy kultury, Vyp. 110. *Arkheologicheskie raboty na novostroykakh*, Tom II. Leningrad.

1956   Voprosy istorii irrigatsii drevney Fergany. *Kratkie Soobshcheniya o Dokladakh i Polevykh Issledovaniyakh Instituta Istorii Materialnoy Kultury AN SSSR*, Vyp. 64. Moskva.

1957   Voprosy istorii irrigatsii drevney Fergany. *Kratkie soobshcheniya Instituta etnografii AN SSSR*, Vyp. XXVI. Moskva.

1959   Nekotorye voprosy metodiki izucheniya istorii irrigatsii Sredney Azii. *Sovetskaya Arkheologiya*, No. 3.

1962   Voprosy istorii irrigatsii i oroshaemogo zemledeliya drevney Fergany. *Obobshchayushny doklad po rabotam, predstavlennym kak dissertatsiya na soiskanie uchenoy stepeni doktora istoricheskikh nauk.* Leningrad.

Latyshev, V. V.

1904–1906   *Izvestiya drevnikh pisateley, grecheskikh i latinskikh o Skifii i Kavkaze, Chast I–II.* Sankt-Peterburg.

Legostaev, V. M., and B. K. Konkov

1953   *Meliorativnoe rayonirovanie Sredney Azii.* Tashkent.

Letunov, P. A.

1958   *Pochvenno-meliorativnye usloviya v nizovyakh Amu-Dari.* Moskva.

Levin, M. G., and N. N. Cheboksarov

1955   Khozyaystvenno-kulturnye tipy i istoriko-geograficheskie oblasti. *Sovetskaya Etnografiya*, No. 4.

Levina, L. M.

1966   Keramika i voprosy khronologii pamyatnikov djety-asarskoy kultury. *Materialnaya kultura narodov Sredney Azii i Kazakhstana*, Moskva.

1967   *Keramika Nijney i Sredney Syr-Dari v pervom tysyacheletii n. e.* Moskva. (avtoref. kand. diss.)

Levshin, A. I.

1832   *Opisanie kirgiz-kazachikh ili kirgiz-kaysatskikh ord i stepey, Chast I–III.* Sankt-Peterburg.

Lips, Yu

1954   Proiskhojdenie veshchey. *Iz istorii kultury chelovechestva.* Moskva.

Lisitsyna, G. N.

1963   *Drevnie zemledeltsy v delte Tedjena.* Priroda, No. 10.

1964   Rastitelnost Yujnoy Turkmenii v epokhu eneolita po paleobotanicheskim dannym. *Kratkie Soobshcheniya o Dokladakh i Polevykh Issledovaniyakh Instituta Arkheologii AN SSSR*, Vyp. 98.

1965   *Oroshaemoe zemledelie epokhi eneolita na yuge Turkmenii.* Moskva.

1966   Izuchenie geoksyurskoy orositelnoy seti v Yujnoy Turkmenii v 1964 g. *Kratkie Soobshcheniya o Dokladakh i Polevykh Issledovaniyakh Instituta Arkheologii AN SSSR*, Vyp. 108.

Lisitsyna, G. N., V. M. Masson, V. I. Sarianidi, and I. N. Khlopin

1965   Itogi arkheologicheskogo i paleogeograficheskogo izucheniya Geoksyurskogo oazisa. *Sovetskaya Arkheologiya*, No. 1.

Litvinskiy, B. A.

1952   Namazga-Depe. *Sovetskaya Etnografiya*, No. 1.

1956   Ob arkheologicheskikh rabotakh v Vakhshskoy doline i v Isfaganskom rayone (v Vorukhe). *Kratkie Soobshcheniya o Dokladakh i Polevykh Issledovaniyakh Instituta Istorii Materialnoy Kultury AN SSSR*, Vyp. 64.

1959   Pamyatniki epokhi bronzy i rannego jeleza v Kayrak-Kumakh. *Materialy Vtorogo Soveshchaniya Arkheologov i Etnografov Sredney Azii.*

1963   Sotsialno-ekonomicheskiy stroy drevney Sredney Azii. Istoriya Tadjikskogo naroda. Tom I, *S drevneyshikh vremen do V v. n. e.* Moskva.

1968   Orujie naseleniya Pamira I Fergany v sakskoe vremya. *Materialnaya kultura Tajikistana*, Vyp. 1. Dushanbe

Litvinskiy, B. A., and E. A. Davidovich

1954   Predvaritelnyy otchet o rabotakh Khuttalskogo otryada na territorii Vakhshskoy doliny v 1953 g. *Doklady AN Tadjskoy SSR*, Vyp. II.

Livshits, V. A.

1962   Sogdiyskie dokumenty s gory Mug, vyp. II: yuridicheskie dokumenty i pisma. *Chtenie, perevod i kommentarii.* Moskva.

1963   Drevneyshie gosudarstvennye obrazovaniya. Istoriya Tadjikskogo naroda. Tom I, *S drevneyshikh vremen do V v. n. e.* Moskva.

Lokhovits, V. A.

1968   Novye dannye o podboynykh pogrebeniyakh v Turkmenii. *Istoriya, arkheologiya i etnografiya Sredney Azii i Kazakhstana*, A. V. Vinogradov, M. G. Vorobeva, T. A. Jdanko, M. A. Itina, L. M. Levina, Yu. A. Rapoport. Moskva.

Lopatin, G. V.

1952   *Nanosy rek (obrazovanie i perenos)*. Moskva.

Makeev, P. S.

1952   Zemli drevnego orosheniya na sukhikh ruslakh Kunya-Dari i Djany-Dari. *Izvestiya Vsesoyuznogo Geograficheskogo Obshchestva*, Tom 84, Vyp. 5.

Maksimov, D. N.

1929   *Nakanune zemledeliya. Rossiyskoy assotsiatsii nauchno-issledovatelskikh institutov obshchestvennykh nauk*, Tom 3. Moskva.

Maksimova A. G., M. S. Mershchiev, B. I. Vaynberg, and L. M. Levina

1968   *Drevnosti Chardary (Arkheologicheskie issledovaniya v zone chardarinskogo vodokhranilishcha)*. Alma-Ata.

Mandelshtam, A. M.

1964   K istorii Baktrii: Tokharistana. *Kratkie Soobshcheniya o Dokladakh i Polevykh Issledovaniyakh Instituta Arkheologii AN SSSR*, Vyp. 98.

Manylov, Yu.

1962   Jeleznye orudiya VIII veka s gorodishcha Tok-kala. *Vestnik Kara-Kalpaskogo Filiala AN Uzbeskoy SSR*, No. 3(9).

Margulan, A. X., K. A. Akishev, M. K. Kadyrbaev, and A. M. Orazbaev

1966   *Drevnyaya kultura Tsentralnogo Kazakhstana*. Alma-Ata.

Markov, K. K.

1951a   *Paleogeografiya*. Moskva.

1951b   Vysykhaet li Srednyaya i Tsentralnaya Aziya?. *Voprosy Geografii*. Sbornik, Sb. 24.

Marushchenko, A. A.

1939   Anau. Istoricheskaya spravka. *Arkhitekturnye pamyatniki Turkmenii*, Ashkhabad – Moskva.

1956a   Staryy Serakhs. Otchet o raskopkakh 1953 g. Trudy Instituta Istorii, *Arkheologii i Etnografii AN Turkmenskoy SSR*, Tom II. Ashkhabad.

1956b   Khosrov-kala (Otchet o raskopkakh 1953 g.). *Trudy Instituta Istorii, Arkheologii i Etnografii AN Turkmenskoy SSR*, Tom II. Ashkhabad.

1956c   Itogi polevykh arkheologicheskikh rabot 1953. Instituta istorii, arkheologii i etnografii AN TurkmSSR. *Trudy Instituta Istorii, Arkheologii i Etnografii AN Turkmenskoy SSR*, Tom II.

Masson, M. E.

1940   Termezskaya arkheologicheskaya kompleksnaya ekspeditsiya (TAKE). *Kratkie Soobshcheniya o Dokladakh i Polevykh Issledovaniyakh Instituta Istorii Materialnoy Kultury AN SSSR*, Vyp. VIII.

1945   Raboty Termezskoy arkheologicheskoy kompleksnoy ekspeditsii (TAKE) 1937 i 1938 gg. *Trudy AN Uzbeskoy SSR*, Ser. I, Tom II. Tashkent

1949   Gorodishcha Nisy v selenii Bagir i ikh izuchenie. *Trudy Yujno-Turkmenistanskoy Ekspeditsii AN SSSR. Tom I*, M. E. Masson. Ashkhabad.

1950   Nekotorye novye dannye po istorii Parfii. *Vestnik Drevney Istorii*, No. 3.

1951   Nekotorye itogi raboty YuTAKE i perspektivy arkheologicheskogo izucheniya Yujnogo Turkmenistana. *Izvestiya AN Turkmenskoy SSR*, No. 1.

1954   O rabotakh Yujno-Turkmenistanskoy arkheologicheskoy ekspeditsii. *Tezisy dokladov na sessii Otdeleniya istoricheskikh nauk i plenuma IIMK*. Moskva.

1955   Kratkaya khronika polevykh rabot YuTAKE za 1948-1952 gg. *Trudy Yujno-Turkmenistanskoy Ekspeditsii AN SSSR. Tom V*, M. E. Masson. Ashkhabad.

1956   Kratkiy ocherk istorii izucheniya Sredney Azii v arkheologicheskom otnoshenii (ch. I, do pervoy poslevoennoy pyatiletki). *Trudy Sredneaziatskogo Gosudarstvennogo Universiteta*, Vyp. 81. Istoricheskie nauki, Kn. 12.

Masson, V. M.

1954a *Drevnyaya kultura Dakhistana (istoriko-arkheologicheskie ocherki)*. Moskva - Leningrad. (avtoref. kand. diss.)

1954b Misrianskaya ravnina v epokhu pozdney bronzy i rannego jeleza. *Izvestiya AN Turkmenskoy SSR*, No. 2.

1956a Izuchenie drevnezemledelcheskikh poseleniy v delte Murgaba. *Izvestiya AN Turkmenskoy SSR*, No. 2.

1956b Pamyatniki kultury arkhaicheskogo Dakhistana v Yugo-Zapadnoy Turkmenii. *Trudy Yujno-Turkmenistanskoy Ekspeditsii AN SSSR. Tom VII, Pamyatniki kultury kamennogo i bronzovogo veka yujnogo turkmenistana*, M. E. Masson. Ashkhabad.

1956c Raspisnaya keramika Yujnoy Turkmenii po raskopkam B. V. Kuftina. *Trudy Yujno-Turkmenistanskoy Ekspeditsii AN SSSR. Tom VII, Pamyatniki kultury kamennogo i bronzovogo veka yujnogo turkmenistana*, M. E. Masson. Ashkhabad.

1957a Djeytun i Kara-Depe. *Sovetskaya Arkheologiya*, No. 1.

1957b Izuchenie eneolita i bronzovogo veka Sredney Azii. *Sovetskaya Arkheologiya*, No. 4.

1957c Izuchenie drevnezemledelcheskikh poseleniy v delte Murgaba. *Kratkie Soobshcheniya o Dokladakh i Polevykh Issledovaniyakh Instituta Istorii Materialnoy Kultury AN SSSR*, Vyp. 69.

1958   Problema drevney Baktrii i novyy arkheologicheskiy material. *Sovetskaya Arkheologiya*, No. 2.

1959   Drevnezemledelcheskaya kultura Margiany. *Materialy i Issledovaniya po Arkheologii SSSR*, No. 73.

1961a Yujnoturkmenistanskiy tsentr rannezemledelcheskikh kultur. *Trudy Yujno-Turkenistanskoy ekspeditsii AN SSSR. Tom X, Pamyatniki kultury kamennogo i bronzovogo veka yujnogo turkmenistana II*, M. E. Masson. Ashkhabad.

1961b Djeytunskaya kultura. *Trudy Yujno-Turkenistanskoy ekspeditsii AN SSSR. Tom X, Pamyatniki kultury kamennogo i bronzovogo veka yujnogo turkmenistana II*, M. E. Masson. Ashkhabad.

1962a Novye dannye o Djeytune i Kara-Depe. *Sovetskaya Arkheologiya*, No. 3.

1962b Eneolit yujnykh oblastey Sredney Azii: Pamyatniki razvitogo eneolita yugo-zapadnoy Turkmenii. *Arkheologiya CCCP: Svod Arkheologicheskikh Istochnikov, BZ-8. Chast II*. Moskva - Leningrad.

1963   Srednyaya Aziya i Iran v III tysyacheletii do nashey ery. *Kratkie Soobshcheniya o Dokladakh i Polevykh Issledovaniyakh Instituta Arkheologii AN SSSR*, Vyp. 93.

1964   *Srednyaya Aziya i Drevniy Vostok*. Moskva - Leningrad.

1965   Zemledelie pervobytnogo i ranneklassovogo obshchestva. *Materialy sessii, posvyashchennoy itogam arkheologicheskikh i etnograficheskikh issledovaniy 1964 goda v SSSR (tezisy dokladov)*. Baku.

1966a Ot vozniknoveniya zemledeliya do slojeniya ranneklassovogo obshchestva (etapy kulturnogo i khozyaystvennogo razvitiya po materialam Aziatskogo materika). *Doklady i soobshcheniya arkheologov SSSR. VII Mejdunarodnyy kongress istorikov i protoistorikov*. Moskva.

1966b Zemledelcheskiy neolit yugo-zapada Sredney Azii. *Srednyaya Aziya v Epokhu Kamnya i Bronzy*, V. M. Masson. Moskva - Leningrad.

1967a Stanovlenie ranneklassovogo obshchestva na Drevnem Vostoke. *Voprosy Istorii*, No. 5.

1967b Protogorodskaya kultura Altyn-depe. *Arkheologicheskie Otkrytiya 1966 goda*. Moskva.

1968   K voprosu ob obshchestvennom stroe drevney Sredney Azii. *Istoriya, arkheologiya i etnografiya Sredney Azii i Kazakhstana*, A. V. Vinogradov, M. G. Vorobeva, T. A. Jdanko, M. A. Itina, L. M. Levina, Yu. A. Rapoport. Moskva.

Mechnikov, L. P.

1924   *Tsivilizatsiya i velikie istoricheskie reki (Geograficheskaya teoriya progressa i sotsialnogo razvitiya).* Moskva.

Menabde, V. L.

1948   *Pshenitsy Gruzii.* Tbilisi.

Mikhaylov, V. Ya.

1959   *Aerofotografiya i obshchie osnovy fotografii.* Moskva.

Minashina, N. G.

1960   Mikromorfologicheskoe issledovanie lessa i ego izmeneniy pri pochvoobrazovanii. *Doklady Sovetskikh pochvovedov k VII Mejdunarodnoy kongressi v SSHA.* Moskva.

1962   Drevneoroshaemye pochvy Murgabskogo oazisa. *Pochvovedenie,* No. 8. Moskva.

1963   Raspredelenie soley v pochvakh i gruntovykh vodakh na massive drevnego orosheniya v tsentralnoy chasti Murgabskoy delty. *Vliyanie orosheniya na pochvy oazisov Sredney Azii.* Moskva.

Miroshnichenko, V. P.

1957   *Materialy k ispolzovaniyu aerometodov pri izuchenii pochv i rastitelnosti Severnogo Kazakhstana.* Moskva - Leningrad.

1960   Takyry kak indikatory noveyshikh tektonicheskikh dvijeniy v peschanykh pustynyakh Sredney Azii. *Trudy Laboratorii aerometodov AN SSSR,* Tom IX.

1961   Primenenie aerometodov pri izuchenii zonalnykh i regionalnykh zakonomernostey landshafta. *Primenenie aerometodov v landshaftnykh issledovaniyakh.* Moskva - Leningrad.

Monakhova, V. N.

1946   K voprosu o zonalnom i nezonalnom landshafte. *Voprosy Geografii.* Sbornik, No. 1.

Mongayt, A. L.

1962   Zadachi i vozmojnosti arkheologicheskoy kartografii. *Sovetskaya Arkheologiya,* No. 1.

1963   *Arkheologiya i sovremennost.* Moskva.

Murzaev, E. M.

1940   K geomorfologii yugo-vostochnoy okrainy Sarykamyshskoy kotloviny. *Trudy Instituta geografii AN SSSR,* Vyp. XXXV.

1955   Sovremennoe sostoyanie problemy izmeneniy klimata Aziatskikh pustyn. *Pamyati akademika L. S. Berga.* Moskva - Leningrad.

1957   *Srednyaya Aziya. Fiziko-geograficheskiy ocherk.* Izd. 2. Moskva.

Narimanov, I. G.

1966   Drevneyshaya zemledelcheskaya kultura Zakavkazya. Doklady i soobshcheniya arkheologov SSSR. *VII Mejdunarodnyy kongress istorikov i protoistorikov.* Moskva

Negmatov, N. N.

1959   *Usrushana v VII-IX vv. n. e.* Dushanbe.

Negmatov, N. N., and T. I. Zeymal

1959   Usrushanskiy zamok v Shakhristane. *Sovetskaya Arkheologiya,* No. 2.

Negrul, A. M.

1938   *Evolyutsiya kulturnykh form vinograda.* Moskva.

Nerazik, E. E.

1958   Arkheologicheskoe obsledovanie gorodishcha Kunya-Uaz v 1952 g. *Trudy Khorezmskoy arkheologo-etnograficheskoy ekspeditsii. Tom II, Arkheologicheskie i Etnograficheskie raboty Khorezmskoy Ekspeditsii 1949-1953,* S. P. Tolstov i T. A. Jdanko. Moskva.

1959a   Raskopki v Berkut-kalinskom oazise. *Materialy Khorezmskoy Ekspeditsii. Vyp. 1, Polevye issledovaniya Khorezmskoy ekspeditsii v 1954-1956 gg.,* S. P. Tolstov i M. G. Vorobeva. Moskva.

1959b Keramika Khorezma afrigidskogo perioda. *Trudy Khorezmskoy Arkheologo-Etnograficheskoy Ekspeditsii. Tom IV, Keramika Khorezma*, S. P. Tolstov i M. G. Vorobeva. Moskva.

1963   Raskopki Yakke-Parsana. *Materialy Khorezmskoy Ekspeditsii. Vyp. 7, Polevye issledovaniya Khorezmskoy ekspeditsii v 1958-1961 gg.*, S. P. Tolstov i E. E. Nerazik. Moskva.

1966   *Selskie poseleniya afrigidskogo Khorezma (Po materialam Berkut-kalinskogo oazisa)*. Moskva.

1968   O nekotorykh napravleniyakh etnicheskikh svyazey naseleniya Yujnogo i Yugo-Vostochnogo Priaralya. *Istoriya, arkheologiya i etnografiya Sredney Azii i Kazakhstana*, A. V. Vinogradov, M. G. Vorobeva, T. A. Jdanko, M. A. Itina, L. M. Levina, Yu. A. Rapoport. Moskva.

Nerazik, E. E. and M. S. Lapirov-Sklobo

1959   Raskopki Barak-tama-I v 1956 g. *Trudy Khorezmskoy arkheologo-etnograficheskoy ekspeditsii. Tom I, Arkheologicheskie i Etnograficheskie Raboty Khorezmskoy Ekspeditsii 1945-1948*, S. P. Tolstov i T. A. Jdanko. Moskva.

Nesteruk, F. Ya.

1955   Vodnoe khozyaystvo Kitaya. *Iz istorii nauki i tekhniki Kitaya*. Moskva.

Nikiforov, V. N.

1968   Logika diskussii i logika v diskussii. *Voprosy Istorii*, No. 2.

Nikolskiy, G. V., D. V. Radakov, and V. D. Lebedev

1952   Ostatki ryb iz neoliticheskoy stoyanki Djanbas-kala 4. *Trudy Khorezmskoy arkheologo-etnograficheskoy ekspeditsii. Tom I, Arkheologicheskie i Etnograficheskie Raboty Khorezmskoy Ekspeditsii 1945-1948*, S. P. Tolstov i T. A. Jdanko. Moskva.

Novikov, Yu. F.

1959   O vozniknovenii zemledeliya i ego pervonachalnykh formakh. *Sovetskaya Arkheologiya*, No. 4.

1964   Ocherki iz istorii razvitiya tekhniki zemledeliya v drevnem mire. *Vestnik Drevney Istorii*, No. 1.

Okladnikov, A. P.

1953   Izuchenie pamyatnikov kamennogo veka v Turkmenii. *Izvestiya AN Turkmenskoy SSR*, No. 2.

1956a Drevneyshee proshloe Turkmenistana (Drevnie okhotniki i sobirateli v stepyakh i pustynyakh Turkmenistana). *Trudy Instituta Istorii, Arkheologii i Etnografii AN Turkmenskoy SSR*, Tom I.

1956b Peshchera Djebel: pamyatnik drevney kultury prikaspiyskikh plemen Turkmenii. *Trudy Yujno-Turkmenistanskoy Ekspeditsii AN SSSR. Tom VII, Pamyatniki kultury kamennogo i bronzovogo veka yujnogo turkmenistana*, M. E. Masson. Ashkhabad.

1959   Kamennyy vek Tadjikistana. Itogi i problemy. *Materialy vtorogo oveshchaniya arkheologov i etnografov Sredney Azii*. Moskva - Leningrad.

1962   O nachale zemledeliya za Baykalom i v Mongolii. *Drevniy mir*, N. V. Pigulevskaya, D. P. Kallistov, I. S. Katsnelson, M. A. Korostovtsev. Moskva.

1966a K voprosu o mezolite i epipaleolite v Aziatskoy chasti SSSR. Sibir i Srednyaya Aziya. U istokov drevnikh kultur. *Materialy i Issledovaniya po Arkheologii SSSR*, No. 126. Moskva.

1966b Verkhnepaleoliticheskoe i mezoliticheskoe vremya. *Srednyaya Aziya v Epokhu Kamnya i Bronzy*, V. M. Masson. Moskva - Leningrad.

Olderogge, D. A., and I. I. Potekhin

1954   *Narody Afriki*. Moskva.

Oranskiy, I. M.

1963   Pismennye pamyatniki na iranskikh yazykakh narodov Sredney Azii (do VII-VIII vv. n. e.). *Istoriya Tadjikskogo naroda. Tom I, C drevneyshikh vremen do V v. i. e.* Moskva.

Orlov, M. A.

1952a Barak-tam: Novye pamyatniki pozdneantichnoy arkhitektury severo-vostochnogo Khorezma. *Trudy Khorezmskoy arkheologo-etnograficheskoy ekspeditsii. Tom I, Arkheologicheskie i Etnograficheskie Raboty Khorezmskoy Ekspeditsii 1945-1948, S. P. Tolstov i T. A. Jdanko. Moskva.*

1952b Pamyatniki sadovo-parkovogo iskusstva srednevekovogo Khorezma. *Trudy Khorezmskoy arkheologo-etnograficheskoy ekspeditsii. Tom I, Arkheologicheskie i Etnograficheskie Raboty Khorezmskoy Ekspeditsii 1945-1948, S. P. Tolstov i T. A. Jdanko. Moskva.*

Ostrovskiy, S. F.

1907 *Irrigatsionnaya sistema Indii.* Sankt-Peterburg.

Parkhomenko, L. M.

1949 K voprosu ob obvodnenii nekotorykh takyrov. *Pochvovedenie,* No. 9.

Passek, T. S.

1949 Periodizatsiya tripolskikh poseleniy. *Materialy i Issledovaniya po Arkheologii SSSR,* No. 10.

Pavlov, A. P.

1920 *Predstavlenie o vremeni v istorii, arkheologii i geologii.* Pg.

Pavlov, S. P.

1934 Primenenie aerosemki v arkheologii. *Problemy Istorii Dokapitalisticheskikh Obshchestv,* No. 11-12. Moskva - Leningrad.

Pavlov, V.

1950 *Predislovie k knige E. Martona "Aerogeografiya".* Moskva.

Pazukhiny, B., and S. Pazukhiny

1894 Nakaz Borisu i Semenu Pazukhinym, poslannym v Bukharu, Balkh i Yurgench v 1669. *Russkaya istoricheskaya biblioteka,* Tom 15.

Pechirka, Ya.

1967 Zamechaniya po povodu diskussii o rabovladelcheskoy formazii i aziatskom sposobe proizvodstva. *Antichnoe obshchestvo: trudy Konferentsii po izucheniyu problem antichnosti.* Moskva.

Pelt, I. N.

1951 Zemli drevnego orosheniya Djanadarinskoy drevnealluvialnoy ravniny. *Izvestiya Vsesoyuznogo geograficheskogo obshchestva,* No. 3.

Perikhanyan, A. G.

1952 K voprosu o rabovladenii i zemlevladenii v Irane parfyanskogo vremeni. *Vestnik Drevney Istorii,* No. 4.

1968 Agnaticheskie gruppy v drevnem Irane. *Vestnik Drevney Istorii,* No. 3.

Petek, L.

1965 "Derevyannyy vol". O pervykh tekhnicheskikh izobreteniyakh. *Kurer,* may.

Petrushevskiy, I. P.

1960 *Zemledelie i agrarnye otnosheniya v Irane XIII-XIV vv.* Moskva - Leningrad.

Pokrovskiy, M. V.

1937 Gorodishche i mogilnik Srednego Prikubanya. *Trudy Krasnodarskogo Gosudarstvennogo Pedagogicheskogo Instituta,* Tom VI, Vyp. I.

Polynov, B. B.

1956 *Uchenie o landshaftakh. Izbrannye trudy.* Moskva.

Popov, A. N.

1853 Snoshenie Rossii s Khivoy i Bukharoy pri Petre Velikom. *Zapiski Russkogo Geograficheskogo Obshchestva,* Kn. IX. Sankt-Peterburg.

Popov, I. P.

1925 Problemy navodneniy v Sredney Azii i Severnoy Amerike. *Vestnik irrigatsii*, No. 2.

Postovskaya, N. M.

1947 Nachalnaya stadiya razvitiya gosudarstvennogo apparata v drevnem Egipte. *Vestnik Drevney Istorii*, No. 1.

1952 Skorpion i ego vremya. *Vestnik Drevney Istorii*, No. 1.

Ptitsyn, G. V.

1947 K voprosu o geografii "Shakh-Name". *Trudy Otdela Vostoka*, Tom IV. Leningrad.

Pyankov, I. V.

1963 Doklad "Voprosu o sfere vliyaniya doakhemenidskogo Khorezma". *Vestnik Drevney Istorii*, No. 1.

1964 K voprosu o marshrute pokhoda Kira II na massagetov. *Vestnik Drevney Istorii*, No. 3.

Radlov, V. V.

1893 *Opyt slovarya tyurkskikh narechiy*, Tom I. Sankt-Peterburg.

1905 *Opyt slovarya tyurkskikh narechiy*, Tom III. Sankt-Peterburg.

Rakhmanova, R. M.

1963 Khozyaystvo drevney Sredney Azii po antichnym istochnikam. *Nauchnye Trudy Tashkentskogo Gosudarstvennogo Universiteta*, Vyp. 200.

1964a Srednyaya Aziya V-IV vv. do n. e. i pokhod Aleksandra Makedonskogo. Leningrad. (avtoref. kand. diss.)

1964b *K voprosu o sotsialnom stroe Sredney Azii v IV v. do n. e.* Sbornik Nauchnykh Rabot Aspirantov Tashkentskogo Gosudarstvennogo Universiteta.

Ranov, V. A.

1960 Raskopki paleoliticheskoy peshchery v Afganistane. Izvestiya AN Tadjskoy SSR. *Otdelenie obshchestvennykh nauk*, No. 1 (22).

1961 Issledovanie pamyatnikov kamennogo veka Vostochnogo Pamira v 1958 g. *Trudy Instituta Istorii AN Tadjskoy SSR*, Tom XXVII.

Rapoport, Yu. A.

1958 Raskopki gorodishcha Shakh Senem v 1952. *Trudy Khorezmskoy arkheologo-etnograficheskoy ekspeditsii. Tom II, Arkheologicheskie i Etnograficheskie raboty Khorezmskoy Ekspeditsii 1949-1953*, S. P. Tolstov i T. A. Jdanko. Moskva.

1962 Khorezmiyskie astodany (K istorii religii Khorezma). *Sovetskaya Etnografiya*, No. 4.

Rapoport, Yu. A., and M. S. Lapirov-Skoblo

1963 Raskopki dvortsovogo zdaniya na gorodishche Kalalygyr v 1958. *Materialy Khorezmskoy Ekspeditsii*. Vyp. 6, *Polevye Issledovaniya Khoresmoskoy Ekspeditsii v 1958-1961 gg*, S. P. Tolstov i A. V. Vinogradov. Moskva.

Reder, D. G.

1960 *Ekonomicheskoe razvitie Nijnego Egipta (Delty) v arkhaicheskiy period (V-IV tysyacheletiya do n. e.). Drevniy Egipet.* Moskva

1962 *Poyavlenie motygi i pluga v drevnem Egipte i Shumere.* Moskva.

1965 *Mify i legendy Drevnego Dvurechya.* Moskva.

Reyzer, P. Ya.

1959 *Aerometody i ikh primenenie. Bibliograficheskiy ukazatel 1836-1955.* Moskva - Leningrad. (vvodnaya istoriogr. Glava)

1963 *Razvitie aerometodov v Rossii i Sovetskom Soyuze.* Moskva.

Rodin, L. E.

1961 *Dinamika rastitelnosti pustyn na primere Turkmenii.* Moskva - Leningrad.

1964 Pastbishcha i geobotanicheskoe rayonirovanie Siriyskoy Arabskoy Respubliki. *Geobotanika*, XVI. Moskva - Leningrad.

Rogov, M. M.

1957   Gidrologiya delty Amu-Dari. *Geogr.-gidrol.* Kharakteristika. Leningrad.

Roslyakov, A. A.

1951   Pervye seldjukidy. *Izvestiya Turkmenskogo filiala AN SSSR*, No. 3.

Rubinshteyn, N. L. [translated from the original]

1938   *Angliyskie puteshectvenniki v Moskovskom gosudarstve v XVI veke.* Moskva, Sozekgiz. Translated from English by
         Yu. V. Gote.

Rybakov, B. A.

1957   *O korpuse arkheologicheskikh istochnikov SSSR.* Moskva. (Tezisy doklada na plenume Instituta Istorii
         Materialnoy Kultury v marte 1957 g.)

1965   Kosmogoniya i mifologiya zemledeltsev eneolita. *Sovetskaya Arkheologiya*, No. 1.

Rychkov, P. I.

1887   *Topografiya Orenburgskoy gubernii. 1762 god. Izdano na sredstva F. I. Bazilevskogo / P. I. Rychkov - Orenburg;*
         Orenburgskiy otd. Imperatorskogo Russkogo georg. O-va.

Ryjkov, P. A.

1957   Noveyshie i sovremennye tektonicheskie dvijeniya v Fergane. *Trudy Komissii po Izucheniyu Chetvertichnogo
         Perioda*, XIII.

Salnikov, K. V.

1951a   Andronovskoe poselenie Zauralya. *Sovetskaya Arkheologiya*, XX.

1951b   Bronzovyy vek Yujnogo Zauralya. *Materialy i Issledovaniya po Arkheologii SSSR*, No. 21. Moskva.

1965   Istoriya Yujnogo Urala v epokhu bronzy. *Doklad po opublikovannym rabotam, predstavlennym na soiskanie
         uchenoy stepeni doktora istoricheskikh nauk.* Moskva.

Samoylov, I. V.

1952   *Ustya rek.* Moskva.

Sapojnikova, S. A.

1951   Nekotorye osobennosti klimata oazisov v usloviyakh Sredney Azii. *Izvestiya Russkogo Geograficheskogo
         Obshchestva*, Tom 83, Vyp. 3.

Sarianidi, V. I.

1961   Eneoliticheskoe poselenie Geoksyur. Trudy Yujno-Turkenistanskoy ekspeditsii AN SSSR. Tom X,
         Pamyatniki kultury kamennogo i bronzovogo veka Yujnogo Turkmenistana II, M. E. Masson. Ashkhabad.

1964   Khapuz-depe kak pamyatnik epokhi bronzy. *Kratkie Soobshcheniya o Dokladakh i Polevykh Issledovaniyakh
         Instituta Arkheologii AN SSSR*, Vyp. 98.

1965   Pamyatniki pozdnego eneolita Yugo-Vostochnoy Turkmenii. *Arkheologiya CCCP: Svod Arkheologicheskikh
         Istochnikov, BZ-8.* Moskva.

Saushkin, Yu. G.

1946   Kulturnyy landshaft. *Voprosy Geografii.* Sbornik, Sb. 1.

1947   *Geograficheskie ocherki prirody i selskokhozyaystvennoy deyatelnosti naseleniya v razlichnykh rayonakh Sovetskogo
         Soyuza.* Moskva.

Saveleva, T. N.

1962   *Agrarnyy stroy Egipta v period Drevnego tsarstva.* Moskva.

Semenov, S. A.

1957   *Pervobytnaya tekhnika. Materialy i Issledovaniya po Arkheologii SSSR*, No. 51. Moskva - Leningrad.

1968   *Razvitie tekhniki v kamennom veke.* Leningrad.

Semenov, Yu. I.

1957   *K voprosu o pervoy forme klassovogo obshchestva.* Krasnoyarsk.

1965   Problemy sotsialno-ekonomicheskogo stroya drevnego Vostoka. *Narody Azii i Afriki*, No. 4.

Senigova, T. N.

1953   *Keramika gorodishcha Altyn-Asar (opyt khronologicheskoy periodizatsii).* Moskva. (avtoref. kand. diss.)

Serebryannyy, L. R.

1968   Primenenie radiouglerodnogo metoda v chetvertichnoy geologii. *Proceedings of the VII Congress International Union for Quaternary Research (Boulder-Denver, Colorado, 1965). Vol. 14, Glaciation of the Alps*, G. M. Richmond. University of Colorado Studies. Series in Earth Sciences, No. 7. University of Colorado Press, Boulder.

Severtsov, N. A.

1873 (1947) *Puteshestviya po Turkestanskomu krayu.* Sankt-Peterburg - Moskva.

Shcherbak, A. M.

1959   Znaki na keramike i kirpichakh iz Sarkela - Beloy Veji (K voprosu o yazyke i pismennosti pechenegov). *Materialy i Issledovaniya po Arkheologii SSSR*, No. 75.

Shishkin, V. A.

1940   *Arkheologicheskie raboty 1937 g. v zapadnoy chasti Bukharskogo oazisa.* Tashkent.

1956   Nekotorye itogi arkheologicheskikh rabot na gorodishche Varakhsha (1947-1953). *Trudy Instituta Istorii i Arkheologii AN Uzbekskoy SSR*, Vyp. VIII. Tashkent.

1957   Aerorazvedka i aerosemka v arkheologii Uzbekistana. *Izvestiya AN UzbSSR, ser. obshchestv, nauk*, No. 3.

1963   *Varakhsha.* Moskva.

Shkapskiy, O.

1900   *Amu-Darinskie ocherki.* Tashkent.

Shnitnikov, A. V.

1949   Obshchie cherty tsiklicheskikh kolebaniy urovnya ozer i uvlajnennosti territorii Evrazii v svyazi s solnechnoy aktivnostyu. *Byulleten Kom. po issledovaniyu Solntsa*, No. 3-4 (17-18).

1957   Izmenchivost obshchey uvlajnennosti materikov severnogo polushariya. *Zapiski GO*, Tom 16, Novaya Seriya, Moskva - Leningrad.

1961   Dinamika vodnykh resursov Arala. *Ozera Nijnego Povoljya i Aralo-Kaspiyskoy nizmennosti.* Moskva - Leningrad.

Sholpo, N. A.

1941   Irrigatsiya v Drevnem Egipte. *Uchenye zapiski Leningradskogo gosudarstvennogo universiteta*, No. 78, Ser. Istor. Nauk, Vyp. 9.

Shramko, B. A.

1961   K voprosu o tekhnike zemledeliya. *Sovetskaya Arkheologiya*, No. 1.

Shternberg, L. Ya.

1936   *Pervobytnaya religiya v svete etnografii.* Leningrad.

Shults, V. L.

1945   Nekotorye rezultaty ekspeditsii Energeticheskogo instituta AN UzbSSR v deltu Amu-Dari. *Byulleten AN Uzbekskoy SSR*, No. 9-10.

Shuvalov, S. A.

1950   Pochvennyy pokrov Kara-Kalpakskoy ASSR i ego izuchennost. *Materialy po proizvoditelnym silam Uzbekistana*, Tom I. Tashkent.

Sinitsyn, V. M.

1949   Tektonicheskiy faktor v izmenenii klimata Tsentralnoy Azii. *Byulleten Moskovskogo Ob-va estestvoispytateley prirody*, Otd. geol., Tom XXIX.

Sinskaya, E. N.

1955   Proiskhojdenie pshenitsy. *Problemy botaniki*, Vyp. 2. Moskva - Leningrad.

1966   Uchenie N. I. Vavilova ob istoriko-geograficheskikh ochagakh razvitiya kulturnoy flory. *Voprosy geografii kulturnykh rasteniy i N.V. Vavilov.* Moskva - Leningrad.

Skvortsov, Yu. A.

1959   K voprosu ob izuchenii geomorfologii i chetvertichnykh otlojeniy. *Materialy po proizv. silam Uzbekistana,* Vyp. 10.

Smirnov, E. A.

1933   Ochistka irrigatsionnoy seti v Sredney Azii i ee ratsionalizatsiya. *Trudy Sredneaziatskogo nauchno-issledovannogo instituta irrigatsii,* Vyp. XIV.

Smirnova, O. I.

1950   Arkheologicheskie razvedki v basseyne Zeravshana. *Materialy i Issledovaniya po Arkheologii SSSR,* No. 15. Moskva - Leningrad.

1961   Karta verkhovev Zeravshana pervoy chetverti VIII veka. *Strany i narody Vostoka,* Vyp. II.

Snesarev, G. P.

1960   Obryad jertvoprinosheniya vode u uzbekov Khorezma, geneticheski svyazannyy s drevnim kultom plodorodiya. *Materialy Khorezmskoy Ekspeditsii.* Vyp. 4, *Polevye issledovaniya Khorezmskoy ekspeditsii 1957 g,* S. P. Tolstov i M. A. Itina. Moskva.

Sokolov, N. A.

1952   *Aerofotosemka gorodov.* Moskva.

Sokolov, S. N.

1961   *Avestiyskiy yazyk.* Moskva.

Solntsev, N. A.

1948   Prirodnyy geograficheskiy landshaft i nekotorye obshchie ego zakonomernosti. *Trudy Vtorogo Geograficheskogo Sezda,* Tom I. Moskva.

Sorokin, V. S.

1962   Jilishche poseleniya Tasty-Butak. *Kratkie Soobshcheniya o Dokladakh i Polevykh Issledovaniyakh Instituta Arkheologii AN SSSR,* Vyp. 91.

Sousa, Ahmed

1948   *Raiy Samarra' fi' akhd al-khilafa al-'abbasiyya,* Chast 1-2. Bagdad (na arab. yaz.).

Sprishevskiy, V. I.

1954   Chustskaya stoyanka epokhi bronzy. *Sovetskaya Etnografiya,* No. 3.

1957   Chustskoe poselenie epokhi bronzy (Raskopki 1955 g.). *Kratkie Soobshcheniya o Dokladakh i Polevykh Issledovaniyakh Instituta Istorii Materialnoy Kultury AN SSSR,* Vyp. 71.

Staviskiy, B. Ya.

1961a   Osnovnye etapy osvoeniya zemledelcheskim naseleniem gornykh rayonov verkhnego Zeravshana (Kukhistana). *Materialy etnograficheskogo otdeleniya VGO,* Vyp. 1. Leningrad.

1961b   O severnykh granitsakh Kushanskogo gosudarstva. *Vestnik Drevney Istorii,* No. 1.

Stetkevich

1889   *Materialy dlya statisticheskogo opisaniya Khorezmskogo oazisa.* Tashkent.

Struve, V. V.

1934   Problema zarojdeniya, razvitiya i razlojeniya rabovladelcheskikh obshchestv Drevnego Vostoka. *Izvestiya Gosudarstvennoy Akademii Istorii Materialnoy Kultury,* Vyp. 77.

1946   Pokhod Dariya I na sakov-massagetov. *Izvestiya AN SSSR, Ser. istor. i filos.,* Tom III, No. 3.

1948   Rodina zoroastrizma. *Sovetskoe Vostokovedenie,* No. 5.

1949   Retsenziya na kn. S. P. Tolstov "Drevniy Khorezm". *Vestnik Drevney Istorii,* No. 4.

1965a Nekotorye aspekty sotsialnogo razvitiya Drevnego Vostoka. *Voprosy Istorii*, No. 5.

1965b Ponyatie "aziatskiy sposob proizvodstva". *Narody Azii i Afriki*, No. 1.

Stuchevskiy, I. A.

1966   *Zavisimoe naselenie Drevnego Egipta*. Moskva.

Suvorov, N. I.

1955   Zaleji drevnego orosheniya u razvalin Asanasa v doline Syr-Dari. *Uchenye zapiski Almaatinskogo Pedagogicheskogo instituta*, Tom 6.

Talitskaya, I. A.

1952   Materialy k arkheologicheskoy karte basseyna r. Kamy. *Materialy i Issledovaniya po Arkheologii SSSR*, No. 27.

1953   Materialy k arkheologicheskoy karte Nijnego i Srednego Priobya. *Materialy i Issledovaniya po Arkheologii SSSR*, No. 35.

Ter-Akopyan, N. B.

1965   Razvitie vzglyadov K. Marksa i F. Engelsa na aziatskiy sposob proizvodstva i zemledelcheskuyu obshchinu. *Narody Azii i Afriki*, No. 2-3.

Terenojkin, A. I.

1940a Arkheologicheskie razvedki v Khorezme. *Sovetskaya Arkheologiya*, VI.

1940b O drevnem goncharstve v Khorezme. *Izvestiya Uzbeskogo Filiala AN*, No. 6.

1947   Arkheologicheskaya razvedka na gorodishche Afrasiab v 1945 g. *Kratkie Soobshcheniya o Dokladakh i Polevykh Issledovaniyakh Instituta Istorii Materialnoy Kultury AN SSSR*, Vyp. XVII.

1950   Sogd i Chach. *Kratkie Soobshcheniya o Dokladakh i Polevykh Issledovaniyakh Instituta Istorii Materialnoy Kultury AN SSSR*, Vyp. XXXIII.

1965   Osnovy khronologii predskifskogo perioda. *Sovetskaya Arkheologiya*, No. 1.

Titov, V. S.

1962   Pervoe obshchestvennoe razdelenie truda. Drevneyshie zemledelcheskie i skotovodcheskie plemena. *Kratkie Soobshcheniya o Dokladakh i Polevykh Issledovaniyakh Instituta Arkheologii AN SSSR*, Vyp. 88.

1965a Period neolita v Gretsii. *Novoe v sovetskoy arkheologii*. Moskva.

1965b Rol radiouglerodnykh dat v sisteme khronologii neolita i bronzovogo veka Peredney Azii i Yugo-Vostochnoy Evropy. *Arkheologiya i estestvennye nauki*.

Tokarev, S. A., and S. P. Tolstov

1956   *Narody Avstralii i Okeanii*. Moskva.

Tolstov, S. P.

1932   Ocherki pervonachalnogo islama. *Sovetskaya Etnografiya*, No. 2.

1935a Voennaya demokratiya i problemy "geneticheskoy revolyutsii". *Problemy Istorii Dokapitalisticheskikh Obshchestv*, No. 7-8.

1935b Perejitki totemizma i dualnoy organizatsii u turkmen. *Problemy Istorii Dokapitalisticheskikh Obshchestv*, No. 9-10.

1938a Osnovnye voprosy drevney istorii Sredney Azii. *Vestnik Drevney Istorii*, No. 1.

1938b Tiraniya Abruya. *Istoricheskie zapiski*, Tom III.

1939   Drevnekhorezmiyskie pamyatniki Karakalpakii. *Vestnik Drevney Istorii*, No. 3(8).

1940a Khorezmskaya ekspeditsiya 1939 g. (predvaritelnyy otchet). *Kratkie Soobshcheniya o Dokladakh i Polevykh Issledovaniyakh Instituta Istorii Materialnoy Kultury AN SSSR*, Vyp. VI.

1940b Cherty obshchestvennogo stroya Vostochnogo Irana i Sredney Azii po Aveste. *Istoriya SSSR s drevneyshikh vremen do obrazovaniya drevnerusskogo gosudarstva (maket)*, Chast I-II. Moskva - Leningrad.

1941   Drevnosti Verkhnego Khorezma (Osnovnye itogi rabot Khorezmskoy ekspeditsii IIMK 1939 g.). *Vestnik Drevney Istorii*, No. 1.

1945a Osnovnye itogi i ocherednye zadachi izucheniya istorii i arkheologii Karakalpakii i karakalpakov. *Byulleten AN Uzbekskoy SSR*, No. 9-10.

1945b K istorii khorezmiyskikh Siyavushidov. *Izvestiya AN SSR*, 1945, No. 4.

1946a Drevniy Khorezm (Tezisy doktorskoy dissertatsii s prilojeniem tablits klassifikatsii pamyatnikov drevnego Khorezma). *Kratkie Soobshcheniya o Dokladakh i Polevykh Issledovaniyakh Instituta Istorii Materialnoy Kultury AN SSSR*, Vyp. XIII.

1946b Khorezmskaya arkheologo-etnograficheskaya ekspeditsiya AN SSSR 1945 g. *Izvestiya AN SSSR*, Ser. istor. i filos., No. 1.

1946c Novye materialy po istorii kultury Drevnego Khorezma. *Vestnik Drevney Istorii*, No. I.

1947a Drevnekhorezmiyskaya tsivilizatsiya v svete noveyshikh arkheologicheskikh otkrytiy (1937-1945 gg.). *Obshchee Sobranie AN SSSR 1-4 iyulya 1946. Moskva – Leningrad.*

1947b Gorod guzov (istoriko-etnograficheskie etyudy). *Sovetskaya Etnografiya*, No. 3.

1947c Khorezmskaya arkheologo-etnograficheskaya ekspeditsiya AN SSSR v 1946 g. *Izvestiya AN SSSR*, Ser. istor. i filos., Tom IV, No. 2.

1947d K voprosu o proiskhojdenii kara-kalpakskogo naroda. *Kratkie Soobshcheniya Instituta Etnografii AN SSSR*, Vyp. II.

1948a Drevniy Khorezm. *Opyt istoriko-arkheologicheskogo issledovaniya*. Moskva.

1948b *Po sledam drevnekhorezmiyskoy tsivilizatsii*. Moskva - Leningrad.

1949   Khorezmskaya arkheologo-etnograficheskaya ekspeditsiya AN SSSR 1948 g. *Izvestiya AN SSSR*, Ser. istor. i filos., Tom VI, No. 3.

1950a Khorezmskaya arkheologo-etnograficheskaya ekspeditsiya AN SSSR v 1949 g. *Izvenstiya AN SSSR*, Ser. istor. i filos., No. 6.

1950b Oguzy, pechenegi, more Daukara (Zametki po istoricheskoy etnonimike vostochnogo Priaralya). *Sovetskaya Etnografiya*, No. 4.

1952a Khorezmskaya arkheologo-etnograficheskaya ekspeditsiya AN SSSR (1945-1948 gg.). *Trudy Khorezmskoy arkheologo-etnograficheskoy ekspeditsii. Tom I, Arkheologicheskie i Etnograficheskie Raboty Khorezmskoy Ekspeditsii 1945-1948*, S. P. Tolstov i T. A. Jdanko.. Moskva.

1952b Arkheologicheskie razvedki na trasse Glavnogo Turkmenskogo kanala. *Kratkie Soobshcheniya Instituta Etnografii AN SSSR*, Vyp. XIV.

1953a Arkheologicheskie raboty Khorezmskoy ekspeditsii AN SSSR v 1952 g. *Vestnik Drevney Istorii*, No. 2.

1953b Khorezmskaya arkheologo-etnograficheskaya ekspeditsiya AN SSSR v 1950 g. *Sovetskaya Arkheologiya*, XVIII.

1954   Arkheologicheskie raboty Khorezmskoy arkheologo-etnograficheskoy ekspeditsii AN SSSR v 1951. *Sovetskaya Arkheologiya*, XIX.

1955a *Raboty Khorezmskoy ekspeditsii AN SSSR v 1954. Sovetskoe Vostokovedenie.*

1955b Rabota Khorezmskoy arkheologo-etnograficheskoy ekspeditsii 1951-1954 godov. *Voprosy Istorii*, No. 3.

1955c Itogi rabot Khorezmskoy arkheologo-etnograficheskoy ekspeditsii AN SSSR v 1953 g. *Vestnik Drevney Istorii*, No. 3:192-205.

1957a Itogi dvadtsati let raboty Khorezmskoy arkheologo-etnograficheskoy ekspeditsii (1937-1956 gg.). *Sovetskaya Etnografiya*, No. 4.

1957b Biruni i ego "Pamyatniki minuvshikh pokoleniy". *Abureykhan Biruni (973-1048), Izbrannye proizvedeniya*, I.

1958   Raboty Khorezmskoy arkheologo-etnograficheskoy ekspeditsii AN SSSR v 1949-1953 gg. *Trudy Khorezmskoy arkheologo-etnograficheskoy ekspeditsii. Tom II, Arkheologicheskie i Etnograficheskie raboty Khorezmskoy Ekspeditsii 1949-1953*, S. P. Tolstov i T. A. Jdanko. Moskva.

1959a Rabota Khorezmskoy arkheologo-etnograficheskoy ekspeditsii v 1954-1956. *Materialy Khorezmskoy Ekspeditsii. Vyp. 1, Polevye issledovaniya Khorezmskoy ekspeditsii v 1954-1956 gg.*, S. P. Tolstov i M. G. Vorobeva. Moskva.

1959b Varvarskie plemena periferii antichnogo Khorezma po noveyshim arkheologicheskim dannym. Materialy Vtorogo Soveshchaniya Arkheologov i Etnografov Sredney Azii. Moskva - Leningrad.

1960   Osnovnye teoreticheskie problemy sovremennoy sovetskoy etnografii. *Sovetskaya Etnografiya*, No. 6.

1961a Priaralskie skify i Khorezm (k istorii zaseleniya i osvoeniya drevney delty Syr-Dari). *Sovetskaya Etnografiya*, No. 4.

1961b Drevnyaya irrigatsionnaya set i perspektivy sovremennogo orosheniya (Po issledovaniyu drevney delty Syr-Dari). *Vestnik AN SSSR*, No. 11.

1961c O zemlyakh drevnego orosheniya v nizovyakh Amu-Dari i Syr-Dari i vozmojnosti ikh osvoeniya v sovremennykh usloviyakh. *Obshchestvennye nauki v Uzbekistane*, No. 81.

1961d Datirovannye dokumenty iz dvortsa Toprak-kaly i problema "Ery Shaka" i "Ery Kanishki". *Problemy Vostokovedeniya*, No. 1.

1962a *Po drevnim deltam Oksa i Yaksarta.* Moskva.

1962b Rezultaty istoriko-arkheologicheskikh issledovaniy 1961 t. na drevnikh ruslakh Syr-Dari (v svyazi s problemoy ikh osvoeniya). *Sovetskaya Arkheologiya*, No. 4.

Tolstov, S. P., and B. V. Andrianov

1957   Novye materialy po istorii razvitiya irrigatsii Khorezma. *Kratkie soobshcheniya Instituta etnografii AN SSSR*, Vyp. XXVI. Moskva.

1969   Zemli drevnego orosheniya v Priarale. *Gidrotekhnika i melioratsiya*, No. 1.

Tolstov, S. P., and M. A. Itina

1960a *Polevye Issledovaniya Khorezmskoy Ekspeditsii v 1957 godu. Materialy Khorezmskoy Ekspeditsii*, Vyp. 4, S. P. Tolstov i M. A. Itina. Moskva.

1960b Problema suyarganskoy kultury. *Sovetskaya Arkheologiya*, No. 1.

1964   Mogilnik Tagisken (V svyazi s voprosom o proiskhojdenii skifskoy kultury). *Tezisy doklada na zased., posvyashch. itogam polevykh issled. 1963 g.* Moskva.

1966   Saki Nizovev Syr-Dari. *Sovetskaya arkheologiya*, No. 2.

Tolstov, S. P., and A. S. Kes

1956   Istoriya pervobytnykh poseleniy na protokakh drevnikh delt Amu-Dari i Syr-Dari. *Sbornik statey dlya XVIII Mejdunarodnogo geograficheskogo kongressa.* Moskva - Leningrad.

1960   Nizovya Amu-Dari, Sarykamysh, Uzboy: Istoriya formirovaniya i zaseleniya. *Materialy Khorezmskoy Ekspeditsii*, Vyp. 3, S. P. Tolstov i A. S. Kes. Moskva.

Tolstov, S. P., and M. A. Orlov

1948   Opyt primeneniya aviatsii v arkheologicheskikh rabotakh Khorezmskoy ekspeditsii. *Vestnik AN SSSR*, No. 6.

Tolstov, S. P., and V. A. Shishkin

1942   *Arkheologiya. 25 let sovetskoy nauki v Uzbekistane.* Tashkent.

Tolstov, S. P. and B. I. Vaynberg

1967   *Trudy Khorezmskoy arkheologo-etnograficheskoy ekspeditsii. Tom V, Koy-Krylgan-Kala: pamyatnik kultury Drevnego Khorezma IV v. do n.e.* S. P. Tolstov i B. I. Vaynberg. Moskva.

Tolstov, S. P., B. V. Andrianov, and N. I. Igonin

1962   Ispolzovanie aerometodov v arkheologicheskikh issledovaniyakh. *Sovetskaya Arkheologiya*, No. 1.

Tolstov, S. P., M. A. Itina, and A. V. Vinogradov

1967   Khorezmskaya arkheologo-etnograficheskaya ekspeditsiya. *Arkheologicheskie Otkrytiya 1966 goda.* Moskva.

Tolstov, S. P., T. A. Jdanko, and M. A. Itina

1963   Raboty Khorezmskoy arkheologo-etnograficheskoy ekspeditsii AN SSSR v 1958-1961 gg. *Materialy Khorezmskoy Ekspeditsii. Vyp. 6, Polevye Issledovaniya Khoresmoskoy Ekspeditsii v 1958-1961 gg*, S. P. Tolstov i A. V. Vinogradov. Moskva.

Tolstov, S. P., A. S. Kes, and T. A. Jdanko

1954   Istoriya Sarykamyshskogo ozera v srednie veka. *Izvestiya AN SSSR, Ser. geogr.*, No. 1.

1955   Istoriya srednevekovogo Sarykamyshskogo ozera. *Voprosy geomorfologii i paleogeografii Azii.* Moskva.

Tolstov, S. P., M. G. Vorobeva and Yu. A. Rapoport

1960   Raboty Khorezmskoy arkheologo-etnograficheskoy ekspeditsii v 1957 g. *Materialy Khorezmskoy Ekspeditsii. Vyp. 4, Polevye issledovaniya Khorezmskoy ekspeditsii 1957 g*, S. P. Tolstov i M. A. Itina. Moskva.

Tolstova, L. S.

1959   *Karakalpaki Ferganskoy doliny.* Nukus.

Torn, D. and Peterson, Kh.

1952   *Oroshaemye zemli.* Moskva.

Trever, K. V.

1940   Gopatshakh pastukh-tsar. *Trudy Otdela Istorii Kultury i Iskusstva Vostoka Gosudarstvennogo Ermitaja*, Tom II.

Trofimova, T. A.

1964   Naselenie Sredney Azii v epokhu eneolita i bronzovogo veka i ego svyazi s Indiey. *Trudy Moskovskogo obshchestva ispytateley prirody*, Tom XIV.

Trudnovskaya, S. A.

1958   Steklo s gorodishcha Shakh-Senem. *Trudy Khorezmskoy arkheologo-etnograficheskoy ekspeditsii. Tom II, Arkheologicheskie i Etnograficheskie raboty Khorezmskoy Ekspeditsii 1949-1953*, S. P. Tolstov i T. A. Jdanko. Moskva.

Tsalkin, V. I.

1966   *Drevnee jivotnovodstvo plemen Vostochnoy Evropy i Sredney Azii.* Moskva.

Tsinzerling, V.

1927   Oroshenie na Amu-Dare. Obshchie osnovaniya orositelnogo stroitelstva. Plan vodnogo khozyaystva. *Pervoocherednye raboty.* Moskva.

Tyumenev, A. I.

1956   *Gosudarstvennoe khozyaystvo Drevnego Shumera.* Moskva - Leningrad.

Vakturskaya, N. N.

1952   O raskopkakh 1948 g. na srednevekovom gorode Shemakha-kala Turkmenskoy SSR. *Trudy Khorezmskoy arkheologo-etnograficheskoy ekspeditsii. Tom I, Arkheologicheskie i Etnograficheskie Raboty Khorezmskoy Ekspeditsii 1945-1948*, S. P. Tolstov i T. A. Jdanko. Moskva.

1959   Khronologicheskaya klassifikatsiya srednevekovoy keramiki Khorezma (IX-XVII vv.). *Trudy Khorezmskoy Arkheologo-Etnograficheskoy Ekspeditsii. Tom IV, Keramika Khorezma*, S. P. Tolstov i M. G. Vorobeva. Moskva.

1963   O srednevekovykh gorodakh Khorezma. *Materialy Khorezmskoy Ekspeditsii. Vyp. 7, Polevye issledovaniya Khorezmskoy ekspeditsii v 1958 – 1961 gg.*, S. P. Tolstov i E. E. Nerazik. Moskva.

Vakturskaya, N. N., and O. A. Vishnevskaya

1959   Pamyatniki Khorezma epokhi Velikikh Khorezmshakhov (XII- nachalo XIII v.). *Materialy Khorezmskoy Ekspeditsii. Vyp. 1, Polevye issledovaniya Khorezmskoy ekspeditsii v 1954-1956 gg.*, S. P. Tolstov i M. G. Vorobeva. Moskva.

Vasilev, L. S.

1961   *Agrarnye otnosheniya i obshchina v drevnem Kitae.* Moskva.

1962   Zemledelie v Drevnem Kitae. *Vestnik istorii mirovoy kultury, mart-aprel,* No. 2.

1965   Obshchee i osobennoe v istoricheskom razvitii stran Vostoka. *Narody Azii i Afriki,* No. 6.

Vasileva, G. P.

1954   Turkmeny-nokhurli. *Sredneaziatskiy etnograficheskiy sbornik.* Moskva.

Vavilov, N. I.

1962   *Pyat kontinentov.* Moskva.

1965   *Izbrannye trudy,* Tom V. Moskva – Leningrad.

1967   *Izbrannye proizvedeniya v dvukh tomakh,* I–II. Leningrad.

Vavilov, N. I., and D. D. Bukinich

1929   *Zemledelcheskiy Afganistan.* Leningrad.

Vayman, A. A.

1961a   Dva klinopisnykh dokumenta o provedenii orositelnogo kanala. *Trudy Gosudarstvennogo Ermitaja,* Tom V.

1961b   *Shumero-vavilonskaya matematika.* Moskva.

1962   Ob izuchenii shumero-vavilonskoy prikladnoy matematiki. *Trudy XXV Mejdunarodnogo kongressa vostokovedov,* Tom I. Moskva.

1963   Issledovanie shumerskikh piktograficheskikh tekstov. *Tezisy dokladov nauchnoy sessii GE v 1962 g.*

Vaynberg, B. I.

1959   K istorii turkmenskikh poseleniy XIX v. v Khorezme. *Sovetskaya Etnografiya,* No. 5.

1960   Turkmenskie poseleniya po Daryalyku. *Materialy Khorezmskoy Ekspeditsii. Vyp. 4, Polevye issledovaniya Khorezmskoy ekspeditsii 1957 g,* S. P. Tolstov i M. A. Itina. Moskva.

1961   Pozdnie turkmenskie poseleniya i jilishcha "zemel drevnego orosheniya" Levoberejnogo Khorezma. Moskva. (avtoref. kand. diss.)

Veber, M.

1925   Agrarnaya istoriya drevnego Mira. Moskva.

Veselovskiy, N. I.

1877   *Ocherk istoriko-geograficheskikh svedeniy o Khivinskom khanstve ot drevneyshikh vremen do nastoyashchego.* Sankt-Peterburg.

Veselovskiy, N. N.

1945   *Fotogrammetriya.* Moskva.

Viktorov, S. V.

1955   *Ispolzovanie geobotanicheskogo metoda pri geologicheskikh i gidrologicheskikh issledovaniyakh.* Moskva.

Vilyams, V. R.

1951   *Sobranie sochineniy,* Tom 6. Moskva.

Vinogradov, A. V.

1957a   K voprosu o yujnykh svyazyakh kelteminarskoy kultury. *Sovetskaya Etnografiya,* No. 1.

1957b   Kelteminarskaya kultura. Moskva (avtoref. kand. diss.)

1958   Rannekelteminarskaya stoyanka Kunyak I. *Kratkie Soobshcheniya Instituta Etnografii AN SSSR,* Vyp. 30.

1959   Arkheologicheskaja razvedka v rayone Aralska-Saksaulskoy v 1955 g. *Trudy Instituta Istorii Arkheologii Etnografii AN Kazakskoy SSR,* Tom 7. Alma-ta.

1960   Novye neoliticheskie nakhodki Khorezmskoy ekspeditsii AN SSSR 1957 g. *Materialy Khorezmskoy Ekspeditsii. Vyp. 4, Polevye issledovaniya Khorezmskoy ekspeditsii 1957 g,* S. P. Tolstov i M. A. Itina. Moskva.

1963   Novye materialy dlya izucheniya kelteminarskoy kultury. *Materialy Khorezmskoy Ekspeditsii. Vyp. 6, Polevye Issledovaniya Khoresmoskoy Ekspeditsii v 1958-1961 gg,* S. P. Tolstov i A. V. Vinogradov. Moskva.

1968a Neoliticheskie pamyatniki Khorezma. *Materialy Khorezmskoy Ekspeditsii, Vyp. 8*, S. P. Tolstov i A. V. Vinogradov, Moskva.

1968b Neolit Kyzylkumov. Problemy arkheologii Sredney Azii. *Tezisy dokladov i soobshcheniy k soveshchaniyu po arkheologii Sredney Azii (1-7 aprelya 1968).* Leningrad.

Vinogradov, B. V.

1961 Opyt krupnomasshtabnogo landshaftnogo deshifrirovaniya i kartirovaniya klyuchevykh uchastkov v aridnykh i subaridnykh zonakh Sredney Azii i Kazakhstana. *Primenenie aerometodov v landshaftnykh issledovaniyakh.* Moskva - Leningrad.

1962 Geograficheskie zakonomernosti dalneyshey ekstrapolyatsii priznakov deshifrirovaniya landshaftov-analogov. *Primenenie aerometodov dlya izucheniya gruntovykh vod.* Moskva - Leningrad.

Vishnevskaya, O. A.

1963 Arkheologicheskie razvedki na srednevekovykh poseleniyakh levoberejnogo Khorezma. *Materialy Khorezmskoy Ekspeditsii. Vyp. 7, Polevye issledovaniya Khorezmskoy ekspeditsii v 1958-1961 gg.,* S. P. Tolstov i E. E. Nerazik. Moskva..

Voeykov, A. I.

1915 Retsenziya na st. P. A. Tutkovskogo "O geograficheskikh prichinakh nashestviy varvarov". *Meteorologicheskiy vestnik,* No. 9-10.

1963 *Vozdeystvie cheloveka na prirodu. Izd. 2.* Moskva.

Volin, S.

1941 K istorii drevnego Khorezma. *Vestnik Drevney Istorii,* No. 1.

Vorobeva, M. G.

1955 Keramika Khorezma rabovladelcheskoy epokhi. *Kratkie Soobshcheniya Instituta Etnografii AN SSSR, XXII.*

1958 Keramika s gorodishcha Kyuzeli-gyr. *Trudy Khorezmskoy arkheologo-etnograficheskoy ekspeditsii. Tom II, Arkheologicheskie i Etnograficheskie raboty Khorezmskoy Ekspeditsii 1949-1953,* S. P. Tolstov i T. A. Jdanko. Moskva.

1959a Keramika Khorezma antichnogo perioda. *Trudy Khorezmskoy arkheologo-etnograficheskoy ekspeditsii. Tom IV, Keramika Khorezma,* S. P. Tolstov i M. G. Vorobeva. Moskva.

1959b Raskopki arkhaicheskogo poseleniya bliz Dingildje. *Materialy Khorezmskoy Ekspeditsii. Vyp. 1, Polevye issledovaniya Khorezmskoy ekspeditsii v 1954-1956 gg.,* S. P. Tolstov i M. G. Vorobeva. Moskva.

1961 Opyt kartografirovaniya goncharnykh pechey dlya istoriko-etnograficheskogo atlasa Sredney Azii i Kazakhstana. *Materialy k Istoriko-Etnograficheskomu Atlasu Sredney Azii i Kazakhstan.* Moskva - Leningrad.

Vorobeva, M. G., M. S. Lapirov-Skoblo, and E. E. Nerazik

1963 Arkheologicheskie raboty v Khazaraspe v 1958-1960 gg. *Materialy Khorezmskoy Ekspeditsii. Vyp. 6, Polevye Issledovaniya Khoresmoskoy Ekspeditsii v 1958-1961 gg,* S. P. Tolstov i A. V. Vinogradov. Moskva.

Voronets, M. E.

1951 Arkheologicheskaya rekognostsirovka 1950 goda po Namanganskoy oblasti. *Izvestiya AN Uzbekskoy SSR,* No. 5.

Vulf, E. V.

1932 *Vvedenie v istoricheskuyu geografiyu rasteniy.* Leningrad.

Vyzgo, M. S.

1947 *Vodozabornye uzly.* Tashkent.

Yagodin, V. N.

1963 Poselenie amirabadskoy kultury Kavat 2. *Materialy Khorezmskoy Ekspeditsii. Vyp. 6, Polevye Issledovaniya Khoresmoskoy Ekspeditsii v 1958-1961 gg,* S. P. Tolstov i A. V. Vinogradov. Moskva.

Yakubovskiy, A. Yu.

1929   Razvaliny Sygnaka (Sugnaka). *Soobshcheniya Gosudarstvennoy Akademii Istorii Materialnoy Kultury*, Tom II.

1940   GAIMK-IIMK i arkheologicheskoe izuchenie Sredney Azii za 20 let. *Kratkie Soobshcheniya o Dokladakh i Polevykh Issledovaniyakh Instituta Istorii Materialnoy Kultury AN SSSR*, Vyp. VI.

1947   Voprosy etnogeneza turkmen. *Sovetskaya Etnografiya*, No. 3.

1950   Itogi rabot Sogdiysko-Tadjikskoy arkheologicheskoy ekspeditsii v 1946-1947 gg. *Materialy i Issledovaniya po Arkheologii SSSR*, No. 15. Moskva - Leningrad.

Yakubtsiner, M. M.

1956   K istorii kultury pshenitsy v SSSR. *Materialy po istorii zemledeliya SSSR*, Vyp. II. Moskva - Leningrad.

Yamnov, A. A., and V. N. Kunin

1953   Nekotorye teoreticheskie itogi noveyshikh issledovaniy v rayone Uzboya, v oblasti paleogeografii i geomorfologii. *Izvestiya AN SSSR*, ser. geogr., No. 3.

Zadneprovskiy, Yu. A.

1960   *Arkheologicheskie pamyatniki yujnykh rayonov Oshskoy oblasti.* Frunze.

1962   Drevnezemledelcheskaya kultura Fergany. *Materialy i Issledovaniya po Arkheologii SSSR*, No. 118.

1966   Chustskaya kultura v Ferganskoy doline. *Srednyaya Aziya v Epokhu Kamnya i Bronzy*, V. M. Masson. Moskva - Leningrad.

Zadneprovskiy, Yu. A., and G. N. Kislyakova

1965   O kompleksnom metode izucheniya prirodnykh usloviy golotsena vo vnelednikovykh rayonakh. *Arkheologiya i estestvennye nauki.*

Zadneprovskaya, T. I.

1966   Literatura po pervobytnoy arkheologii Sredney Azii na yazykakh narodov SSSR. *Srednyaya Aziya v Epokhu Kamnya i Bronzy*, V. M. Masson. Moskva - Leningrad.

Zadykhina, K. L.

1952   Uzbeki delty Amu-Dari. *Trudy Khorezmskoy arkheologo-etnograficheskoy ekspeditsii. Tom I, Arkheologicheskie i Etnograficheskie Raboty Khorezmskoy Ekspeditsii 1945-1948*, S. P. Tolstov i T. A. Jdanko. Moskva.

Zakhoder, B. N.

1945   Khorasan i obrazovanie gosudarstva seldjukidov. *Voprosy Istorii*, No. 5-6.

Zelin, K. K.

1960   *Issledovanie po istorii zemelnykh otnosheniy v Egipte II-I vekov do nashey ery.* Moskva.

Zeymal, E. V.

1964   Problema kushanskoy khronologii i monetii. *Tezisi dokladov na yubileynoy nauchnoy sessii. Gosudarstvennyy Ermitaj. 1764-1964 (sektsionnye zasedaniya).* Leningrad.

1965   Kushanskoe tsarstvo po numizmaticheskim dannym. Avtoreferat dissertatsii. Leningrad.

1968   *Kushanskaya khronologiya (Materialy po probleme).* Moskva.

Original references in other languages

Adams, R. McC.

1958   Survey of Ancient Water Courses and Settlement in Central Iraq. *Sumer. A journal of Archaeology and History in Iraq* 14 (1-2): 101-103.

1965   *Land behind Baghdad: A History of Settlement on the Diyala Plains.* University of Chicago Press, Chicago - London.

1966   *The Evolution of Urban Society: Early Mesopotamia and Prehispanic Mexico.* Aldine, Chicago.

Amiran, R.

1966   The Beginning of Pottery-Making in the Near East. In *Ceramics and Man,* edited by Frederick R. Matson, pp. 240-247. Viking Fund Publications in Anthropology No. 41. Methuen, London.

Andersson, J. G.

1923   *An Early Chinese Culture.* Bulletin of the Geological Survey of China No. 5, Peking.

1943   Researches into the Prehistory of the Chinese. *Bulletin of the Museum of Far Eastern Antiquities* No. 15. Stockholm.

Arkell, A. J.

1953   *Shaheinab. An Account of the Excavation of a Neolithic Occupation Site Carried out for the Sudan Antiquities Service in 1949-50.* Oxford University Press, London.

Arkell, A. J. and P. J. Ucko

1965   Review of Predynastic Development in the Nile Valley. *Current Anthropology* 6: 145-166.

Armillas, P.

1962   *The Native Period in the History of the New World.* Instituto Panamericano de Geografia e Historia, Mexico.

Armillas, P., A. Palerm and E. R. Wolf

1956   A Small Irrigation System in the Valley of Teotihuacán. *American Antiquity* 21: 396-399.

Audebeau, C. Bey

1932   Les Irrigations dans le Monde Antique. Causes de Leur Décadence. *Revue Générale des Sciences Pures et Appliquées* 43: 272-282.

Barois, J.

1904   *Les Irrigations en Egypte.* Libraire Polytechnique Ch. Beranger, Paris.

Bartholomae, C.

1905   *Die Gatha's des Awesta: Zarathustra's Verspredigten.* Verlag von Karl J. Trubner, Strasburg.

Barton, G. A.

1913   *The Origin and Development of Babylonian Writing.* Part I, *A Genealogical Table of Babylonian and Assyrian Signs with Indices.* Beiträge zur Assyriologie und Semitischen Sprachwissenschaft 9. J. C. Hinrichs'sche Buchhandlung, Leipzig. 1929 *The Royal Inscriptions of Sumer and Akkad.* American Oriental Society, Library of Ancient Semitic Inscriptions, No. 1. Yale University Press, New Haven.

Baumgartel, E. J.

1947   *The Cultures of Prehistoric Egypt.* Oxford University Press, London.

1955   *The Cultures of Prehistoric Egypt,* Vol. I. Oxford University Press, London.

1960   *The Cultures of Prehistoric Egypt,* Vol. II. Oxford University Press, London.

1965   *Predynastic Egypt.* Cambridge University Press, London.

Beazeley, G. A.

1919   Air Photography in Archæology. *The Geographical Journal* 53: 330-335.

1920   Surveys in Mesopotamia during the War. *The Geographical Journal* 55: 109-123.

Beloch, J.

1886   *Die Bevölkerung der griechisch-römischen Welt.* Duncker & Humblot, Leipzig.

Bennett, W. C.

1946   The Atacameño. In *Handbook of South American Indians.* Vol. I, *The Marginal Tribes,* edited by J. H. Steward, pp. 599-618. Smithsonian Institution, Bureau of American Ethnology, Bullettin No. 143. United States Government Printing Office, Washington.

Benveniste, E.

1934   L'Ērān-vēž et l'Origine Légendaire des Iraniens. *Bulletin of the School of Oriental Studies, University of London* 7 (3): 265-274.

Beresford, M. W.

1954   *The lost villages of England.* Lutterworth Press, London.

Beresford, M. W. and J. K. S. St Joseph

1958   *Medieval England: An Aerial Survey with 117 air photographs.* Cambridge University Press, Cambridge.

Bhandarkar, D. R.

1902a   A Kushan stone-inscription and the question about the origin of the Saka Era. *Journal of the Bombay Branch of the Royal Asiatic Society* 20: 269-302.

1902b   A Peep into the Early History of India from the Foundation of the Maurya Dynasty to the Downfall of the Imperial Gupta Dynasty, BC 322-circa 500 AD. *Journal of the Bombay Branch of the Royal Asiatic Society* 20: 356-408.

Boak, A. E. R.

1926a   Notes on Canal and Dike Work in Roman Egypt. *Aegyptus* 7: 215-219.

1926b   Irrigation and Population in the Faiyum, the Garden of Egypt. *Geographical Review* 16: 353-364.

Borchardt, L.

1906   *Nilmesser und Nilstandsmarken.* Verlag der Konigl. Akademie der Wissenschaften, Berlin.

1934   *Nachtrage zu Nilmesser und Nilstandsmarken.* Akademie der Wissenschaften, Berlin.

Bosworth, C. E. and G. Clauson

1965   Al-Xwārazmī on the Peoples of Central Asia. *The Royal Asiatic Society of Great Britain and Ireland* 97: 2-12.

Bowden, W. M.

1959   The Taxonomy and Nomenclature of the Wheats, Barleys and Ryes and Their Wild Relatives. *Canadian Journal of Botany* 37: 657-684.

Bowen, R. LeB. Jr.

1958   Irrigation In Ancient Qatabān (Beihān). In *Archaeological Discoveries in South Arabia,* edited by R. Bowen LeB. Jr. and F. P. Albright, pp. 43-112. Johns Hopkins Press, Baltimore.

Bowen, R. LeB. Jr. and F. P. Albright

1958   *Archaeological Discoveries in South Arabia.* Johns Hopkins Press, Baltimore.

Bradford, J.

1957   *Ancient Landscapes. Studies in Field Archaeology.* Bell, London.

Braidwood, R. J.

1952   *The Near East and the Foundations for Civilization.* Condon Lectures, Eugene [Oregon, USA].

1958   Near Eastern Prehistory. *Science* 127: 1419-1430.

Braidwood, R. J and B. Howe (editors)

1960   *Prehistoric Investigations in Iraqi Kurdistan.* Studies in Ancient Oriental Civilization No. 31. The University of Chicago Press, Chicago.

Braidwood, R. J. and G. R. Willey (editors)

1962   *Courses toward Urban Life: Archeological Considerations of Some Cultural Alternates.* Viking Fund Publications in Anthropology No. 32. Aldine Publishing Company, Chicago.

Breasted, J. H. (editor)

1906   *Ancient records of Egypt: Historical Documents from the Earliest Times to the Persian Conquest.* 4 vols. The University of Chicago Press, Chicago.

Brothwell, D. R. and E. Higgs (editors)

196   *Science in Archaeology. A Comprehensive Survey of Progress and Research.* Basic Books, New York.

Brunton, G.

1937   *British Museum Expedition to Middle Egypt. First and Second Years, 1928, 1929. Mostagedda and the Tasian Culture.* Quaritch, London.

Brunton, G. and G. Caton-Thompson

1928   *The Badarian Civilisation and Predynastic Remains near Badari.* British School of Archaeology in Egypt, London.

Bryan, K.

1929   Flood-Water Farming. *Geographical Review* 19: 444-456.

Buck, J. L.

1930   *Chinese Farm Economy: a Study of 2866 Farms in Seventeen Localities and Seven Provinces in China.* University of Nanking, Shanghai; University of Chicago Press, Chicago.

Butzer, Karl W.

1964   *Environment and Archeology. An Introduction to Pleistocene Geography.* Aldine Pub., Chicago.

Campbell, A. H.

1965   Elementary Food Production by the Australian Aborigines. *Mankind* 6 (5): 206-211.

Casanova, E.

1946   The Cultures of the Puna and the Quebrada de Humahuaca. In *Handbook of South American Indians.* Vol. I, *The Marginal Tribes*, edited by J. H. Steward (ed.), pp. 619-632. Smithsonian Institution, Bureau of American Ethnology, Bullettin No. 143. United States Government Printing Office, Washington.

Castetter E. F. and W. H. Bell

1942   *Pima and Papago Indian Agriculture.* The University of New Mexico Press, Albuquerque.

Caton-Thompson G. and E. W. Gardner

1934   *The Desert Fayum.* The Royal Anthropological Institute of Great Britain and Ireland, London.

1939   Climate, Irrigation and Early Man in the Hadhramaut. *The Geographical Journal* 93: 18-35.

Chang, Kwang-chih

1963   *The Archaeology of Ancient China.* Yale University Press, New Haven [Connecticut, USA].

Chevallier, R.

1957   *Bibliographie des Applications Archéologiques de la Photographie Aérienne.* Bulletin d'Archéologie Marocaine, Tome II (Supplément). EDITA, Casablanca.

1961a   Un Document Fondamental pour l'Histoire et la Géographie Agraires. La Photographie Aérienne. *Études Rurales: Revue Trimestrielle d'Histoire, Géographie, Sociologie et Économie des Campagnes* 1: 70-80.

1961b   La Centuriation et les Problèmes de la Colonisation Romaine. *Études Rurales: Revue Trimestrielle d'Histoire, Géographie, Sociologie et Économie des Campagnes* 3: 54-80.

1963   *Archéologie Aérienne et Techniques Complémentaires.* Inventaire et Sauvegarde du Patrimoine Historique. 4 juillet - 9 novembre 1963. Institut Pédagogique National, Paris.

1964a   Les Applications de la Photographie Aérienne aux Problèmes Agraires. *Études Rurales: Revue Trimestrielle d'Histoire, Géographie, Sociologie et Économie des Campagnes* 13-14: 120-124.

1964b *L'Étude des Modes Anciens d'Utilisation des Terres (Archéologie Agraire) par la Photographie Aérienne et son Intérêt Pratique.* UNESCO document, Toulouse [France].

Chevallier, R. and J. Soyer

1962   Cadastres Romains d'Algérie. *Bulletin de la Société Française de Photogrammetrie* 5: 43-48. Childe, V. Gordon

1950   The Urban Revolution. *Town Planning Review* 21: 3-17.

1952   *New Light on the Most Ancient East.* Routledge & Paul, London.

1953   Old World Prehistory: Neolithic. In *Anthropology Today: An Encyclopedic Inventory,* edited by A. L. Kroeber, pp. 193-210. University of Chicago Press, Chicago.

Christian, V.

1940   *Altertumskunde des Zweistromlandes: von der Vorzeit bis zum Ende der Achamenidenherrschaft.* Hiersemann, Leipzig.

Clark, J. C. D.

1965   Radiocarbon Dating and the Spread of Farming Economy. *Antiquity* 39: 45-48.

Coghlan H. H.

1943   The Evolution of the Axe from Prehistoric to Roman Times. *The Journal of the Royal Anthropological Institute of Great Britain and Ireland* 73 (1-2): 27-56.

Contenau, G.

1954   *Everyday life in Babylon and Assyria.* Translation by K. R. and A. R. Maxwell-Hyslop. E. Arnold, London.

Coon, C. S.

1957   *The Seven Caves: Archaeological Explorations in the Middle East.* J. Cape, London.

Crawford, O. G. S.

1921   *Man and His Past.* H. Milford, New York; Oxford University Press, London.

1923   Celtic Britain From the Air. *The Observer* 8 July: p. 9, cols B-D. London.

1924   *Air survey and archaeology.* Ordnance Survey, Southampton.

1929a   Desert Markings Near Ur. *Antiquity* 11:342.

1929b   Air Photographs of the Middle East. *The Geographical Journal* 73: 497-509.

1953   *Archaeology in the Field.* Phoenix House, London.

1954   A Century of Air-Photography. *Antiquity* 28: 206-210.

Crawford, O. G. S. and A. Keiller

1928   *Wessex from the Air.* Clarendon Press, Oxford.

Cressey, G. B.

1958   Qanats, Karez, and Foggaras. *Geographical Review* 48: 27-44.

Curwen E.C. and G. Hatt

1953   *Plough and Pasture: the Early History of Farming.* Schuman, New York.

Dale, T. and Carter V. G.

1955   *Topsoil and Civilization.* University of Oklahoma Press, Norman [Oklahoma, USA].

Damm, H.

1954   Form und Anwendung der Feldgerate beim pfluglosen Anbau der Ozeanier. *Ethnographisch-archaologische Forschungen* 2: Xxx.

Daniel, G. E.

1950   *A Hundred Years of Archaeology.* Duckworth, London.

Darby, H. C.

1956   The Clearing of the Woodland in Europe. In *Man's Role in Changing the Face of the Earth,* edited by William L. Thomas, pp. 183-216. University of Chicago Press, Chicago - London.

Darlington, C. D.
  1963   *Chromosome Botany and the Origins of Cultivated Plants.* Hafner and Allen, New York; Unwin, London.

Darmesteter, J.
  1893   *Etudes Iraniennes.* 2 vols. F. Vieweg, Paris.

  1880-1887 *The Zend-Avesta.* 3 vols. Clarendon Press, Oxford.

Das Gupta, P. C.
  1964   *The Excavations at Pandu Rajar Dhibi.* Directorate of Archaeology, West Bengal.

de Goeje, M. J.
  1875   *Das alte Belt del Oxus.* Leiden.

  1906   *Bibliotheca Geographorum Arabicorum.* Pars Tertia, *Descriptio Imperii Moslemici,* by Al-Moqaddasi. 2ⁿᵈ ed. E. J. Brill, Lugduni Batavorum.

Deimel, A.
  1924   *Ausgrabungen der Deutschen Orientgesellschaft in Fara und Abu Hatab.* Vol. III, *Wirtschaftstexte aus Fara.* J. C. Hinrichs, Leipzig.

Delattre, A. J.
  1888   Les Travaux Hydrauliques en Babylonie. *Revue des Questions Scientifiques* 24: 451-507.

Deshayes, J.
  1960   *Les Outils de Bronze de l'Indus au Danube (IVe au IIe millénaire).* 2 vols. Impr. Catholique, Paris.

  1963   Haches-Herminettes Iraniennes. *Syria* 40: 273-276.

Drioton, É.
  1953   Les Origines Pharaoniques du Nilomètre de Rodah. *Bulletin de l'Institut d'Egypte* 34: 291-316.

Drower, M. S.
  1954   Water-Supply, Irrigation and Agriculture. In *A History of Technology.* Vol. I, *From Early Times to Fall of Ancient Empires,* edited by Charles Singer, E. J. Holmyard and J. M. Donaldson, pp. 520-557. Clarendon Press, Oxford.

Dupree, L.
  1964   Prehistoric Archeological Surveys and Excavations in Afghanistan: 1959-1960 and 1961-1963. *Science* 146: 638-640.

Ehrich, R. W. (editor)
  1965   *Chronologies in Old World Archaeology.* University of Chicago Press, Chicago - London.

El-Wailly F. and Abu es-Soof B.
  1965   The Excavations at Tell-as-Sawwan, First Preliminary Report. *Sumer: a Journal of Archaeology and History in Iraq* 21: 17-32.

Engel, F.
  1963   *A Preceramic Settlements on the Central Coast Peru: Asia, Unit 1.* Transactions of the American Philosophical Society New Series Vol. 53, Part 3. The American Philosophical Society, Philadelphia.

Evans-Pritchard, E. E.
  1940   *The Nuer: a Description of the Modes of Livelihood and Political Institutions of a Nilotic People.* Clarendon Press, Oxford.

Falkenstein, A.
  1936   *Archaische Texte aus Uruk.* Ausgrabungen der Deutschen Forschungsgemeinschaft in Uruk-Warka No. 2. Deutsche Forschungsgemeinschaft, Berlin.

Fergusson, J.
  1870   On Indian Chronology. *The Journal of the Royal Asiatic Society of Great Britain and Ireland, New Series* 4: 81-137.

1880   On the Saka, Samvat, and Gupta Eras. A Supplement to His Paper on Indian Chronology. *Journal of the Royal Asiatic Society of Great Britain and Ireland* 12: 259-285.

Field, H.

1960   *North Arabian Desert Archaeological Survey, 1925-50.* Papers of the Peabody Museum of Archaeology and Ethnology Vol. 45, No. 2. Harvard University, Cambridge [Massachusetts, USA].

Fish, T.

1935   Aspect of Sumerian Civilization during the Third Dynasty of Ur III: Rivers and Canals. *Bulletin of the John Rylands Library, Manchester* 19: 90-101.

Flannery, K. V.

1965   The Ecology of Early Food Production in Mesopotamia. Prehistoric Farmers and Herders Exploited a Series of Adjacent but Contrasting Climate Zones. *Science* 147: 1247-1256.

Forbes, R. J.

1955   *Studies in Ancient Technology,* Vol. II. E. J. Brill, Leiden.

1964   *Studies in Ancient Technology,* Vol. VIII. 2nd ed. E. J. Brill, Leiden.

Forde, D. C.

1963   *Habitat, Economy and Society: a Geographical Introduction to Ethnology.* Methuen, London.

Fox, C. F.

1943   *The Personality of Britain: its Influence on Inhabitant and Invader in Prehistoric and Early Historic Times.* The National Museum of Wales, Press Board of the University of Wales, Cardiff.

Frankfort, H.

1939   *Cylinder Seals: a Documentary Essay on the Art and Religion of the Ancient Near East.* Macmillan, London.

Gadd, C. J.

1962   *The Cities of Babylonia.* Cambridge University Press, Cambridge.

Gardiner, A. H.

1957   *Egyptian Grammar: Being an Introduction to the Study of Hieroglyphs.* Griffith Institute, Ashmolean Museum, Oxford.

Garrod, D. A. E.

1930   The Palaeolithic of Southern Kurdistan. Excavations in the Caves of Zarzi and Hazar Merd. *Bulletin of the American School of Prehistoric Research* 6: 9-43.

1932   A New Mesolithic Industry: The Natufian of Palestine. *The Journal of the Royal Anthropological Institute of Great Britain and Ireland* 62: 257-269.

1957   The Natufian Culture. The Life and Economy a Mesolithic People in the Near East. *Proceedings of the British Academy* 43: 211-227.

Garrod, D. A. E. and D. M. A. Bates

1937   *The Stone Age of Mount Carmel.* Vol. I, *Excavations at the Wady el-Mughara.* Clarendon, Oxford.

Garstang, J.

1953   *Prehistoric Mersin: Jumuk Tepe in Southern Turkey: the Neilson Expedition in Cilicia.* Clarendon Press, Oxford.

Geiger, W.

1882   *Ostiranische Kultur im Altertum.* A. Deichert, Erlangen.

Gershevitch, I.

1959   *The Avestan Hymn to Mythra.* University Press, Cambridge.

Ghirshman, R.

1938   *Fouilles de Sialk près de Kashan 1933, 1934, 1937.* Vol. I. Librairie Orientaliste Paul Geuthner, Paris.

Giles, H. A.

1927   *A History of Chinese Literature.* D. Appleton, New York - London.

Gladwin, H. S.

1948   *Excavations at Snaketown.* Vol. IV, *Reviews and Conclusions.* Medallion Papers No. 38. Gila Pueblo, Globe [Arizona, USA].

1957   *A History of the Ancient Southwest.* Bond Wheelwright, Portland [Maine, USA].

Gladwin, H. S., E. W. Haury, E. B. Sayles and N. Gladwin

1937   *Excavations at Snaketown.* Vol. I, *Material Culture.* Medallion Papers No. 25, Gila Pueblo, Globe [Arizona].

Glueck, N.

1959   *Rivers in the Desert: a History of the Negev.* Farrar, Straus and Cudahy, New York.

Goetze, A.

1955   Archaeological Survey of Ancient Canals. *Sumer: a Journal of Archaeology in Iraq* 11: 127-128.

Gompertz, M.

1927   *Corn from Egypt.* G. Howe ltd., London.

Goodwin, A. J. H.

1947   The Bored Stones of South Africa. *Annals of South African Museum* 37. Neil & Co., Edimburg.

Gopal, L.

1962   Irrigation-tax in Ancient India. *The Indian Historical Quarterly* 38 (1): 65-70.

Gothein, M. L. S.

1928   *A History of Garden Art,* 2 vols. Edited by Walter P. Wright, translated from the German by M. A. Archer-Hind. Dutton, New York.

Gray, R. F.

1963   *The Sonjo of Tanganyika: an Anthropological Study of an Irrigation-based Society.* International African Institute by the Oxford University Press, London - New York.

Gruber, J. W.

1948   Irrigation and Land Use in Ancient Mesopotamia. *Agriculture History* 22 (2): 69-77.

Gulley, J. L. M.

1960   Some Problems of Archaeologic Mapping. *Revue Archéologique* 1960, Tome I: 141-159.

Gutkind, E. A.

1956   Our World from the Air: Conflict and Adaptation. In *Man's Role in Changing the Face of the Earth,* edited by William L. Thomas, pp. 1-44. University of Chicago Press, Chicago - London.

Halseth, O. S.

1936   Prehistoric Irrigation in the Salt River Valley. In *Symposium on Prehistoric Agriculture,* pp. 42-47. University of New Mexico Bulletin No. 296; University of New Mexico Anthropological Series Vol. 1, 5. University of New Mexico Press, Albuquerque.

Hansen, D. P.

1965   The Relative Chronology of Mesopotamia. Part II, The Pottery Sequence at Nippur from Middle Uruk to the end of the Old Babylonian Period (3400-1600 BP). In *Chronologies in Old World Archaeology,* edited by Robert W. Ehrich, pp. 201-213. University of Chicago Press, Chicago.

Harlan J. K. and D. Zohary

1966   Distributions of Wild Wheat and Barley. *Science* 153: 1074-1080.

Harris, D. R.

1967   New Light on Plant Domestication and the Origins of Agriculture: a Review. *Geographical Review* 57: 90-107.

Harrisson, T.

1964   Inside Borneo: The Dickson Asia Lecture. *The Geographical Journal* 130: 329-336.

Haury, E. W.

1936   The Snaketown Canal. In *Symposium on Prehistoric Agriculture*, pp. 48-50. University of New Mexico Bulletin No. 296; University of New Mexico Anthropological Series Vol. 1, 5. University of New Mexico Press, Albuquerque.

1967   The Hohokam. First Masters of the American Desert. *National Geographic* 131: 670-695.

Hayes, W. S.

1964   Most Ancient Egypt. Chapter III, The Neolithic and Chalcolithic Communities of Northern Egypt. *Journal of Near Eastern Studies* 23: 217-272.

Heichelheim, F. M.

1938   *Wirtschaftsgeschichte des Altertums vom Paläolithikum bis zur Völkerwanderung der Germanen, Slaven und Araber.* 2 vols. Sijthoff, Leiden.

Heiser, C. B.

1965   Cultivated Plans and Cultural Diffusion in Nuclear America. *American Anthropologist* 67: 930-949.

Heizer, R. F.

1953   Long-Range Dating in Archaeology. In *Anthropology Today: an Encyclopedic Inventory*, edited by A. L. Kroeber, pp. 3-42. University of Chicago Press, Chicago.

Helbaek, H.

1955   Ancient Egyptian Wheats. *Proceedings of the Prehistoric Society* 21: 93-95.

1959a   Domestication of Food Plants in the Old World. *Science* 130: 365-372.

1959b   Ecological Effects of Irrigation in Ancient Mesopotamia. *Iraq* 21: 186-196.

1960   Paleoethnobotany of the Near East and Europe. In *Prehistoric investigations in Iraqi Kurdistan*, edited by Braidwood, Robert J. and Bruce Howe, pp. 99-118. Studies in Ancient Oriental Civilization No. 31. The University of Chicago Press, Chicago.

1964a   First Impressions of the Çatal Hüyük Plant Husbandry. *Anatolian Studies* 14: 121-123.

1964b   Early Hassunan Vegetable Remains at es-Sawwan near Samarra. *Sumer: a Journal of Archaeology in Iraq* 20 (1-2): 45-48.

Henning, W. B.

1951   *Zoroaster, Politician or Witch-doctor?*. Oxford University Press, London.

Henshaw, H. W.

1887   Perforated Stones from California. *Bureau of American Ethnology Bulletin* 2: 5-34.

Herrmann, A.

1914   Alte Geographie des unteren Oxusgebietes. *Abhandlungen der Gesellschaft der Wissenschaften in Göttingen: Philologisch-historische Klasse* 15 (4): 1-35.

Herzfeld, E.

1909   Uber die Historische Geographie von Mesopotamien: Ein Programm. *Petermanns Mitteilungen* 55: Xxx.

1947   *Zoroaster and his World.* Vol. 2. Princeton University Press, Princeton.

Hole, F.

1962   Archaeological Survey and Excavation in Iran, 1961. *Science* 137: 524-526.

Hole, F. and R. F. Heizer

1965   *An Introduction to Prehistoric Archaeology.* Holt, Rinehart and Winston, New York.

Hole, F., K. Flannery and J. Neely

1965   Early Agriculture and Animal Husbandry in Deh Luran, Iran. *Current Anthropology* 6 (1): 105-106.

Holmes, W. H.

1910   Perforated Stones. In *Handbook of American Indians north of Mexico*, Part 2, edited by Hodge, Frederick Webb, pp. 231-232. Smithsonian Institution, Bureau of American Ethnology, Bulletin No. 30. Rowman and Littlefield, Totowa [New Jersey, USA].

Hopfen, H. J.

1960   *Farm Implements for Arid and Tropical Regions.* FAO Agricultural Development Papers No 67. FAO, Rome.

Hopkins, S. W.

1883   *Life among the Piutes: their Wrongs and Claims.* Cupples, Upham, Boston.

Horst, H.

1964   *Die Staatsverwaltung der Grosselgügen und Hōrazmšahs (1038-1231).* F. Steiner, Wiesbaden.

Huntington, E.

1905   The Climate and History. In *Explorations in Turkestan, with an Account of the Basin of eastern Persia and Sistan. Expedition of 1903, under the direction of Raphael Pumpelly*, edited by Raphael Pumpelly, pp. 302-317. Carnegie Institution of Washington, Washington.

1906   The Rivers of Chinese Turkestan and the Desiccation of Asia. *The Geographical Journal* 28: 352-367.

1907   The Historic Fluctuations of the Caspian Sea. *Bulletin of the American Geography Society* 39: 577-596.

1908a   The Climate of Ancient Palestine. Part I. *Bulletin of the American Geography Society* 40: 513-522.

1908b   The Climate of Ancient Palestine. Part II. *Bulletin of the American Geography Society* 40: 577-586.

1908c   The Climate of Ancient Palestine. Part III (Conclusion). *Bulletin of the American Geography Society* 40: 641-652.

Hutchinson, J. B. (editor)

1965   *Essays on Crop Plant Evolution.* University Press, Cambridge.

Ionides, M. G.

1938   Two Ancient Irrigation Canals in Northern ʻIraq. *The Geographical Journal* 92: 351-354.

Izumi, S. and T. Sono

1963   *Andes 2: Excavations at Kotosh, Peru, 1960.* Kadokawa, Tokyo.

Jackson, A. V. W.

1899   *Zoroaster: The prophet of Ancient Iran.* Macmillan, London.

Jacobsen, T.

1958   *Salinity and Irrigation Agriculture in Antiquity.* Diyala Basin Archaeological Project, report on essential results June 1, 1957, to June 1, 1958 (mimeographed). Baghdad.

1960   The Waters of Ur. *Iraq* 22: 174-185.

Jacobsen, T. and R. M. Adams

1958   Salt and Silt in Ancient Mesopotamian Agriculture. *Science* 128: 1251-1258.

Jacobsen, T. and S. Lloyd

1935   *Sennacherib's aqueduct at Jerwan.* The University of Chicago Oriental Institute publications Vol. 24. The University of Chicago Press, Chicago.

Jelinek, A. J.

1962   An Index of Radiocarbon Dates Associated with Cultural Materials. *Current Anthropology* 3: 451-477.

Jennings, J. D. and E. Norbeck (editors)

1964   *Prehistoric Man in the New World.* University of Chicago Press, Chicago.

Johnson, G. R.

1930   *Peru from the Air.* American Geographical Society, Special Publication No. 12. American Geographical Society, New York.

Joseph, J. K. St.

1945   Air Photography and Archaeology. *The Geographical Journal* 105: 47-59.

Junker, H.

1945   Merimda-Benisalama: Fouilles de l'Académie des Sciences de Vienne (1939). *Chronique d'Égypte* 39-40: 74-75.

Kaplony, P.

1963   *Die Inschriften der Ägyptischen Frühzeit.* 3 vols. O. Harrassowitz, Wiebaden.

Kempton, J. H.

1936   Maize as a measure of Indian skill, In *Symposium on Prehistoric Agriculture*, pp. 19-28. University of New Mexico Bulletin No. 296; University of New Mexico Anthropological Series Vol. 1, 5. University of New Mexico Press, Albuquerque.

Kenyon, K. M.

1960   *Archaeology in the Holy Land.* Praeger, London.

Kihara, H., K. Yamashita and M. Tanaka

1965   Morphological, Physiological, Genetical and Cytological Studies in Aegilops and Triticum Collected in Pakistan, Afghanistan and Iran. In *Cultivated Plants and their Relatives*, edited by Kosuke Yamashita, pp. 4-41. Kyoto University, Kyoto.

King, F. H.

1911   *Farmers of Forty Centuries; or, Permanent Agriculture in China, Korea and Japan.* Madison.

Kirkbride, D.

1966   Beidha. An Early Neolithic Village in Jordan. *Archaeology* 19 (3): 199-207.

Klíma, B.

1955   Přínos nové paleolitické stanice v Pavlové k problematice nejstarších zeměděských nástrojů. *Památky Archeologické* 46: 7-25.

Kosok, P.

1965   *Life, Land and Water in Ancient Peru.* Long Island University, New York.

Kramer, F. L.

1967   Eduard Hahn and the End of the "Three Stages of Man". *Geographical Review* 57: 73-89.

Kroeber, A. L.

1948   *Anthropology: Race, Language, Culture, Psychology, Prehistory.* Harcourt, Brace and Company, New York.

1953   *Anthropology Today: an Encyclopedic Inventory.* University of Chicago Press, Chicago.

Laessøe, J.

1953   Reflexions on Modern and Ancient Oriental Water Works. *Journal of Cuneiform Studies* 7 (1): 5-26.

Lamberg-Karlovsky, C. C.

1967   Archeology and Metallurgical Technology in Prehistoric Afghanistan, India and Pakistan. *American Anthropologist* 69: 145-162.

Lampert, R. J.

1967   Horticulture in the New Guinea Highlands - C_{14} Dating. *Antiquity* 41: 307-309.

Laufer, B.

1919   *Sino-Iranica: Chinese Contributions to the History of Civilization in Ancient Iran, with special reference to the History of Cultivated Plants and Products.* Field Museum of Natural History. Publication 201. Anthropological Series Vol. 15, No. 3. Chicago.

Le Strange, G.

1905   *The Lands of the Eastern Caliphate: Mesopotamia, Persia and Central Asia from the Moslem conquest to the time of Timur.* University Press, Cambridge.

Lees, G. M. and N. L. Falcon
1952   The Geographical History of the Mesopotamian Plains. *The Geographical Journal* 118: 24-39.

Leser, P.
1931   *Entstehung und Verbreitung des Pfluges.* Aschendorffsche Verlagsbuchhandlung, Munster.

Lewis, R. A.
1966   Early Irrigation in West Turkestan. *Annals of the Association of American Geographers* 56: 467-491.

Libby, W. F.
1963   Accuracy of Radiocarbon Dates. *Science* 140: 278-280.

Licent, E. and P. Teilhard de Chardin
1925   Note sur deux Instruments Agricols du Neolithique de China. *L'Anthropologie* 35: 63-74.

Liddell, H. G., R. Scott, F. Passow, and H. Drisler
1846   *A Greek-English Lexicon.* 1st ed. Harper & brothers, New York.

Lips, J. E.
1953   *Vom Ursprung der Dinge: Eine Kulturgeschichte des Menschen.* Bibliographisches Institut, Leipzig.

Lloyd, S. and Fuad Safar
1947   Eridu: A Preliminary Communication on the First Season's Excavation January - March 1947. *Sumer: a Journal of Archaeology and History in Iraq* 3 (2): 84-111.

1948   Eridu: A Preliminary Communication on the Second Season's Excavation 1947-1948. *Sumer: a Journal of Archaeology and History in Iraq* 4 (2): 115-127.

Lloyd, S., Fuad Safar and R. J. Braidwood
1945   Tell Hassuna. Excavations by the Iraq Government Directorate General of Antiquities in 1943 and 1944. *Journal of Near Eastern Studies* 4 (4): 255-289.

Lorenzo, J. L.
1961   *La Revoluciòn Neolìtica en Mesoamèrica.* Publicaciones Instituto Nacional de Antropología e Historia, Departamento de Prehistoria Vol. 11. Instituto Nacional de Antropologìa e Historia, Mexico.

Lowdermilk, W. C. and D. R. Wickes
1942   Ancient Irrigation in China Brought up to Date. *The Scientific Monthly* 55: 209-225.

Mackay, D.
1945   Ancient River Beds and Dead Cities. *Antiquity* 19: 135-144.

MacNeish, R. S.
1958   *Preliminary Archaeological Investigations in the Sierra de Tamaulipas, Mexico.* American Philosophical Society Transactions, New Series Vol. 48, Pt. 6. American Philosophical Society, Philadelphia.

1965   The Origins of American Agriculture. *Antiquity* 39: 87-94.

Mallon, A., R. Koeppel and N. René
1934   *Teleilaāt Ghassūl.* Vol. I, *Compte rendu des fouilles de l'Institut Biblique Pontifical, 1929-1932.* Institut Biblique Pontifical, Rome.

Mallowan, M. E. L.
1964   Noah's Flood Reconsidered. *Iraq* 26: 62-82.

Markwart, J.
1938   *Wehrot und Arang: Untersuchungen zur mythischen und geschichtlichen Landeskunde von Ostiran.* Brill, Leiden.

Marquart, J.
1901   *Erānšahr, nach der Geographie des Ps. Moses Xorenac'i. Abhandlungen der Gesellschaft der Wissenschaften in Göttingen: Philologisch-historische Klasse* 3 (2). Weidmannsche Buchhlandlung, Berlin.

1903   *Osteuropäische und ostasiatische Streifzüge: ethnologische und historisch-topographische Studien zur Geschichte des 9. und 10. Jahrhunderts (ca. 840-940).* Weicher, Leipzig.

Mason, J. A.

1957   *The Ancient Civilizations of Peru*. Penguin Books, Harmondsworth [Middlesex, USA].

McGovern, W. M.

1939   *The Early Empires of Central Asia*. The University of North Carolina Press, Chapel Hill [North Carolina, USA].

Mellaart, J.

1964   Excavations at Çatal Hüyük, 1963, Third Preliminary Report. *Anatolian Studies* 14: 39-119.

1965a Çatal Hüyük West. *Anatolian Studies* 15: 135-156.

1965b *Earliest Civilizations of the Near East*. Thames and Hudson, London.

Millon, R.

1957   Irrigation Systems in the Valley of Teotihuacan. *American Antiquity* 23: 160-166.

Minorsky, V.

1939   A "Soyūrghāl" of Qāsim b. Jahāngīr Aq-qoyunlu (903/1498). *Bulletin of the School of Oriental Studies, University of London* 9: 927-960.

Mortensen, P.

1964   Additional Remarks on the Chronology of Early Village-Farming Communities in the Zagros Area. *Sumer: a Journal of Archaeology in Iraq* 20 (1-2): 28-36.

Mouterde, R. and A. Poidebard

1945   *Le Limes de Chalcis: Organisation de la Steppe en Haute Syrie Romaine. Documents Aèriens et Èpigraphiques*. 2 vols. P. Geuthner, Paris.

Müller, F. M.

1879   *The Sacred Books of the East*. Vol. I, *The Upanishads*. Part I. M. Clarendon Press, Oxford.

1880   *The Sacred Books of the East*. Vol. V, *Pahlavi Texts*. Part I, edited by E. W. West. Clarendon Press, Oxford.

1882   *The Sacred Books of the East*. Vol. XVIII, *Pahlavi Texts*. Part II, edited by E. W. West. Clarendon Press, Oxford.

Murray, G. W.

1951   The Egyptian Climate: An Historical Outline. *The Geographical Journal* 117: 422-434.

Narr, K. J.

1956   Early Food-producing Populations. In *Man's Role in Changing the Face of the Earth*, edited by William L. Thomas, pp. 134-151. University of Chicago Press, Chicago - London.

Nelson, H. S.

1962   An Abandoned Irrigation System in Southern Iraq. *Sumer: a Journal of Archaeology in Iraq* 18 (1-2): 67-72.

Nilles, J.

1942-45 Digging-Sticks, Spades, Hoes, Axes, and Adzes of the Kuman People in the Bismarck Mountains of East-Central New Guinea. *Anthropos: internationale Zeitschrift für Völker- und Sprachenkunde* 37-40 (1-3): 205-212.

Nyberg, H. S.

1938   *Die Religionen des alten Iran*. J. C. Hinrichs verlag, Leipzig.

Ohlin, G.

1965   Historical Outline of World Population Growth. In *Proceedings of the World Population Conference: Belgrade, 30 August - 10 September 1965*, edited by World Population Conference. Background Paper. General, No. 486. United Nations Organization, New York.

Oldenberg, H.

1881   Über die Datierung der ältern indischen Münz- und Inschriftreihen. *Zeitschrift für Numismatik* 8: 289-328.

Owen, T. R. H.

1937   The Hadendowa. *Sudan Notes and Records* 20: 183-208.

Patrick, H. R.

1903   *The Ancient Canal Systems and Pueblos of the Salt River Valley, Arizona, with notes and charts.* Phoenix Free Museum, Phoenix [Arizona].

Pearl, O. M.

1950   Ἐξάθυφος: Irrigation Works and Canals in the Arsinoite Nome. *Aegyptus* 31: 223-230.

Perkins, A. L.

1949   *The Comparative Archaeology of Early Mesopotamia.* Studies in Ancient Oriental Civilization No. 25. University of Chicago Press, Chicago.

Perry, W. J.

1916   The Geographical Distribution of Terraced Cultivation and Irrigation. *Memoirs and proceedings of the Manchester Literary & Philosofical Society* 60 (6): 1-25.

1924   *The Growth of Civilization.* Methuen & Co. Ltd., London.

Petrie, W. M. Flinders

1917   *Tools and Weapons illustrated by the Egyptian Collection in University College, London and 2,000 Outlines from Other Sources.* British School of Archaeology in Egypt, London.

1920   *Prehistoric Egypt, illustrated by over 1,000 Objects in University College, London.* British School of Archaeology in Egypt, London.

Platts, J. T.

1884   *A Dictionary of Urdū: Classical Hindī and English.* W.H. Allen & Co, London.

1959   *A Dictionary of Urdū: Classical Hindī and English.* Moscow.

Poidebard, A.

1928   Mission Archéologique en Haute Djezireh (automne 1927). *Syria* 9 (3): 216-223.

1929a   Résultats de sa Mission en Haute Djéziré en automne 1928. *Comptes-rendus des Séances de l'Académie des Inscriptions et Belles-Lettres* 73 (2): 91-94.

1929b   Les Révélations Archéologiques de la Photographie Aérienne. Une Nouvelle Méthode de Recherches et d'Observations en Région de Steppe. *L'Illustration*, 24 Mars, 1929.

1932   Méthode Aérienne de Recherches en Géographie Historique. *Terre, Air, Mer: La Géographie* 1: 1-16.

1934   *La trace de Rome dans le Désert de Syrie: Le limes de Trajan à la Conquête Arabe, Recherches Aérienne (1925-1932).* 2 vols. Librairie Orientaliste Paul Geuthner, Paris.

Popper, W.

1951   *The Cairo Nilometer: Studies in Ibn Taghrî Birdî's chronicles of Egypt.* Vol. I. University of California Press, Berkeley.

Porada, E.

1965   The Relative Chronology of Mesopotamia. Part I, Seals and Trade (6000-1600 BP). In *Chronologies in Old World Archaeology*, edited by Robert W. Ehrich, pp. 133-200. University of Chicago Press, Chicago.

Pumpelly, R.

1908   *Explorations in Turkestan. Expedition of 1904. Prehistoric civilizations of Anau: Origins, Growth, and Influence of Environment.* 2 vols. Carnegie Institution, Washington.

Reder, D. G.

1958   Ancient Egypt, a Centre of Agriculture. *Cahiers d'Histoire Mondiale* 4 (4): 801-817.

Reed, C. A.

1959   Animal Domestication in the Prehistoric Near East. *Science* 130: 1629-1639.

1960   A Review of the Archaeological Evidence on Animal Domestication in the Prehistoric Near East. In *Prehistoric investigations in Iraqi Kurdistan*, edited by Braidwood, Robert J. and Bruce Howe, pp. 119-145. Studies in Ancient Oriental Civilization No. 31. The University of Chicago Press, Chicago.

Reisner, G. A.

1910   *The Archaeological Survey of Nubia.* Archaeological report for 1907-1908 Vol. 1. National Printing Department, Cairo.

Roth, H. L.

1887   On the Origin of the Agriculture. *The Journal of the Anthropological Institute of Great Britain and Ireland* 16: 102-136.

Russell, J. G.

1958   Late Ancient and Medieval Population. *Transactions of the American Philosophical Society* 48 (3). American Philosophical Society, Philadelphia.

Ryder, M. L.

1958   Follicle Arrangement in Skin from Wild Sheep, Primitive Domestic Sheep and in Parchment. *Nature* 182: 781-783.

Sachau, E.

1873   Zur Geschichte und Chronologie von Khwârism. *Sitzungsberichte der Philosophisch-historischen Klasse der Kaiserlichen Akademie der Wissenschaften* 73: 471-506.

Safar, F.

1947   Sennacherib's Project for Supplying Erbil with Water. *Sumer: a Journal of Archaeology in Iraq* 3: 23-25.

Saloen, A.

1965   *Die Hausgeräte der alten Mesopotamier nach Sumerischakkadischen Quellen: eine Lexikalische und Kulturgeschichtliche Untersuchung.* Vol. I. Suomalainen Tiedeakatemia, Helsinki.

Sankalia, H. D.

1964   *Stone Age Tools. Their Techniques, Names and Probable Functions.* Deccan College, Poona.

Sauer, C. O.

1952   *Agricultural Origins and Dispersals.* American Geographical Society, New York.

1956   The Agency of Man on the Earth. In *Man's Role in Changing the Face of the Earth,* edited by William L. Thomas, pp. 49-69. University of Chicago Press, Chicago - London.

Schiemann, E.

1943   Entstehung der Kulturpflanzen. *Ergebnisse der Biologie* 19: 409-552.

Schmidt, E. F.

1940   *Flights over Ancient Cities of Iran.* University of Chicago Press, Chicago.

Schneider, N.

1931   *Die Drehem- und Dioha-Urkunden der Strassburger Universitäts- und Landesbibliothek.* Analecta Orientalia No. 1. Pontificio Istituto Biblico, Roma.

1932   *Die Drehem- und Dioha-Texte im Kloster Montserrat (Barcelona).* Analecta Orientalia No. 7. Pontificio Istituto Biblico, Roma.

Schove, D. J.

1965   Solar Cycle and Equatorial Climates. *Geologische Rundschau* 54 (1): 448-477.

Schroeder, A. H.

1943   Prehistoric Canals in the Salt River Valley, Arizona. *American Antiquity* 8: 380-386.

1951   Snaketown IV vs. The Facts. *American Antiquity* 16: 263-265.

Schuchhardt, C.

1918   *Die sogenannten Trajanswälle in der Dobrudscha.* Abhandlungen der Preussischen Akademie der Wissenschaften: Philosophisch-historische Klasse, Vol. 1918, No. 12. G. Reimer, Berlin.

Scollar, I.

1965   *Archäologie aus der Luft. Arbeitsergebnisse d. Flugjahre 1960 u. 1961 im Rheinland.* Rheinland-Verlag, Dusseldorf.

Semple, E. C.

1929   Ancient Mediterranean Pleasure Gardens. *Geographical Review* 19: 420-443.

Serjeant, R. B.

1964   Some Irrigation Systems in Ḥaḍramawt. *Bulletin of the School of Oriental and African Studies, University of London* 27: 33-76.

Shetrone, H. C.

1945   A Unique Prehistoric Irrigations Projects. *Annual Report of the Board of Regents of the Smithsonian Institution:* 379-386.

Shippee, R.

1932   The "Great Wall of Peru" and Other Aerial Photographic Studies by the Shippee-Johnson Peruvian Expedition. *Geographical Review* 22: 1-29.

Solecki, R. S.

1955a   Shanidar Cave, a Paleolithic Site in Northern Iraq and its Relationship to the Stone Age Sequence of Iraq. *Sumer: a Journal of Archaeology and History in Iraq* 11: 14-38.

1955b   Shanidar Cave, a Late Pleistocene Site in Northern Iraq. *Annual Report of the Smithsonian Institution* 1954: 389-425.

1963   Prehistory in Shanidar Valley, Northern Iraq. *Science* 139: 179-193.

Steensberg, A.

1964   A Bronze Age Ard Type from Hama in Syria Intended for Rope Traction. *Berytus, Archaeological Studies* 15: 111-139.

Steward, J. H.

1933   Ethnography of the Owens Valley Paiute. *University of California publications: American Archaeology and Ethnology* 33 (3): 233-250.

1953   Evolution and Process. In *Anthropology Today: an Encyclopedic Inventory,* edited by A. L. Kroeber, pp. 313-326. University of Chicago Press, Chicago.

1955   *Irrigation Civilizations: a Comparative Study. A Symposium on Method and Result in Cross-cultural Regularities.* Social Science Monographs No. 1. Pan American Union, Washington.

Stow, G. W.

1930   *Rock-paintings in South Africa from Parts of the Eastern Province and Orange Free State.* Methuen & Co, London.

Taubenschlag, R.

1955   *The Law of Greco-Roman Egypt in the Light of the Papyri: 332 B.C. - 640 A.D.* 2nd ed. Państwowe Wydawnictwo Naukowe, Warszawa.

Tenney, F.

1936   *An Economic Survey of Ancient Rome.* Vol. II, *Roman Egypt to the Reign of Diocletian,* edited by A. C. Johnson. The Johns Hopkins Press, Baltimore.

Thomas, W. J. (editor)

1956   *Man's Role in Changing the Face of the Earth.* University of Chicago Press, Chicago - London.

Titiev, M.

1944   *Old Oraibi: a Study of the Hopi Indians of Third Mesa.* The Museum, Cambridge [Massachusetts, USA].

Tolstov, S. P.

1963   Dated documents from the Toprak Kala Palace and the problem of the Saka Era and Kaniṣka Era. *Bibliotheca Orientalis,* 20.

Towle, M. A.
1961   *The Ethnobotany of Pre-Columbian Peru.* Viking Fund Publications in Anthropology No. 30. Aldine, Chicago.

Turney, O. A.
1929   Prehistoric Irrigation. *Arizona Historical Review* 2 (3): 9-45.

Unger, E.
1935   Ancient Babylonian Maps and Plans. *Antiquity* 9: 311-322.

Usher, A. P.
1930   The History of Population and Settlement in Eurasia. *Geographical Review* 20: 110-132.

Van der Meer, P. E.
1955   *The Chronology of Ancient Western Asia and Egypt.* E. J. Brill, Leiden.

Van Liere, W. J.
1963   Capitals and Citadels of Bronze-Iron age Syria in their Relationship to Land and Water. *Les Annales Archéologiques de Syrie: Revue Archéologique et d'Historie Syriennes* 13: 109-122.

Van Liere, W. J. and J. Lauffray
1954-1955   Nouvelle Prospection Archéologiques dans la Haute Jesireh Syriennes. *Les Annales Archéologiques de Syrie: Revue Archéologique et d'Historie Syriennes* 4-5: 129-148.

Van Zeist, W. and H. E. Wright
1963   Preliminary Pollen Studies at Lake Zeribar, Zagros Mountains, Southwestern Iran. *Science* 140: 65-67.

Van Zeist, W. and W. A. Casparie
1968   Wild Einkorn Wheat and Barley from Tell Mureibit in Northern Syria. *Acta Botanica Neerlandica* 17 (1): 44-53.

Velàzquez, P. F.
1945   *Códice Chimalpópoca: Anales de Cuauhtitlan y Leyenda de los Soles.* Publicación del Instituto de Historia de la Universidad Nacional Autónoma de México, Series 1, No. 1. Imprenta Universitaria, México.

Von Li, H.
1931   Die Geschichte des Wasserbaues in China. *Beiträge zur Geschichte der Technik und Industrie* 21: 59-73.

Von Wissmann, E., H. Poech, G. Smolla and F. Kussmaul
1956   On the Role of Nature and Man in Changing the Face of the Dry Belt of Asia. In *Man's Role in Changing the Face of the Earth*, edited by William L. Thomas, pp. 278-303. University of Chicago Press, Chicago - London.

Voyevodsky, M.
1938   A Summary Report of a Khwarizm Expedition. Bulletin of the American Institute for Iranian Art and Archaeology 5 (3): 235-244.

White, L. A.
1943   Energy and Evolution of Culture. *American Anthropologist* 45: 335-356.

Whyte, R. G.
1961   Evolution of Land use in South-Western Asia. A History of Land use in Arid Regions. In *A History of Land use in Arid Regions*, edited by L. Dudley Stamp, pp. 57-118. Arid Zone Research No. 17. UNESCO, Paris.

1963   The Significance of Climatic Change for Natural Vegetation and Agriculture. In *Changes of Climate: Proceedings of the Rome Symposium organized by Unesco and the World Metereological Organization*, pp. 381-386. Arid Zone research No. 20. UNESCO, Paris.

Willcocks, W.
1889   *Egyptian Irrigation.* Spon, London.

1903   *The Restoration of the Ancient Irrigation Works on the Tigris or the Recreation of Chaldea.* National Printing Department, Cairo.

1917   Irrigation of Mesopotamia. 2nd ed. Spon, London.

Willis, E. H.

1963   Radiocarbon Dating. In *Science in Archaeology: a Comprehensive Survey of Progress and Research*, edited by Don R. Brothwell and Eric Higgs, pp. 35-46. Thames & Hudson, London.

Wilson, J. A.

1962   Water and Ancient Egypt. *Trudy dvadtsat pyatogo mejdunarodnogo kongressa vostokovedov. Moskva 9 - 6 avgusta 1960. Tom 1, Obshchaya chast zasedaniya sektsii I-V*, B. G. Gafurov. Moskva

Wittfogel, K. A.

1956   The Hydraulic Civilizations. In *Man's Role in Changing the Face of the Earth*, edited by William L. Thomas, pp. 152-164. University of Chicago Press, Chicago - London.

1957   *Oriental Despotism: a Comparative Study of Total Power*. Yale University Press, New Haven.

1967   Review of *The Evolution of Urban Society: Early Mesopotamia and Prehispanic Mexico*, edited by R. McC. Adams. *American Anthropologist* 69 (1): 90-92.

Woolley, C. L.

1938   *Ur of the Chaldees: a Record of Seven Years of Excavation*. Penguin, Harmondsworth.

1954   *Excavations at Ur. a Record of Twelve Years Work*. E. Benn, London.

Wreszinski, W.

1923-1936   *Atlas zur Altaegyptischen Kulturgeschichte*. 3 vols. J. C. Hinrichs, Leipzig.

Wright, H. E. Jr.

1968   Natural Environment of Early Food Production North of Mesopotamia. *Science, New Series* 161: 334-339.

Yeivin, S.

1930   The Ptolemaic System of Water Supply in the Fayyûm. *Annales du Service des Antiquités de l'Égypte* 30: 27-30.

Zeuner, F. E.

1955   The Goats of Early Jericho. *Palestine Exploration Quarterly*: 70-86.

1963   The History of the Domestication of Cattle. In *Man and Cattle: Proceedings of a Symposium on Domestication at the Royal Anthropological Institute, 24-26 May 1960*, edited by A. E. Mourant and F. E. Zeuner, pp. 9-19. Royal Anthropological Institute. Occasional Paper No. 18. Royal Anthropological Institute of Great Britain & Ireland, London.